Fluid Mechanics and Its Applications

Founding Editor

René Moreau

Volume 140

The purpose of this series is to focus on subjects in which fluid mechanics plays a fundamental role. As well as the more traditional applications of aeronautics, hydraulics, heat and mass transfer etc., books will be published dealing with topics, which are currently in a state of rapid development, such as turbulence, suspensions and multiphase fluids, super and hypersonic flows and numerical modelling techniques. It is a widely held view that it is the interdisciplinary subjects that will receive intense scientific attention, bringing them to the forefront of technological advancement. Fluids have the ability to transport matter and its properties as well as transmit force, therefore fluid mechanics is a subject that is particulary open to cross fertilisation with other sciences and disciplines of engineering. The subject of fluid mechanics will be highly relevant in such domains as chemical, metallurgical, biological and ecological engineering. This series is particularly open to such new multidisciplinary domains. The median level of presentation is the first year graduate student. Some texts are monographs defining the current state of a field; others are accessible to final year undergraduates; but essentially the emphasis is on readability and clarity.

Springer and Professor Thess welcome book ideas from authors. Potential authors who wish to submit a book proposal should contact Dr. Mayra Castro, Senior Editor, Springer Heidelberg, e-mail: mayra.castro@springer.com

Indexed by SCOPUS, EBSCO Discovery Service, OCLC, ProQuest Summon, Google Scholar and SpringerLink.

Valery Rudyak

Thermophysical Properties of Nanofluids

Springer

Valery Rudyak
Novosibirsk State University
of Architecture and Civil Engineering
Novosibirsk, Russia

ISSN 0926-5112 ISSN 2215-0056 (electronic)
Fluid Mechanics and Its Applications
ISBN 978-3-031-97994-1 ISBN 978-3-031-97995-8 (eBook)
https://doi.org/10.1007/978-3-031-97995-8

This Springer imprint is published by the registered company Springer Nature Switzerland AG
The registered company address is: Gewerbestrasse 11, 6330 Cham, Switzerland

Preface

Transport processes in gases and liquids are
visually observed effect of molecular microfluctuations
taking place in these media

The term *nano* has become one of the markers of the transition to a new economic order. Nanotechnology is no longer a new and widely used term that integrates a huge number of modern technologies for various purposes. Nanofluids have appeared in a series of such technologies. Although the term was introduced over a quarter of a century ago, nanofluids have been studied before. The author of this book had constructed the kinetic theory of gases and liquids with submicron particles since the early 1990s. It turned out that such gas suspensions have unusual properties. The kinetic theory could be constructed only for rarefied finely dispersed gas suspensions. In all other cases, a strict kinetic theory cannot be constructed. This is not surprising, since it is principally impossible to construct a kinetic theory even for simple liquids.

Suspensions containing coarse dispersed particles have long been known and widely used. More than half a century ago, the idea was born to use suspensions with micron-sized metal particles as a working fluid for multiple thermophysical applications, in particular, as a cooling fluid. The motivation for this was quite obvious. Metal particles possess a very high thermal conductivity, and their addition to the carrier fluid should have significantly enhanced its heat transfer characteristics. It soon became clear, however, that this was impossible, the particles sedimented quite quickly. In addition, they were an abrasive for the channel walls and significantly damaged them. The development of technologies for producing large quantities of submicron particles marked the rebirth of this idea. The pioneering experiments, conducted at the Argonne National Laboratory, have shown encouraging results. The term *nanofluid* was also born there, referring to a suspension of liquid and nanoparticles with a characteristic size from one to one hundred nanometers. This idea has since been taken up by numerous research laboratories worldwide, with a primary focus on investigating thermal conductivity and viscosity of nanofluids. Surprisingly, the experimental data obtained turned out to be extremely heterogeneous and often contradictory. Furthermore, it quickly became clear that the thermophysical properties of nanofluids were not described by classical theories. This was another very important irritant for researchers, and it was necessary, on the one hand, to systematically experimentally study the properties of nanofluids, and on the other, to explain their non-standard behavior. The challenge was so strong that the number of relevant publications grew exponentially quickly. To date, thousands of articles have been published and their number continues to grow. This indicates the relevance of the topic. However, this also signifies that numerous issues yet remain unresolved, and the obtained data frequently contradict each other.

Generally speaking, the term nanofluid still irritates many people, since it is claimed that long before this term appeared, the properties of such dispersed liquids were actively studied and in chemistry they were called colloidal solutions. There is some truth in these words. Colloids usually include dispersed particles with characteristic sizes from one to a thousand nanometers, i.e. up to a micrometer and even several micrometers. In Perrin's famous experiments on Brownian particles, their minimum sizes were equal to several hundred nanometers. Therefore, at the upper boundary, colloidal particles are ordinary Brownian particles. On the other hand, it turned out that suspensions particles with a size not exceeding one hundred nanometers have completely unusual properties. It is precisely such dispersed fluids that were called nanofluids. However, one more explanation is necessary. In the early 1990s, carbon tubes were synthesized. Their properties were so unusual that it was natural to anticipate that dispersed fluids containing carbon tubes would exhibit similarly remarkable characteristics. These fluids are also called nanofluids. Generally speaking, these dispersed fluids do not fit into the Procrustean bed of nanofluids. One size (diameter) of these tubes is indeed on the nanometer scale, but the other (length) can be several micrometers or even tens of micrometers, which places them in the macroscopic scale. Despite this size dissimilarity, the properties of these nanofluids were found to be similar to those of nanofluids containing conventional spherical particles. However, the significant differences in the sizes of carbon tubes, of course, determined that their properties of the corresponding nanofluids turned out to be quite heterogeneous.

The complexity of the topic is that the spectrum of nanofluids is extremely broad. For a long time, it was unclear which nanofluids could be compared with each other and which properties were most significant and distinctive. Since the time of J. C. Maxwell and A. Einstein, who have created the first theories of thermal conductivity and viscosity of dispersed liquids, it was believed that these properties of suspensions were determined only by the concentration of dispersed particles. However, as it turned out, this is not the case for nanofluids. Over the past two decades of studying nanofluids, numerous good monographs and extensive reviews have been dedicated to delineating their properties. However, the unstoppable growth of publications revealed more and more new nuances, which required making appropriate adjustments to the understanding of these properties. It was particularly difficult to understand and explain the causes why the thermophysical properties of nanofluids were found to be non-standard. The reasons for this are clear. The transport processes are of a molecular nature, and all experiments on the study of transport coefficients are macroscopic and interpreted through well-known macroscopic theories, usually hydrodynamic. Other tools were needed. Awareness of this problem came quite a long time ago, and already about ten years after the beginning of the study of nanofluids, attempts were made to answer these questions using molecular dynamic modeling. The complexity of implementing this program is well understood, its implementation required significant computing resources and time, time

Moreover, it was necessary to combine systematic experimental study of thermophysical properties of nanofluids and their molecular dynamic simulation. The latter is in fact an ideal experiment as it enables the consequent investigation of the impact of different factors on the transport processes. Additionally, simulating these processes permits the examination of their mechanisms and the contribution of different interactions, such as intermolecular effect, interaction of molecules with nanoparticles, and nanoparticles with one another. Implementing this program has been a goal of the author and his research team for many years. The present day, many important issues can be already answered quite clearly. Consequently, it is possible to predict the properties of nanofluids even without possessing a comprehensive theory describing their transport processes. In fact, this makes it possible to create nanofluids with controlled and predetermined properties. Today it is already possible to formulate a range of important and meaningful conclusions regarding the properties of nanofluids. Confidence in such ability prompted the author to write this book.

The volume of this book is necessarily quite large, since I wanted to touch upon all the significant processes of transference. Nevertheless, it should not be seen as a comprehensive overview of current knowledge. The author focused mainly on his studies and that of his research team. However, each section includes an extensive review of data available in literature.

The book consists of eight chapters. Chapter 1 provides a definition of nanofluids and presents various techniques of nanofluid preparation, as well as their characterization, and stability control. In addition, it briefly addresses the production methods for ready-made nanoparticle powders, which are used in one of the primary methods for preparing nanofluids.

Chapter 2 is devoted to the study of diffusion of nanoparticles and carbon nanotubes. Since it presents not only experimental data but also the results of molecular dynamic simulation, the molecular dynamics method is briefly discussed beforehand. This chapter begins with the formulation of the fluctuation-dissipation theorems, which establish the theoretical foundation of the entire gas and liquid transport theory. Subsequently, this chapter provides an in-depth analysis of the diffusion of carbon nanotubes, with a specific focus on their rotational diffusion that is exceedingly challenging to investigate experimentally.

Chapter 3 analyzes the viscosity and rheology of nanofluids. The discussion begins with a preliminary analysis of the viscosity of coarse dispersed fluids, followed by a review of the theories of A. Einstein, G. K. Batchelor, and others, as well as the underlying viscosity mechanisms. Finally, this chapter presents experimental data on the viscosity and rheology of nanofluids with both conventional nanoparticles and carbon nanotubes. To understand the momentum transfer mechanisms in nanofluids, which determines their viscosity, the molecular dynamics method is employed. This approach entails calculating the viscosity coefficient of various nanofluids and analyzing their rheology. Additionally, since some nanofluids exhibit non-Newtonian properties, their microrheology is investigated. Further, the rheology of nanofluids is comprehensively examined using the molecular dynamics method.

Chapter 4 is critical for utilizing nanofluids in diverse thermophysical applications. It focuses on nanofluid thermal conductivity, examining both conventional spherical particles and carbon nanotubes. The chapter begins with an objective examination of Maxwell's classical theory. The subsequent analysis focuses on the heat conduction mechanisms, discussing both longstanding explanations and new data from molecular dynamic simulation.

Chapter 5 explores the electrical conductivity of nanofluids. This chapter, having a structure similar to the previous one, investigates the electrical conductivity of nanofluids containing both conventional nanoparticles and carbon nanotubes. The electrical conductivity mechanisms are systematically analyzed and compared to thermal conduction mechanisms.

Chapter 6 focuses on the convective heat transfer in nanofluids and presents systematic experimental data. The data show that the heat transfer efficiency is primarily determined by the flow regime of nanofluids, whether laminar or turbulent.

Chapter 7 deals with the nanofluid flow simulations. It is shown that the simulation presented requires utmost caution when applying traditional approaches due to its non-standard nature. This is particularly evident when using various similarity parameters. Further, the simulation data on heat transfer of nanofluids with both conventional nanoparticles and carbon nanotubes are provided. In all cases, experimentally measured transport coefficients are used. This chapter ends with a discussion of the laminar-turbulent transition in nanofluid flows. The data obtained for a number of model flows are presented, which are then confirmed by experimental data.

The concluding Chap. 8 examines the topic with which most of the articles begin, namely, application areas of nanofluids. This discussion has been relegated to the end of the book to demonstrate first extraordinary properties and immense potential of nanofluids. After reading the book, reader can expect to identify additional prospective applications of nanofluids that are not addressed in this book. Finally, this chapter discusses some important aspects of the properties of nanofluids that were not previously described. These properties were not considered mainly due to the fact that there is insufficient clarity in the study of these properties of nanofluids and further systematic studies are required. Such properties include the optical properties of nanofluids, phase transitions in them, transport properties of nanofluids in confined conditions and boiling in nanofluids. Finally, a new type of nanofluids, hybrid nanofluids, has been shortly discussed.

The concluding section provides a summary of the research findings. Here, the features of nanofluids are again fixed. The focus is on discussing transport mechanisms in nanofluids.

Novosibirsk, Russia Valery Rudyak

Acknowledgements All the results discussed in this book, where the author's name is mentioned, were obtained with the valuable assistance of the colleagues, students, and graduate students, who were or are currently members of the author's research team. These are A. A. Belkin, E. G. Bord, A. A. Gavrilov, D. Geotto, D. V. Guzey, G. R. Dashapilov, S. V. Dimov, V. V. Egorov, D. A. Ivanov, E. B. Isakov, S. L. Krasnolutskii, V. V. Kuznetsov, E. V. Lezhnev, D. N. Lyubimov, A. S. Lobasov, A. V. Minakov, M. I. Pryazhnikov, T. A. Rafalskaya, E. A. Tomilina, S. N. Tretyakov, and A. A. Shupik. I am grateful to all of them; without them this book would not have appeared.

This work was supported by the Russian Science foundation (Agreements No. 20-19-00043 and 23-19-45002).

Competing Interests The author has no competing interests to declare that are relevant to the content of this manuscript.

Contents

Nomenclature

a	Perturbation propagation velocity or velocity of sound
B	Magnetic field induction
b	Lattice constant
C_p	Specific heat capacity
C_{pp}	Specific heat capacities of nanoparticles
C_{pf}	Specific heat capacities of base fluid
C_v	Specific heat capacity
c	Thermal velocity of the base molecules
c_p	Thermal velocity of the particle
$\boldsymbol{D}$	Strain rate tensor
D	Diffusion coefficient
D_E	Einstein's diffusion coefficient
D_r	Rotational diffusion coefficient
D_T	Translational diffusion coefficient
D_{Th}	Thermodiffusion coefficient
d	Diameter of nanoparticle or CNT
d_c	Channel diameter
d_e, d_i	External and internal diameters of the heated channel, respectively
d_h	Hydraulic diameter of the channel
d_m	Diameter of the molecule
d_p	Diameter of the particles
$\boldsymbol{E}$	Vector of the electric field strength
E_α	Partial energy of component α
$\boldsymbol{e}$	Unit vector
$\boldsymbol{F}_i$	Force acting on the molecule *i*
$\boldsymbol{F}_{ij}$	Interaction force of particles *i* and *j*
F_s	Stokes force
$\boldsymbol{f}(t)$	Langevin random force
$f(R_0)$	Size distribution of nanoparticles
f_1	Single-particle distribution function

$f_2(\boldsymbol{r}_1, \boldsymbol{r}_2)$	Binary distribution function
f_d	Size distributions of nanoparticles
f_M	Maxwell one-particle distribution function
G	Mass flow rate
G'	Elastic modulus
g	Acceleration of gravity
g_2	Radial distribution function
H	Characteristic linear dimension of the flow
h	Enthalpy
h_K	Kapitza radius
Ha	Hartmann number
I	Moment of inertia of the particle
$\boldsymbol{J}$	Current density vector
$\boldsymbol{J}_i$	Fluxes of macroscopic characteristics
$\boldsymbol{J}_2$	Stress tensor
$\boldsymbol{J}_3$	Heat flux
$\boldsymbol{j}_{D2}$	Diffusion flux of nanoparticles
$\boldsymbol{j}_Q$	Heat flux
$\boldsymbol{j}_{2\alpha}$	Interfacial interaction strength
$j_{3\alpha}$	Energy production due to the inter-component interaction
k	Boltzmann constant
k_1 and k_2	Slip and thermal slip coefficients
k_v	Consistency parameter
Kn	Knudsen number
L	Length of CNT
L_{ch}	Length of the channel
L_s	Characteristic linear scale of the system
l	Mean free path of gas molecules
l_c	Length of the cell edge
l_i	Mobility of ions
l_{mp}	Average distance between particles
l_{np}	Mobility of nanoparticle
l_{ph}	Mean free path of phonons
M	Particle mass
m	Molecule mass
m_i	Mass of particle i
N	Number of molecules in the system
N_{hf}	Number of molecules in the hydrodynamic physically infinitesimal volume
$\boldsymbol{n}$	Normal
n	Magnetic moment of the particle
n_f	Numerical density of liquid molecules
n_M	Numerical density of MWCNTs
n_p	Numerical density of dispersed particles
n_S	Numerical density of SWCNTs

Nu	Nusselt number
p	Pressure of the flow
p_0	Unperturbed pressure in the flow
p'	Pressure disturbance
Pe	Peclet number
Pr	Prandtl number
Pr_r	Relative Prandtl number
q	Heat flux
R_0	Radius of nanoparticle
R_{cnt}	Radius of CNT
R_h	Effective hydrodynamic radius
R_K	Kapitza resistance
Re	Reynolds number
Re_{cr}	Critical Reynolds numbers
$\boldsymbol{r}_i$	Radius-vector of the particle i
$\mathbf{r}_{ij}$	Radius-vector between the centers of the interacting particles
r_{hf}	Hydrodynamic physically infinitesimal length scale of the carrier fluid
r_{hp}	Hydrodynamic physically infinitesimal length scale of the particles pseudogas
r_{hnf}	Hydrodynamic physically infinitesimal scale of nanofluid
$\mathbf{S}$	Strain rate tensor
$\boldsymbol{S}$	Principal moment of the random forces
S	Cross-sectional area of the channel
S_{cnt}	Surface area of a CNT
S_{sp}	Surface area of spherical nanoparticles
$\tilde{S}$	Specific surface area of the powder
T	Temperature
$T_f(x)$	Liquid temperatures
T_o, T_i	Liquid temperatures at the channel inlet and outlet, respectively
T_w	Local temperature of the channel wall
$\bar{T}$	Mean liquid temperature
T_α	Partial temperature of the component α
$T_{\alpha\beta}$	Diffusion temperatures
T_m	Fusion temperature of the material
t_p	Plateau time
$\boldsymbol{U}$	Unit tensor of the second rank
U	Potential energy
U_a	Activation energy
U_G	Velocity determined by the flow rate
U_s	Sedimentation velocity
$\boldsymbol{u}$	Flow velocity
$\boldsymbol{u}'$	Velocity disturbance
$\boldsymbol{u}_\alpha$	Partial velocity of component α
$\boldsymbol{u}_{\alpha\varphi} = \boldsymbol{u}_\alpha - \boldsymbol{u}_\varphi$	Diffusion rates

u_p	Macroscopic velocity of the particle pseudogas
u_w	Velocity of the channel wall
V	Particle velocity
$\boldsymbol{V}_c$	Velocity of the particle's center of mass
V	Volume of the system
V_{np}	Volume of the nanoparticle
V_{CNT}	Volume of the CNT
$V_p,\ V_s$	Volumes of nanoparticle and adsorbed layer
$\boldsymbol{v}$	Velocity of the gas molecules
$\boldsymbol{v}_i$	Molecule velocity
v_{hf}	Hydrodynamic physically infinitesimal volume
v_s	Sedimentation rate
w	Weight concentration
α	Thermal diffusivity
α_c	Local heat transfer coefficient
α_{cw}	Local heat transfer coefficient of water
α_i	Lagrange multipliers
α_r	Relative local heat transfer coefficient
$\bar{\alpha}_c$	Average heat transfer coefficient
$\bar{\alpha}_r$	Average relative heat transfer coefficient
β	Volume expansion coefficient
γ	Drag coefficient
$\dot{\gamma}$	Shear rates
$\dot{\gamma}_{cr}$	Critical shear rate
$\boldsymbol{\gamma}_r$	Tensor drag coefficient
γ_r	Rotational drag coefficient
δ	Steric parameter
ε_p	Virial parameter of the dispersed particles pseudogas
ϵ_r, ϵ_0	Relative dielectric constants of particles and vacuum, respectively
ε_{ij}	Depth of the potential well
$\varsigma(t)$	Drag coefficient of nanoparticle
ζ	Coefficient of the thermal-hydraulic efficiency
ι	Root-mean-square deviation of the weight fraction of the mixture component from its mean value
$\kappa = \lambda_f/\lambda_p$	Relative thermal conductivity
Λ	Dimensionless wavelength of unstable disturbances in a cylindrical channel
λ	Thermal conductivity coefficient
λ_{br}	Thermal conductivity coefficient contribution caused by Brownian motion
λ_f	Thermal conductivity coefficient of based fluid
λ_p	Thermal conductivity coefficient of dispersed particle
λ_r	Relative thermal conductivity coefficient
μ	Viscosity coefficient

μ_{app}	Apparent viscosity coefficient
μ_b	Viscosity coefficient of the base fluid
μ_e	Effective viscosity coefficient
μ_f	Water viscosity coefficient corresponding to the temperature at the channel axis
μ_i	Transport coefficients
μ_r	Relative viscosity coefficient
μ_w	Water viscosity coefficient at the wall temperature
υ	Kinematic viscosity coefficient
ξ	Hydraulic resistance coefficient.
ρ	Density of the carrier fluid
ρ_f	Liquid densities
ρ_p	Density of the particle material
ρ_{eff}	Effective density of a nanoparticle
ρ_α	Partial density of component α
ρ_s	Densities of surfactant
σ_{bf}	Specific electrical conductivity of the base fluid
σ_{ij}	Effective size of the particle
σ_K	Kapitza conductivity
σ_p	Specific electrical conductivity of dispersed particles
σ_r	Relative electrical conductivity coefficient
$\boldsymbol{\tau}, \tau$	Stress tensor
τ_0	Interaction time of molecules
τ_p	Time to reach the plateau value
τ_y	Yield stress
$\Phi(\lvert\boldsymbol{r}_k - \boldsymbol{r}_m\rvert)$	Interaction potential of atoms
Φ_{LJ}	Lennard-Jones potential
Φ_{mp}	Interaction potential of molecule and nanoparticle
Φ_{pp}	Interaction potential of nanoparticles
$\chi_i(\boldsymbol{r}, t, t+t_1)$	Two-time correlation functions
χ_1	Velocity autocorrelation function
χ_{VV}	Velocity autocorrelation function of Brownian particle
$\chi_{\omega\omega}$	Angular velocity autocorrelation function
χ_μ	Stress autocorrelation function
$\chi_i(\boldsymbol{r}, t, t+t_1)$	Two-time correlation functions
Ψ	Stream function
$\boldsymbol{\Omega}$	Vorticity tensor
Ω	Cross section of the particles interaction
Ω	Angular velocity of the cylinder
$\boldsymbol{\omega}$	Angular velocity
ω_0	Magnetic permeability of vacuum
ω_i	Disturbance growth rate

Chapter 1
Nanofluids: Definition, Creation, and Characterization

1.1 Introduction Definition of Nanofluids

Nanofluids are a two-phase system consisting of a carrier fluid and nanoparticles. Nanoparticles are particles with characteristic sizes ranging from 1 to 100 nm. The term nanofluids first appeared in [13]. Later in our papers [64–66, 70, 71] gas nanosuspensions were also called nanofluids. This was done for several reasons. First, both gas nanosuspensions and nanosuspensions are commonly used in practical applications. Secondly, the transport properties of neither of them are usually described by classical theories. Thirdly, numerous properties of nanofluids and gas nanosuspensions are similar, especially if the carrier gas in the latter is dense enough. Finally, transport processes in both cases are often studied using similar methods and models. For example, the procedure for implementing the molecular dynamics method is the same in both cases. Nevertheless, in the literature, the term nanofluid is commonly understood as a dispersed fluid, where the carrier component is a liquid, i.e., a suspension. This book will focus on discussing such nanofluids specifically. Any reference to gas nanosuspensions will be limited to special notes if mentioned at all.

Carrier fluids typically include water, ethylene glycol, machine oils, some organic liquids, etc. The spectrum of nanoparticles is also quite broad. These are metals, different oxides, and some exotic particles, such as nanodiamonds, carbon particles, nitride, sulfide, and polymer particles, etc. The smallest nanoparticle is considered to be fullerene, with a diameter of approximately one nanometer. Viruses occupy an intermediate position, with sizes on the order of tens of nanometers. In general, particles made from various materials can be considered, as long as their sizes fall within the specified range. Typically, these particles are spherical or nearly spherical (see Fig. 1.1a). As will be demonstrated later, many of these particles are formed through the condensation of corresponding droplets, which explains their spherical shape. Thus, the material of possible nanoparticles has quite different properties. For this reason, one could try to classify nanofluids by the type of material of the nanoparticles. Although the thermophysical properties of nanofluids depend on nanoparticle

V. Rudyak, *Thermophysical Properties of Nanofluids*, Fluid Mechanics and Its Applications 140, https://doi.org/10.1007/978-3-031-97995-8_1

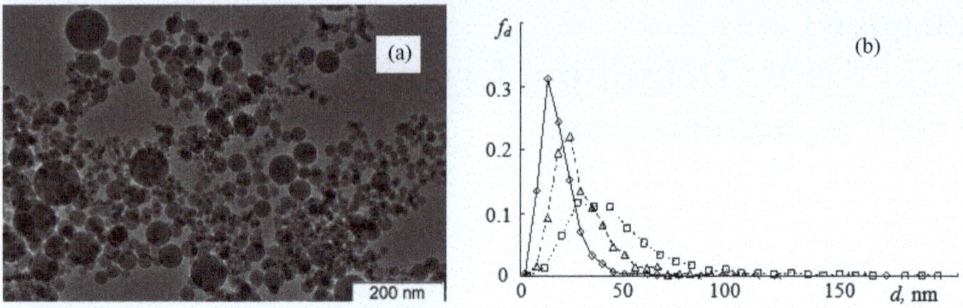

Fig. 1.1 Electron micrograph of silicon oxide nanoparticles with an average size of 28.3 nm (**a**) and differential size distributions of oxide nanoparticles (**b**)

material, the introduction of such a classification seems unnecessary, despite such attempts have been made in literature.

Nanofluids have been actively studied since the mid-nineties of the last century. Their defining feature is that the majority of dispersed particles (both in terms of volume and weight) are nanoparticles. This enables determining an average particle size, which is the most crucial characteristic of nanofluids. Therefore, when studying any property of nanofluids, it is necessary to strictly control the size of the nanoparticles, including their size distributions. Of course, it is usually not possible to create monodisperse nanofluids in real conditions. A typical electron micrograph of silicon oxide nanoparticles with an average size of 28.3 nm in an ethylene glycol-based nanofluid is shown in Fig. 1.1a [69]. The size distribution range of nanoparticles should to be quite narrow. Figure 1.1b displays typical nanoparticle size distributions f_d in nanofluids based on ethylene glycol containing silicon oxide particles of three various sizes (here d is the particle diameter).

On the other hand, in the literature, fluids containing dispersed CNTs are also commonly referred to as nanofluids. However, it is important to note that these are special type of nanofluids. Depending on the production technology, CNTs can be single-walled (SWCNT) or multi-walled (MWCNT) [29, 34]. SWCNTs have a diameter of approximately one to two nanometers, while MWCNTs have diameter that can reach tens of nanometers. The former range in length from hundreds of nanometers to a few micrometers, whereas the latter can be up to tens of micrometers in length. Thus, one of the charecteristic sizes of CNTs falls within the definition of nanofluids, while the latter may be quite macroscopic. Nanofluids containing CNTs merit individual consideration. In certain situations, they exhibit conventional dispersed fluid behavior, while in others, they at least structurally resemble polymer solutions or even liquid crystals. The latter can be realized in external fields, for example, in a shear flow. The presence of nanotubes in a fluid, particularly at significant concentrations, essentially alters the structure of the fluid. Therefore, studying the properties of such fluids without examining their structure is not feasible. Thus, when discussing nanofluids, it is important to specify what nanoparticles are used as dispersed component.

In ordinary coarse dispersed fluids, at sufficiently low particle concentrations, the structure of the carrier fluid remains practically unchanged. In fact, macroscopic particles are commonly streamlined by fluid. However, the situation is distinctly different in nanofluids. Here, even at low nanoparticle concentrations, their number per unit volume is very large. This leads to the fact that, firstly, the particles actively interact with each other through the medium, secondly, they can interact directly, and thirdly, the dynamics of the molecules of the carrier fluid in the vicinity of each nanoparticle is correlated with its dynamics. This leads to a change in the structure of the carrier fluid, not just in the nanoparticle's proximity, but throughout the entire volume. Further details on this matter will be discussed in subsequent chapters.

The particles shown in Fig. 1.1 have a spherical or close to spherical shape. In general, however, their shape may not be spherical. Nanoparticles with noticeable asymmetry are widely distributed. Usually, they can almost always be modeled by spheroids whose longitudinal and transverse dimensions differ. If these dimensions are close, the properties of nanofluids will almost match those of nanofluids with spherical particles. Otherwise, these properties may differ significantly. In particular, the rheology of nanofluids can change. Other important consequences may also be present. In these types of fluids, nanoparticles possess both translational and rotational degrees of freedom, which can significantly impact their transport properties. Further discussion on this matter will be addressed later.

The nanoparticle size is a fundamental factor determining the structure of conventional nanofluids with spherical particles. The characteristic sizes of nanoparticles vary by two orders of magnitude. The smallest ones are only two to three times larger than conventional inorganic molecules while the largest ones are similar in size to Brownian particles (see Fig. 1.2). To better understand this, it is useful to classify the situations that arise.

Nanofluids with CNTs are radically different from conventional nanofluids. CNTs, unlike spherical particles, have two characteristic dimensions: diameter d and length L. The structure of such fluids should depend essentially on the properties of CNTs and on their aspect ratio, i.e. the ratio of CNT length to its diameter. This ratio for SWCNTs varies from hundreds to several thousands, and for MWCNTs it varies from tens to thousands. Thus, nanofluids containing ordinary spherical particles and CNTs have notably different sizes of internal structural elements, ultimately resulting

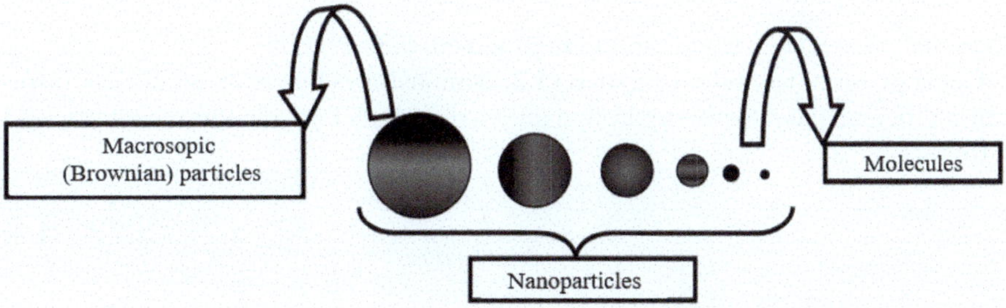

Fig. 1.2 Schematic diagram illustrating the characteristic size spectrum of nanoparticles

in significant variations in their overall structure. On the other hand, the thermophysical properties of fluids are determined by their structure. This becomes particularly apparent when comparing the properties of gases, liquids, and ordinary coarse dispersed fluids. The dimensions of the internal structural elements of nanofluids are a crucial factor in studying their properties. Therefore, they should be thoroughly examined in all experiments aimed at measuring any nanofluid property.

Can the characteristic sizes of conventional spherical nanoparticles and CNTs be compared? The answer is yes. In fact, most methods for measuring the size of particles or CNTs in nanofluids use the concept of their effective size. There are multiple approaches to determine effective size. The determination of particle size is commonly achieved through the dynamic light scattering (DLS) method, as will be outlined in Sect. 1.3. This method enables the determination of the diffusion coefficient D of nanoparticles using the Stokes–Einstein formula [20]

$$D = \frac{kT}{6\pi \mu R_h}, \tag{1.1}$$

where μ is the viscosity coefficient of the carrier fluid, T is the temperature, k is the Boltzmann constant, and R_h is the effective hydrodynamic radius of the particles under study. In practice, there is typically a particle size distribution present in nanofluids (see Fig. 1.1). Therefore, in fact, formula (1.1) accounts for an effective hydrodynamic radius of the nanoparticles (even if they are all spherical) at which the experimentally measured diffusion coefficient is attained. Formula (1.1) is also used to determine the effective hydrodynamic radius of CNTs. However, for nonspherical particles the relation (1.1) is modified. In particular, for cylindrical tubes of length L, when their length is much greater than the diameter d it takes the form [18]

$$D = \frac{kT \ln(L/d)}{3\pi \mu L}. \tag{1.2}$$

Often in practice, when the aspect ratio is not very large, a slightly modified formula is used

$$D = \frac{kT \ln[(L/d) + \zeta]}{3\pi \mu L}, \tag{1.2a}$$

where the parameter ζ takes values ranging from 0.32 to 0.86.

Comparison of formulas (1.1) and (1.2) demonstrates that at an equivalent diffusion coefficient, the effective hydrodynamic size and CNT parameters are interrelated as follows

$$R_h = \frac{L}{2 \ln(L/d)}. \tag{1.3}$$

If the diameter of the CNT is known, this relation becomes quite definite. Otherwise, it determines the relationship between the effective hydrodynamic radius of the

CNT and its length, which has been reduced by a factor of $2\ln(L/d)$. For small aspect ratios, the denominator in formula (1.3) should be $2\ln[(L/d)+\zeta]$. When the aspect ratio is of the order of one, the effective hydrodynamic radius exceeds the length of CNT. On the contrary, in cases where the aspect ratio is large, the CNT length is significantly greater than R_h. Thus, at $L/d \sim 10^3$, which is typical for SWCNTs, the hydrodynamic radius is smaller than the CNT length by about twenty times.

For spherical particles, the average radius of nanoparticles can be defined using relation (1.1). Their average effective volume can be expressed as $V_{np} = (4/3)\pi R^3$. This quantity is important since it has a direct correlation to the particles' volume concentration. On the other hand, the average effective volume of CNT is $V_{CNT} = (\pi/4)d^2L$. These two volumes can be equal when the following relation is fulfilled

$$R = \left(\frac{3d^2L}{16}\right)^{1/3}. \tag{1.4}$$

It is evident that SWCNTs and MWCNTs have essentially different effective hydrodynamic radii. This, in particular, means that at the same effective hydrodynamic radius they will have essentially different number concentrations at the same weight concentration.

1.2 Methods of Nanofluid Preparation

Nanofluids are complex two-phase media, requiring extensive efforts and continual monitoring during their preparation. Nanofluids should maintain their stability and properties for at least a considerable duration. Furthermore, it is preferable that they are chemically neutral. The stability and thermophysical properties of nanofluids can be significantly altered depending on the preparation method used. This alteration is particularly affected by the size of particles. Due to their small size and developed active surface area, nanoparticles can form agglomerates. The latter change the structure and size distribution of dispersed nanoparticles at a certain concentration. Sufficiently large nanoparticle agglomerates, having macroscopic sizes, can sediment, thus changing the concentration of nanoparticles. Therefore, it is necessary to use special techniques when creating nanofluids with a given concentration and average particle size. An important aspect to consider is the lifetime of the nanofluid, i.e. the period during which its properties remain unchanged. Therefore, when discussing about the use of nanofluids or the study of their properties, it is always necessary to pay due attention to their preparation. To date, the preparation of nanofluids is still mainly the domain of research teams, although there are already firms in several countries offering certain nanofluids on the market. The manufacturer guarantees quite certain properties and lifetime of nanofluid, but still these parameters need to be checked. Additionally, the manufacturer typically does not disclose the precise composition of the nanofluid they produce since it is company's know-how. This creates a situation in which purchasers are essentially buying a black box, making it extremely challenging, if not impossible, to investigate properties of these nanofluids. For instance, it is often unclear what base fluid was used. For research purposes, it is essential that the nanofluid has a controlled composition to enable further monitor of its properties. This section briefly describes the main techniques for preparing nanofluids, which significantly impact their properties.

1.2.1 Single-Step Method

There exist two main techniques for preparing nanofluids, namely, single-step and two-step [17, 50, 52, 93, 101]. In the first method, the nanofluid is concurrently produced while synthesising the nanoparticles. Various techniques are employed, with the most common being the generation of nanoparticles through melting and vaporization of the target material based on the physical vapor deposition technique, as well as different chemical methods, such as the liquid chemical method. One of the first single-step physical methods was described in [3, 99], which is the vacuum evaporation onto a running oil substrate (VEROS-method). This method has been modified multiple times [10, 19, 43, 44, 61, 87, 88]. Usually, this method has been mainly used to create nanofluids with highly reactive particles, such as rapidly oxidizable metallic nanoparticles (Cu, Ag, Al). Some oxides show a high tendency to aggregate unless they are surrounded by a dissolved surfactant.

A significant benefit of the single-step physical method is the low nanoparticle agglomeration. Besides, the resultant particles are almost similar in size. On the other hand, this technology has its drawbacks. It is extremely difficult to separate nanoparticles here and create sufficiently noticeable and controlled concentrations of them in the fluid. Furthermore, particle condensation and their immersion in the fluid (i.e., the production of nanofluid) occurs in a vacuum chamber. This method is unsuitable for producing significant quantities of nanofluids due to the high cost of necessary equipment.

A much simpler and more economical is a single-step chemical method. This method was first used in Zhu et al. [105]. A schematic diagram of the method is shown in Fig. 1.3. The nanofluid is produced through the reaction between liquid 1 and liquid 2, and subsequent treatment of the reaction products in an ultrasonic or microwave bath. Zhu et al. [105] have obtained 10 nm copper nanoparticles by reducing inorganic copper salts with sodium hypophosphite in various base fluids. This method was employed for obtaining mineral oil-based nanofluids with silver nanoparticles. The single-step chemical method has also been used to produce nanofluids with other particles [25, 40, 102], including MWCNTs [51]. The primary benefit of the single-step chemical method is the ability to acquire uniformly-sized nanoparticles. Nevertheless, a challenge lies in eliminating hard to remove reagents from the nanofluid. Additionally, the base fluid becomes poorly predictable, its thermophysical properties are usually unknown.

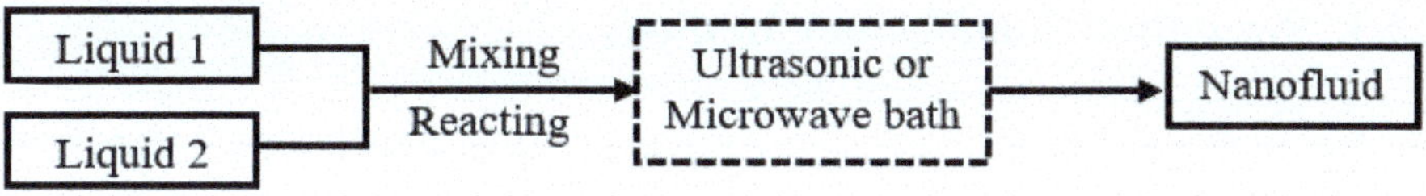

Fig. 1.3 Schematic diagram of a single-step chemical method for preparing nanofluids

1.2.2 Two-Step Method

The idea of the second method is extremely simple. First, a base fluid and a nanopowder are selected. Then, to create a given weight concentration of nanoparticles, an appropriate amount of nanopowder is added to a certain volume of the base fluid (Fig. 1.4). The resulting dispersed fluid is stirred mechanically.

However, a nanopowder with a specified mean particle size commonly includes a significant number of agglomerates. To eliminate these agglomerates, the nanofluid undergoes ultrasonic treatment (UST) at a second step (see, for example, [14, 27]). The dispersion mechanism of nanoparticle or CNT agglomerates during UST is assosiated with the creation of cavitation bubbles in the fluid. When these bubbles collapse, localized shock waves are generated, which destroy the aggregates. The process of splitting CNT bundles occurs in multiple stages and is aptly defined as the "unzipping mechanism" [80]. The processing time generally ranges from a few tens of minutes to several hours. Sometimes, ultrasonic processing is replaced with intense magnetic force agitation or shear flow homogenization [92]. However, the latter method is technologically more intricate and not universally applicable.

The two-step method is highly advanced technologically as it incorporates a pre-prepared nanoparticle powder. The process of preparing the nanofluid can be continuous, resulting in mass production and cost-effectiveness. This method is suitable for the industrial production of nanofluids. It is convenient for investigating the properties of nanofluids. This book presents the results of studying the thermophysical properties of nanofluids, prepared using this particular two-step method. An important benefit of the approach is the ability to monitor the weight (and volume) concentration of nanoparticles and the average size of particles in nanofluids. This is extremely important because, as will become clear later, all thermophysical properties of nanofluids are dependent on these two factors.

Fig. 1.4 The scheme of the two-step method of nanofluid preparation

1.2.3 Nanofluid Stabilization Methods

One disadvantage of the two-step method for preparing nanofluids is the initial usage of nanopowders, which may contain a large number of agglomerates of varying sizes. In the second step, UST can usually cleave most of these agglomerates, resulting in a nanofluid with a fairly uniform distribution of nanoparticles. However, nanoparticles and CNTs are constantly in thermal motion and can agglomerate directly in the nanofluid. This is a distinct characteristic of CNTs due to the significant van der Waals forces acting between them. For instance, 100 nm long SWCNTs exhibit an interaction energy of 2.9 keV [16]. The presence of such interactions may lead to the formation of agglomerates. At the same time, it is clear that from a practical point of view, nanofluids should be stable and not change their properties for quite a long time. Moreover, agglomeration of nanoparticles or CNTs at a particular concentration significantly alters the fluid's structure and the size distribution of dispersed nanoparticles. Therefore, the thermophysical properties of nanofluid will change. Large enough nanoparticle agglomerates, which have already reached macroscopic sizes, can sediment and alter the nanoparticle concentration.

Nanoparticles possess high surface energy, making them highly active and thus increasing the probability of their agglomeration. As previously noted, agglomeration is associated with van der Waals attraction forces acting at large distances between nanoparticles. The characteristic scale of these forces is about a dozen nanometers (or even dozens of nanometers). On the other hand, for example, the average distance between nanoparticles of diameter $d \sim 10$ nm at a volume concentration $\varphi \sim 10^{-2}$ is of the order of $l_{mp} \sim d/[3]\varphi \sim 45$ nm. Consequently, nanoparticles largely remain within the range of action of van der Waals forces and actively interact with one another. The probability for possible agglomeration rises as the volume concentration of nanoparticles increases.

There are several techniques to preventing the agglomeration of nanoparticles or CNTs, with the use of surfactants (SUR) being one of the primary methods. The first step when using a surfactant is to create an initial base fluid (IBF), followed by the addition of the necessary amount of surfactant to the IBF. As a result, the base liquid (BF) is described by the formula: (BF) = (IBF) + (SUR). The resulting nanofluid is characterized by a certain weight concentration of surfactant and nanoparticles or CNTs. The weight concentration w is converted to the volume concentration φ using the formula

$$\varphi = \frac{w\rho}{w\rho + (1-w)\rho_p}. \tag{1.5}$$

where ρ_p and ρ are the density of the particle material and the carrier fluid, respectively.

In addition to the existing Brownian motion of nanoparticles, which prevents their agglomeration (however, Brownian motion also promotes agglomeration, because it is due to this motion nanoparticles are brought to the distance of the van der Waals forces), three primary stabilization mechanisms are distinguished, namely, electrostatic, steric, and electrosteric. In the first case, electrostatic forces, resulting from the presence of similar electric charges on the nanoparticles, prevent their approach to each other. Another mechanism preventing agglomeration is weak electrostatic repulsion of the nanoparticles associated with the appearance of a double electric layer on the surface of moving nanoparticles, formed by ions adsorbed from the fluid. The potential difference, resulting between the dispersed fluid and the stationary fluid layer surrounding the particle, is called the zeta potential (ζ-potential). It arises from particle movement and the disruption of the double electric layer. The separation plane between solid and liquid phases as they move relative to each other is called the slip plane, and the potential at the slip plane is the ζ-potential. The electrostatic interaction between particles is measured by the ζ-potential, which is determined at the slip plane and is a primary factor affecting the stability of disperse systems. The higher the absolute value of the ζ-potential, the greater are the repulsive forces arising between the particles, and the more stable the nanofluid is.

Brownian motion and electrostatic repulsion typically have a weak effect on preventing nanoparticle agglomeration. In contrast, surfactants are highly effective in preventing nanoparticle agglomeration, with even small amounts causing significant changes to the system's surface characteristics. Some of the most commonly used surfactants include sodium dodecylbenzenesulfate (SDBS), sodium dodecyl sulfate (SDS), salt and oleic acid, dodecyltrimethylammonium bromide (DTAB), hexadecyltrimethylammonium bromide (HCTAB), polyvinylpyrrolidone (PVP), gum arabic (GA), and some others. Surfactants commonly consist of a hydrophobic tail and a hydrophilic head group positioned along the hydrocarbon chain. The different compositions of these groups lead to the categorization of surfactants into four distinct classes [101]: (i) nonionic surfactants that have no charged ion in the head group (for example, alcohols, polyethylene oxide, etc.; (ii) anionic surfactants with negatively charged head group; (iii) cationic surfactants with positively charged head group; (iv) amphoteric surfactants, whose charge depends on the pH of the fluid.[1] Roughly speaking, surfactants can be classified as either ionic or steric surfactants. The choice of surfactant is key to creating a nanofluid (in case when a surfactant is used). In the case of a polar base fluid, either a water-soluble or oil-soluble surfactant must be chosen. For non-ionic surfactants, the solubility can be determined by the hydrophilic-lipophilic balance (HLB). This indicator is commonly found in reference books. Fatty acids and their salts are frequently used for creating stable nanofluids with inorganic particles. Among the effective stabilizers, oleic acid stands out due to its ability to chemisorb onto the surface of zinc oxide nanoparticles [48].

The idea of using steric surfactants is quite simple. Surfactant molecules (see Fig. 1.5) adsorb on the surface of the nanoparticle, resulting in the creation of a protective layer. The adsorption of surfactant can occur through various ways. There

[1] pH is a hydrogen index indicating the intensity of medium acidity, determined by the concentration of hydrogen ions present. In perfectly pure distilled water, these ions balance each other. In such cases, the water is neutral. When different chemicals are dissolved in water or dispersed particles are added, this balance can be disturbed, resulting in a change in pH.

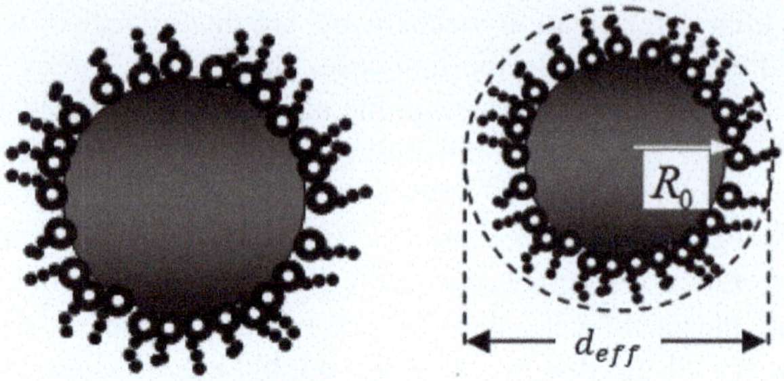

Fig. 1.5 Nanoparticle of radius R_0 with surfactant molecules adsorbed on its surface

are several surfactant adsorption models: (a) encapsulation into a surfactant cylindrical micelle; (b) hemimicellar adsorption of surfactant; and (c) random adsorption of surfactant molecules [103]. The so-called steric repulsion force arises between the layers of molecules absorbed on different nanoparticles. This force's repulsion mechanism is quite simple [63, 83] and effectively obstructs nanoparticle coagulation. The energy of such repulsion is described according to Rosensweig's model by the following relation [63, 83]

$$U_s = \begin{cases} 2\pi R_0^2 NkT\left[2 - \left(\frac{h+2R_0}{\delta}\right)\ln\left(\frac{\delta+R_0}{h+2R_0}\right) - \frac{h}{R_0}\right], & h \le 2\delta \\ 0, & h > 2\delta \end{cases}, \tag{1.6}$$

where R_0 is the radius of the nanoparticle, N is the number of surfactant molecules adsorbed on a unit surface of the nanoparticle, h is the distance between the surfaces of spherical nanoparticles, δ is a steric parameter equal, roughly speaking, to the thickness of the adsorbed layer.

Adsorption layers create a potential barrier that prevents particles in a nanofluid from approaching each other and coagulating. The width of this barrier should be greater than the characteristic scale of the zone in which the attractive forces act. Since this scale is about ten nanometres, it is clear that stabilizing a nanofluid is all the more difficult the smaller the particle size. This is because as the particle size decreases (at the same volume concentration), the average distance between the particles also decreases. Thus, using finer nanoparticles results in a lower concentration at which a stable nanofluid can be created.

An important class of nanofluids is represented by the so-called magnetic nanofluids, which include dispersed ferromagnetic particles, predominantly single-domain magnetic particles (metals Fe, Co and Ni, as well as rare-earth metals Gd, Tb, Dy, Ho, Er have ferromagnetic properties). Due to the fact that such particles are magnetised to saturation, such particles experience substantial magnetic attraction forces. The potential energy U of this interaction between two spherical particles of radius R_0 is inversely proportional to the cube of the distance between them [83]

$$U = -\frac{n^2}{2\pi\omega_0(2R_0 + h)^3}, \tag{1.7}$$

where n is the magnetic moment of the particle, ω_0 is the magnetic permeability of vacuum, and h is the distance between the surfaces of interacting spheres.

Depending on the nanoparticle material, whether oxide, magnetite, or metal, magnetic nanofluids are subdivided sometimes into oxide and metal nanofluids. Iron, nickel, cobalt, magnetite (Fe_3O_4), ferrite are the typical particles used to prepare magnetic nanofluids, with particle sizes ranging from approximately 2.6 nm for cobalt to 11 nm for nickel. Along with water, complex liquids such as kerosene, mineral oils, heptane, decane, alkylnaphthalene, perfluoropolyether, etc., are used as a base fluid. One notable feature of magnetic fluids in practical use is the requirement for high particle concentrations yet unlike conventional nanofluids. Therefore, volume concentrations can reach up to tens of percent, with a limiting concentration of approximately 0.65 [7]. In this case, distance between particles is smaller than their size. For this reason, particles actively interact with each other. In addition to the Van der Waals attractive forces, magnetic forces also act between dispersed magnetic nanoparticles. As a result, the agglomeration capacity of these particles far surpasses that of ordinary nanoparticles. Consequently, tiny nanoparticles, measuring around 1–10 nm, are continuously exposed to these forces even at low concentrations (for example, less than one percent). As a result, stable nanofluids cannot be created with these nanoparticles since they agglomerate and sediment quickly. Larger nanoparticles also undergo agglomeration. Therefore, creating stable magnetic nanofluids always requires using special methods.

The standard approach for generating stable magnetic nanofluids is using surfactants. If the carrier fluid is nonpolar (e.g. oil) a solitary layer of surfactant is sufficient. In this case the polar head of the surfactant adheres to the particle surface, while the carbon chain contacts with the carrier fluid. This is exactly the case shown in Fig. 1.5. On the other hand, if the carrier fluid is polar, such as water, a dual layer of surfactant molecules is produced, as shown in Fig. 1.6.

The stabilization of magnetic particles is achieved if there is compatibility between the surfactant phase and the carrier fluid. Usually, the best stabilizers are substances that are highly soluble in the base liquid. As a rule, these substances are comprised of a lengthy hydrocarbon chain (C_{10}–C_{20}), particularly those with double unsaturated bonds between carbon atoms or aromatic groups, and have functional groups (–OH, $-NH_2$,–COOH, $-OSO_3H$). The stabilization is attributed to a decrease in the surface

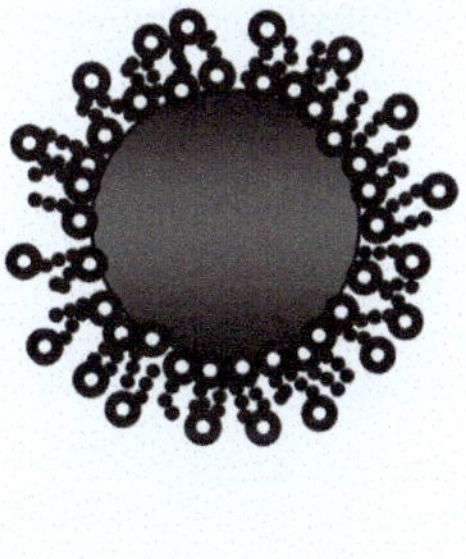

Fig. 1.6 Nanoparticle of radius R_0 with two layers of surfactant molecules adsorbed on its surface

energy of the nanoparticles and an increase in the absolute value of their electrokinetic potential, as well as due to the steric factor.

When selecting a surfactant, it should be borne in mind that the ideal situation is when the stabilizer is related to the base fluid (for example, using decane when the base fluid is kerosene or heptanes). If a non-polar magnetite nanofluid is mixed with a polar surfactant, such as ethanol or acetone, the particles will aggregate and rapidly sediment. If the settled particles are then placed in a pure base fluid (heptane, decane or kerosene), peptization of the particles occurs again. This suggests that the particles are flocculating along with the surfactant layer that is covering each particle (for a simple explanation of flocculation, refer to monograph of Michaelides [49]). However, certain fluids, when added to the magnetic fluid, can dissolve the surfactant, resulting in irreversible flocculation. It is worth noting that with an increase in stabilizer thickness the volume fraction of the magnetic material decreases rapidly, resulting in a decrease in saturation magnetization.

In addition to typical nonpolar magnetic nanofluids, there exist ionic counterparts that require no surfactants for stabilization. Electrostatics mechanism of stabilization is realized in these fluids. Magnetite or ferrite particles for these fluids are prepared through chemical precipitation, which generates an electrical charge on their surface through an acid-alkaline reaction with their surroundings. In case, if water is used as the base fluid, the pH of the solution can vary from 2 to 12 depending on the sign of the surface charge [73]. Acidic magnetic nanofluids ($pH < 7$) contain positively charged particles, while alkaline magnetic nanofluids ($pH > 7$) have negatively charged particles (see Fig. 1.7). The surface charge density depends on the pH of the fluid.

As mentioned earlier, besides steric surfactants, ionic surfactants also exist and can be categorized as either anionic (for example, SDS and SDBS) or cationic surfactants (for example, cetyltrimethylammonium bromide, CTAB; cetylpyridinium chloride, CPC; benzalkonium chloride, BAC; dioctadecyldimethylammonium bromide, DODAB). The former possess a negatively charged head group while the latter feature a positive charge at the hydrophilic tail. Adsorption of certain surfactants leads to the creation of electrostatic repulsion among nanoparticles.

The addition of surfactants to nanofluid is a simple and quite effective method for enhancing its stability. However, it should be noted that surfactants can only be used at temperatures that are not too high as high temperatures can cause deterioration of

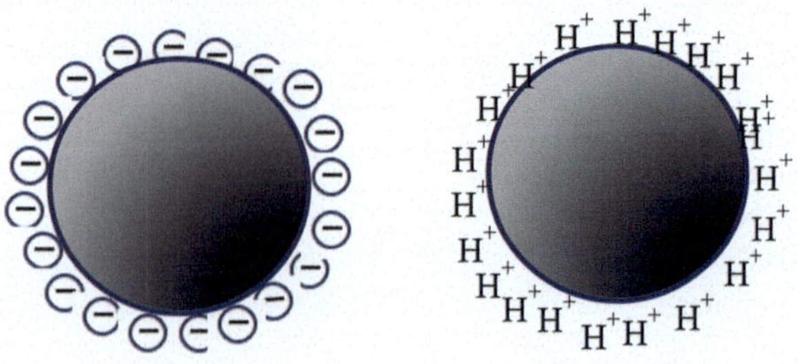

Fig. 1.7 Particle charges in ionic acidic (right) and alkaline (left) nanofluids

the bond between the surfactant and nanoparticle, as well as to surfactant degradation. Commonly, used surfactants degrade significantly at temperatures above 60 °C [26]. Significant foaming occurs with some surfactants. Additionaly, intensive UST may also result in surfactant degradation [4, 82]. Finally, it should be noted that the use of surfactants typically results in noticeable changes in the properties of base fluids. Specifically, their viscosity, thermal conductivity, and electrical conductivity may be changed. Therefore, when studying the thermophysical properties of nanofluids, it is always necessary to initially examine the corresponding properties of the base fluid obtained with the use of surfactants.

Surfactants alter the surface area of nanoparticles, their size (this is especially significant for small size particles), and density. Figure 1.5 illustrates that the effective diameter d_{eff} of a nanoparticle with surfactant increases by the size of the adsorbed layer of surfactant molecules. Typically, the thickness of this layer is 1–3 nm. Thus, small particle size can change by tens of percent or even several times. On the other hand, the density of such a composite nanoparticle significantly differs from that of the nanoparticle material. The expression defining the effective density ρ_{eff} of a nanoparticle with an adsorbed surfactant layer is

$$\rho_{eff} = \frac{\rho_p V_p + \rho_s V_s}{V_p + V_s}, \tag{1.8}$$

where ρ_p, ρ_s are the densities of the nanoparticle material and surfactant, respectively, and V_p, V_s are the volumes of the nanoparticle and adsorbed layer. It is easy to see that for particles with diameters up to ten to fifteen nanometers, the effective density (1.8) can be significantly different from that of the nanoparticle material.

The aggregation ability of CNTs is much higher than that of conventional nanoparticles; in addition, they are hydrophobic. Nevertheless, using organic solutions, it is possible to create nanofluids with CNTs without surfactants. However, the number of such solutions is limited. Alcohols such as ethylene glycol and isopropyl alcohol, as well as some intricate organic solvents, are examples of such solutions [5, 30, 38, 86].

The significant impact of surfactants on nanofluid properties has initiated numerous attempts to stabilize them without the use of such substances using the properties of the nanoparticles surface. Thus, the stability of water-based nanofluids (and also of ethylene glycol-based nanofluids, which very actively absorbs water from the atmosphere) will be significantly different for hydrophilic and hydrophobic particles. This approach was previously utilized by Chen and Xie [11] when adding a hydroxyl group onto the surface of carbon tubes. The chemical modification of CNT surface is a commonly employed method to generate stable nanofluids. This method is widely used when creating nanofluids with CNTs, which, as already mentioned, are hydrophobic and thus it is impossible to create a water-based nanofluid with CNTs. In this case, surfactants are used or CNTs are doped with some molecules [98]. Various CNT surface functionalizing methods are known. When the base fluid is polar (water or polar organic solvents, etc.), liquid-phase or gas-phase oxidation of the CNTs with various reagents, including nitric acid, nitrogen dioxide, ammonium persulfate, potassium permanganate, sodium hypochlorite, hydrogen peroxide, ozone, etc. are employed [41, 74, 75, 78, 90, 97].

Creating nonpolar fluid-based nanofluids with CNTs necessitates the utilization of complex functionalization methods or special additives. Some approaches can be found in [22, 77]. Functionalizing CNTs can alter their properties and consequently the properties of corresponding nanofluids. For instance, electrical and optical properties of nanofluids can be significantly modified [36, 86]. The objective of this book is to examine the thermophysical properties of nanofluids using conventional particles and CNTs. The author strived to study only completely natural CNTs, and thus, nanofluids with functionalized CNTs are beyond the scope of this book.

Finally, it is worth noting some less conventional methods of producing nanofluids. For instance, Yu et al. [100] employed plasma treatment of a mixture of methane and carbon gas to alter the surface of diamond nanoparticles. The introduction of a polar group on the nanoparticles' surface led to an improvement in their dispersion characteristics in water. In fact, this is also a specific method of functionalization. Other methods of nanoparticle surface modification have been utilized as well to minimize coagulation in nanofluids, as described in reviews [50, 89, 101]. However, it should be noted that these techniques are significantly labor-intensive and unsuitable for large-scale implementation.

When discussing the general principles of creating stable nanofluids, several circumstances should be noted. As demonstrated through various experiments, a stability of nanofluid is inherently associated with its electrokinetic properties. Primarily, it has been long established that the van der Waals attraction strongly correlates with the dielectric constant of the carrier fluid [39, 81]. As this constant increases, the stability of the nanofluid decreases. Another crucial aspect to consider is the pH level of the fluid. However, there is still a lack of systematic experimental data that clearly indicates the nature of this dependence.

The presence of identical electric charges on the surface of nanoparticles, of course, prevents their coagulation due to the action of repulsive Coulomb forces. This can usually be achieved by adsorption of the ion onto the surface of the metal particle. However, implementing electrostatic stabilization is technically arduous. Furthermore, this typically alters the pH of the solution, which might produce a contrary effect.

1.3 Characterization of Nanofluid and Control of Its Stability

Certainly, in the process of working with nanofluids, it is of the utmost importance to control its stability, including the average size of nanoparticles, their size distribution, concentration, etc. Although this remains a challenging issue, some clear and implicit recommendations can be made. The sedimentation method represents the simplest and most obvious method to determine nanofluid stability. Sedimentation is the process by which particles of a dispersed phase in a liquid or gas deposit under the action of a gravitational field or centrifugal forces. A constant sedimentation rate, for example in a gravity field, takes place when the gravity force acting on the particle is balanced by a fluid drag force. As the latter, the Stokes force acting on a spherical particle of radius R_0 which is $F_s = 6\pi \mu R_0 U_s$ can be taken for estimates. Then the sedimentation velocity U_s is described by the formula

$$U_s = \frac{gV_p(\rho_p - \rho_f)}{6\pi \mu R_0} = \frac{2gR_0^2(\rho_p - \rho_f)}{9\mu}, \tag{1.9}$$

where ρ_f is the liquid densities, g is the acceleration of gravity.

Simple estimates indicate that in laboratory or industrial installations (with characteristic dimensions ranging from one centimeter to several meters) for reasonable times the sedimentation of relatively small nanoparticles is practically absent. However, for particles with a characteristic size of the order of hundreds of nanometers, it becomes noticeable. Therefore, if the agglomeration processes are active enough, then, at some point, the dispersed particles will begin to actively sediment. By observing this process, it will be possible to infer the degree of nanofluid stability. This is a fairly typical method of monitoring the stability of nanofluids. In fact, it is used in one form or another by almost all research teams. However, to obtain precise measurements, it is necessary to employ videography. Then, by determining the sedimentation rate of the particles, the dynamics of the agglomeration process can be established [91]. Quantitative determination of the nanofluid concentration can be obtained using UV spectrometry. While using this technique simultaneously with sedimentation analysis, the nanoparticle agglomeration dynamics and nanofluid concentration variation can be studied.

Sedimentation in a gravitational field is a time-consuming process, making it challenging to use for testing particle agglomeration during nanofluid production. Sedimentation can be studied in the field of centrifugal forces. Such a method was applied, for example, in the work [76]. By adjusting the rotational speed of the centrifuge, the sedimentation rate can be increased. This method offers the advantage of being able to calculate the required centrifuge rotation speed for monitoring agglomerate sedimentation of a specific size. It should be noted, however, that implementing this method requires additional quite complex equipment.

The preparation of nanofluid is complex due to the lack of a universal recipe for its ultrasonic treatment time, which is dependent on the type and concentration of nanoparticles or CNTs used. The treatment should be continued until the characteristic size of the particles or CNTs no longer changes. Relevant examples will be provided below. The effectiveness of the UST during the second step of the nanofluid preparation is well determined using the Turbiscan colloidal stability analyzer, which operates on the principle of multiple light scattering. The sample being examined, consisting of 20 ml, is contained in a glass vial. In study [67], a near-infrared laser diode (880 nm) was used as the radiation source. Two synchronous optical sensors measured the radiation intensity pass through the sample, while the radiation scattered in the opposite direction was also recorded. The analyzer scanned the sample along the vial height (up to 55 mm), taking transmittance and backscatter measurements every 40 μm along the way. This method enables comprehension of the level of colloidal stability of the nanofluid. As an example, Fig. 1.8 shows the backscatter profiles of an isopropyl alcohol-based nanofluid with SWCNTs at a weight concentration of 0.05%, both without UST (a) and after being subjected to UST for 90 min (b). In the latter case, the measured signal appears to be stable and uniform, which indicates the homogeneity and stability of the produced nanofluid. This is also indicated by the photographs of the vials with nanofluids on the right.

One of the commonly used methods to determine the dimensions of nanoparticles and CNTs is electron microscopy. This approach also allows one to found the size distribution of nanoparticles (see Fig. 1.1b). The two commonly used methods in practice are TEM (transmission electron microscopy) and SEM (scanning electron microscope), each with its own advantages. SEM offers a 3D image, while TEM provides a 2D image. Therefore, the former method furnishes comprehensive information regarding the particle's shape and relative placement, whereas TEM offers higher resolution and amplification capabilities. This method identifies nanoparticles of about one nanometer in size [72]. It should be noted, however, that the measurements conducted by TEM or SEM are indirect. A sample is taken from the nanofluid, which is then analyzed. Therefore, in order to obtain adequate information, it is necessary to obtain an ensemble of micro-photographs, the averaging of

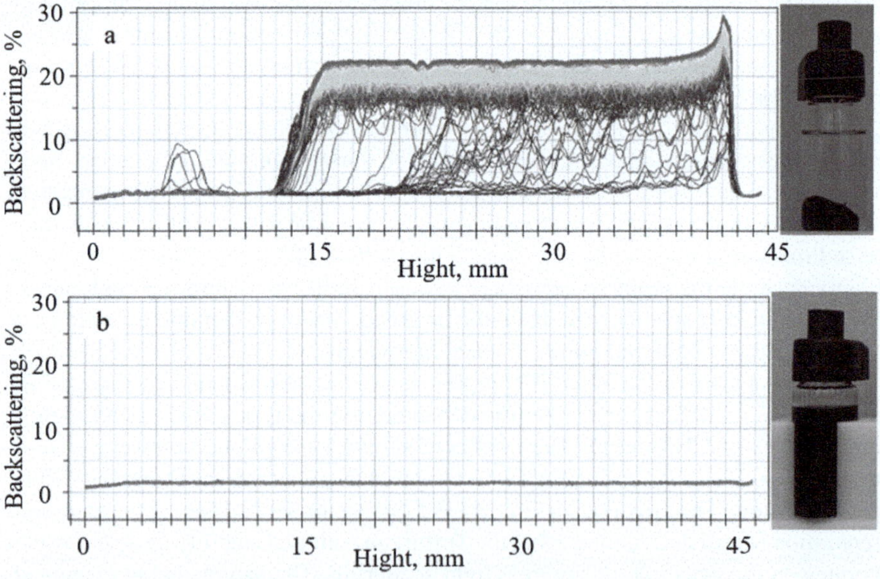

Fig. 1.8 Backscatter signal of IPA-based nanofluid with 0.05% weight concentration of SWCNTs without UST (**a**) and after 90 min of UST (**b**)

Fig. 1.9 TEM micrograph of SWCNTs in isopropyl alcohol-based nanofluid

which will allow us to obtain information on the average size of nanoparticles, their size distribution and the size of agglomerates [68]. With a sufficiently large sample of electron micrographs, objective information on nanoparticle size distribution in nanofluids can be obtained. However, nanofluids containing CNTs present a much more complex situation.

As an example, Fig. 1.9 shows a micrograph of SWCNTs in an isopropyl alcohol-based nanofluid, with a weight concentration of 0.05%. The micrographs reveal two distinct types of nanotubes, with sizes measuring approximately 5 and 20–40 nm. Nevertheless, the obtained microphotographs do not actually capture isolated nanotubes; their diameter determined by Raman-scattering spectroscopy is 1.6 ± 0.4 nm. The second circumstance that should be pointed out is that SWCNTs form a complex percolation structure.

When obtaining micrographs of the type shown in Fig. 1.9, a typical pre-evaporation technique of the base fluid was used. However, such microscopy does not allow us to understand the spatial distribution structure of CNTs in the base fluid, nor does it enable determination of the length and length distribution of nanotubes. Directly in the fluid itself, the average size of nanoparticles or CNTs can be measured utilizing several methods. One of these techniques is the electroacoustic method. It is based on the measurement of the attenuation degree of the ultrasonic signal and is used to determine the size of particles, including in opaque and concentrated media. Ultrasonic waves are scattered on particles, resulting in changes in the spectrum of the ultrasonic signal. The average particle size and particle size distribution are calculated from the attenuated signal spectrum and sound velocity measurements. This approach enables monitoring of the size of nanoparticles or CNTs during UST process. An example illustrating the alteration in the particle size distribution (PSD) of SWCNTs as a result of UST is shown in Fig. 1.10a. Here, the distributions of SWCNTs in an isopropyl alcohol-based nanofluid with a weight concentration of

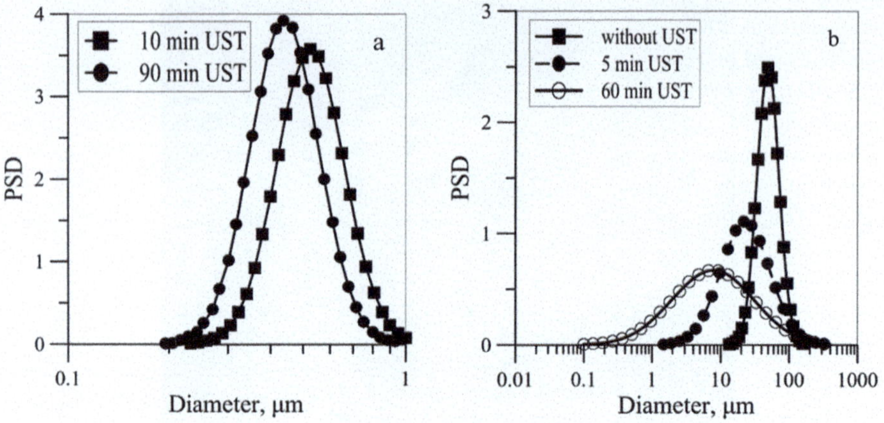

Fig. 1.10 Size distributions of: **a** SWCNTs at a concentration of 0.1% in isopropyl alcohol-based nanofluid at different UST times, **b** Taunit-MD SWCNTs in W + SDS-based nanofluid at the concentration of SWCNTs and surfactant equal to 0.2%

0.1% after 10 and 90 min of UST time are compared. The average size of CNTs after a 10-min UST is 0.52 μm, whereas after 90-min UST session it is 0.43 μm.

When using UST, it is always necessary to keep in mind that there is a certain processing time, starting from which the nanotube (or nanoparticle) size distribution, and therefore their average size, does not change This limit state should be achieved when creating nanofluids for research. The above is illustrated in Fig. 1.10b, which shows the effective diameter distributions of MWCNTs (Taunit-MD) in W + SDS-based nanofluids at different UST times. The weight concentration of MWCNTs and SDS was equal to 0.2%. The UST time was altered between 0 and 60 min. After 60 min of UST the size distributions remained practically unchanged, but the maximum of size distributions decreased by nearly five-fold. The average size of MWCNTs also changed significantly. Figure 1.11 illustrates the impact of UST on the average size of these MWCNTs.

Another important parameter, ζ-potential, is measured using the electroacoustic technique. The reason for its appearance has been explained previously. Measuring ζ-potential enhances the comprehension and monitoring of dispersion and aggregation mechanisms when preparing nanofluids. The greater the ζ-potential's absolute value, the stronger the repulsive forces between particles. Consequently, increased ζ-potential reduces nanoparticle coagulation. It is safe to assert that nanofluids are considerably stable when they have a ζ-potential between 40 and 60 millivolts; surpassing 60 millivolts results in high stability levels [101].

Finally, the distribution of nanoparticles and CNTs in nanofluid can be measured using different light scattering techniques. The dynamic light scattering (DLS) with a Malvern Zetasizer Nano ZS analyzer is very commonly used method. This instrument can measure particles ranging in size from 0.3 nm to 10.0 μm. The measurement scheme is quite simple and is shown in Fig. 1.12. The intensity of light scattered

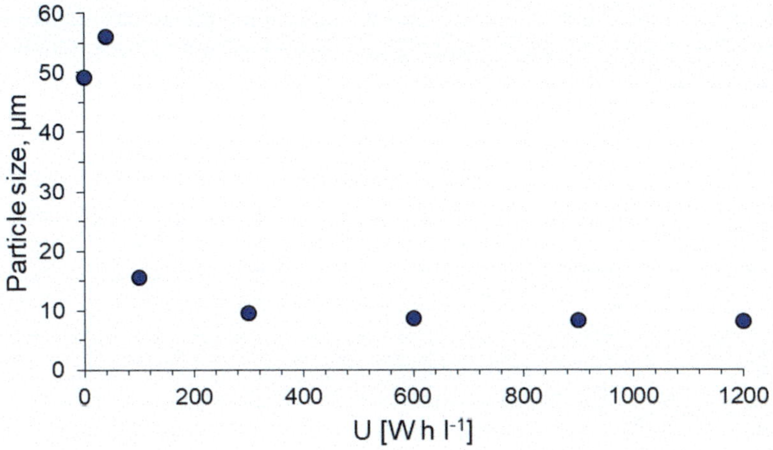

Fig. 1.11 Effect of ultrasonic treatment on the average size of Taunit-MD nanotubes in W + SDS-based nanofluid

by particles is measured and the corresponding two-time correlation function is constructed. According to this function, the relaxation time is determined, which is then used to determine the diffusion coefficient of the particles under study. Using the Stokes–Einstein formula (1.1) and the given coefficient, the effective hydrodynamic radius (or diameter) of nanoparticles is calculated and the resultant size distribution is constructed. For spherical particles, this radius corresponds to their actual radius, resulting in an accurate particle size distribution. The DLS method is also applied to CNTs, but in this case only the effective hydrodynamic radius of CNTs is obtained, i.e. the effective radius of a spherical nanoparticle, which would have the same diffusion coefficient as the measured one (see formula (1.4)). Certainly, the effective hydrodynamic radius of the CNTs obtained in this way does not agree well with its actual size. The adequacy of these measurements can be improved by utilizing formula (1.2) to relate the diffusion coefficient to the CNT size. However, it is necessary to have information on the CNT diameter in this case.

When utilizing DLS method, it is crucial to exercise caution, as the data retrieved may overestimate the average nanoparticle size by a significant margin (sometimes

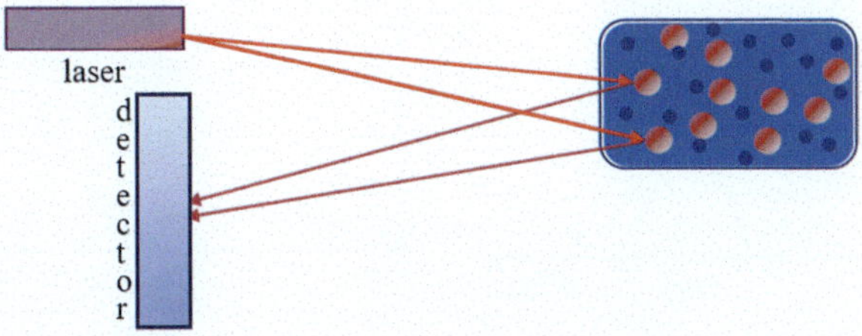

Fig. 1.12 Schematic diagram of the dynamic light scattering method

multiple times) [84]. This is attributed to the fact that the maximum scattered light intensity is produced by the largest particles, and in the case of CNTs, by their bundles (the scattered light intensity is proportionate to the particle size). This method, however, allows one to obtain information about the volume concentration of such large particles and the initial information obtained can be corrected.

Complex surfactants can be used in the preparation of nanofluids, making it imperative to investigate the structure of the corresponding base fluids using DLS. Figure 1.13 presents the size distributions of SDBS (Fig. 1.13a) and PVP (Fig. 1.13b) molecules in water. Formula (1.1) was utilized to reconstruct the particle sizes (hydrodynamic diameter d). The size distributions of SDBS molecules (Fig. 1.13a) have a distinctive feature where two separated maxima of the reflected light intensity I can be observed. This indicates the presence of particles with significantly different sizes in the solution. The first peak maximum for SDBS molecules corresponds to a particle size of 4.19 nm. According to molecular dynamic simulation of SDBS micelles, their diameter is 4 nm [56], which is consistent with experimental data. The presence of the second peak corresponds to larger cylindrical micelle structures. In addition to being a micellar solution, this surfactant also exhibits polymorphism. In the first scenario, SDBS molecules form nearly spherical aggregates, while in the second situation, they create cylindrical micellar structures. It is these structures that are responsible for generating the second peak in Fig. 1.13a with its maximum height of 132 nm, which in this case is the hydrodynamic diameter of these micelles. While assuming a diameter of micelles equal to 4.19 nm, it is easy to calculate the characteristic length using relation (1.2). However, it becomes apparent that the length of the polymicella, obtained from its hydrodynamic size, is greatly underestimated. In this case, its length is about 1 μm. It is worth noting that the formation of micelles in a SDBS solution is an analogue of some phase transition and begins at the critical concentration of molecules.

The size distributions of PVP molecules are presented in Fig. 1.13b. The average size of PVP molecules is 282 nm, which, as already mentioned, corresponds to the effective hydrodynamic dimeter. However the PVP molecule (C_6H_9NO) is close in shape to an elongated ellipsoid of rotation, the major axis of which is approximately 8.1 nm, and the minor axis is approximately 4.3 nm. So, their ratio is approximately

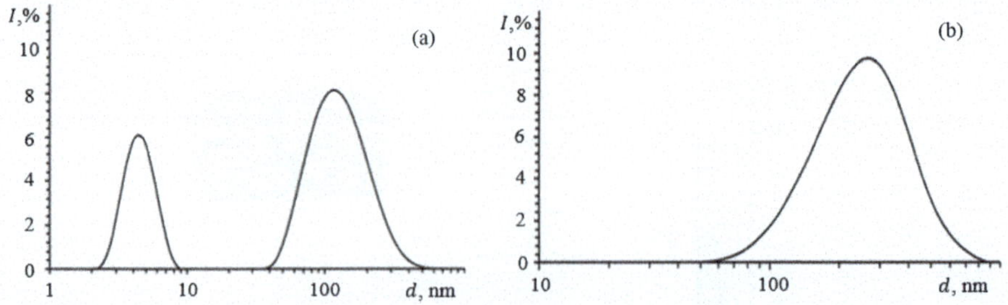

Fig. 1.13 Size distributions of SDBS (**a**) and PVP (**b**) molecules in aqueous solution

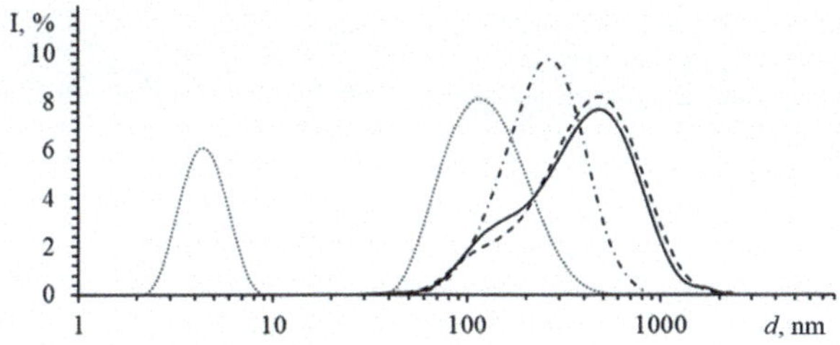

Fig. 1.14 Distributions of SWCNTs as well as PVP and SDBS particles in nanofluids based on W + PVP and W + SDBS solutions

1.9 [85]. The surfactant used in the experiment features the formula $(C_6H_9NO)_n$, and roughly speaking can be described as a chain of *n* similar molecules.

The DLS method effectively determines the sizes of surfactant molecules. However, it is crucial to determine the presence of corresponding peaks of surfactant molecules on DLS analysis curves acquired for CNT-containing nanofluids. Figure 1.14 demonstrates a relevant data example. Here data for two nanofluids with a weight concentration of SWCNTs at 0.01% are presented, with the concentration of both surfactants twice as high. The solid and dashed lines represent the size distributions of SWCNTs in nanofluids based on W + PVP and W + SDBS. The dotted and dashed-dot lines display the size distributions of SDBS and PVP molecules. As before, relation (1.1) was used to reconstruct the size distributions from the diffusion data. Analysis of the distributions presented in Fig. 1.14 shows that the characteristic sizes of SWCNTs in nanofluids at such low CNT concentrations appear to be almost the same with both surfactants. The maximum hydrodynamic diameters of SWCNTs are around 480 nm, with slightly lower average sizes. On the other hand, the size distributions of PVP and SDBS molecules are localized in their proper places (compare with Fig. 1.13). The maxima of their distributions do not coincide with those of SWCNTs. Consequently, the DLS method precisely identifies CNTs in both nanofluids, irrespective of the surfactant used.

As the concentration of SWCNTs increases, the effectiveness of the surfactants differs. The size distribution of SWCNTs in water-based nanofluids utilizing two distinct surfactants is illustrated in Fig. 1.15. The nanofluid with SDBS is represented by the solid line, while the nanofluid with PVP is illustrated by the dashed line. The weight concentration of SWCNT and curfactants in both cases is the same and equals 0.1%, and 0.2% respectively. The formula (1.2) was used to reconstruct the size of SWCNTs from the measured values of diffusion coefficient, assuming an average diameter of the SWCNT bundles equal to 8 nm. It is worth reiterating that the characteristic diameter of an isolated SWCNT is approximately two nanometers. However, it is practically impossible to achieve the presence of only isolated CNTs at the concentrations of the surfactants used. Consequently, a significant portion of nanotubes form bundles in nanofluids at the studied concentrations, resulting in an effective radius substantially greater than that of isolated SWCNTs.

Regarding the distributions shown in Fig. 1.15, two key observations should be made. Firstly, there is a significant difference in the average sizes of SWCNTs when using different surfactants. If SDBS is used for the preparation of the base fluid, the

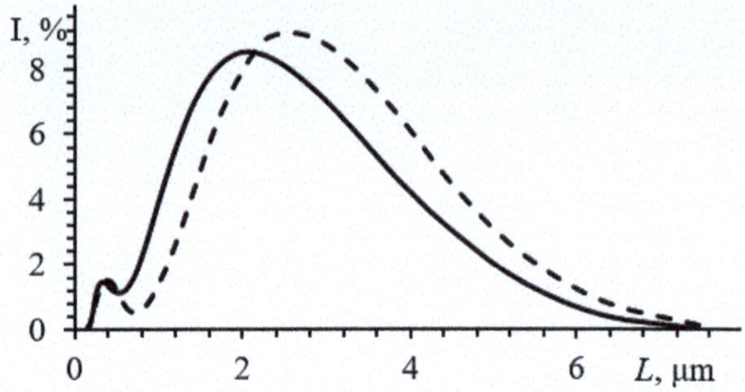

Fig. 1.15 Distributions of SWCNTs in water-based nanofluids with PVP (dashed line) and SDBS (solid line) μm

average size of the tube bundles is approximately 2.6 μm, while in nanofluids with PVP, the average size of the tubes exceeds 3 μm. The second important circumstance is that the size distributions in all cases have two maxima. The first one corresponds to the low intensity of reflected light and identifies individual SWCNTs, nearly always present in nanofluids, unless their concentrations are overly high. Both nanofluids contain nanotubes of identical characteristic size. However, their average length is nearly ten times shorter than that of the tube bundles.

1.4 Nanopowder Production Methods

The production of nanofluids using the two-step method assumes availability of nanopowder, a substance composed of nanoparticles. Currently, nanopowders for various purposes are produced industrially. The nanopowder market has become quite noticeable and already exceeds several tens of billions of dollars. Nanopowders can be classified based on the types of nanoparticles they contain: (i) metallic nanoparticles (copper, gold, silver, platinum, titanium, aluminum, etc.); (ii) oxide nanoparticles (copper, gold, silver, platinum, titanium, aluminum, silicon oxides, etc.); (iii) other nanoparticles (polymeric, carbides, nitrides, silicides, composite, etc.); (iv) SWCNTs, and (v) MWCNTs. Nnanopowders have already found wide scope of application, the best known ones are summarized in Table 1.1.

It is rather difficult to determine the particle size in nanopowders due to significant particle aggregation. Therefore, traditionally, the average particle size in nanopowders is determined using the Brunauer–Emmett–Teller (BET) method [9]. This method is based on low-temperature adsorption of inert gases (usually nitrogen) and used to directly measure the specific surface area of the powder $\tilde{S}$, i.e., the surface area of the particles in one gram of powder; expressed in m^2/g. The specific surface area can then be used to determine the characteristic size of the nanoparticles. For spherical particles it is equal to $d = 6/(\rho\tilde{S})$.

Direct methods of nanoparticle size measurement have shown that the size distribution of nanoparticles $f(R_0)$ is generally lognormal

Table 1.1 Scope of application of the best known nanopowders

Formula	Scope of application
Ag	Production of cosmetics, textiles, coatings; catalysis
Al	Biomedical technology, catalysis
Au	Electronics, biomedical technology, catalysis
C (nanodiamond)	Production of coatings, semiconductors, abrasives
Cu	Biomedical technology, optics, electronics
Fe	Optics, production of magnets and magnetic storage devices, ecology
Mo	Production of coatings and polymers; catalysis
Ni	Production of hydrogen fuel cells, polymers, coatings, textiles; catalysis, biomedical technology, optics, electronics
Pt	Electronics, production of hydrogen fuel cells, fiberglass; biomedical technology, catalysis
Si	Electronics; production of ceramics, abrasives, building materials, pyrotechnics
W	Production of polymers and coatings
Zn	Production of polymers, textiles, hydrogen fuel cells, solar cells; biomedical technologies
Al_2O_3	Production of abrasives, ceramics; optics, electronics, ecology
CeO_2	Optics; production of hydrogen fuel cells and glass
CuO	Optics, electronics, biomedical technologies
Fe_2O_3, Fe_3O_4	Production of ceramics, glass, magnets; electronics, ecology, catalysis
MgO	Biomedical technologies, production of metal alloys, textiles, polymers; optics, electronics
SiO_2	Optics, electronics, abrasives; production of paint and varnish, cosmetics
TiO_2	Production of paint and varnish, cosmetic materials, abrasives; optics, ecology
Y_2O_3	Production of hydrogen fuel cells, ceramics, fluorescent lamps, monitors
ZnO	Production of polymers, hydrogen fuel cells, textiles, cosmetics, solar cells
ZrO_2	Production of hydrogen fuel cells, ceramics
CNT	Production of advanced materials and batteries; microelectronics, biomedical technologies

$$f(R_0) = \frac{3}{4\sqrt{2\pi}\sigma R_0^3} \exp\left[-\frac{9\left(\ln R_0 - \ln \overline{R}_0\right)^2}{2\sigma^2}\right], \tag{1.10}$$

where $\overline{R}_0$ is the average particle size, σ is the variance.

Typical particle size distributions of the three nanopowders are presented in Fig. 1.16. The aluminum nanopowder (L-ALEX™) was treated with palmitic acid $CH_3(CH_2)_{14}CO_2H$ for its passivation. A histogram of the size distribution of aluminum powder particles is shown in Fig. 1.16a. According to Fig. 1.16a, the

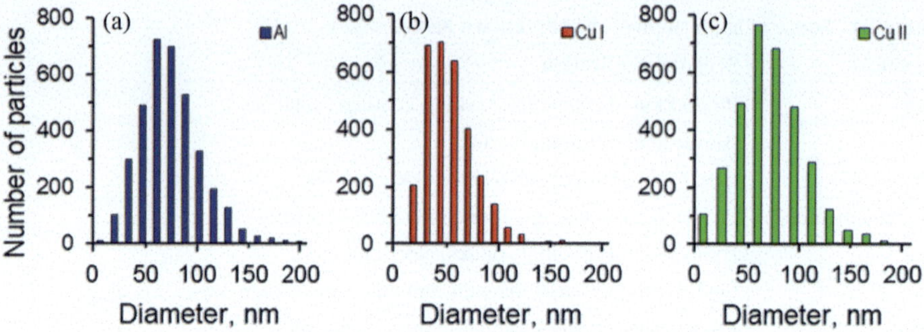

Fig. 1.16 Distribution histograms of Al (**a**), Cu I (**b**), and Cu II (**c**) particles in nanopowders

average size of the nanoparticles was 75.2 nm. Copper nanopowder contained the particles of two sizes. The corresponding histograms of the copper powder particle distribution by size are shown in Fig. 1.16b, c, respectively. The average particle size of these powders was 51.5 nm (Fig. 1.16b) and 79.4 nm (Fig. 1.16c). These particles are further referred to as Cu I and Cu II, respectively.

A systematic review of nanopowder and CNT manufacturing methods is beyond the scope of this book, and can be found in specialized reviews and handbooks. In this section, only a brief description of the main methods of their fabrication is presented.

1.4.1 Chemical Methods

Chemical methods for producing nanopowders usually involve a range of techniques, including chemical vapour deposition or chemical solution deposition, thermal decomposition, high-energy synthesis (detonation, plasma chemistry, gas dispersion), pyrolysis, gas-phase chemical reactions (reduction, hydrolysis), and electrodeposition [28, 32]. Metal hydroxides are precipitated from salt solutions through the use of precipitants, specifically sodium and potassium alkali solutions. The crystallization rate can be adjusted and highly dispersed hydroxide can be formed by manipulating the pH level and temperature of the solution. The gel method is employed in producing powders of different metals, involving the precipitation of insoluble metal compounds from aqueous gel solutions. In addition, reduction and thermal decomposition methods are utilized to obtain powders of varying shapes such as spherical, needle-shaped, and flake-shaped particles. In particular, nanopowders of Fe, W, Ni, Co, Cu, and other metals are produced by reduction of their oxides with hydrogen. Carbon, metals, and their hydrides can function as solid reducing agents. Typically, metal salts are reduced by reducing agents such as alumohydrides, borohydrides, hypophosphites, formaldehyde, or oxalic and tartaric acid salts. The particle size of the resulting material can be adjusted by altering the reducing agent used. This technique leads to the creation of nearly monodisperse nanoparticles.

Silver and gold nanoparticles have widespread usage in a variety of applications. The production of silver nanoparticles primarily relies on chemical methods that reduce silver ions in solution using various reducing agents. This typically transpires in the liquid phase. These particles tend to aggregate actively in the presence of sulfuric and hydrochloric acids. The resistance of silver particles to aggregation also depends on the pH of the medium. Gold nanoparticles are often obtained by reduction of gold halogenides (for example, $NAuCl_4$) using chemical reducing agents and/or ultrasonic or UV irradiation. Various organic and inorganic compounds (sodium citrate, ascorbic acid, ethylenediaminetetraacetic acid, etc.) serve as reducing agents in this process.

Chemical methods are commonly used in the production of magnetic nanoparticles. Specifically, the interplay of compounds containing divalent and trivalent iron salts is applied, resulting in the formation of magnetite nanoparticles. In fact, it is possible to create a magnetic nanofluid concurrently [95]. Many other similar examples can be found. Chemical methods enable the production of magnetic fluids based also on MnZn and NiZn ferrites. The process is multi-stage. The initial chemical reactions are as follows [12]

$$MnSO_4 + ZnSO_4 + Fe_2(SO_4)_3 + \text{(aqueous solution NaOH)}$$

and

$$NiSO_4 + ZnSO_4 + Fe_2(SO_4)_3 + \text{(aqueous solution NaOH)}.$$

The products produced during the reactions are mixed and treated with oleic acid, resulting in paste-like briquettes, from which magnetic liquid is obtained by peptization in cyclohexane.

Chemical methods include thermal methods, in which a chemically unstable organometallic substance is subjected to thermal decomposition. As a result, magnetic nanoparticles are formed, which can be dispersed into the corresponding base liquid to obtain a magnetic fluid. A similar example is presented in Kilner et al. [37]. In certain circumstances, ultraviolet (or other high-energy) radiation is used for the decomposition of metal carbonyls instead of the application of heat [31]. After ultraviolet irradiation, nickel nanoparticles are synthesized. The particles are then peptized in toluene, creating a magnetic fluid with the nickel nanoparticles.

In the past two decades, radiation-chemical reduction techniques for many metal ions in aqueous solutions with stabilizers were developed to produce magnetic nanofluids. Unlike many other methods, these nanofluids are obtained without any contamination from additives introduced in the solution, which is always the case when using any reducing agents. Furthermore, the reducing radicals are uniformly generated through the solution volume during radiation. The radiation-chemical method is a useful complement to other methods for producing metallic nanoparticles, including photochemical and electrochemical methods. As an example, one can note the recovery of spherical nanoparticles of cobalt or nickel through γ-irradiation of deaerated solutions of $Co(ClO_4)_2$ and $Ni(ClO_4)_2$ (containing Co^{2+} and Ni^{2+}) in the presence of alkali and PVA [21]. The metal particles settle on the bottom and walls of the vessel over time. The nanofluid is then produced by adding a stabilizer (polyethylenimine, polyvinyl alcohol, polyacrylate, or polyvinyl sulfate).

Nanoparticles of various ferrites are typically synthesized through co-precipitation. The research [23] provides an example of how to obtain $MnFe_2O_4$ spinel nanoparticles with an average size of 40 nm. These nanoparticles are formed by adding an aqueous solution of stoichiometric amounts of Mn^{2+} and Fe^{3+} chlorides to an alkali solution (the process is carried out with constant mechanical stirring). This process can be used to produce particles of other ferrites as well, with sizes ranging from 2 to 50 nm.

Chemical methods for producing metal nanoparticles are widely used due to their simplicity and availability. Nonetheless, nanoparticles generated through reduction or ion exchange reactions tend to contain ions and reaction products, which can cause challenges in separation. Additionally, these methods usually have low productivity.

1.4.2 Physical Methods

These methods can be conditionally divided in three groups: physical vapor deposition, melt blowing, and mechanical grinding. The first two procedures are based on metal, alloy, or oxide evaporation followed by their condensation to form nanoparticles, usually in a vacuum or under reduced pressure [35, 57]. The substance is heated intensely to evaporate before being fed into the reaction space by means of a carrier gas, where it rapidly cools. Nanoparticles are formed by condensation of the vapors. The condensation of individual molecules of the evaporated substance occurs spontaneously since the free energy of the system decreases when the molecules or atoms combine. One of the earliest methods utilized was spark discharge between electrodes, and cooling took place in a liquid bath. Currently, various heating techniques are employed, such as utilizing plasma, laser radiation, electric arc, and induction currents. By monitoring the temperature and the environment (gas flow rate, gas pressure, and type of gas) where evaporation and condensation take place, the average particle size and reactor performance can be regulated. Nanoparticle condensation can occur in both the reactor volume and on the cooled substrate (reactor walls). Evaporation and condensation can occur in a vacuum, gas atmosphere, or plasma jet. This method allows obtaining powders of various metals and oxides with nanoparticle sizes from several nanometers to tens of nanometers. It is also possible to produce nanoparticles of nearly all metals, alloys, and compounds using this method. The particles obtained typically have a nearly spherical shape and a narrow size distribution. The closed loop flow facilitates a continuous dispersion process. The productivity of the gas-phase method increases with increasing metal evaporation temperature, but, in general, is not too high, as a rule, not exceeding a kilogram per day.

An ingenious heating method was suggested in Bardakhanov et al. [6]. To achieve evaporation, an industrial linear electron accelerator with a power capacity of up to

1.4 MeV operating continuously was used providing the release of a focused beam into the atmosphere. The equipment can achieve material temperatures high enough to evaporate any currently known refractory substance, with a heating rate of over 1000 deg/s. Nanopowders of SiO_2, SiO, MgO, Al_2O_3, TiO_2, Y_2O_3, Gd_2O_3, Cu_2O, WO_3 Bi_2O_3, ZnO etc., as well as various metals, including W, Ta, Mo, Co, Al, Fe, Ni, and Ag, etc. were successfully synthesized using this method. With increasing power (current) of the electron beam, an increase in the average particle size is observed. This is explained by the increase in the concentration of evaporating substances. On the contrary, the average particle size can be reduced by increasing the flow rate of the transport gas.

The vacuum chamber has also been extensively utilized for evaporating the initial material. An interesting method, for example, was developed by Nakatani et al. [53], where the vapors of molten metal were employed to obtain nanoparticles. The procedure involved placing a crucible filled with molten metal inside a cooled cylinder rotating around a horizontal axis. The cylinder's inner surface was wetted with liquid oil containing surfactant. The process was implemented under high vacuum. Cobalt, iron, and nickel were evaporated using the described method, resulting in uniform-sized particles approximately two nm in diameter. Altering the heating temperature can increase their size multiple times. Thermal evaporation can also be carried out in an inert gas flow. In this case, the atoms of the evaporated metal, colliding with the atoms of the gas, lose kinetic energy and cluster, and then condense on the cooled substrate. This method enables producing not only pure metal particles but also composite particles, specifically FeNi, FeMn, FePt, and FeCr composites [59].

The well known galvanic method can be utilized to prepare mercury-based magnetic fluid. Another physical method, which can also be considered a chemical method, is electrolytic method. Actually, the electrolytic method can also be classified as a condensation method. This method involves the process known as electrolysis, which has been long used to obtain metals, including aluminum, as well as other substances. The method's operation principle is extremely simple: when an electric current passes through a solution or electrolyte melt, the components of dissolved or other substances that are the result of secondary reactions on the electrodes are released on the electrodes immersed in them [8]. The process described in the paper [94] was carried out in a two-layer electrolytic bath with a rotating cathode, where the lower layer is an electrolyte solution and the upper layer is a stabilizer solution. The cathode rotated rapidly and the metal nanoparticles condensing on its surface immediately fell into the dispersed fluid and were covered with a layer of surfactant (for example, oleic acid). The size of the resulting particles depended on the rotation speed of the cathode, the depth of its immersion, the electrolyte concentration, the temperature regime, and the material of the cathode. This method was employed to obtain magnetic liquids with cobalt and magnetite particles. The best results were attained with the use of toluene as a carrier fluid and an aqueous solution of cobalt chloride or iron sulfate as an electrolyte. The magnetic nanoparticles generated had a size ranging from 30 to 80 nm. The average size of the formed particles was inversely proportional to the current density. The disadvantage of this method is its poor performance.

1.4.3 Plasma-Chemical Method

Plasma-chemical synthesis of nanopowders is indeed a physical method, although some authors consider it chemical. The synthesis process is carried out in a low-temperature plasma flow, for example, in an arc discharge. In the later case, the nanoparticles arise from the evaporation of fluid metal particles and their oxidation in oxygen-containing plasma once there is a sharp drop in temperature. By varying the cooling rate, the size and formation rate of condensed phase particles can be altered. The procedure is exceptionally productive and enables the production of particle sizes in the range of tens of nanometers, given certain conditions. However, along with this, a significant amount of substandard powder is formed, having a particle size of tenths of a micron or even more.

1.4.4 Mechanical Methods

Various mills (ball, planetary, centrifugal, and vibrating), gyroscopic devices, attritors, and simoloyers are utilized when applying mechanical methods, which are based on grinding of materials. The degree of grinding is reliant on the properties of the materials. For example, the size of tungsten and molybdenum oxides particles was 5–10 nm, and iron particles were 10–20 nm in size. The method is extremely simple and highly productive. At the same time, using this method, it is impossible to produce nanopowders with a specific average particle size because the particle size distribution is always broad. Additionally, the resulting powder is typically contaminated with abrasive material. Mechanical methods of nanopowder production are sometimes combined with simultaneous chemical processes [58].

1.4.5 Production of Composite Nanoparticles

In addition to uniform nanoparticles, various composite and nonuniform nanoparticles are of significant practical interest. The last chapter will discuss their uses, however, the need for composite nanoparticles stems also from several other reasons. Special coatings for nanoparticles, for instance, prevent their agglomeration or protect them from oxidation. For example, to protect copper nanoparticles from oxidation, they are coated with graphene layers [45]. In this case copper nanoparticles were formed in a burner flame during the decomposition of the organometallic precursor in nitrogen environment. The presence of acetylene resulted in the formation of a stabilizing graphene shell on the surface of the nanoparticles. These copper particles have the potential to be an alternative to those made of silver or gold in various applications.

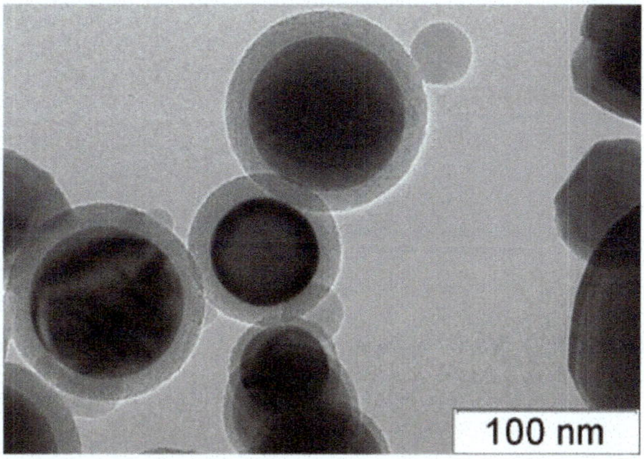

Fig. 1.17 Dioxide-coated copper nanoparticles

Figure 1.17 depicts the silicon dioxide-coated copper nanoparticles [54]. Composite nanoparticles, like uniform ones are produced by evaporation–condensation method. For instance, in [46] anodic sputtering of a metal-graphite composite electrode within a low-pressure arc reactor was used. This process concurrently synthesizes metal particles and a carbon matrix, which serves to retain the particles and prevent their coagulation. In a similar manner, the synthesis of platinum, palladium and nickel nanoparticles was carried out. Helium was used as the carrier gas at pressures of 5–50 Torr. With this method, along with encapsulated (composite) nanoparticles, homogeneous particles are also formed, of course, and their separation is a certain problem.

A more complex technology was used in research [54, 55] to obtain Ag/Si composite powders, which is briefly explained below. The schematic diagram of the setup used is presented in Fig. 1.18. A pre-prepared silver ingot (2 in Fig. 1.18) was placed in a graphite crucible, and pieces of pure monocrystalline silicon were placed on top. The mass ratio of Si to Ag was 1:10. The monocrystalline silicon pieces were melted by an electron beam (the energy of the irradiating electrons was 1.4 MeV, the current in the beam varied from 3 to 10 mA, the maximum power density of the electron beam was 106 W/cm^2). After the termination of the electron beam, the silicon with the silver underneath solidified. Then the two substances Ag and Si were heated again by an electron beam to the melting temperature, resulting in a homogeneous composite melt. Further, the beam current was increased to a level that led to the intense evaporation of the composite liquid. Argon was employed as a transport gas within the condensation zone. The resulting nanocomposite particles were gathered on a designated filter.

Shells may form on the surfaces of small metallic particles during their production or as a result of surface segregation, which involves the enrichment of near-surface regions with atoms of components that differ from the main component in atom size or surface energy. Surface segregation has been extensively researched in numerous papers. The nature and the character of segregation of binary melts is attributed to

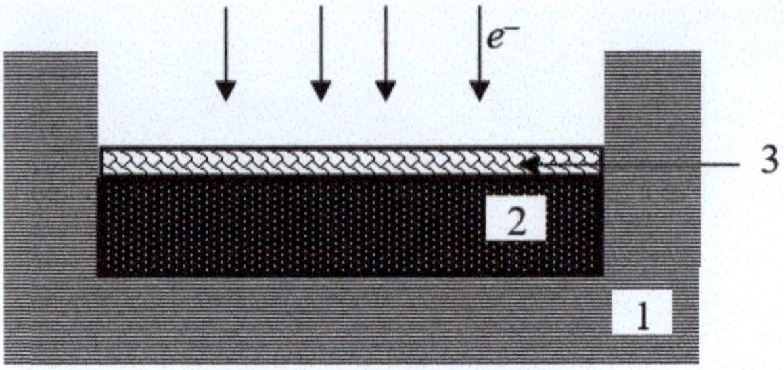

Fig. 1.18 Schematic diagram of producing composite Ag/Si nanoparticles: 1 is graphite crucible, 2 is solid Ag sample, 3 is Si pieces, e^- is electron beam

the difference in surface energy and size of the two elements, as well as their ability (inability) to mix uniformly throughout the volume.

A method of synthesizing Sb/Sb_2O_3 spherical composite nanoparticles by hybrid induction and laser heating is described in Zeng et al. [104]. For this purpose, a pure Sb ingot was placed in a graphite crucible and inserted into a chamber where argon and oxygen (with an oxygen partial pressure of 500 Pa) were injected under a pressure of around 10 Pa. The final particle size averaged at around 80 nm. Highly dispersed magnetic-containing polymer nanoparticles with narrow particle size distribution and high content of magnetic material were synthesized in [15]. Styrene magnetic liquid obtained by chemisorption onto the surface of sodium oleate nanoparticles served as a base fluid. Acetic acid was added to the system to hydrophobize the surface of the nanoparticle, subsequently, styrene was injected under ultrasound exposure to the system. A stable suspension of magnetite nanoparticles in styrene was added to an aqueous solution of sodium dodecyl sulfate under simultaneous and constant ultrasonic treatment. This process facilitated the creation of a mini-emulsion, followed by polymerization, resulting in nanoparticles with an average diameter of 80–100 nm.

In addition to composites, hollow nanoparticles are of great practical importance. One simple process to create these particles is as follows. Polymer nanoparticles are bio- or chemically coated with a solution of a silicon containing substance. Then the core of the obtained composite particle, consisted of polymer/SiO_2, is dissolved in base fluid. An analogous process was performed by Romanov et al. [62], in which the authors initially synthesized core–shell Cu/SiO_2 nanoparticles for the production of hollow particles. After heating, the molten copper core exited through defects in the shell and was eliminated chemically. This resulted in hollow particles that possess holes, which is very significance for accommodating active substances.

Finally, Janus-like particles, in addition to conventional composite nanoparticles, are of significant interest [79]. These particles are composite, thus featuring a range of distinct properties, including a dipole moment. In the $TaSi_2$/Si particle, for instance, the first part exhibits metallic conductivity, while the second part (silicon) lacks free electrons. This results in the spatial separation of electrons, the emergence of charge polarization, and subsequently, the presence of a dipole moment. These nanoparticles were produced using the same procedure depicted in Fig. 1.18, but with a silicon bottom layer and a tantalum top layer [54]. This is because the saturated vapor pressure of silicon is higher than that of tantalum at all temperatures.

1.4.6 The CNT Creation Methods

Today, several different technologies for the production of CNTs are known. These technologies were discovered during the creation of fullerenes by electric arc evaporation of graphite electrodes. It is generally accepted that CNTs were first clearly identified using electron microscopy by Iijima [33]. During electric arc evaporation of a graphite anode, cylindrical bundles of CNTs are formed on the cathode perpendicular to the flat surface of its face, which then form a honeycomb structure. Typically, an arc discharge is created in an inert gas at atmospheric pressure. This arc discharge method remains one of the most popular today [1, 2, 29, 60, 106]. The gaps between the bundles are filled with other carbon materials, the CNT content in this forming sediment is about 60%. The components of this precipitate are then separated usually by centrifugation using various chemical procedures.

Also, CNTs can be synthesized through laser ablation. This approach has been used, for example, in paper [24], where graphite target was placed in a furnace at 1200 °C is irradiated with a laser beam that scans its entire surface. The carbon, knocked out by the laser beam, is deposited on the cold substrate, forming CNTs. By varying the catalysts used in the process (nickel, cobalt) and the wavelength of laser radiation, different CNTs can be obtained. The process can be carried out in helium, hydrogen, and nitrogen atmosphere.

The use of metal catalysts has sparked the development of a novel low-temperature method known as chemical vapor deposition (CVD). Essentially, it involves the pyrolysis of hydrocarbons in the presence of metal catalysts [41, 42, 47, 96]. This process is organized quite simply. A mixture of hydrocarbon gas, acetylene, methane or ethylene, and nitrogen is introduced into the reaction chamber. Pyrolysis is carried out at atmospheric pressure and temperatures ranging from 700 to 900 °C. The CNTs are formed on a substrate where catalyst particles are deposited. The catalyst can be applied by chemical deposition from a solution containing it, thermal evaporation, etc.

References

1. Ahmad A, Kholoud MM, Abou E et al (2012) Carbon nanotubes, science and technology part (I) structure, synthesis and characterization. Arab J Chem 5:1–23
2. Ajayan PM (1999) Nanotubes from carbon. Chem Rev 99:1787–1799
3. Akoh H, Tsukasaki Y, Yatsuya S et al (1978) Magnetic properties of ferromagnetic ultrafine particles prepared by vacuum evaporation on running oil substrate. J Crystal Growth 45:495–500

4. Antti G, Pentti P, Hanna K (2008) Ultrasonic degradation of aqueous carboxymethylcellulose: effect of viscosity, molecular mass, and concentration. Ultrasonics Sonochem 15(4):644–648
5. Backes C, Hirsch A (2010) Noncovalent functionalization of carbon nanotubes. In: Akasaka T, Wudl F, Nagas S (eds) Chemistry of nanocarbons. Wiley, Chichester, pp 1–48
6. Bardakhanov SP, Korchagin AI, Kuksanov NK et al (2006) Nanopowder production based on technology of solid raw substances evaporation by electron beam accelerator. Mat Sci Eng B 132(1–2):204–208
7. Bica D, Vékás L, Raşa M (2002) Preparation and magnetic properties of concentrated magnetic fluids on alcohol and water carrier liquids. J Magn Magn Mat 252:10–12
8. Bloom EYa, Mayorov MM, Tsebers AO (1989) Magnetic fluids. Znanie, Riga
9. Brunauer S, Emmett PH, Teller E (1938) Adsorption of gases in multimolecular layers. J Am Chem Soc 60(2):309–319
10. Chang H, Tsung TT, Chen LC et al (2005) Nanoparticle suspension preparation using the arc spray nanoparticle synthesis system combined with ultrasonic vibration and rotating electrode. Int J Adv Manuf Technol 26:552–558
11. Chen L, Xie H (2010) Surfactant-free nanofluids containing double- and single-walled carbon nanotubes functionalized by a wet-mechanochemical reaction. Thermochim Acta 497(1–2):67–71
12. Chikazumi S, Taketomi S, Ukita M et al (1987) Physics of magnetic fluids. J Magn Magn Mat 65(2–3):245–251
13. Choi S (1995) Enhancing thermal conductivity of fluids with nanoparticles. In: Siginer DA, Wang HP (eds) Developments applications of non-Newtonian flows, FED-vol. 231/MD-vol. 66. ASME, New York, pp 99–105
14. Chung SJ, Leonard JP, Nettleship I et al (2009) Characterization of ZnO nanoparticle suspension in water: effectiveness of ultrasonic dispersion. Powder Tech 194:75–80
15. Cherkasov VR, Gervald AYu, Prokopov NI et al (2006) Synthesis of magnetically filled polymer microspheres for immunomagnetometric analysis. High-Molecular Compounds A 48(4):572–579
16. Cwirzen A, Habermehl-Cwirzen K, Penttala V (2008) Surface decoration of carbon nanotube sand mechanical properties of cement/carbon nanotube composites. Adv Cem Res 20(2):65–73
17. Das D, Kulkarni D (2012) Nanofluids properties and their applications. Lambert Academic Publilishing
18. Dhont JKG, Briels WJ (2004) Rod-like Brownian particles in shear flow. WILEY-VCH Verlag Berlin GmbH, Berlin
19. Eastman JA, Choi SUS, Li S et al (2001) Anomalously increased effective thermal conductivities of ethylene glycol-based nanofluids containing copper nanoparticles. Appl Phys Lett 78:718–720
20. Einstein A (1905) Über die von der molekularkinetischen Theorie der Wärme geforderte Bewegung von in ruhenden Flüssigkeiten suspendierten Teilchen. Ann Phys 17:549–560
21. Ershov BG (2001) Metal nanoparticles in aqueous solutions: electronic, optical and catalytic properties. Russ Chem J 45(3):20–30
22. Esfe MH, Saedodin S, Mahian O et al (2014) Thermophysical properties, heat transfer and pressure drop of COOH-functionalized multi walled carbon nanotubes/water nanofluids. Int Commun Heat Mass Transfer 58:176–183
23. Fang Q, Lin Y, Yin P et al (2001) Magnetic properties and formation of Sr ferrite nanoparticles and Zn, Ti/Ir substituted phases. J Magn Magn Mat 234(3):366–370
24. Gamly E (2000) Single-walled carbon NT formation with a continuous CO_2-laser: experiments and theory. Appl Phys A 70:161–168
25. Garg J, Poudel B, Chiesa M et al (2008) Enhanced thermal conductivity and viscosity of copper nanoparticles in ethylene glycol nanofluid. J Appl Phys 103:074301
26. Ghadimi A, Saidur R, Metselaar HSC (2011) A review of nanofluid stability properties and characterization in stationary conditions. Int J Heat Mass Transfer 54:4051–4068

27. Goharshadi E, Ding Y, Jorabchi M et al (2009) Ultrasound-assisted green synthesis of nanocrystalline ZnO in the ionic liquid. Ultrason Sonochem 16:120–123
28. Gubin SP, Koksharov YuA, Khomutov GB et al (2005) Magnetic nanoparticles: preparation, structure and properties. Russ Chem Rev 74(6):489–520
29. Hahm MG, Hashim DP, Vajtai R et al (2011) A review: controlled synthesis of vertically aligned carbon nanotubes. Carbon Lett 12(4):185–193
30. Ham HT, Choi YS, Chung IJ (2005) An explanation of dispersion states of single-walled carbon nanotubes in solvents and aqueous surfactant solutions using solubility parameters. J Colloid Interface Sci 286:216–223
31. Hoon SR, Kilner M, Russel GJ, Tanner BK (1983) Preparation and properties of nickel ferrofluids. J Magn Magn Mat 39:107–110
32. Hyeon T (2003) Chemical synthesis of magnetic nanoparticles. Chem Commun 8:927–934
33. Iijima S (1991) Helical microtubules of graphitic carbon. Nature 354(6348):56–58
34. Journet C, Maser WK, Bernier P et al (1997) Large-scale production of single-walled carbon nanotubes by the electric-arc technique. Nature 388:756–761
35. Kato M (1997) Preparation of ultrafine particles of refractory oxides by gase vaporation method. JJAP 15(5):757–760
36. Kharissova OV, Kharisov BI, de Casas Ortiz EG (2013) Dispersion of carbon nanotubes in water and non-aqueous solvents. RSC Adv 3:24812–24852
37. Kilner M, Hoon SR, Lambrock DB et al (1984) Preparation and properties of metallic iron ferrofluids. IEEE Trans Magn 20(5):1735–1737
38. Kim KK, Yoon SM, Choi JY et al (2007) Design of dispersants for the dispersion of Carbon nanotubes in an organic solvent. Adv Funct Mater 17:1775–1783
39. Krishnakumar S, Somasundaran P (1996) ESR investigations on the stabilization of alumina dispersions by Aerosol-OT in different solvents. Coll Surfaces A 117:37–44
40. Kumar SA, Meenakshi KS, Narashimhan B et al (2009) Synthesis and characterization of copper nanofluid by a novel one-step method. Mater Chem Phys 113:57–62
41. Lee G-W. Jagannathan S, Chae HG et al (2008) Carbon nanotube dispersion and exfoliation in polypropylene and structure and properties of the resulting composites. Polymer 49(7):1831–1840
42. Li WZ, Xie SS, Qian LX et al (1996) Large-scale synthesis of aligned carbon nanotubes. Science 274:1701–1703
43. Lo C-H, Tsung T-T, Chen L-C (2005) Shaped-controlled synthesis of Cu-based nanofluid using submerged arc nanoparticle synthesis system (SANSS). J Crystal Growth 277:636–642
44. Lo C-H, Tsung T-T, Chen L-C et al (2005) Fabrication of copper oxide nanofluid using submerged arc nanoparticle synthesis system (SANSS). J Nanoparticle Res 7:313–320
45. Luechinger NA, Athanassiou EK, Stark WJ (2008) Graphene-stabilized copper nanoparticles as an air-stable substitute for silver and gold in low-cost ink-jet printable electronics. Nanotechnology 19(44):445201
46. Maltsev VA, Nerushev OA, Novopashin SA et al (2007) Synthesis of metal nanoparticles on a carbon matrix. Russian Nanotech 2(5–6):85–89
47. Manawi MY, Ihsanullah SA et al (2018) A review of carbon nanomaterials' synthesis via the chemical vapor deposition (CVD) method. Materials 11:822
48. Matei A, Cernica I, Cadar O et al (2008) Synthesis and characterization of ZnO–polymer nanocomposites. Int J Mater Form 1:767–770
49. Michaelides EE (2014) Nanofluidics. Thermodynamic and transport properties. Springer, Switzerland
50. Mukherjee S, Paria S (2013) Preparation and stability of nanofluids—a review. IOSR J Mech Civil Eng 9(2):63–69
51. Munkhbayar B, Tanshen MR, Jeoun J et al (2013) Surfactant-free dispersion of silver nanoparticles into MWCNT-aqueous nanofluids prepared by one-step technique and their thermal characteristics. Ceram Int 39:6415–6425
52. Murshed SMS, de Castro CAN (2014) Nanofluids: synthesis, properties and applications. NOVA Science Publications Inc., New York

53. Nakatani I, Furubayahi T, Takahashi T et al (1987) Preparation and magnetic properties of colloid ferromagnetic metals. JMMM 65(2/3):261–264
54. Nomoev AV (2012) Mechanisms of formation, structural features and physical properties of nano-sized structures obtained by electron beams. Doctoral dissertation, University Ulan-Ude
55. Nomoev AV, Bazarova DZh, Bardakhanov SP (2009) Synthesis and study of copper-containing nanopowders. News Universities. Phys 12/3:228–232
56. Palazzeci F, Calvareci M, Zerbetto F (2011) A molecular dynamics investigation of structure and dynamics of SDS and SDBS micelles. Soft Matter 7(19):9148–9156
57. Ramsey JDF, Avery RG (1974) Ultrafine oxide powders prepared by electron beam evaporation. J Mater Sci 9:1681–1688
58. Paskevicius M, Webbb J, Pitt MP, Blach TP et al (2009) Mechanochemical synthesis of aluminium nanoparticles and their deuterium sorption properties to 2 kbar. J Alloys Compounds 481:595–599
59. Petrov YuI, Shafranovsky EA, Krupyanskii YuF et al (2001) Specific features Of the structure and local magnetic order in nanoparticles of Fe-Cr alloy. Doklady Phys Chem 379:194–199
60. Ren ZF, Huang ZP, Xu JW (1998) Synthesis of large array of well aligned carbon nanotubes on glass. Science 282:1105
61. Romano JM, Parker JC, Ford QB (1997) Application opportunities for nanoparticles made from condensation of physical vapors. Adv Powder Metall Partic Mat 2:12–13
62. Romanov NA, Kalashnikov SV, Nomoev AV (2012) Mechanisms of formation of composite nanoparticles (hollow nanoparticles). Young Sci 8(43):11–13
63. Rosensweig RE (1985) Ferrohydrodynamics. Cambridge University Press, Cambridge
64. Rudyak VYa (2004) Statistical aerohydromechanics of homogeneous and heterogeneous media, vol. 1, Kinetic Theory. NSUACE, Novosibirsk
65. Rudyak VYa (2005) Statistical aerohydromechanics of homogeneous and heterogeneous media, vol. 2, Hydromechanics. NSUACE, Novosibirsk
66. Rudyak VYa, Belkin AA, Krasnolutsky SL (2005) Statistic theory of nanoparticle transport processes in gases and liquids. Thermophys Aeromech 10(4):489–507
67. Rudyak VYa, Dashapilov GR, Minakov AV et al (2023) Comparative characteristics of viscosity and rheology of nanofluids with multi-walled and single-walled carbon nanotubes. Diamond Related Mat 132:109616
68. Rudyak VYa, Dimov SV, Kuznetsov VV (2013a) On the dependence of the viscosity coefficient of nanofluids on particle size and temperature. Tech Phys Lett 39(9):779–782
69. Rudyak VY,a Dimov SV, Kuznetsov VV et al (2013) Measurement of the viscosity coefficient of an ethylene glycol–based nanofluid with silicon dioxide particles. Doklady Phys 58(5):173–176
70. Rudyak VY, Krasnolutskii SL (2003) About viscosity of rarefied gas suspensions with nanoparticles. Doklady Phys 48(10):583–586
71. Rudyak VYa, Krasnolutskii SL (2003) On kinetic theory of diffusion of nanoparticles in a rarefied gas. Atm Oceanic Opt 16(5–6):468–471
72. Rudyak VYa, Krasnolutsky SL, Nasibulin AG et al (2002) Methods of measuring the diffusion coefficient and sizes of nanoparticles in rarefied gas. Doklady Phys 47(10):758–761
73. Scherer C, Naeto AMF (2005) Ferrofluids: properties and applications. Braz J Phys 35(3A):718–727
74. Schierz A, Zanker H (2009) Aqueous suspensions of carbon nanotubes: surface oxidation, colloidal stability and uranium sorption. Environ Pollut 157(4):1088–1094
75. Shieh Y-T, Liu G-L, Wu H-H et al (2007) Effects of polarity and pH on the solubility of acid-treated carbon nanotubes in different media. Carbon 45(9):1880–1890
76. Singh AK, Raykar VS (2008) Microwave synthesis of silver nanofluids with polyvinylpyrrolidone (PVP) and their transport properties. Colloid Polymer Sci 286(14–15):1667–1673
77. Sluzarenko N, Heurtefeu B, Maugey M et al (2006) Diblock copolymer stabilization of multiwall carbon nanotubes in organic solvents and their use in composites. Carbon 44(15):3207–3212

78. Soltanimehr M, Afrand M (2016) Thermal conductivity enhancement of COOH functionalized MWCNTs/ethylene glycol–water nanofluid for application in heating and cooling systems. Appl Therm Eng 105:716–723
79. Sotiriou GA, Hir AM, Lozach P-Y et al (2011) Hybrid, silica-coated, Janus-like plasmonic-magnetic nanoparticles. Chem Mat 23(7):1985–1992
80. Strano MS, Moore VC, Miller MK et al (2003) The role of surfactant adsorption during ultrasonication in the dispersion of single-walled carbon nanotubes. J Nanosci Nanotech 3:81–86
81. Studart AR, Amstad E, Gauckler LJ (2007) Colloidal stabilization of nanoparticles in concentrated suspensions. Langmuir 23(5):1081–1090
82. Taghizadeh MT, Abdollahi R (2015) Ultrasonic degradation of polyvinyl pyrrolidone (PVP): effect of power of ultrasound, temperature and concentration. Am Chem Sci J 9(6508):1–11
83. Taketomi S, Chikazumi S (1993) Magnetic fluids. Mir, Moscow
84. Timofeeva EV, Smith DS, Yu W et al (2010) Particle size and interfacial effects on thermo-physical and heat transfer characteristics of water-based α-SiC nanofluids. Nanotechnology 21:215703
85. Tsvetkov VN (1986) Rigid-chain polymer molecules. Nauka, Leningrad
86. Vaisman L, Wagner H, Marom G (2006) The role of surfactants in dispersion of carbon nanotubes. Adv Colloid Inter 128:37–46
87. Wagener M, Murty BS, Gűnther B (1997) Preparation of metal nanosuspensions by high-pressure DC-sputtering on running liquids. In: Komarnenl S, Parker JC, Wollenberger HJ (eds) Nanocrystalline and nanocomposite materials II. Materials Research Society, Pittsburgh, pp 149–157
88. Wagener M, Gűnther B (1999) Sputtering in liquids: a versatile process for the production of magnetic suspensions. J Magn Magn Mat 201:41–52
89. Wang L, Quintard M (2009) Nanofluids of the future. In: Wang LQ (ed) Advances in transport phenomena. Springer, Berlin, pp 179–243
90. Wang Z, Liu Q, Zhu H et al (2007) Dispersing multi-walled carbon nanotubes with water–soluble block copolymers and their use as supports for metal nanoparticles. Carbon 45(2):285–292
91. Wei X, Zhu H, Kong T et al (2009) Synthesis and thermal conductivity of Cu_2O nanofluids. Int J Heat Mass Transfer 52(19–20):4371–4374
92. Wen D, Ding Y (2005) Formulation of nanofluids for natural convective heat transfer applications. Int J Heat Fluid Flow 26(6):855–864
93. Williams WC, Bang IC, Forrest E et al (2006) (2006) Preparation and characterization of various nanofluids. NSTI-Nanotech 2:408–411
94. Windle PL, Popplewell J, Charles SW (1975) The long-term stability of mercury based ferromagnetic liquids. IEEE Trans Magnetic 11(5):1367–1369
95. Wu J-H, Ko SP, Liu H-L et al (2007) Sub 5 nm magnetite nanoparticles: synthesis, microstructure, and magnetic properties. Mat Lett 61:3124–3129
96. Xie S, Li W, Pan Z et al (2000) Carbon nanotube arrays. Mater Sci Eng A 286:11
97. Xing M, Yu J, Wang R (2015) Thermo-physical properties of water-based single-walled carbon nanotube nanofluid as advanced coolant. Appl Therm Eng 87:344–351
98. Yadav P, Gupta SM, Sharma SK (2022) A review on stabilization of carbon nanotube nanofluid. J Therm Anal Calorim 147:6537–6561
99. Yatsuya S, Tsukasaki Y, Mihama K et al (1978) Preparation of extremely fine particles by vacuum evaporation onto a running oil substrate. J Cryst Growth 45:490–497
100. Yu Q, Kim YJ, Ma H (2008) Nanofluids with plasma treated diamond nanoparticles. Applied Phys Lett 92(10):103111
101. Yu W, Xie H (2012) A review on nanofluids: preparation, stability mechanisms, and applications. J Nanomaterials 2012:1–17
102. Yu W, Xie H, Chen L et al (2009) Synthesis and characterization of monodispersed copper colloids in polar solvents. Nanoscale Res Lett 4:465–470

103. Yurekli K, Mitchell CA, Krishnamoorti R (2004) Small-angle neutron scattering from surfactant-assisted aqueous dispersion of carbon nanotubes. J Am Chem Soc 126(32):9902
104. Zeng DW, Chen X, Jiang R et al (2006) Thermal stability of Sb/Sb_2O_3 nanoparticles. Mat Chem Phys 96(2/3):454–456
105. Zhu HT, Lin YS, Yin YS (2004) A novel one-step chemical method for preparation of copper nanofluids. J Coll Interface Sci 277:100–103
106. Zhu HW, Xu CL, Wu DH et al (2002) Direct synthesis of long single-walled carbon nanotube strand. Science 296:884–886

Chapter 2
Diffusion of Nanoparticles and CNTs in Fluids

2.1 Fluctuation–Dissipation Theorems

The transport processes in liquids and dispersed fluids are molecular in nature. In particular, therefore, the transport coefficients can be calculated only from the first principles. The first success in this direction was achieved in the kinetic theory of rarefied gases. Dynamics of such gases are defined by the Boltzmann kinetic equation, which is a nonlinear integro-differential equation. This equation can be solved when Knudsen numbers are small ($Kn = l/H \ll 1$) [12, 20]. Here l is the mean free path of gas molecules, and H is the characteristic linear dimension of the flow (such as channel height, pipe diameter, etc.). The obtained solution corresponds to the hydrodynamic description of gas dynamics as described by the Navier–Stokes equations. This description corresponds to a weakly nonequilibrium gas, wherein the fluxes $\boldsymbol{J}_i$ of its macroscopic characteristics (such as particle number, momentum, and energy) are proportional to the gradients of their corresponding variables φ_i (density, velocity, and temperature)

$$\boldsymbol{J}_i(\boldsymbol{r}, t) = -\mu_i \nabla \varphi_i(\boldsymbol{r}, t), i = 1, 2, 3. \tag{2.1}$$

The coefficients of proportionality in these constitutive ratios μ_i are the transport coefficients that have been explicitly calculated for a rarefied gas [8, 12]. Thus, the constitutive relations (2.1) describe weak deviations of the gas state from equilibrium. The consequence of this is that they are calculated by mean of the quasi-equilibrium Maxwell one-particle distribution function f_M

$$\mu_i = \int d\boldsymbol{v} A_i(\boldsymbol{v}) f_M(\boldsymbol{v}), \quad i = 1, 2, 3 \tag{2.2}$$

V. Rudyak, *Thermophysical Properties of Nanofluids*, Fluid Mechanics and Its Applications 140, https://doi.org/10.1007/978-3-031-97995-8_2

Here $\mu_1 = D$ is the self-diffusion coefficient, $\mu_2 = \mu$ is the viscosity coefficient, $\mu_3 = \lambda$ is the thermal conductivity coefficient of the gas, $\boldsymbol{v}$ is the velocity of the gas molecules, and $A_i(\boldsymbol{v})$ is some functions of these velocities.

For liquids, and especially for dispersed liquids, it is fundamentally impossible to construct a kinetic theory. The dynamics of a liquid cannot be described in terms of a single-particle distribution function. It is widely recognized that to calculate even the total energy of a liquid in equilibrium, knowledge of at least the binary distribution function is required [56, 58]. Nevertheless, it is feasible to establish constitutive relations for weakly nonequilibrium processes (2.1) using first principles. This is achieved through the application of the Liouville equation for the N-particle distribution function of liquid molecules (here N is the number of molecules in the system). Generally speaking, they turn out to be non-local and retarded, but for the fluid transport processes usually considered they reduce to relation (2.1), where the transport coefficients are now determined by the so-called fluctuation–dissipation theorems (FDT) [59, 61, 89]

$$\mu_i(t) = \int_0^{t_p} dt_1 \chi_i(\boldsymbol{r}, t, t + t_1), \quad i = 1, 2, 3. \tag{2.3}$$

Here $\chi_i(\boldsymbol{r}, t, t + t_1)$ are the two-time correlation functions of the system's corresponding dynamic variables. The definition for each case will be provided later. It is important to note that the correlation functions χ_i are calculated using the quasi-equilibrium N-particle distribution function of the system being analyzed. It is also important to realize that the transport coefficients are thus determined by the transfer of the number of particles, momentum, and energy in the system, and this transfer is caused by thermal (equilibrium) molecular fluctuations.

In the literature, relations of the type (2.3) are referred to as Green–Kubo formulas [4, 22, 38, 39]. It is important to note that the transport coefficient μ_i in relation (2.3) is a function of time. A real measurable value of the transport coefficient is only attainable when the corresponding correlation function decays. The coefficient μ_i then reaches its plateau, which is determined by time t_p. In the Green–Kubo formulas, it is assumed that $t_p = \infty$. However, the quasi-periodicity of the phase trajectories[1] leads to the system's reversible motion. Therefore, in particular, the strict computation of transport coefficients using the Green–Kubo formulas yield zero. This problem was first observed by Kirkwood while calculating the friction coefficient of a Brownian particle [34]. To achieve a non-zero result, he had to truncate the infinite time limit in the integration process with a finite value, resulting in the emergence of the so-called plateau problem for correlation function integrals. Subsequently, it was demonstrated that the correlation functions of a finite system are nearly periodic functions of time.

[1] The quasiperiodicity of phase trajectories in a conservative Hamiltonian system with a finite number of molecules is determined by the renowned Poincaré recurrence theorem (see, e.g., [5]). According to this theorem, the phase trajectory of the system will return to its initial state as closely as possible after a finite time.

The first strict results were obtained in [7, 50]. Assuming the presence of mixing phase trajectories in the system, the equality to zero of the Green–Kubo formulas was established by Lebowitz [41]. Somewhat different arguments can be found in the monograph [55].

Finally, it should be noted that externally relation (2.3) substantially differs from the corresponding formulas of the kinetic theory of gases (2.2). Nevertheless, their equivalence has been rigorously proven in Ernst [17], Khon'kin [32].

The simplest fluctuation–dissipation theorem (FDT) determines the self-diffusion coefficient of liquid molecules as follows

$$D(t) = \frac{1}{3}\int_0^{t_p} dt_1 \chi_1(\boldsymbol{r}, t, t + t_1), \tag{2.4}$$

where its evolution is determined by the velocity autocorrelation function (VACF) of the molecule velocity $\boldsymbol{v}_i$

$$\chi_1(t) = \frac{1}{3}\sum_{i=1}^{N} \langle \boldsymbol{v}_i(0) \cdot \boldsymbol{v}_i(t) \rangle. \tag{2.5}$$

Here, the angular brackets denote averaging over the quasi-equilibrium N-particle distribution function. Again, the observed value of the self-diffusion coefficient is obtained only after time reaches the plateau value [66].

FDTs are universal in nature; they determine the transport processes in both simple and dispersed fluids and in nanofluids [59, 61]. The diffusion coefficient of the nanoparticle or CNT will also be determined by the relation (2.4), in which the VACF of the nanoparticle now has the form

$$\chi_{vv}(0, t) = \frac{1}{3}\langle \boldsymbol{V}_c(0) \cdot \boldsymbol{V}_c(t) \rangle. \tag{2.6}$$

Here $\boldsymbol{V}_c$ is the velocity of the particle's center of mass, and the VACF averaging is already performed over the $(N + 1)$-particle distribution function of the fluid-particle system.

For a Brownian particle, its diffusion coefficient can also be calculated by setting the corresponding law of its interaction with molecules of the carrier fluid[2] according to Eq. (2.4). However, it is useful to determine the evolution of the VACF by proceeding from the Langevin description of its dynamics. P. Langevin believed that the interaction of a Brownian particle with the molecules of a carrier liquid can be represented as a superposition of a systematic drug force and some random force that changes over time. He designated the systematic drag force as the Stokes force

$$\boldsymbol{F}_s = 6\pi\mu R\boldsymbol{V}, \tag{2.7}$$

[2] Generally speaking, the law of interaction does not need to be potential.

where R and $\boldsymbol{V}$ are the particle radius and velocity, and μ is the viscosity coefficient of the carrier fluid. As a result, the motion equation of the particle takes the form [40]

$$\dot{\boldsymbol{V}} = -\gamma \boldsymbol{V} + \boldsymbol{f}(t), \tag{2.8}$$

here $\gamma = (6\pi\mu R)/M$ is the drag coefficient, M is the particle mass, and $\boldsymbol{f}(t)$ is the Langevin random force having the following properties

$$\langle \boldsymbol{f}(t)\rangle = 0, \langle \boldsymbol{f}(t)\cdot \boldsymbol{f}(t')\rangle = \Gamma\delta(t'-t), \tag{2.9}$$

where, as noted above, the angular brackets denote averaging.

The velocity of the Brownian particle is obtained by integrating Eq. (2.8)[3]

$$\boldsymbol{V}(t) = \boldsymbol{V}_0 e^{-\gamma t} + \int_0^t e^{-\gamma(t-t')}\boldsymbol{f}(t')dt', \tag{2.10}$$

where $\boldsymbol{V}_0$ is the initial velocity of the particle. Utilizing Eq. (2.10), deriving the VACF of the Brownian particle becomes quite easy

$$\boldsymbol{V}(t)\cdot\boldsymbol{V}(t+\tau) \equiv \chi_{VV}(t, t+\tau) = (3kT/M)e^{-\gamma\tau}, \tag{2.11}$$

When deriving (2.11), the following initial condition is taken into account: $\chi_{VV}(t,t) = \langle V^2(t)\rangle = 3kT/M$.

Thus, the VACF of a Brownian particle exhibits an exponential decay with a relaxation time determined by the Stokes force, and equal to γ^{-1}. This exponential decay implies the existence of a sole velocity relaxation mechanism. By substituting VACF (2.11) into the FDT (2.4), we derive the diffusion coefficient of a Brownian particle, previously derived by Einstein [16]

$$D_E = \frac{kT}{6\pi\mu R}. \tag{2.12}$$

Another important circumstance in the study of diffusion is the need in the general case to also take into account the presence of rotational diffusion. This is especially critical for CNTs. The rotational diffusion of nanoparticles can also be described by the corresponding Langevin equation

$$\frac{d}{dt}(\boldsymbol{I}\cdot\boldsymbol{\omega}) = -\boldsymbol{\gamma}_r\cdot\boldsymbol{\omega} + \boldsymbol{S}, \tag{2.13}$$

[3] Equation (2.8) is a stochastic equation, and the second term in (2.10) is a stochastic integral operator. The rules for calculating such integrals were formulated by K. Itô [28] and later generalized by R. Stratonovich [79].

where $\boldsymbol{I}$ is the moment of inertia of the particle, $\boldsymbol{\omega}$ is its angular velocity, $\boldsymbol{\gamma}_r$ is the tensor drag coefficient, and $\boldsymbol{S}$ is the principal moment of the random forces acting on the particle with respect to its center of mass.

Similarly to translational Brownian motion, the principal moment of random forces can be simulated by white noise. Then (compare with formula (2.9)),

$$\langle \boldsymbol{S}(t) \rangle = 0, \left\langle \boldsymbol{S}(t) \cdot \boldsymbol{S}\left(t'\right) \right\rangle = \Gamma_r \delta\left(t' - t\right). \tag{2.14}$$

Equation (2.13) takes into account that the moment of inertia tensor can vary in the general case. However, if it doesn't, the equation simplifies. It has a particularly simple form for a spherical top, when the principal values of the moment of inertia tensor with respect to the three axes of the Cartesian coordinate system coincide $\left(I_{xx} = I_{yy} = I_{zz} = I\right)$

$$I \dot{\boldsymbol{\omega}} = -\gamma_r \boldsymbol{\omega} + \boldsymbol{S}, \tag{2.13a}$$

where the drag coefficient is equal to [59]

$$\gamma_r = 8\pi \mu R_h^3. \tag{2.15}$$

Here R_h is the hydrodynamic radius of the CNT (1.1). While utilizing the FDT of the form (2.9),

$$D_r = \int_0^{t_p} \chi_{\omega\omega}(0, t) dt, \; \chi_{\omega\omega}(0, t) = \frac{1}{3} \langle \boldsymbol{\omega}_c(0) \cdot \boldsymbol{\omega}_c(t) \rangle, \tag{2.16}$$

we obtain the rotational diffusion coefficient equal to

$$D_r = kT / \left(8\pi \mu R_h^3\right). \tag{2.17}$$

This formula also applies to describe the rotational diffusion of CNTs due to their hydrodynamic radius being in the denominator. However, in this case, the value of the coefficient D_r will be significantly overestimated. A more accurate estimation requires transitioning from Eq. (1.1) to the one for CNTs, which can be closely simulated by long solid rods with a diameter d and length L. For these particles, the rotation around the nanotube axis may be disregarded. The inertia tensor components, related to the rotation around the cylinder's long axis, are negligible in comparison to the other components. The angular velocity, in this case, refers to the angular velocity component that is perpendicular to the axis of symmetry of the nanotube. In this case $\boldsymbol{I} \cdot \boldsymbol{\omega} = ML^2 \boldsymbol{\omega}/12$, and Eq. (2.13) becomes

$$\frac{ML^2}{12} \dot{\boldsymbol{\omega}} = -\gamma_t \boldsymbol{\omega} + \boldsymbol{S}, \tag{2.13b}$$

where the drag coefficient γ_t is given as

$$\gamma_t = \frac{\pi \mu L^3}{3\ln(L/d)}. \tag{2.18}$$

As a result, it can be established that the translational and rotational diffusion coefficients are determined respectively by the following formulas

$$D = \frac{kT\ln(L/d)}{3\pi \mu L}, \quad D_r = \frac{3kT\ln(L/d)}{\pi \mu L^3} \tag{2.19}$$

2.2 Molecular Dynamics Method

The FDTs (2.3), which determine the transport coefficients, require the computation of the corresponding correlation functions. To do this, it is necessary to know the dynamic variables of the system, the coordinates and velocities of all its molecules at successive moments in time. These dynamic variables can be determined by solving the system of Newton's equations

$$m_i \frac{d\boldsymbol{v}_i}{dt} = \sum_{i=1}^{N} \boldsymbol{F}_{ij}, \boldsymbol{v}_i = \frac{d\boldsymbol{r}_i}{dt}, \quad i = 1, 2, \ldots, N, \tag{2.20}$$

describing the evolution of particles of the considered system. Here $\boldsymbol{r}_i$ is the radius-vector of the particle, m_i is its mass, $\boldsymbol{v}_i$ is its velocity, and $\boldsymbol{F}_{ij}$ is the interaction force of particles i and j.

Of course, we would like to have analytical solutions of Eq. (2.20), which, unfortunately, is impossible. With the advent of sufficiently fast computers, it became possible to solve such a problem numerically. The MD method is used to obtain the numerical solution of Eq. (2.20), sequentially determine the dynamic variables of the system, and find all the necessary observables with their aid. The MD method was first developed by Alder and Wainwright in their works [1–3]. Initially, the method was applied to a system of hard spheres. Later, Gibson et al. [21] extended it to systems of molecules with real interaction potentials. The forthcoming discussion will show the fundamental difference between these two approaches.

Today, a new direction of research has formed around the MD method, which is called computational physics. The MD method and its various variants are extremely successfully used in physics, mechanics, chemistry, biology. Over the past sixty years, MD method has been used to predict a number of data that were only later confirmed experimentally (or still have not been confirmed). In certain cases, the use of MD method is essentially the only option. In particular, only with the help of the MD method can one try to understand the mechanisms of transport processes in nanofluids.

2.2.1 MD Method: Continuous Potentials

The starting point of the MD method is Newton's Eq. (2.20) applied to the particles under consideration. One defining element of this microscopic simulation of the system's evolution is setting interparticle forces. In practice, continuous potentials, such as the Lennard–Jones potential or the potential of hard spheres (see Sect. 2.2.5), are typically utilized. The application of the method varies between these two cases. The algorithm construction schemes for both situations will be briefly outlined below. For a more systematic understanding of the MD method, it is recommended to refer to specialized sources (for example, [53]).

Usually, when utilizing the MD method, the intermolecular interaction potentials are presumed to be pairwise and additive.[4] The system's dynamics is then described by system of Newton's Eq. (2.20), and the task is to solve this equation system using a computer. To achieve this, a cell is chosen, commonly cubic wherein from several hundred to many thousand particles are positioned. Each particle receives an initial velocity according to a specified law, and then particles are homogeneously distributed in the cell according to the density of the simulated fluid. Further for the system of Eq. (2.20) it is necessary to solve the Cauchy problem to determine the coordinates and impulses for all particles at a given moment in time, provided that the relevant data was known at a prior moment. This can be achieved by replacing Eq. (2.20) with finite-difference equations. For each molecule i, determining dynamic variables $\boldsymbol{r}_i^{(n+1)}$ and $\boldsymbol{v}_i^{(n+1)}$ at the moment in time $t_{n+1} = t_n + \delta t$ is necessary, based on their known values at a prior moment t_n. Many different algorithms have been developed to compute the considered dynamic variables. The simplest discretization of Eq. (2.20) has the following form

$$\boldsymbol{r}_i^{(n+1)} = \boldsymbol{r}_i^{(n)} + \boldsymbol{v}_i^{(n)}\delta t, \quad \boldsymbol{v}_i^{(n+1)} = \boldsymbol{v}_i^{(n)} + m^{-1}\delta t\sum_{j\neq i}^{N}\boldsymbol{F}_{ij}\left(\boldsymbol{r}_i^{(n)}, \boldsymbol{r}_j^{(n)}\right).$$

This explicit scheme, known as the Euler scheme, is first-order accurate with respect to δt. However, there exist more complex approximations. The method of central differences assumes that the coordinate increments are determined by the velocity values at the middle of the time step

[4] This assumption greatly simplifies the computations, although not fundamentally. Moreover, the MD method is actually the only reliable tool that allows studying the effect of non-additive forces on the transport processes. In defense of using pair potentials, it is noteworthy that most of the potentials used in practice are effective and at least partially take into account the non-additivity effects.

$$\boldsymbol{r}_i(t+\delta t) = \boldsymbol{r}_i(t) + \delta t \boldsymbol{v}_i(t+\delta t/2), \quad \boldsymbol{v}_i(t+\delta t/2) = \boldsymbol{v}_i(t-\delta t/2) + m^{-1}\delta t \sum_{j \neq i}^{N} \boldsymbol{F}_{ij}(t).$$

Many algorithms use a series expansion of the $\boldsymbol{r}_i(t \pm \delta t)$ function: $\boldsymbol{r}_i(t \pm \delta t) = \boldsymbol{r}_i(t) \pm \dot{\boldsymbol{r}}_i(t)\delta t + \ddot{\boldsymbol{r}}_i(t)\left(\delta t^2/2\right) + \cdots$ The addition of these two equations to the third order terms gives: $\boldsymbol{r}_i(t+\delta t) = 2\boldsymbol{r}_i(t) - \boldsymbol{r}_i(t-\delta t) + \ddot{\boldsymbol{r}}_i(t)\delta t^2$. The velocity is obtained by subtracting these two equations. With the same accuracy we have: $\dot{\boldsymbol{r}}_i(t) = [\boldsymbol{r}_i(t+\delta t) - \boldsymbol{r}_i(t-\delta t)]/2\delta t$. The described algorithm was proposed by Verlet [83]. Numerous other schemes have also been applied, such as predictor–corrector scheme, Runge–Kutta methods of different order of accuracy, etc. In general, the choice of numerical approximation is largely dictated by the problem to be solved and should be discussed in each particular case. The δt interval must be carefully selected to ensure stable integration of the Eq. (2.20) while maintaining minimal deviation from the exact solution over time. Conservation laws are typically used to monitor the solution's stability.

All particle coordinates and velocities obtained at different time points are preserved in computer memory and can be utilized to derive specific properties of the system. Multiple computations are performed, resulting in the generation of an ensemble of distinct representations of the system's evolution in the phase space.

2.2.2 MD Method: Elastic Hard Spheres

When considering a system with molecules simulated by elastic hard spheres, the application process of MD method is notably simplified. In this case, there is no need to solve Newton's equations at all. Here, the procedure involves identifying a sequence of time intervals between molecule collisions within the system. The minimum interval at each step will identify a real colliding pair of molecules. Then all particles are shifted along their trajectories by this minimum time interval, while the velocities of the colliding particles are recalculated. The computation procedure is then repeated. Exactly such an algorithm was proposed by Alder and Wainwright [2, 3] and subsequently modified many times (see for example, [18]).

Let us examine how the time before two particles collide is determined. So, let two particles be given with velocities $\boldsymbol{v}_1$ and $\boldsymbol{v}_2$, respectively, located at points in space with coordinates $x_i^0,\ y_i^0,\ z_i^0,\ i = 1,\ 2$ at the initial instant. Assuming their collision occurred after time δt, the coordinates of the particles became equal to $x_i^1,\ y_i^1,\ z_i^1,\ i = 1,\ 2$. At the moment of collision the following condition is satisfied

$$(x_1 - x_2)^2 + (y_1 - y_2)^2 + (z_1 - z_2)^2 = (R_1 + R_2)^2, \tag{2.21}$$

where $R_1$and R_2 are the radii of the hard spheres. During the time δt (before the collision) the particles moved with constant velocities $\boldsymbol{v}_1$ and $\boldsymbol{v}_2$, therefore

$$x_1 = x_1^0 + v_{1x}\delta t, \quad y_1 = y_1^0 + v_{1y}\delta t, \quad z_1 = z_1^0 + v_{1z}\delta t,$$

$$x_2 = x_2^0 + v_{2x}\delta t, \quad y_2 = y_2^0 + v_{2y}\delta t, \quad z_2 = z_2^0 + v_{2z}\delta t.$$

Then a quadratic equation is obtained to determine the time before collision

$$a_{12}\delta t^2 + b_{12}\delta t + c_{12} = 0,$$

where $a_{12} = (v_{1x} - v_{2x})^2 + (v_{1y} - v_{2y})^2 + (v_{1z} - v_{2z})^2$, $b_{12} = 2\left(\boldsymbol{r}_1^0 - \boldsymbol{r}_2^0\right) \cdot (\boldsymbol{v}_{1x} - \boldsymbol{v}_{2x})$, $c_{12} = (x_1^0 - x_2^0)^2 + (y_1^0 - y_2^0)^2 + (z_1^0 - z_2^0)^2 - (R_1 + R_2)^2$. The real roots of this equation exist if

$$b_{12} < 0, \quad b_{12}^2 - a_{12}c_{12} > 0, \tag{2.22}$$

in which case they are equal to

$$\delta t_{1,2} = \left[-b_{12} \pm \sqrt{b_{12}^2 - a_{12}c_{12}}\right] a_{12}^{-1}.$$

It is necessary to choose the minimum root, since it corresponds to the collision. The second root fulfills condition (2.21) when one particle passes through another. If the conditions (2.22) are not satisfied, a collision does not occur and $\delta t = \infty$.

After two particles collide, it is essential to recalculate their velocities. These velocities are determined by the momentum and energy conservation laws and have the form

$$\boldsymbol{v}_1' = \boldsymbol{v}_1 + 2\frac{m_2}{m_1 + m_2}(\boldsymbol{v}_{21} \cdot \boldsymbol{e})\boldsymbol{e}, \quad \boldsymbol{v}_2' = \boldsymbol{v}_2 - 2\frac{m_1}{m_1 + m_2}(\boldsymbol{v}_{21} \cdot \boldsymbol{e})\boldsymbol{e},$$

where $\boldsymbol{e}$ is a unit vector directed from the first particle to the second along the normal to the surfaces of the colliding hard spheres and $\boldsymbol{v}_{21} = \boldsymbol{v}_2 - \boldsymbol{v}_1$.

2.2.3 Boundary and Initial Conditions

Limited computer resources necessitate simulating systems with only a comparatively small number of molecules. For such a system to have the properties of a system consisting of a large number of particles simulating a liquid or gas, the so-called periodic boundary conditions are most often applied. For this purpose, the analyzed volume V is partitioned into cells, for example, cubic cells with edge length l_c containing N particles. It is assumed that the cells form a periodic lattice. Thus, when a molecule escapes through a cell face with momentum $\boldsymbol{p}_i$, another molecule with the same momentum replaces it from the opposite face. Figure 2.1 illustrates this concept for a planar system.

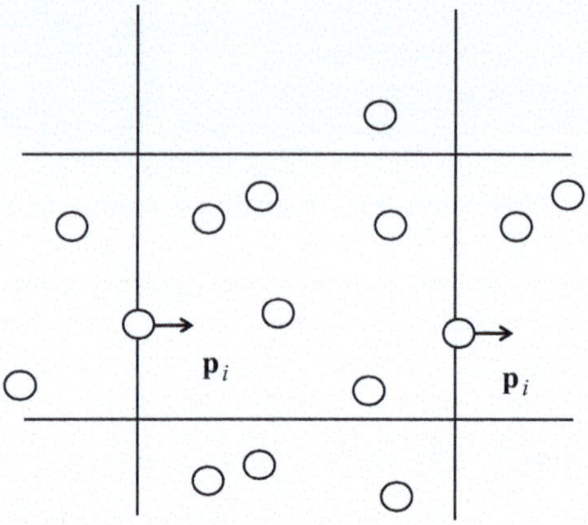

Fig. 2.1 Periodic boundary conditions used in the MD method

The potential for intermolecular interactions is assumed to be truncated, with the effective radius of the intermolecular forces significantly smaller than l_c.. To reduce computation time, the periodic boundary conditions can be modified by explicitly taking into account the motions of the molecules at the cell boundaries. It is quite clear that under periodic boundary conditions, the number of particles in each cell is conserved along with energy and momentum (certainly, angular momentum is not conserved). The corresponding ensemble for such a system is the microcanonical ensemble, where no fluctuations in the number of particles occur. The absence of these fluctuations leads to statistical errors and molecule correlations associated with the finite number of particles, which are not typical of real systems. Statistical errors can be reduced, firstly, by increasing the statistics of the computations performed, and secondly, by allowing the existence of particle number fluctuations in the cell. To achieve this, stochastic boundary conditions are implemented where, in such cases, a particle crossing the cell boundary disappears from the system. However, new molecules with random momentum are randomly introduced into the system at random time intervals and locations. At the same time, the law of particle generation must ensure that the system's average density, momentum, and energy remain unchanged. The benefit of this system lies in its ability to simulate the dynamics of a large canonical ensemble. It is only necessary to carefully monitor the fulfillment of the conservation laws of the number of particles, and energy, while allowing fluctuations of these quantities.

The complete problem formulation for the Eq. (2.20) necessitates establishing initial conditions, which includes the coordinates and velocities of all molecules in the cell. The assignment of initial conditions is determined by the physics of the problem. In studies of high density systems such as liquids and solids, the particles arrange themselves in an ordered manner. For instance, argon forms a face-centered cubic lattice when it crystallizes; therefore, its simulation needs using a cubic cell. The molecules in this cell constitute a face-centered cubic lattice and are positioned at the nodes of this lattice. Conversely, when studying the gaseous phase, the molecules' coordinates are randomly set, with an additional condition to preclude overlapping configurations. Various methods can be utilized to set the initial velocities of the molecules. When analyzing the equilibrium states, it is essential to apply certain conditions to the velocities, which demand the total momentum of the system to be equal zero and the total energy of the system to correspond with its temperature. The system's temperature is determined by the average kinetic energy of its molecules. Such arbitrariness in setting the initial conditions is due to the impossibility of determining which microscopic state of the system corresponds to simulated macroscopic state.

2.2.4 On the Accuracy of the Molecular Dynamics Method

Computer simulation accuracy is always very important. The MD method is remarkable since it enables simulation of real scenarios that may be difficult or even impossible to study experimentally. And here the simulation results acquire the predictive power of an experimental fact. In such cases, the accuracy of the results is of utmost importance. When using the MD method, certain systematic errors arise from four distinct sources: (i) rounding errors; (ii) errors stemming from Newton's equations' numerical integration schemes; (iii) errors due to the finite size of the computational cell and periodic boundary conditions; and (iv) errors resulting from particle correlations due to the limited number of particles in the simulated system.

Rounding errors are a crucial element to consider when using computers for simulations. As these errors are pseudo-random,[5] this type of error, when introduced into the computation, makes the motion of the system in phase space irreversible. This phenomenon is confirmed by direct computations. The finite precision of the computer representation of numbers is especially important to consider when solving evolutionary problems, such as calculating the trajectories of molecules. However, the manifestation of this effect is weak when determining integral characteristics such as mean free path, relaxation times, and autocorrelation functions. Thus, when calculating the self-diffusion coefficient of a dense gas, the difference between the values of this coefficient calculated with single and double precision of number representation does not exceed 0.5%, including the statistical error [66].

Errors associated with finite-difference approximation of Newton's equations are unavoidable and the magnitude of these errors increases over time. Therefore, the main criterion for the application of one or another finite-difference scheme is, first of all, the correct determination of the confidence time interval during which these errors remain small enough not to significantly affect the quality of the calculation. If this time is t_e, then the errors in determining the dynamic variables at such times

[5] Strictly speaking, here randomness is excluded. Rounding happens when the computer's digit capacity is exceeded by significant digits following the decimal point. All of these factors contribute to the randomness of the rounding process. Rounding is dependent on various factors, including computer architecture, computation methods, and frequency of occurrence of irrational numbers, etc. However, re-computing using the same computer always yields the same outcome, so the result obtained is deterministic.

should be small. Since finite-difference schemes of one or another order give an error in calculation of dynamic variables proportional to the discretization step h, the latter must fulfill the condition: $h^{k+1} t_e \ll 1$, where k is the order of the scheme. In practice, the accuracy of the solution of the dynamic problem, as already mentioned, can be checked by the fulfillment of conservation laws.

Due to the finite length of the computational cell l_c, another source of errors emerges. In a real gas or liquid, a disturbance arising at some point spreads throughout the volume and then decays due to dissipative effects. Computational cell sizes are usually small enough, and a disturbance that appears at a certain point can reach it again after a time of the order of l_c/a. Here a is the perturbation propagation velocity, which is equal to the speed of sound in gases and liquids. Thus, guaranteed accuracy can only be expected, given all other conditions are equal, only at times of the order of $t \sim l_c/a$.

A finite number of particles N in a simulated system can cause significant errors when the system is infinitely large. The physical reason for such errors is that correlations develop in the final system that are not typical for the simulated system. These correlations are generally poorly understood. The exception is the case of a rarefied gas [57, 58].

The appearance of all the indicated inaccuracies is completely natural. Any modeling of real processes or phenomena is associated with the introduction of certain errors that arise both during model construction and solution. In the MD method the situation is in a certain sense worse. The phase trajectories of the simulated molecular system are unstable with respect to small perturbations. More precisely, there is a local instability (exponential divergence of two trajectories that are arbitrarily close at the initial moment) as well as mixing (see [46, 47, 59] and the literature cited there). This phenomenon, known as dynamical chaos in the literature, is generally typical of many-particle systems, such as for example, vortices [60]. The presence of dynamic chaos leads to the fact that the MD method does not calculate true Newtonian phase trajectories. It is often suggested that this is attributed to the inaccurate solving of the Newtonian equation system (see, for example, [35]), and developing a corrective method for equation solution at each step would lead to true phase trajectories. This is, certainly, a misconception. True trajectories cannot be obtained even for a system of hard spheres, for which there is no need to solve Newton's equations at all [59].

Adequate simulation results of all required observable processes by the MD method can be obtained only by averaging data over a large number of independent phase trajectories. Additionally, it is crucial to consider a substantial number of molecules to ensure sufficiently high accuracy.

2.2.5 Particle Interaction Potentials

The central point of implementing the MD method is to specify intermolecular forces. Typically, potential forces are used for this purpose. Contemporary MD packages provide the option to apply various types of intermolecular potentials. These potentials are actually semi-phenomenological. They are based on reasonable physical representations, but the exact solution to the interaction problem of two molecules cannot be solved even in the simplest cases. For this reason, the potentials used always contain a number of parameters that are determined from certain experimental data. Most often for this aim the various thermophysical or mechanical data (transport coefficients, equation of state, etc.) are used. In this case, depending on the choice of particular thermophysical or mechanical data, the obtained parameters of intermolecular potentials will differ, sometimes significantly.

It was previously stated that this book uses only pair potentials. As it concerns simulation of transport processes in nanofluids, it is necessary to introduce several types of interaction potentials. These include interaction potentials between the base fluid molecules, these molecules and nanoparticles, as well as nanoparticles themselves. Typically, determining the optimal potential to use is a complicated matter and largely dependent on the specific problem being addressed. It is impossible to select a universal potential that can be used to solve any problem. Since this book primarily aims to calculate transport coefficients, we can confine ourselves to relatively simple potentials, due to transport coefficients are fairly rough integral characteristics of the simulated system. The simplest pair interaction potential is the potential of hard spheres

$$U(r) = \begin{cases} \infty, & r \le d \\ 0, & r > d \end{cases}, \tag{2.23}$$

where d is the diameter of the hard sphere.

The advantages of this model are obvious, it is visual and allows calculations to be made very simple. On the other hand, its disadvantages are also clearly visible. The potential (2.23) increases too steeply at short distances, and does not take into account the long-range component of the interaction. It is evident, therefore, that this model is only applicable at adequately high fluid temperatures. Nevertheless, even at such temperatures, the model does not provide an accurate dependency of the transport coefficients on temperature. On the other hand, potential (2.23) is universal. By changing the diameter of the spheres, it can be used to describe the interaction of molecules, as well as nanoparticles and cross-interactions. Denoting the diameter of the molecules as d_m, and that of the particles as d_p, the simplest combination relation for the interaction of base fluid molecules with nanoparticles can be given as: $d_{mp} = \left(d_m + d_p\right)/2$.

The primary drawback of potential (2.23) is the lack of an attractive part of the intermolecular interaction potential, that all real potentials have. The two-parameter Lennard–Jones potential is quite realistic and at the same time simple

$$\Phi_{LJ}(r) = 4\varepsilon_{ij}\left[\left(\sigma_{ij}/r\right)^{12} - \left(\sigma_{ij}/r\right)^{6}\right], \tag{2.24}$$

where $r = |\boldsymbol{r}_i - \boldsymbol{r}_j|$ is the distance between the centers of molecules i and j, σ_{ij} is the effective size of the particle (atom, molecule), and ε_{ij} is the depth of the potential well.

The potential (2.24) has a well-understood domain of applicability. It is not suitable for detailed descriptions of processes such as the excitation of molecules, their ionization or recombination. However, it is appropriate for calculating most thermophysical characteristics of liquids and gases. With its use, it became feasible to determine viscosity coefficient of water with a sufficiently high level of accuracy, which was not achievable for an extended period using even more complex (and more realistic for water) potentials [74].

The combination rules are commonly used for determining the interaction potential parameters amongst various particles of distinct substances. A thorough description of these relations can be found in [36, 59]. In this study, the simplest combination rules are implemented for the potentials (2.23) and (2.24)

$$\sigma_{ij} = \left(\sigma_{ii} + \sigma_{jj}\right)/2 \quad \text{or} \quad \sigma_{ij} = \sqrt{\sigma_{ii}\sigma_{jj}}, \quad \varepsilon_{ij} = \sqrt{\varepsilon_{ii}\varepsilon_{jj}}. \tag{2.25}$$

where σ_{ij}, σ_{ii}, σ_{jj} и ε_{ij}, ε_{ii}, ε_{jj} are the interaction potential parameters of molecule of types i and j, i and i, j and j, respectively.

Nanoparticles are frequently simulated as distinct clusters of varying sizes whose atoms interacted with each other by means of a potential (2.24) (see, for example, [52]). However, a cluster is not a nanoparticle,in a liquid its shape and size can change (and change). Additionally, the cluster atoms can be dislodged by molecules of the carrier fluid. Conversely, in real nanofluids, the shape and size of the nanoparticles remain consistent throughout their evolution.

Contemporary MD packages enable the simulation of nanoparticles and their force interactions with the nanoparticle-containing carrying molecular medium in different ways. First, a nanoparticle can be simulated as a combination of its constituent atoms or molecules. This approach requires setting the interaction potentials of each nanoparticle atom with other atoms, as well as with the molecules of the carrier medium. The selection of potentials relies on the nanoparticles' characteristics to be simulated. To avoid the nanoparticles from breaking down into separate atoms, supplementary potentials are integrated to hold the nanoparticle molecules together, or they can be simulated as a set of atoms in which a constant distance is maintained between them, for instance as a solid body that has a specific weight, center of mass, and corresponding moments of inertia.

Three potentials, namely the Lennard–Jones, Kihara, and Weeks–Chandler–Andersen potentials [86] have been used for simulating nanoparticle dynamics and transport properties. The primary challenge of simulating nanoparticle–molecule interactions using these potentials is the selection of their constants. These potentials were developed to describe intermolecular interactions, and the established constants are relevant solely for this purpose. It remains completely unclear how they should be employed for describing molecule–nanoparticle interactions. Thus, these potentials have been exclusively utilized for studying qualitative features of nanofluids [49].

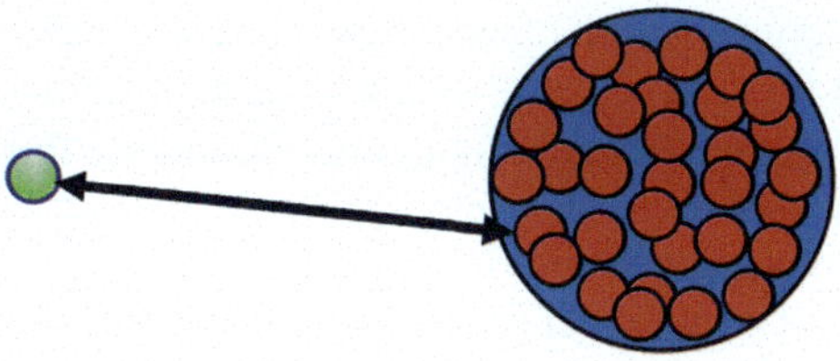

Fig. 2.2 Construction of the interaction potential of a carrier fluid molecule with a nanoparticle

Krasnolutskii and the author of this book have developed potentials that describe how a carrier fluid molecule interacts with different types of nanoparticles, including spherical solid, hollow, and composite one [70–73, 77]. These potentials are constructed using the same principles. As an example, let us briefly consider the derivation of the interaction potential of a carrier fluid molecule with a solid spherical nanoparticle, since this is a potential that is further used in this book.

A solid spherical nanoparticle with a radius R (Fig. 2.2) is considered as an assembly of atoms (or molecules). The potential interaction between a molecule of the carrier medium and the nanoparticle can be described by the following equation

$$\Phi(|\boldsymbol{r}|) = \sum_{k=1}^{N} \Phi(|\boldsymbol{r}_k - \boldsymbol{r}_m|), \tag{2.26}$$

where N is the number of atoms of the nanoparticle, $\Phi(|\boldsymbol{r}_k - \boldsymbol{r}_m|)$ is the interaction potential of atom k of the nanoparticle with the medium molecule, $\boldsymbol{r}_m$ and $\boldsymbol{r}_k$ are the radius-vectors of the carrier medium molecule and the k-th atom of the nanoparticle, respectively.

It is not feasible to summarize series (2.26) analytically. Therefore, integration is used instead of summation. Physically, this means that the nanoparticle is approximated by a continuum model (solid body). Assuming further that $\Phi(|\boldsymbol{r}_k - \boldsymbol{r}_m|)$ is described by the Lennard–Jones potential (2.24) and calculating the resulting integral, we obtain the desired potential

$$\Phi_{mp}(r, R) = \Phi_9(r, R) - \Phi_3(r, R), \tag{2.27}$$

where

$$\Phi_n(r, R) = C_n\left\{\left[\frac{1}{(r-R)^n} - \frac{1}{(r+R)^n}\right] - \frac{n}{(n-1)r}\left[\frac{1}{(r-R)^{(n-1)}} - \frac{1}{(r+R)^{(n-1)}}\right]\right\},$$

$$C_n = \frac{8\pi\varepsilon_{12}\sigma_{12}^{(n+3)}}{n(n+1)V_p}$$

Here r is the distance between the centers of the nanoparticle and the carrier fluid molecule, ε_{12} and σ_{12} are the parameters of the interaction potential between the nanoparticle atoms and the carrier fluid molecule (here index 1 refers to the nanoparticle atom and 2 relates to the carrier molecule). They are determined by the combination rules (2.25).

The potential (2.27) (hereafter it will be called the RK potential) offers a notable benefit: it is a two-parameter potential, similarly to potential (2.24). The parameters C_n are easily determined by the parameters of the potential (2.24) and the combination rules (2.25).

The interaction potential of the spherical particle was proposed by Hamaker [23]. This potential is now commonly used for macroscopic particles [27]. According to Hamaker's theory, van der Waals forces play a crucial role in interparticle interactions. This is because macroscopic particles have extremely low kinetic energy, making their repulsion energy negligible compared to attraction forces. However, nanoparticles have characteristic sizes ranging from 1 to 100 nm and are comparable in size to molecules, and obviously have appreciable thermal velocities. In these circumstances, not only attraction forces but also repulsion forces between nanoparticles should be taken into account in their interaction. The nanoparticle interaction potential, which would take into account not only Van der Waals forces but also repulsive forces, was obtained in the paper [76]. This potential was derived in the same way as the RK potential (2.27). It is defined as the sum of the interaction potentials of the atoms of the first nanoparticle with all the atoms of the second nanoparticle and is described by the potential (2.24). Thus, the interaction potential of two identical nanoparticles (RKI potential) of radius R is defined by the following formula

$$\Phi_{pp}(r, R) = \Phi_{pp,7}(r, R) - \Phi_{pp,1}(r, R), \tag{2.28}$$

$$\begin{aligned}
\Phi_{pp,7}(r, R) = C_7 \Bigg\{ & \frac{R^2}{r}\left[\frac{1}{(r-2R)^7} + \frac{2}{r^7} + \frac{1}{(r+2R)^7}\right] \\
& - \frac{R}{3r}\left[\frac{1}{(r-2R)^6} - \frac{1}{(r+2R)^6}\right] \\
& - \frac{1}{30r}\left[\frac{1}{(r-2R)^5} + \frac{2}{r^5} + \frac{1}{(r+2R)^5}\right]\Bigg\}, \\
\Phi_{pp,1}(r, R) = C_1 \Bigg\{ & \ln\left[\frac{r^2 - 4R^2}{r^2}\right] + 2R^2\left[\frac{1}{r^2 - 4R^2} + \frac{1}{r^2}\right]\Bigg\}, \\
C_7 = \frac{\pi^2}{315}\frac{\varepsilon\sigma^{12}}{V_p^2}, & \quad C_1 = \frac{2\pi^2}{3}\frac{\varepsilon\sigma^6}{V_p^2},
\end{aligned}$$

where r is the distance between the centers of the nanoparticles.

It is easy to show that the attractive part of the potential (2.28) $\Phi_{pp,1}$ coincides with the Hamaker potential; however, the parameters in the potential (2.28) are defined through the parameters of the interaction potential (2.24) of the nanoparticle atoms. It is known that the Hamaker constant inadequately describes the experimental data.

The Lifshitz theory, based on continuum electrodynamics [42] provides a more accurate description of van der Waals forces between macroscopic particles. However, this theory is only applicable to macroscopic particles. Conversely, potential (2.28) is applicable to sufficiently small particles. Here, it can be assumed that the constant C_1, which determines the value of the attractive forces in formula (2.28), may also correspond to the Hamaker constant for such particles.

The potentials described in this study differ significantly in their action radius. As an example, Fig. 2.3 illustrates the Lennard–Jones interaction potential between zinc atoms (line 1), the RK potential (2.27) corresponds to the interaction between a Zn atom and a Zn nanoparticle with a diameter of 2 nm (curve 2), and the RKI potential (2.28) demonstrates the interaction between a 2 nm diameter Zn nanoparticles (curve 3). The aforementioned potentials vary not only in the well depth of their attraction forces, but also in their action radius. The action radius of nanoparticles' van der Waals forces is at least the same as the diameter of the nanoparticles themselves.

Since the diffusion of CNTs will also be discussed below, it is necessary to mention corresponding models used. Carbon nanotubes are intricate structures made up of large number of atoms. A SWCNT is a rolled-up sheet of graphene in which the carbon atoms are arranged at the vertices of a regular hexagon. The sheet can be rolled-up at various angles, which affects the SWCNT chirality, determined by two chirality indices (n and m). The SWCNT's diameter is determined by these indices and is equal to

$$d = \frac{\sqrt{3}a_0}{\pi}\sqrt{m^2 + mn + n^2}, \quad (2.29)$$

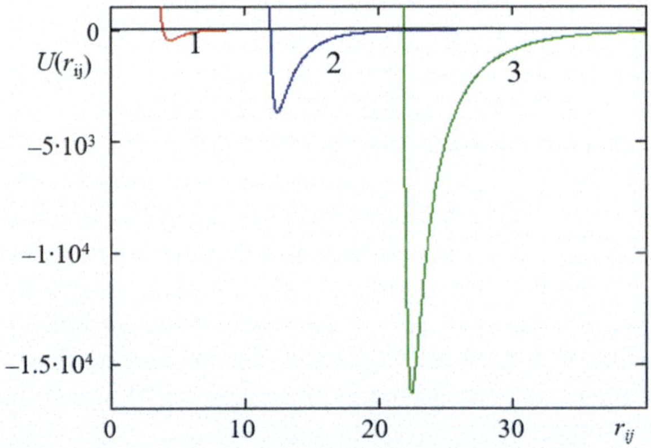

Fig. 2.3 Comparisons of the Lennard–Jones interaction potential between zinc atoms (1), the RK interaction potential between a Zn atom with a 2 nm diameter Zn nanoparticle (2), and the RKI interaction potential between a 2 nm diameter Zn nanoparticles (3); here the interparticle distance r_{ij} is given in angstroms

where $a_0 = 0.142$ nm is the distance between neighboring carbon atoms of the graphene sheet.

Armchair-type SWCNTs with chirality indices ($m = 8$, $n = 8$) and ($m = 6$, $n = 6$) and diameters of 1.1 and 0.818 nm were utilized in the study. Two distinct models of SWCNTs were employed. The lengths of the SWCNTs varied from 1.1 to 32 nm. The first model simulated the SWCNTs as a solid structure consisting of a set of carbon atoms with a time-constant distance between them, i.e. the nanotube was characterized by its weight, position of the center of mass, and its corresponding moments of inertia. The second was modeled by a set of carbon atoms interacting via the AIREBO potential [80]. Thus, this latter model allows the tube to bend during evolution, it has a known elasticity. Additionally, the interaction between CNT atoms and the carrier fluid molecules was described by the Lennard–Jones potential.

2.3 MD Simulation of Diffusion of Spherical Nanoparticles

Diffusion of nanoparticles is a key transport process in nanofluid and its defining features of the relaxation of the velocity of nanoparticles in the carrier fluid will affect both the viscosity and thermal conductivity of nanofluids. To date, there is no rigorous theory that can predict the diffusion of nanoparticles in dense gases and liquids. Typically, the diffusion of nanoparticles is thought to be the same as that of ordinary Brownian particles. However, the relation (2.12) is modified by introduction of a new drag coefficient γ_S

$$D = kT/\gamma_S, \quad \gamma_S = C\pi\mu R, \tag{2.30}$$

where C is the constant, determined by boundary conditions and equal to 6 for no-slip conditions. In this case (2.30) is reduced to formula (2.12).

Nonetheless, experiments measuring the diffusion coefficient of fullerenes [25, 31], large organic molecules [19, 37], and metallic nanoparticles [87] have yielded highly dissimilar results. Various explanations have been proposed for these discrepancies. The papers [25, 87] contend that non-slip boundary conditions must be changed by slip conditions for small objects, resulting in a modification of the constant in (2.30). The authors [19, 31] suggest using an empirical correlation $D = A/\mu^p$ instead of relation (2.30), where the parameters A and p depend on temperature and the ratio of R and the molecule radius. However, the mechanisms that could explain this relationship are not analyzed.

There are studies on MD simulation of nanoparticle diffusion. However, in these cases, the validity of the law (2.30) is assumed a priori, and the simulation data only serve to determine the constant C. The study [48] has demonstrated that with this approach, the constant value heavily relies on the ratio of particle and molecule radii and generally does not align with either slip or non-slip conditions. In paper [44] it is shown that for a fixed radius of the nanoparticle the constant C increased with increasing weight of the nanoparticle. The authors determine the diffusion coefficient by the law (2.30) with slip conditions; however, for this they had to introduce some effective radius of the nanoparticle.

According to FDT (2.4), the law determining the diffusion of nanoparticles in liquids and dense gases will be determined by the evolution of the VACF (2.5). In [68, 69] using the MD method for a system of hard spheres it was established that in dense gases and in liquids the relaxation of the VACF of small-sized nanoparticles occurs in two stages and is described by the superposition of two exponentials

$$\chi_{vv}(t) = a_1 \exp(-t/\tau_1) + a_2 \exp(-t/\tau_2), \tag{2.31}$$

where a_1, a_2, τ_1, τ_2 are some constants, which depend on the characteristics of the nanoparticles.

The presence of two exponential relaxation branches of the nanoparticle VACF (2.31) indicates the existence of two different velocity relaxation mechanisms. The first mechanism relates to individual interactions of the carrier fluid molecule with nanoparticle and is characterized by time τ_1. This time is of the same order as the interaction time of the molecule with the nanoparticle, $\tau_1 \sim R/c$, where c is the thermal velocity of the molecule. Certainly, the momentum of a heavy nanoparticle changes slightly when it collides with a molecule. Therefore, this diffusion mechanism becomes increasingly ineffective as particle mass increases. For this reason, the exponent contributions in formula (2.31) changes with the ratio of nanoparticle mass to carrier fluid molecule mass, and as this ratio increases, the ratio a_2/a_1 also increases.

The second term in the Eq. (2.31) is due to the collective effect of the interaction of nanoparticles with microfluctuations of macroscopic fields (density and velocity) of the carrier medium, which are formed around the nanoparticle during its movement. Once the concentration of nanoparticles reaches a certain critical threshold, their diffusion occurs primarily through interactions with microfluctuations (the study of these microfluctuations by MD method can be found in papers of [62–65]). Microfluctuations constantly arise in the medium even being at equilibrium state. A trivial example of such a microfluctation is a local change in the density of the medium when molecules collide. This fluctuation has a characteristic scale of the order of the size of molecules and is usually much smaller than the size of nanoparticles. These fluctuations do not significantly impact the nanoparticle's evolution. It can become noticeable only if the characteristic size of the fluctuation is of the order of the size of the nanoparticle (at least for not too large nanoparticles). The interaction time of the nanoparticle with this microfluctuation (time τ_2) can then be estimated as follows

$$\tau_2 \sim \frac{R}{c_p} \sim \frac{R}{c}\sqrt{\frac{M}{m}} \sim \tau_1\sqrt{\frac{M}{m}}, \tag{2.32}$$

where c_p and c are the thermal velocities of the nanoparticle and molecule, M and m are the masses of nanoparticle and molecule, respectively. The estimate (2.32) shows that the time τ_2 for even the smallest nanoparticles is at least an order of magnitude larger than the time τ_1. As nanoparticles increase in size and mass, this difference also increases.

MD modeling of the VACF of a nanoparticle based on the model of hard spheres allowed us to establish an important qualitative character of nanoparticle velocity relaxation (see formula (2.31)). However, it is important to understand what type of changes can occur for a system with real interaction potentials. Such an analysis was conducted by Rudyak et al. [75]. Here, the diffusion of lithium and aluminum nanoparticles with sizes from 1 to 4 nm in argon was considered. Figure 2.4 illustrates the corresponding data, representing the VACF $\chi' = \chi/V^2(0)$ of 2 nm lithium particles in argon (solid curve). The argon molecule interaction was represented by potential (2.24), whereas the interaction between the carrier medium molecules and the nanoparticle was represented via the RK potential (2.27). The time t' is measured in units of $\tau = \sigma/c$. The density of argon is $n\sigma^3 = 0.707$, where n denotes the number density of argon molecules and σ is the parameter of the potential (2.24) that determines the effective size of a molecule. The temperature of the carrier fluid was varied from 248 to 420 K.

For comparison, the dashed curve in the Fig. 2.4 displays the VACF of the nanoparticle in the hard sphere fluid. In both cases, two exponential branches of the evolution of the VACF are recorded. Although both curves are quite similar quantitatively, some qualitative differences exist. Specifically, the VACF for RK potential displays a distinctive non-monotonic region at the initial stage of relaxation. This region is absent if the carrier medium is rarefied. In a dense carrier medium, the nanoparticle is in the mean field of its interaction forces with the molecules of the medium. Thus, a collective effect occurs. Such an average self-consistent field is typical for systems

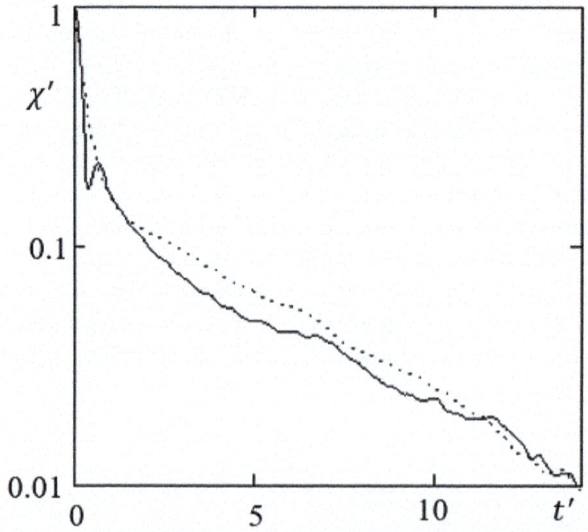

Fig. 2.4 The evolution of the normalized VACF for a Li nanoparticle in Ar; solid and dashed lines correspond to the Lennerd–Jonse (2.24) and hard sphere (2.23) potentials, respectively

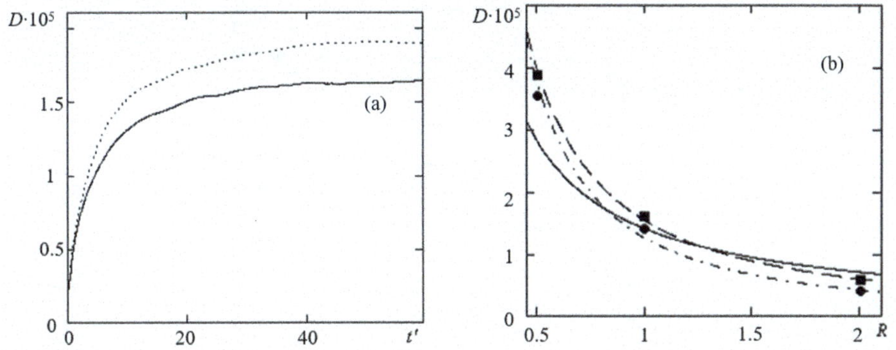

Fig. 2.5 Time-dependent diffusion coefficient of a lithium nanoparticle in argon (**a**), diffusion coefficient of lithium and aluminum depending on the nanoparticle radius (nm) (**b**)

with long-range forces, such as plasma. However, this phenomenon also occurs in ordinary molecular liquids, but on significantly different scales. The potential (2.27) is also quite long-range, and in this case there is also an analog of the self-consistent force (more detailed explanations can be found in study by [75]).

The diffusion coefficient of the nanoparticle is determined using formula (2.4). A typical behavior of function (2.4) over time is demonstrated in Fig. 2.5a for the diffusion coefficient of a lithium nanoparticle in argon. the VACF of which is shown in Fig. 2.4. Time t' is measured in units of $\tau = \sigma/c$, and the diffusion coefficient is measured in sm^2/s. The dashed line represents the hard sphere system. The resulted diffusion coefficients differ by 8%. The presence of an attractive branch in the RK potential and a relatively weaker repulsive component inherently decreases the mobility of the nanoparticle, resulting as a consequence in a lower diffusion coefficient for the nanoparticle with RK potential compared to that of the hard sphere potential.

According to Einstein's formula (2.12) the diffusion coefficient of a particle is inversely proportional to its radius. However, this law does not apply to small nanoparticles. To illustrate, Fig. 2.5b presents the calculated diffusion coefficients of lithium and aluminum nanoparticles with diameters ranging from 1 to 4 nm in argon at a temperature of 322.5 K depending on the nanoparticle radius. Here, the dashed line with squares represents lithium nanoparticles, while the dash-dotted curve with circles represents aluminum nanoparticles. The solid curve represents Einstein's formula (2.12). However, the observed diffusion coefficients cannot be accurately described by the formula (2.12) in any case. Additionally, the diffusion coefficients are dependent on the material of the particles. In general, the dependence of the diffusion coefficient on the nanoparticle radius is described by the function

$$D = aR^{-k}, \tag{2.33}$$

where $k > 1$. For lithium nanoparticles, $k_{Li} = 1.37$, and for aluminum nanoparticles, $k_{Al} = 1.59$.

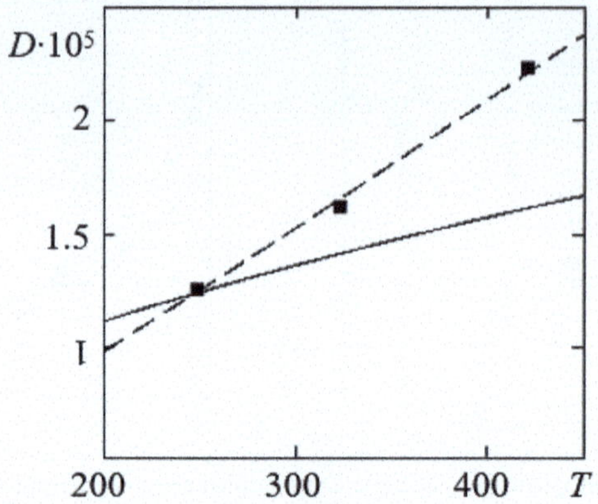

Fig. 2.6 Temperature dependence of the diffusion coefficient of lithium nanoparticle in argon; the solid curve corresponds to Einstein's formula (2.12)

In the previous section it was already noted that the correct temperature dependence of the diffusion coefficient can only be obtained by using a realistic molecule-nanoparticle interaction potential. This dependence is described by a following power function

$$D = aT^n, \tag{2.34}$$

where the exponent n is not universal and depends on the nanoparticle material and size.

As an example, Fig. 2.6 shows the temperature dependence of the diffusion coefficient of 2 nm lithium nanoparticles in argons. Here the marks correspond to the MD calculation, and the dotted curve is its approximation by relation (2.34) with $n = 1.1$. The corresponding dependence, determined by Einstein's formula (solid curve), indicates a significantly lower increase over temperature, and this variation becomes more prominent as the temperature increases.

The relation (2.31) indicates that there are two relaxation processes that are significantly separated in time. This means that the autocorrelation function of the random force (RFACF) of the nanoparticle $\psi_n(t, t')$ cannot be δ-correlated. As a result, the presence of an additional term is imperative. It is clear that this function is damped, and the damping process of the RFACF is fast enough, otherwise the diffusion coefficient would not exist. On the other hand, due to the homogeneity of time, the following condition must be satisfied: $\psi_n(t, t') = \psi_n(t - t')$. It is most simple (and reasonable) to simulate this second branch of the RFACF by an exponent, so that the RFACF of the nanoparticle will have the following form

$$\psi_n(t, t') = \Gamma\delta(t' - t) + ae^{-\beta(t' - t)}. \tag{2.35}$$

Using the solution (2.10), the mean square of the nanoparticle velocity can be calculated

$$\langle v^2(t)\rangle = v_0^2(t)e^{-2\gamma t} + \frac{\Gamma}{2\gamma}\left(1 - e^{-2\gamma t}\right) + \frac{a}{\gamma^2 - \beta^2}\left[1 - e^{(\beta-\gamma)t} - e^{-(\beta+\gamma)t} + e^{-2\gamma t}\right].$$

This expression has a finite limit at $t \to \infty$ only when $\beta < \gamma$, and this limit is equal to

$$\langle v^2(t \to \infty)\rangle = \frac{\Gamma}{2\gamma} + \frac{a}{\gamma^2 - \beta^2} = \frac{3kT}{M}, \tag{2.36}$$

where it is taken into account that $\langle v^2(t)\rangle = 3kT/M$.

Similarly, the RFACF of the nanoparticle can also be calculated, taking into account (2.36). It has the following form

$$\langle v(0) \cdot v(t)\rangle = \frac{\Gamma}{2\gamma}e^{-\gamma t} + \frac{a}{\gamma^2 - \beta^2}e^{-\beta t}, \tag{2.37}$$

or

$$\chi_n(0, t) = \frac{3kT}{M}\left[b_1 e^{-\gamma t} + b_2 e^{-\beta t}\right], \tag{2.37a}$$

The constants b_i, included in (2.37a), are determined naturally (compare (2.37)), and the form of the RFACF of the nanoparticle (2.37a) is obtained when the RFACF is of the form (2.35). While employing now the FDT (2.4), we get the following expression for the diffusion coefficient

$$D = \frac{kT}{M}\left(\frac{b_1}{\gamma} + \frac{b_2}{\beta}\right). \tag{2.38}$$

2.4 Force Acting on the Nanoparticle

The diffusion laws for nanoparticles (2.31) and (2.38) differ significantly from the corresponding law for Brownian particles (2.12). This implies that the force exerted on the nanoparticle by the liquid cannot be described by the classical Stokes law. Conducting experimental studies on the drag force acting on a nanoparticle in a molecular medium is quite challenging for obvious reasons. Several studies of this kind are known. In these studies, the drag force is not directly measured, but is instead calculated using Einstein's relation (2.12) based on the diffusion coefficient. For example, Evans et al. [19] determined the force acting on large molecules using this method, while Haselmeyer et al. [25] and Kato et al. [31] used it to study fullerenes, which are typical nanoparticles of minimal size. In all cases, the force acting on a nanoparticle was found to be significantly different from the Stokes force (2.7). Therefore, the only technique that allows for the direct study of this force is direct MD simulation. Such modeling was performed in [67] for a system of hard spheres.

The evolution of the normalized drag coefficient $\varsigma_*(t) = \varsigma(t)/6\pi\eta R$ for a nanoparticle of radius $R = 4r$ (where r is the radius of the carrier fluid molecules) is presented in Fig. 2.7. The mass ratios $M/m = 200, 100, 50$ correspond to lines 1, 2, and 3, respectively. The time is measured in the mean free path times of the carrier fluid molecules. It is important to note that the drag coefficient is non-stationary. The relaxation zone is quite extended, approximately 30 mean free path times of the carrier gas molecules. The initial value of the drag coefficient $\varsigma_*(0) \equiv \varsigma_{*0}$ is 2–3 times greater than the corresponding value for the Stokes force and depends weakly on the particle weight. As the radius of nanoparticles increases, the deviation of the initial drag force from Stokes' law decreases. The dependence of ς_{*0} on radius is well described by the linear function $\varsigma_{*0} = c_1 R/r + c_2$, where constants c_1 and c_2 are some functions of the density of the carrier fluid and the nanoparticle mass.

The second important circumstance is that the drag coefficient decreases over time and eventually stabilizes at a stationary value $\varsigma_*(\infty)$, which is smaller than the corresponding value of the Stokes force. The value $\varsigma_*(\infty)$ increases as the mass of the nanoparticle increases and approaches the Stokes coefficient for massive particles. The relaxation time of a Brownian particle can be calculated using the formula $\tau_{rs} \sim M/(6\pi\mu R)$, while the drag coefficient of nanoparticles can be determined by the relation $\varsigma_*(\infty) = \tau_{rs}/\tau_2$. Thus, the stationary stage of friction force evolution is influenced by the interaction between nanoparticles and already mentioned microfluctuations of the hydrodynamic fields of the carrier fluid near the nanoparticle.

The microfluctuations generated by the nanoparticle are primarily determined by local variations in the concentration of molecules n. The dimensionless concentration of carrier fluid molecules $g_2(r'/r, \theta, t) = n(r'/r, \theta, t)/n$ (or in other words, the radial distribution function) around a nanoparticle with $R/r = 3$ и $M/m = 100$ is shown

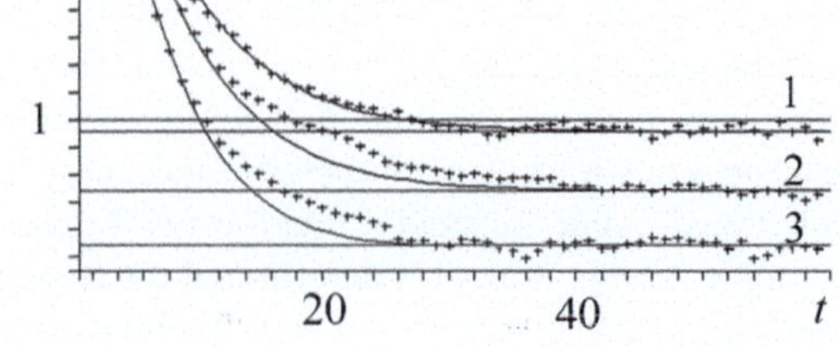

Fig. 2.7 Evolution of the drag coefficient acting on a small nanoparticle

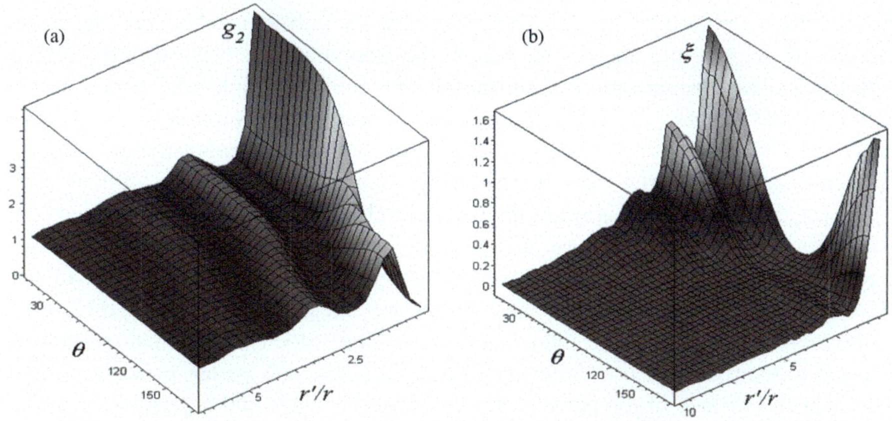

Fig. 2.8 Radial distribution function of the carrier medium molecules near the nanoparticle (**a**), and space structure of the velocity field correlations of the carrier medium and nanoparticle (**b**)

in Fig. 2.8a. The variable r' represents the minimum distance between the surface of the particle and the molecule. The angle θ is defined as the angle between the line connecting the centers of the particle and the molecule, and the direction of the particle's velocity. At the initial instant of time, when the particle is removed from equilibrium state, the function g_2 is initially isotropic but becomes anisotropic over time. The first maximum in the frontal region (at $\theta < 40^o$) increases by 30–50% compared to its equilibrium value and becomes three to five times higher than the mean value of g_2, which is equal to one. A highly rarefied region is formed behind the particle. This anisotropy occurs in the equilibrium state as well. Thus, the relaxation process of the nanoparticle velocity in the liquid is accompanied by an increase in the density of the carrier molecules in the region in front of the particle.

Along with the medium density fluctuations, local fluctuations of momentum play a crucial role, since the efficiency of nanoparticle velocity relaxation is related precisely to the momentum of microfluctuations. The spatial and temporal correlations of the nanoparticle velocity with the mean velocity field of the carrier fluid are described by the following function $\xi\left(r'/r,\theta\right) = \mathbf{v}\left(r'/r,\theta\right)\cdot \mathbf{V}_0/\left(V_0\sqrt{3kT}\right)$, where $\mathbf{v}\left(r'/r,\theta\right)$ is the mean velocity of the carrier fluid. The spatial dependence of this function is shown in Fig. 2.8b; here the system parameters are identical to those in Fig. 2.8a. This function describes the correlations of the nanoparticle velocity and the mean velocity field of the medium at a fixed time instant. It is evident that the particle significantly accelerates the molecules in the frontal region. However there are also layers in the latter region, where the molecule velocities practically do not increase. The reason for this is as follows. A moving particle generates in the medium a perturbation propagating at the speed of sound [63]. This perturbation pushes the molecules closest to the particle in the direction of its motion. These molecules, in turn, transfer the received impulse to the molecules of the next layers. However, these molecules themselves are slowed down, reflected, or even move back towards the particle. Comparing Fig. 2.8a, b reveals that the correlation between the mean velocity of the medium and the particle in the frontal zone is highest in the region with the highest concentrations. Conversely, only very fast molecules penetrate the region behind the particle, where the medium concentration is minimal.

Thus, a moving nanoparticle creates microfluctuations of the density and velocity distributions of the carrier medium. Energy is spent on creating such a fluctuation, which accelerates the relaxation of the particle velocity. Figure 2.7 characterizes the initial stage development of these fluctuations. Then, the velocity field of the carrier medium near the nanoparticle is rearranged. Figure 2.9 shows a fragment of the velocity field of the carrier medium at some instant. The arrows indicate the directions of the medium velocities and their magnitudes. The velocity of the nanoparticle is directed to the right; the arc shows a part of its boundary. A vortex structure can be seen near the surface of the particle. It has a toroidal shape and is positioned in the plane that passes through the center of the particle perpendicularly to its velocity. The structure is formed during several mean free path times of molecules, and the resulting vortex has a diameter of the order of the nanoparticle size.

Figure 2.9 shows that the average velocity of the medium molecules below the center of the vortex is directed towards the particle motion. This is where the velocity of the medium is at its maximum, while the average velocity of the medium far from the particle is zero. The vortex, which is formed as a microfluctuation, pushes the particle, thus slowing down the relaxation process. This is the reason for the appearance of the second relaxation time τ_2. The wave of density disturbances generated in the medium and the appearance of unsteady velocity field with a characteristic vortex structure near the nanoparticle makes its further evolution almost hydrodynamic. A quasi-stationary drag force begins to apply to the nanoparticle (Fig. 2.7), whose magnitude depends on the particle's size and mass. For sufficiently massive, yet still small nanoparticles, this force is close to the Stokes drag force.

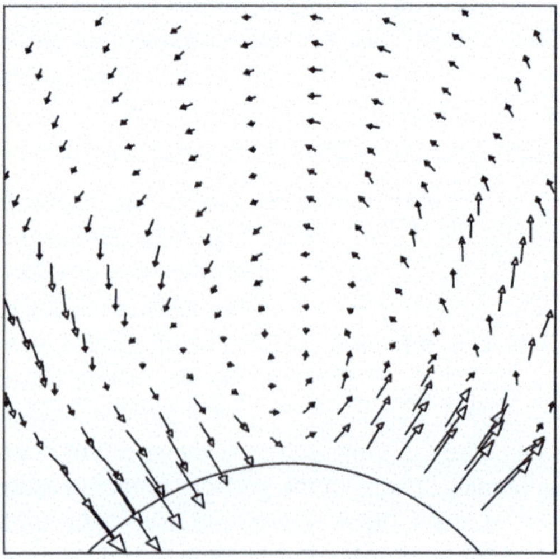

Fig. 2.9 Velocity field of the carrier medium near the nanoparticle

2.5 Diffusion of Carbon Nanotubes

Studying diffusion of the CNTs is very important for various applications. First of all, this includes biomedical technologies, which will be discussed later in this book. Diffusion plays a significant role in creating advanced materials, although there one has to deal with diffusion in more complex condensed materials rather than fluid. Additionally, it is a fundamental transport process that largely influences other transport processes. In this section, we present experimental data on the diffusion of CNTs, followed by their simulation using the MD method.

2.5.1 Experimental Study of CNT Diffusion

Although diffusion plays an important role in the application and comprehension of CNT transport processes, there has been limited research on the diffusion of CNTs in liquids. Only a few studies have been conducted on this topic [30, 43, 78]. The first study examined the diffusion of SWCNTs, treated with two rather complex surfactants (ribonucleic acid or bovine serum albumin), using single-molecule fluorescence microscopy. The SWCNTs had an average diameter of 1.4 nm and a length of 400 nm. The measured diffusion coefficients ranged from 0.374 to 0.442 $\mu m^2/s$. The second paper reports on the measurement of diffusion coefficients of SWCNTs in aqueous solutions of lymph, bile, blood, and cytoplasm. The SWCNTs had a length of 253 nm ($\pm$30.6 nm) and a diameter of 1.6 nm ($\pm$0.34 nm). Fluorescence microscopy was also used here. The estimated diffusion coefficients of SWCNTs ranged from 0.0026 to 0.0145 $\mu m^2/s$. Besides, we can mention studies where the diffusion of CNTs has been examined in various biological systems [33, 82, 85]. The diffusion coefficients were found to be significantly different. This section presents experimental data on the diffusion of SWCNTs and MWCNTs using the DLS method.

Since surfactants are commonly used in the preparation of nanofluids, it is important to study the characteristic sizes of the corresponding molecules (see Fig. 1.13) and their diffusion coefficients. The measurement of the diffusion coefficient distribution of SDBS molecules in aqueous solution is presented in Fig. 2.10a. Here two maxima are present similarly as in the measurements of size of SDBS and PVP molecules (Fig. 1.13a). Since both SDBS micelles and polymicelles with significantly different sizes are present in solution, their diffusion coefficients differ by several tens of times. The diffusion coefficient distributions of PVP molecules are shown in Fig. 2.10b, and corresponding values can be found in Table 2.1. Here M and PM denote the micelles and polymicelles.

The diffusion coefficient is usually determined for fairly dilute suspensions. Strictly speaking, we have to talk about the diffusion coefficient of an isolated dispersed particle, in particular, CNTs. In real experiments, there is always a finite concentration of particles in the base fluid. Simple estimates show that a reasonably accurate value of the diffusion coefficient of SWCNTs is obtained if their weight

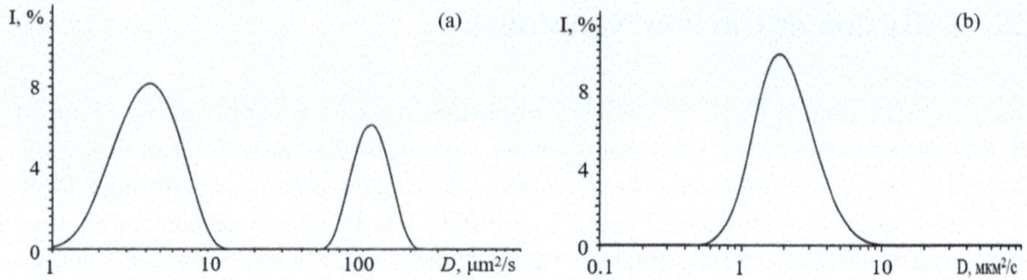

Fig. 2.10 Diffusion coefficient of SDBS (**a**) and PVP (**b**) molecules in aqueous solution

Table 2.1 Diffusion coefficients and effective hydrodynamic sizes of SWCNTs and MWCNTs in different carrier fluids

Liquid	$D \times 10^{-12}$, m^2/s	Size, nm	Liquid	$D \times 10^{-12}$, m^2/s	Size, nm
W + PVP	2.17	18	W + SDBS / SWCNT	2.59	190
W + PVP/ SWCNT	1.07	450	W + SDBS / MWCNT	2.24	220
W + PVP/ MWCNT	0.79	615	EG + SDBS	0.106	255
EG + PVP	1.03	21	EG + SDBS/ SWCNT	0.0158	1700
EG + PVP/ SWCNT	0.051	530	EG + SDBS/ MWCNT	0.0101	2600
EG + PVP/ MWCNT	0.038	712	EG/SWCNT	0.0158	1700
W + SDBS (PM)	3.47	140	EG/MWCNT	0.00756	3500
W + SDBS (M)	123.17	4			

concentration does not exceed 10^{-5}%. The distribution of the diffusion coefficient of SWCNTs in the W + SDBS solution at this concentration is shown by the dash-dotted line in Fig. 2.11. The distribution has two maxima, indicating that even at such low concentrations the size distributions of the SWCNTs are dissimilar, especially in length. The second (smaller) maximum corresponds to single nanotubes and the diffusion coefficient of these nanotubes is many times larger. Thus, the maximum value of the CNT diffusion coefficient D corresponding to the first (larger) maximum is $D = 2.2 \times 10^{-12}$ m^2/s, and that corresponding to the second maximum is $D = 1.02 \times 10^{-11}$ m^2/s. The mean values are somewhat lower.

Naturally, as the concentration of SWCNTs increases, their diffusion coefficient decreases. In Fig. 2.11 the dashed line corresponds to a concentration of 0.002%, and the solid line refers to 0.01%. At these concentrations, the maximum values of the diffusion coefficient are $D = 1.2 \times 10^{-12}$ m^2/s and $D = 0.7 \times 10^{-12}$ m^2/s,

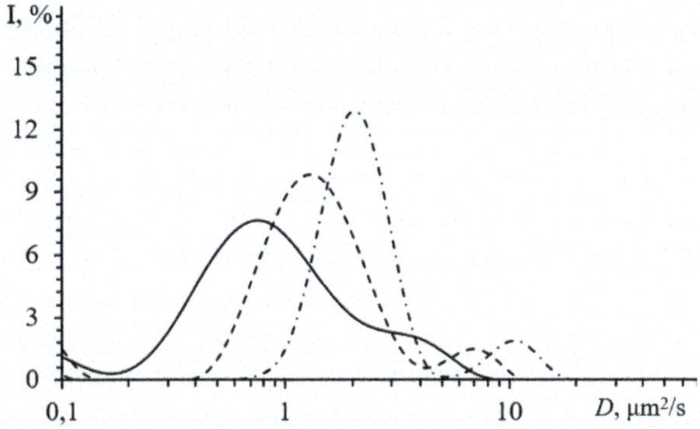

Fig. 2.11 Diffusion coefficient of SWCNTs at varying concentrations:0.00001% (dash-dotted line), 0.002% (dashed line), and 0.01% (solid line)

respectively. The average values are slightly lower. The latter value is in agreement (within 10%) with the data from [43], but in this study, CNTs were coated with either ribonucleic acid polymer or bovine serum albumin. In the study [30], diffusion coefficients were obtained that differed from the data of this work by several orders of magnitude for a concentration of 0.01%. However, they used completely different base fluids that were many times more viscous.

The systematic data on the diffusion coefficients and corresponding effective hydrodynamic sizes of SWCNTs and MWCNTs in different carrier fluids are presented in Table 2.1. The weight concentration of CNTs was 10^{-5}% in all cases. The diffusion coefficients of SWCNTs and MWCNTs were determined in nanofluids using the following carrier fluids: W + PVP, W + SDBS, EG + PVP, EG + SDBS, and EG. The hydrodynamic sizes of CNTs ranged from 190 to 3500 nm. The minimum diffusion coefficient was 0.00756 μm^2/s, obtained for EG-based MWCNTs nanofluids, while the maximum value of 2.59 μm^2/s was obtained for W + SDBS-based SWCNT nanofluid. In all cases the diffusion coefficient increased as the liquid viscosity and CNT length decreased. The table also provides the diffusion coefficients of the surfactant molecules used.

2.5.2 *MD Simulation of Translational Diffusion of CNTs*

The obtained data on the diffusion coefficients of CNTs vary significantly, differing by several orders of magnitude. The reason for this is clear. This is due to the carrier fluids having different viscosities and the length of the considered CNTs differing by more than an order of magnitude. Therefore, experimental data do not provide sufficient information to understand the diffusion laws of CNTs. Furthermore, traditional methods used to measure the diffusion coefficient, such as DLS, fluorescence microscopy, electroacoustic methods, etc. require accurate data on the size of CNTs for proper interpretation of the results. The diffusion of CNTs is a complex phenomenon that involves both translational and rotational diffusion. The rotational diffusion of CNTs will be studied in the next paragraph of this section.

An alternative to the experimental study of CNT diffusion is its modeling using the MD method. Translational diffusion of conventional spherical nanoparticles in liquids has been studied previously (see Sect. 2.3). However, the translational diffusion of CNTs in liquids has not been practically studied. Typically, studies have focused on the tensile properties of nanotubes made of different materials or the relative motion of the nanotube atoms [26, 45]. In thematically related papers [81, 84] studied the drag force acting on CNTs in liquid argon using the nonequilibrium molecular dynamics method. The CNT was simulated as a one-dimensional rod of carbon atoms. It was shown that the force acting on the CNT was significantly higher than that determined by the hydrodynamic formulas. However, this force varied with changes in the shear rate.

It is important to note that carbon nanotubes have a large aspect ratio, meaning their length is much greater than their diameter. As a result, they are anisotropic objects, and their diffusion is also anisotropic. Diffusion can occur along the nanotube axis (longitudinal diffusion) or perpendicular to it (transverse diffusion). However, measuring the corresponding diffusion coefficients is practically impossible. At the same time, this diffusion may prove to be important in actual shear flows [14, 51] or when external fields are present. Researchers have long been interested in the anisotropy of translational diffusion [11, 24, 88]. Mainly, macroscopic theories of Brownian motion of solid rod-shaped particles have been developed with regard to this issue [13].

Belkin et al. [6] studied the longitudinal and transverse diffusion of armchair-type (6, 6) SWCNTs in water. The nanotubes had a diameter of 0.818 nm and lengths varying from 5.25 to 32.23 nm. The volume concentration of nanotubes ranged from 0.0003 to 0.0026. As previously mentioned CNTs exhibit high anisotropy. Therefore, this work presents an algorithm developed to determine and calculate the longitudinal and transverse diffusion coefficients. To achieve this, the velocities of the center of mass of the CNT, $\boldsymbol{V}_0$ and $\boldsymbol{V}_1$, at the initial and current time, t were measured. The corresponding unit vectors, $\boldsymbol{e}_0$ and $\boldsymbol{e}_1$, which are directed along the CNT axis, were determined using the coordinates of the two atoms (center of mass, colored in gray (Fig. 2.12), and at the CNT boundary, colored in black (Fig. 2.12)). Systematic information about the velocities and coordinates of all molecules and CNTs was obtained at successive time instants during the simulation process. The diffusion coefficient of CNTs was determined using the FDT (2.4) with VACF (2.6). To calculate the longitudinal diffusion coefficients $D_{||}$ it is necessary to use longitudinal velocity vectors $\boldsymbol{V}_{||} = \boldsymbol{e}(\boldsymbol{V} \cdot \boldsymbol{e})$ at the corresponding time instants when calculating the corresponding VACF. The diffusion of CNTs in the direction perpendicular to their axis is isotropic. Therefore, the transverse diffusion coefficient $D_{\perp}$ was defined as $D_{\perp} = \left(3D - D_{||}\right)/2$.

The evolution of the diffusion coefficient (2.4) is determined by the attenuation of the VACF (2.6). The nature of the VACF attenuation is largely determined by the length of the CNT. Figure 2.13 shows the evolution of the VACF (10^2 m^2/s^2) of CNTs

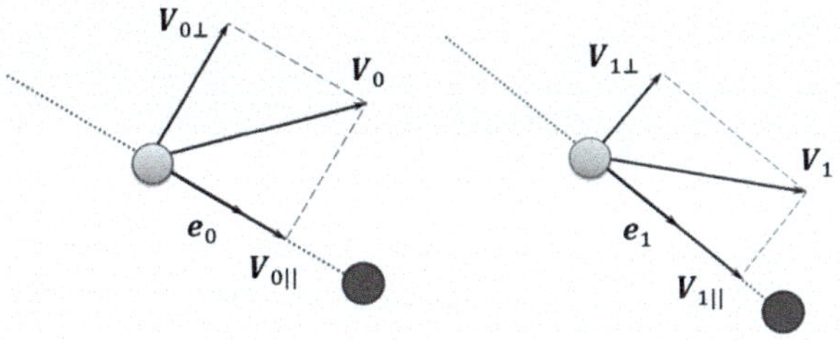

Fig. 2.12 Scheme for determining the longitudinal and transverse velocities of CNTs at the initial and current time points. The left figure shows the initial state of the CNT, while the right shows the state corresponding to the time moment *t*. The dotted line indicates the longitudinal axis of the nanotube

with lengths of 5.25 (solid line) and 10.65 nm (dashed line). A characteristic feature of the evolution of the VACF of rather short CNTs is the presence of a negative branch. For CNTs with a length of 5.25 nm, the negative evolution region of the VACF is small, but it increases as the CNT length decreases. The appearance of this region is due to the low inertness of small CNTs, which results in their motion slowing down and sometimes changing direction to opposite due to interaction with the carrier fluid molecules. The negative region of the VACF describes this reversal. As the CNT length increases, this effect disappears.

The VACF relaxation occurs in two stages. In the first stage, which lasts approximately 1–2 ps (depending on the length), there is an exponential decline with a relaxation time that depends on both the CNT mass and length. This stage of the VACF evolution can be accurately described by the relaxation time of a long rigid cylindrical rod. As previously mentioned, the motion of the rod can be described by the Langevin Eqs. (2.8) and (2.9) of the velocity of its center of mass, however, with

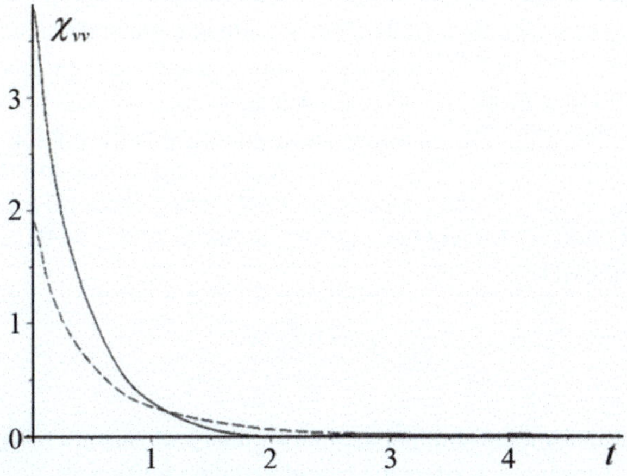

Fig. 2.13 Time evolution of the VACF (10^2 m^2/s^2) of CNTs with lengths 5.25 nm (solid line) and 10.65 nm (dashed line). Time is measured in picoseconds

a tensor drag coefficient [13]

$$\gamma = \frac{4\pi \mu L}{\ln(L/d)}\left(U - \frac{1}{2}ee\right), \tag{2.39}$$

where U is a unit tensor of the second rank, and e is a unit vector directed along the CNT axis. Further, it can be shown in the usual way [59], see also Sect. 2.1) that in the simplest case the relaxation time of CNTs will be equal to

$$\tau_r \sim \frac{M\ln(L/d)}{2\pi \mu L} = \frac{\rho d^2 \ln(L/d)}{8\mu}, \tag{2.40}$$

where ρ is the density of the CNT.

Thus, the relaxation time of the CNT at the first stage increases with its mass but decreases with its length and the viscosity of the carrier fluid. The second estimate (2.40) indicates that the relaxation time increases rapidly with increasing CNT diameter and density. This suggests that the relaxation time of multi-walled nanotubes will be much longer than that of single-walled nanotubes.

The second branch of the VACF of CNT decays much more slowly. It can be also approximated by an exponent with a significantly longer relaxation time. This behavior is similar to the velocity relaxation of spherical nanoparticles, as discussed in Sects. 2.3 and 2.4. The relaxation of CNTs during this stage is also influenced by the interaction with the microfluctuations of hydrodynamic fields of the carrier fluid surrounding them. These microfluctuations propagate through the medium at the speed of sound and are non-isotropic. The volume of the formed microfluctuation is proportional to the surface area of the CNT. Therefore, most of the fluctuations are concentrated in the direction perpendicular to the CNT axis. The second important circumstance is related to the fact that the CNTs shown in Fig. 2.13 have significantly different velocities of their centers of mass. Therefore, their absolute values differ by almost twice as much (see Fig. 2.12). This is indeed expected because the squares of the thermal velocities of CNT centers of mass are inversely proportional to their mass, which are in turn proportional to their lengths.

The plateau values of diffusion coefficients are reached during the damping time of the corresponding VACFs. Table 2.2 summarizes the obtained diffusion coefficients, including the length and aspect ratio (AR). For sufficiently long CNTs, the translational diffusion coefficient (second line in Table 2.2) is well described by the formula proposed by [13]

$$D_T = \frac{kTln(L/d)}{3\pi \eta L}. \tag{2.41}$$

The diffusion coefficients obtained from formula (2.41) are also given in Table 2.2 (third line). This formula describes the calculated data quite well.

Table 2.2 Diffusion coefficient of CNTs in water

L, nm/AR	5.25/6.42	10.65 / 13.02	21.44/26.21	32.23/39.40
D, m²/s	1.7×10^{-10}	1.2×10^{-10}	7.7×10^{-11}	5.2×10^{-11}
D_T, m²/s	1.7×10^{-10}	1.1×10^{-10}	6.8×10^{-11}	5.2×10^{-11}
D_r, rad²/s	4.76×10^7	0.78×10^7	0.18×10^7	–
D_{rt}, rad²/s	5.7×10^7	0.94×10^7	0.15×10^7	

The velocity of the center of mass of the CNT can be represented as a superposition of the longitudinal velocity $\boldsymbol{V}_{\|}$ and the transverse velocity $\boldsymbol{V}_{\perp}$ (see Fig. 2.12). Similarly, the drag coefficient (2.39) can also be expressed as a superposition of two coefficients, $\gamma_{\|}$ and $\gamma_{\perp}$

$$\gamma_{\|} = \frac{2\pi\mu L}{\ln(L/d)}, \gamma_{\perp} = \frac{4\pi\mu L}{\ln(L/d)}. \tag{2.42}$$

For long thin rods in this case it can be shown that the diffusion coefficient is equal to

$$D = \frac{1}{3}\left(D_{\|} + 2D_{\perp}\right) = \frac{kT\ln(L/d)}{3\pi\mu L}, \tag{2.43}$$

which corresponds to the formula (2.41).

Figure 2.14a shows the evolution of the VACF of longitudinal and transverse velocities of 10.65 nm long CNTs. The dashed line represents the averaged normalized VACF of the center of mass, χ_{vv}, while the solid line corresponds to the VACF of the longitudinal velocity $\chi_{\|}$, and the dotted line corresponds to the VACF of the transverse velocity $\chi_{\perp}$. The function χ_{vv} decreases monotonically, and its evolution has already been described above. The evolution of the VACF of longitudinal and transverse velocities of CNTs occurs in two stages, but the process differs qualitatively for each velocity. During the first stage, the relaxation time of the VACF of transverse velocity is much shorter than that of the VACF of longitudinal velocity. During the second stage, the VACF of longitudinal velocity has a noticeable negative branch, while the VACF of transverse velocity has a long weakly damped tail after a sharp decline at the first stage. The presence of this particular section compensates for the negative contribution of the VACF of longitudinal velocity. As a result, the VACF of mean velocity remains positive and is quickly damped.

The exponential damping of all three VACF, observed at the first step of the functions presented in Fig. 2.14, is determined by the systematic drag force (2.39) acting on the CNT from the carrier fluid. This force is similar to the Einstein–Stokes force for ordinary Brownian particles. The drag coefficients of the longitudinal and transverse velocities are described by expressions (2.42). The corresponding relaxation times of the VACF at the first exponential branch are inversely proportional

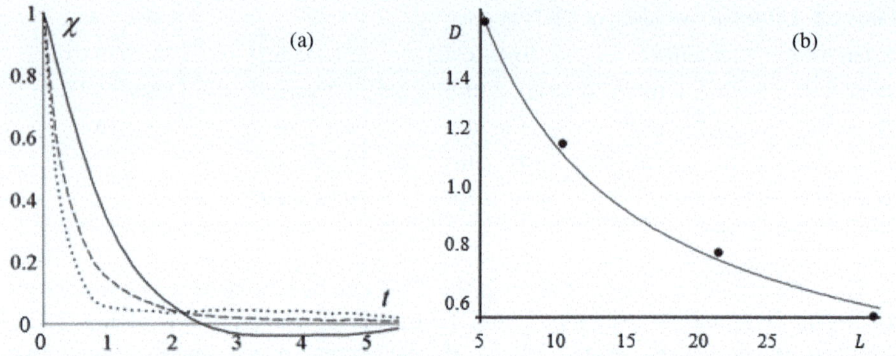

Fig. 2.14 **a** Time-dependent (ps) normalized functions χ (dashed line), $\chi_{||}$ (solid line), and $\chi_{\perp}$ (dotted line); **b** comparison of the translational diffusion coefficient (10^{-10} m^2/s) obtained by the MD method (dark labels) and using formula (2.43) (solid curve) for CNTs of length 10.65 nm

to the drag coefficients (2.42). The relaxation time of the transverse VACF is twice as long (see (2.42)), which is illustrated by Fig. 2.14.

The diffusion coefficients were calculated by integrating the time-dependent VACFs presented in Fig. 2.13. The obtained diffusion coefficients $D_{||}$ and $D_{\perp}$ are compared in Table 2.3. It is observed that the longitudinal diffusion coefficient is consistently larger than the transverse diffusion coefficient.

However, the previous discussion only considered the diffusion of rigid CNTs. In reality, nanotubes are flexible and can change shape while maintaining a fixed diameter during evolution. Despite this flexibility, the diffusion anisotropy remains preserved. To account for this flexibility, the AIREBO potential [80] can be used. Taking into account the flexibility of nanotubes leads to a significant increase in their diffusion coefficient compared to rigid CNTs. This is illustrated by the data given in Table 2.4 for CNTs of three lengths. The diffusion coefficients are given in 10^{-10} m^2/s units. The table also presents the corresponding aspect ratios. The analysis of the data in this table shows that the diffusion coefficient of flexible CNTs is higher than that of rigid CNTs in all cases. The longitudinal diffusion coefficient increases especially strongly.

The evolution of the VACF of flexible CNTs also occurs in two stages. At the first stage, similarly to rigid CNTs, the VACF decays exponentially with a relaxation time determined by systematic drag coefficients (2.42). At the second stage, relaxation significantly slows down. As the size of the CNT increases, a small-amplitude wave-like disturbance becomes a characteristic property of its VACF. The frequency of

Table 2.3 Diffusion coefficients (10^{-10} m^2/s) of CNTs of different lengths

L, nm	5.25	10.65	21.44	32.23
$D_{\|\|}$	2.11	1.65	1.09	0.78
$D_{\perp}$	1.49	0.99	0.61	0.39

Table 2.4 Comparison of diffusion coefficients (10^{-10} m^2/s) of rigid and flexible CNTs, the quantities marked with the subscript E are obtained using Eqs. (2.43) and (2.45)

L, nm/AO	D	D_E	$D_{\|\|}$	$D_{\perp}$	D_r	D_{rE}
5.25/6.42 solid	1.7	1.67	2.11	1.49	4.76	5.7
5.25/6.42 AIREBO	2.42	1.67	3.6	1.83	–	–
10.65 / 13.02 solid	1.2	1.09	1.65	0.99	0.78	0.94
10.65 / 13.02 AIREBO	1.7	1.09	2.6	1.25	–	–
21.44/26.21 solid	0.77	0.68	1.09	0.61	0.18	0.15
21.44/26.21 AIREBO	1.17	0.68	1.89	0.81	–	–

these disturbances is close to the frequency of natural oscillations of CNTs, which affects the diffusion of flexible nanotubes.

The MD simulation data are quite different from the experimental ones. Thus, the diffusion coefficient of the longest CNT is 5.2×10^{-11} m^2/s (Table 2.2). On the other hand, the diffusion coefficient of SWCNTs in the W + SDBS solution is 1.02×10^{-11} m^2/s (Fig. 2.1); however the nanotubes here are many times longer.

2.5.3 *Rotational Diffusion of CNTs*

As previously mentioned, in addition to the usual translational diffusion of CNTs it is necessary to analyze their rotational diffusion. Studying the rotational diffusion of CNTs is not an easy task. Apparently, the first measurements of the rotational diffusion coefficients of SWCNTs in water with surfactant (SDS) were conducted in [15] using fluorescence video microscopy. The diffusion coefficient was found to be in agreement with predictions based on solid rod diffusion theory. Later, a similar method was also employed in a study [54].

The rotation of a short CNT in liquid argon was studied in [81] employing non-equilibrium molecular dynamics method. The CNT had a diameter of 0.81 nm and aspect ratios of 1.44, 1.8, and 2.16. It was found that the rotational relaxation time was significantly higher than that predicted by Jeffery's theory [29], and it increased as the aspect ratio increased. Additionally, the obtained characteristics significantly depended on CNT velocity. It is clear, that the characteristic, obtained in this way, can be hardly associated with rotational diffusion. The rotational diffusion of CNTs in argon was actually studied in [9, 10] using the MD method. A nanotube with an armchair-type (5, 5) configuration was analyzed. The length ranged from 3.1 to 8.1 nm, and the diameter varied from 0.688 to 1.93 nm, resulting in an aspect ratio ranging from 3.5 to 11.8. A comparison of obtained values with known semi-analytical correlations revealed significant discrepancies.

Later, to study rotational diffusion, the MD method was used by Belkin et al. [6]. The data presented below were obtained in this very work. A sufficiently large ratio of the length of the simulated CNT to its diameter determines the character of

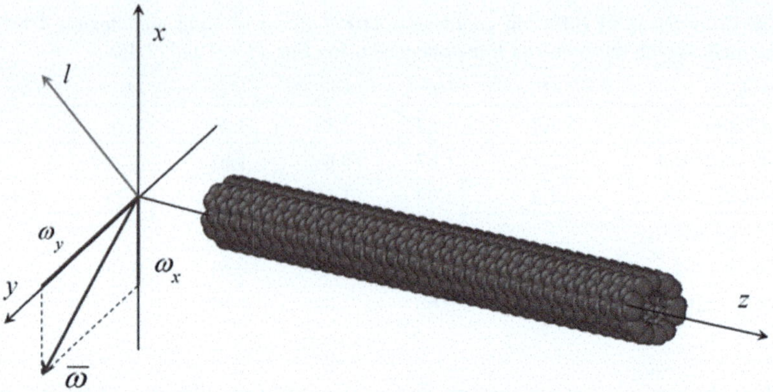

Fig. 2.15 Schematic for determining the rotational diffusion coefficient of SWCNT

rotational diffusion. Indeed, rotation around the longitudinal axis (see Fig. 2.15) does not lead to a change of the nanotube position. In fact, the rotation has two degrees of freedom, and it is necessary to track the evolution of angular velocities perpendicular to the nanotube axis (Fig. 2.15).

The simulation process provides systematic information on the velocities and coordinates of all molecules and CNTs at successive time instants. This information can be used to calculate the rotational diffusion coefficient D_r of CNTs using two different approaches. One can use either the fluctuation–dissipation theorem in the form (2.16) or the Einstein form to relate it to the evolution of the root mean square value of the rotation angle φ with respect to the CNT axis

$$D_r = \varphi^2/4t. \tag{2.44}$$

An example of the evolution of the φ^2 function for CNT of length 10.7 nm is shown in Fig. 2.16a. The corresponding diffusion coefficient is 7.8×10^6 rad^2/s. Naturally, the rotational diffusion coefficient decreases as the CNT length increases. The corresponding values are shown in Fig. 2.16b (dark labels), and the numerical data can be found in Table 2.2. The theory of Brownian diffusion of long rigid rods leads to the following formula [13]

$$D_{rt} = \frac{3kTln(L/d)}{\pi \mu L^3}. \tag{2.45}$$

The solid line in Fig. 2.16b corresponds to the rotational diffusion coefficient calculated using the relation (2.45). The data calculated using Einstein's relation (2.44) agrees qualitatively with the data corresponding to formula (2.45). However, for short nanotubes, the latter formula gives values that exceed those corresponding to (2.44) by about 20%.

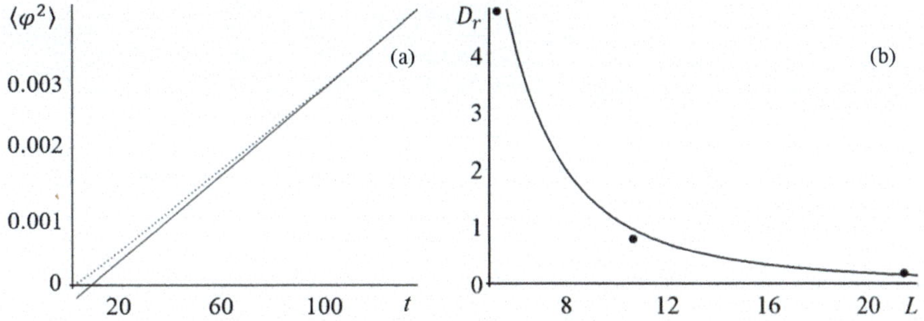

Fig. 2.16 **a** Time dependence (ps) of the root mean square rotation angle of the CNT relative to its axis; **b** rotational diffusion coefficient depending on the CNT length

To calculate the rotational diffusion coefficient using relation (2.16), it is neccessary to analyze the evolution of the corresponding angular velocity autocorrelation function (AVACF). Figure 2.17a shows the evolution of the AVACF for the same CNT as in Fig. 2.16. The AVACF evolution is two-stage, with the first stage characterized by an exponential decay with a relaxation time determined by the systematic drag coefficient. The corresponding diffusion coefficient (Fig. 2.17b) reaches the plateau value over the time of reversal to zero of AVACF, as shown in Fig. 2.17a. The obtained diffusion coefficients are presented in Table 2.2 as well as the data obtained using formula (2.45). The solid straight line in Fig. 2.17b corresponds to the rotational diffusion coefficient obtained using relation (2.44).

The rotational diffusion of CNTs can be quite accurately described by the Langevin Eq. (2.13b) with the drag coefficient (2.18). It can be shown that at the first stage, the AVACF decays exponentially as well as the VACF but with a relaxation time

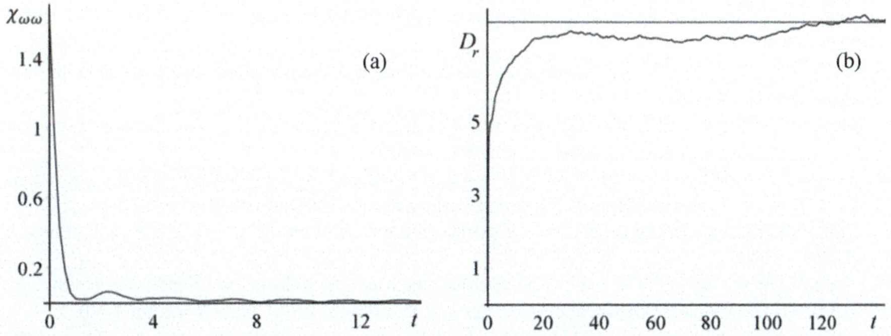

Fig. 2.17 Evolution of AVACF (rad^2/ps^2) (**a**) and diffusion coefficient (2.16) (**b**) for a nanoparticle with length of 10.7 nm

$$\tau_{rot} = \frac{M \ln(L/d)}{4\pi \mu L}. \tag{2.46}$$

It is rather easy to show that the relaxation times of the velocity of center of mass and CNT rotation around the center of mass (2.40) are of the same order, and the relaxation time (2.46) agrees well with the damping time of the first branch of the AVACF presented in Fig. 2.17a.

References

1. Alder BJ, Wainwright TE (1957) Phase transition for hard spheres system. J Chem Phys 27(2):1208–1209
2. Alder BJ, Wainwright TE (1959) Studies in molecular dynamics. I. General method. J Chem Phys 31(2):459–466
3. Alder BJ, Wainwright TE (1960) Studies in molecular dynamics. II. Behavior of a small number of elastic spheres. J Chem Phys 33(5):1439–1451
4. Allen MP, Tildesley DJ (2017) Computer simulation of liquids. Oxford University Press, Oxford
5. Arnold VI (1989) Mathematical methods of classical mechanics. Springer, New York
6. Belkin A, Rudyak V, Krasnolutskii S (2022) Molecular dynamics simulation of carbon nanotubes diffusion in water. Mol Simul 48(9):752–759
7. Bocchiery P, Loinger A (1957) Quantum recurrence theorem. Phys Rev 107(2):337–338
8. Burnett D (1935) The distribution of molecular velocities in a slightly non-uniform gas. Proc London Math Soc 39(6):385–430
9. Cao B-Y, Dong R-Y (2014) Molecular dynamics calculation of rotational diffusion coefficient of a carbon nanotube in fluid. J Chem Phys 140:034703
10. Cao B-Y, Dong R-Y (2015) Investigation of rotational diffusion of a carbon nanotube by molecular dynamics. J Nanosci 15:2984–2988
11. Chakrabarty A, Wang F, Fan CZ, Sun K, Wei Q-H (2013) High-precision tracking of Brownian boomerang colloidal particles confined in quasi two dimensions. Langmuir 29:14396–14402
12. Chapman S, Cowling TG (1990) The mathematical theory of non-uniform gases. Cambridge University Press, Cambridge
13. Dhont JKG, Briels WJ (2004) Rod-like Brownian particles in shear flow. WILEY-VCH Verlag GmbH, Berlin
14. Dong RY, Cao BY (2014) Anomalous orientations of a rigid carbon nanotube in a sheared fluid. Sci Rep 4:6120
15. Duggal R, Pasquali M (2006) Dynamics of individual single-walled carbon nanotubes in water by real-time visualization. Phys Rev Lett 96:246104
16. Einstein A (1905) Über die von der molekularkinetischen Theorie der Wärme geforderte Bewegung von in ruhenden Flüssigkeiten suspendierten Teilchen. Ann Phys 17:549–560
17. Ernst MH (1966) Formal theory of transport coefficients to general order in the density. Physica 32:209–243
18. Erpenbeck JJ, Wood WW (1977) Molecular dynamics techniques for hard-core system. In: Berne BJ (ed) Modern theoretical chemistry, pt B. Plenum Press, New York, pp 1–40
19. Evans DF, Tominaga T, Davis HT (1981) Tracer diffusion in polyatomic liquids. J Chem Phys 74:1299–1305
20. Ferziger JH, Kaper HG (1972) Mathematical theory of transport processes in gases. North Holland Publishing Company, Amsterdam
21. Gibson JB, Goland AN, Milgram M, Vineyard GH (1960) Dynamics of radiation damage. Phys Rev 120(4):1229–1253
22. Green HS (1961) Theories of transport in fluids. J Math Phys 2(2):344–348

23. Hamaker HC (1937) The London–van der Waals attraction between spherical particles. Physica 4(10):1058–1072
24. Han Y, Alsayed A, Nobili M et al (2009) Quasi-two-dimensional diffusion of single ellipsoids: aspect ratio and confinement effects. Phys Rev E 80:11403
25. Haselmeyer R, Holz M, Kappes MM et al (1994) Translational diffusion in C_{60} and C_{70}. Fullerene solutions. Ber Bunsenger Phys Chem 98:878–881
26. Huang J, Fan Y, Guo Y et al (2020) Axial strain effect on translational motion property of carbon nanotube thermal actuator. Mol Simul 46(17):1426–1431
27. Israelachvili J (2011) Intermolecular and surface forces. Academic, New York
28. Itô K, McKean HP (1965) Diffusion processes and their sample paths. Springer
29. Jeffery GB (1922) The rotation of two circular cylinders in a viscous fluid. Proc R Soc London A 101(709):169–174
30. Judkins J, Lee HH, Tung S et al (2013) Diffusion of single-walled carbon nanotube under physiological conditions. J Biomed Nanotech 9(6):1065–1070
31. Kato T, Kikuchi K, Achiba YJ (1993) Measurement of the self-diffusion coefficient of C_{60} in benzene-D_6 using ^{13}C pulsed-gradient spin echo. Phys Chem 97:10251–10253
32. Khon'kin AD (1970) Equations for space-time and time correlation functions and proof of the equivalence of results of the Chapman-Enskog and time correlation methods. Theoret Math Phys 5:1029–1037
33. Kim J-W, Kotagiri N, Kim J-H (2006) In situ fluorescence microscopy visualization and characterization of nanometer-scale carbon nanotubes labeled with 1-pyrenebutanoic acid, succinimidyl ester. Appl Phys Lett 88:213110
34. Kirkwood JG (1946) The statistical mechanical theory of transport processes. I. General Theory. J Chem Phys 14(4):180–201
35. Komatsu N, Abe T (2004) Numerical irreversibility in time reversible molecular dynamics simulation. Physica D 195:391–397
36. Kong CL (1973) Combining rules for intermolecular potential parameters. II. Rules for the Lennard-Jones (12–6) potential and the Morse potential. J Chem Phys 59:2069–2464
37. Kowert RW, Jones JB, Zahm JA et al (2004) Size-dependent diffusion in cycloalkanes. Mol Phys 102:1489–1497
38. Kubo R (1957) Statistical-mechanical theory of irreversible processes. I. General theory and simple applications to magnetic and conduction problems. J Phys Soc Japan 12(6):570–584
39. Kubo R, Yokota M, Nakajima S (1957) Statistical-mechanical theory of irreversible processes. II. Reaction on thermal disturbances. J Phys Soc Japan 12(11):1203–1226
40. Langevin P (1908) Sur la théorie du mouvement brownien. C R Acad Sci Paris 146:530–533
41. Lebowitz JL (1971) Hamiltonian flows and rigorous results in non-equilibrium statistical mechanics. In: Statistical mechanics, new concepts, new problems, new applications. Proceedings of IUPAP conference on statistical mechanics. Chicago University Press, Chicago, pp 41–66
42. Lifshits EM, Dzyaloshinsky IE, Pitaevsky LP (1961) General theory of van der Waals forces. Adv Phys Sci 73:381–422
43. Lu Q, Freedman KO, Rao R et al (2004) Diffusion of carbon nanotubes with single-molecule fluorescence microscopy. J Appl Phys 96:6772–6781
44. McPhie MG, Daivis IK, Snook PJ (2006) Viscosity of a binary mixture: approach to the hydrodynamic limit. Phys Rev E 74:021301
45. Mejía-Rosales S, Sandoval-Salazar SA, Soria-Sánchez A et al (2021) Mechanical properties of MoS_2 nanotubes under tension: a molecular dynamics study. Mol Simul 47(6):471–479
46. Norman GE, Stegailov VV (2002) Stochastic and dynamic properties of molecular dynamics systems: simple liquids, plasma and electrolytes, polymers. Comp Phys Comm 147(4):678–683
47. Norman GE, Stegailov VV (2013) Stochastic theory of the classical molecular dynamics method. Math Mod Comput Simul 5(4):305–333
48. Ould-Caddour F, Levesque D (2000) Molecular-dynamics investigation of tracer diffusion in a simple liquid: test of the Stokes-Einstein law. Phys Rev E 63:011205

49. Ould-Kaddour F, Levesque D (2007) Diffusion of nanoparticles in dense fluids. J Chem Phys 127:154514
50. Percival I (1961) Almost periodicity and the quantum H theorem. J Math Phys 2(2):235–239
51. Pujari S, Rahatekar SS, Gilman JW et al (2009) Orientation dynamics in multiwalled carbon nanotube dispersions under shear flow. J Chem Phys 130(21):214903
52. Puliti G, Paolucci S, Sen M (2012) Thermodynamic properties of gold–water nanofluids using molecular dynamics. J Nanopart Res 14:1296
53. Rapaport DC (2005) The art of molecular dynamics simulation. Cambridge University Press, Cambridge
54. Reuel NF, Dupont A, Thouvenin O et al (2012) Three-dimensional tracking of carbon nanotubes within living cells. ACS Nano 6:5420–5428
55. Résibois P, De Leener M (1977) Classical kinetic theory of fluids. Wiley-Interscience, London
56. Rudyak VYa (1987) Statistical theory of dissipative processes in gases and liquids. Science, Novosibirsk
57. Rudyak VYa (1991) Correlation in a finite number of particles system simulating a rarefied gas. Fluid Dyn 26(6):909–914
58. Rudyak VYa (2004) Statistical aerohydromechanics of homogeneous and heterogeneous media. Vol. 1, Kinetic Theory. NSUACE, Novosibirsk
59. Rudyak VYa (2005) Statistical aerohydromechanics of homogeneous and heterogeneous media. Vol. 2, Hydromechanics. NSUACE, Novosibirsk
60. Rudyak VYa (2011) Mechanics, ttransports processes, fluctuation and irreversibility. NSUACE, Novosibirsk
61. Rudyak VYa (2020) Fluctuation-dissipation theorems and transport coefficients of the gases, liquids and nanofluids. J Phys Conf Series 1560:012002
62. Rudyak VYa, Belkin AA (2003) Nanoparticle velocity relaxation in condensed carrying medium. Tech Phys Lett 29(7):560–562
63. Rudyak VYa, Belkin AA (2004) Mechanisms of collective nanoparticles interaction with condensed solvent. Thermophys Aeromechanics 11(2):54–63
64. Rudyak VYa, Belkin AA (2010) Simulation of transport coefficients. Nanosyst Phys Chem Math 1:156–177
65. Rudyak VYa, Belkin AA (2013) Transport processes of nanoparticles in gases and liquids. Adv Struct Mater 4:135–168
66. Rudyak VYa, Belkin AA, Ivanov DA et al (2008) The simulation of transport processes using the method of molecular dynamics. Self-diffusion Coeff High Temp 46(1):30–39
67. Rudyak VYa, Belkin AA, Tomilina EA (2008) Force acting on nanoparticle in fluid. Tech Phys Lett 34:76–78
68. Rudyak VYa, Kharlamov GV, Belkin AA (2000) The velocity autocorrelation function of nanoparticles in a hard-sphere molecular system. Tech Phys Lett 26:553–556
69. Rudyak VYa, Kharlamov GV, Belkin AA (2001) Diffusion of nanoparticles and macromolecules in dense gases and liquids. High Temp 39:264–271
70. Rudyak VYa, Krasnolutskii SL (1999). The interaction potential of dispersed particles with carrier gas molecules. In: Brun R, Camargue R, Guignol R, Lengrand J-C (eds) Proceedings of 21st international symposium on rarefied gas dynamics, Gépadués-Éditions, Toulouse, vol 1, pp 263–270
71. Rudyak VYa, Krasnolutskii SL (2002) Diffusion of nanoparticles in a rarefied gas. Tech Phys 47:807–813
72. Rudyak VYa, Krasnolutskii SL (2017) Interaction potentials of hollow nanoparticles with each other and with molecules of the carrier medium. Proc Russ High School Acad Sci 2:32–42
73. Rudyak VYa, Krasnolutskii SL (2018) Potentials of composite nanoparticles interaction with each other and with molecules of the carrier medium. Proc Russian Higher School Academy Sci 3:26–37
74. Rudyak VYa, Krasnolutskii SL, Belkin AA et al (2021) Molecular dynamics simulation of water based nanofluids viscosity. J Thermal Anal Calorim 145(6):2983–2990

75. Rudyak VYa, Krasnolutskii SL, Ivanov DA (2011) Molecular dynamics simulation of nanoparticle diffusion in dense fluids. Microfluid Nanofluid 11(4):501506
76. Rudyak VYa, Krasnolutskii SL, Ivanov DA (2012) The interaction potential of nanoparticles. Doklady Phys 57:33–35
77. Rudyak VYa, Krasnolutsky SL, Nasibulin AG et al (2002) Methods of measuring the diffusion coefficient and sizes of nanoparticles in rarefied gas. Dokl Phys 47(10):758–761
78. Rudyak VYa, Tretiakov DS (2020) On diffusion of single-walled carbon nanotubes. Thermophys Aerotech 27:847–855
79. Stratonovich RL (1966) A new representation for stochastic integrals and equations. SIAM J Control 4(2):362–371
80. Stuart SJ, Tutein AB, Harrison JA (2000) A reactive potential for hydrocarbons with intermolecular interactions. J Chem Phys 112(14):6472–6486
81. Tang W, Advani SG (2006) Drag on a nanotube in uniform liquid argon flow. J Chem Phys 125:174706
82. Urbančič I, Brun J, Shrestha D et al (2018) Lipid composition but not curvature is the determinant factor for the low molecular mobility observed on the membrane of virus-like vesicles. Viruses 10:415
83. Verlet L (1967) Computer "experiments" on classical fluids. I. Thermodynamical properties of Lennard-Jones molecules. Phys Rev 159:98–103
84. Walther JH, Werder T, Jaffe RL, Koumoutsakos P (2004) Hydrodynamic properties of carbon nanotubes. Phys Rev E 69(6):062201
85. Wang Y, Bahng JH, Che Q et al (2015) Anomalously fast diffusion of targeted carbon nanotubes in cellular spheroids. ACS Nano 9(8):8231–8238
86. Weeks JD, Chandler D, Andersen HC (1971) Role of repulsive forces in determining the equilibrium structure of simple liquids. J Chem Phys 54:5237–5247
87. Wuelfing WP, Templeton AC, Hicks JF et al (1999) Dispersion measurements of monolayer protected clusters: a physicochemical determination of nanoparticle size. Anal Chem 71:4069–4074
88. Zhang Z-T, Zhao X, Cao B-Y (2019) Diffusion tensors of arbitrary-shaped nanoparticles in fluid by molecular dynamics simulation. Sci Rep 9:18943
89. Zubarev D (1974) Nonequilibrium statistical thermodynamics. Consultants Bureau, New York

Chapter 3
Viscosity and Rheology of Nanofluids

3.1 Viscosity of Liquids

The use of nanofluids is primarily related to their forced flow processes, which require energy expenditure. This expenditure is in turn determined by the viscosity of the fluid being pumped, making viscosity the most important characteristic of nanofluids. The viscosity of conventional coarse dispersed fluids has been systematically studied for over a century. Several fundamental differences between the viscosity of dispersed fluids and the viscosity of the carrier fluid have been established. First of all, the viscosity of the dispersed fluid is always greater than that of the carrier fluid, and this excess is proportional to the volume concentration of dispersed particles. The second important point to consider is that in some cases, dispersed fluids exhibit non-Newtonian behavior. In Newtonian fluids, the stress tensor $\boldsymbol{J}_2$ is directly proportional to the strain rate tensor $\boldsymbol{D}$. For the simplest case of an incompressible fluid, this relationship can be expressed as follows

$$\boldsymbol{J}_2 = \mu \boldsymbol{D}, \quad D_{ij} = \frac{\partial u_i}{\partial x_j} + \frac{\partial u_j}{\partial x_i}. \tag{3.1}$$

where $\boldsymbol{u}$ is the flow velocity, while the viscosity coefficient is just a function of state (density, temperature, etc.).

As previously mentioned, relation (3.1) is not satisfied in many dispersed fluids. The stress tensor appears to be a nonlinear function of the shear rate tensor in these cases. Reducing the corresponding transport law to the Newtonian form (3.1) causes the effective viscosity coefficient to become shear rate dependent. The reasons for this change are not clear yet, because a fluid can become non-Newtonian at relatively low shear rates. Therefore, predicting the transition of a fluid from Newtonian to non-Newtonian rheology is quite challenging.

Already in the first studies of nanofluids, it was discovered that their viscosity is not described by the usual relationships for coarse-dispersed liquids. For a long time,

V. Rudyak, *Thermophysical Properties of Nanofluids*, Fluid Mechanics and Its Applications 140, https://doi.org/10.1007/978-3-031-97995-8_3

the measurement data obtained were contradictory. However, recent advancements have made it possible to formulate corresponding regularities and, moreover, to explain the observed phenomena. The situation, however, is non-trivial. To make the obtained answer clear, it is necessary to consistently consider the processes of momentum transfer in various dispersed liquids: coarsely dispersed, nano-liquids with ordinary spherical particles and nano-liquids with CNTs. On the other hand, to understand the mechanisms of momentum transport in such a system, in addition to experimental data, it is necessary to also provide the result of MD modeling of the viscosity coefficient.

The presence of dispersed particles in a fluid is an additional complicating factor that affects the properties of the dispersed fluid, particularly its viscosity. Therefore, understanding the viscosity mechanisms of the carrier fluid is crucial before discussing the viscosity of the dispersed fluid. Unlike rarefied gases, there is no rigorous theory for liquid viscosity. This is due, on the one hand, to the complex structure of the liquid, and, on the other hand, to the significantly different dynamics of its molecules compared to the dynamics of molecules in both a solid and a gas. Figure 3.1 (left) schematically represents a system of molecules of a not very dense gas. The distance between molecules is much greater than their size. In a rarefied gas, this distance is even larger allowing the molecules to move freely most of the time, and the mean free path l is much larger than the characteristic molecular size σ. The kinetic energy of the molecules kT in this case is much larger than their potential energy of interaction U. As a result, it turns out that the interaction of molecules does not contribute neither to the transport coefficients nor to the equation of state [62, 116, 117]. The viscosity coefficient is proportional to the mean free path. In the simplest case of a hard spheres gas it takes the form

$$\mu = \frac{5}{16}\frac{\sqrt{mkT}}{\sigma^2\sqrt{\pi}}. \tag{3.2}$$

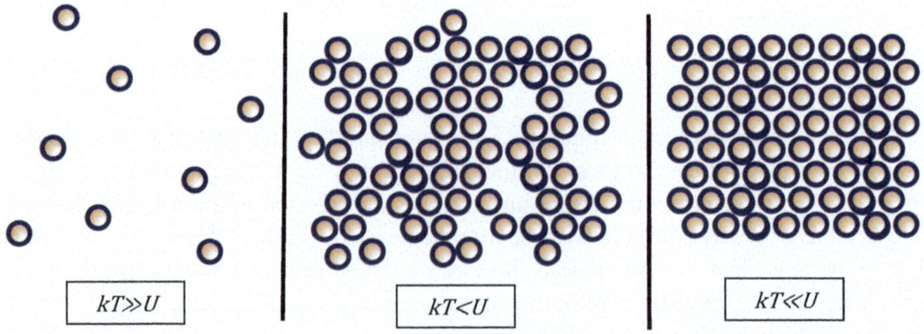

Fig. 3.1 Schematic structure of the gas (left), solid (right) and liquid (middle) molecules

The contribution of interaction to the transport coefficients needs to be considered only for a dense gas. In contrast, in a solid, the molecules are ordered and typically form a crystal lattice of various configurations, with only small oscillations near their equilibrium position (Fig. 3.1, right). This structure is possible because the kinetic energy of molecules is much lower than their potential energy. The redistribution of momentum in such a system occurs, roughly speaking, in a manner similar to that of a system of coupled oscillators.

In the liquid (Fig. 3.1, middle), the molecules are in a disordered state, although there are also ordered fragments separated by nanopores. Each molecule is constantly surrounded by approximately 10–12 other molecules, with which it interacts. It can be said that it is located in the average self-consistent field of intermolecular forces of surrounding environment. The density of a fluid is only about ten percent lower than that of a solid. It is evident that the transport processes in a fluid are determined by the constant evolution of its molecular structure. The periodic change in density, commonly referred to as long-range order, is determined by the ordered arrangement of molecules in a crystalline solid. The structure of a fluid (liquid) can be accurately described by the radial distribution function $g_2(r)$ [4, 118], which defines the configuration correlation of two arbitrarily chosen molecules

$$g_2(r_{12}) = \frac{f_2(\boldsymbol{r}_1, \boldsymbol{r}_2)}{f_1(\boldsymbol{r}_1) f_1(\boldsymbol{r}_2)}, \tag{3.3}$$

where $f_2(\boldsymbol{r}_1, \boldsymbol{r}_2)$ is the binary distribution function of molecules 1 and 2, and $f_1(\boldsymbol{r}_1)$ and $f_1(\boldsymbol{r}_2)$ are their single-particle distribution functions. The normalized radial distribution function is defined by the expression

$$g_2(r) = \frac{1}{4\pi n r^2} \left\langle \frac{\mathrm{d}n(r)}{\mathrm{d}r} \right\rangle, \tag{3.4}$$

where n is the numerical density of the fluid, and the angular brackets denote averaging over the ensemble.

Thus, the radial distribution function determines the average density of fluid molecules around any given molecule. A typical example of the radial distribution function of water molecules at normal pressure and temperature of 25 °C is shown in Fig. 3.2 [124]. The radial distribution function of water has a quasi-periodic damped structure characteristic of a fluid, and describes the so-called short-range order of molecules in the fluid. The radial distribution function is a crucial factor in the study of fluid structure because it is directly measurable. More precisely, the so-called structure factor is measurable, which is the Fourier transform of the radial distribution function [149, 161].

The radial distribution function is an average characteristic of the fluid. However, the transfer of momentum, carried out in the process of interaction of molecules and their movement (in particular, diffusion) continuously changes the local nature of the distribution of molecules near each molecule. Locally, the short-range order undergoes continuous change. Therefore, it can be said that momentum transfer in a

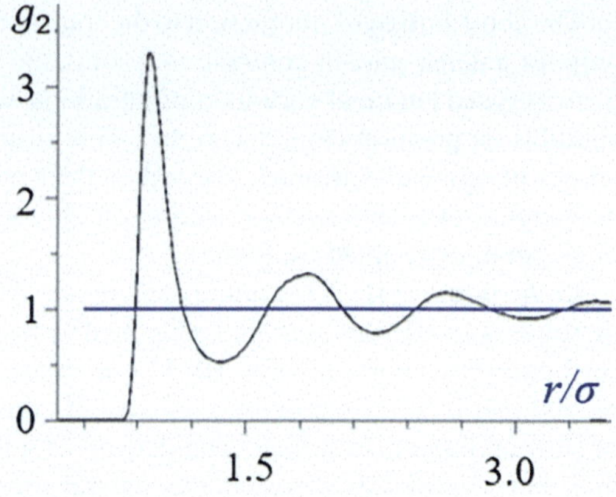

Fig. 3.2 Radial distribution function of water molecules

fluid, and hence its viscosity, is related to the continuous change of the short-range order. The relationship between the viscosity coefficient and the radial distribution function can be directly derived from the Born–Green formula [17]

$$\mu \sim \left(\frac{m}{kT}\right)^{1/2} \rho^2 \int_0^\infty g_2(r) \frac{\mathrm{d}U}{\mathrm{d}r} r^4 \mathrm{d}r. \tag{3.5}$$

The lack of long-range order in fluids is due to the fact that their molecules have significantly lower activation energy U_a than in solids. This means that the potential barrier, which defines this energy and is formed by the self-consistent field of molecules, surrounding this molecule, is significantly lower than that in a solid. Overcoming this barrier determines the local change in the short-range order. The momentum transfer in a fluid is determined by the successive transitions of molecules from one nearest-neighbor cage to another. This mechanism is responsible for the formation of viscosity in a fluid. Frenkel [34] was the first to develop a theory of fluid viscosity that takes this mechanism into account. The starting point for the construction of this theory was the analogy of the transport equations that determine the self-diffusion of fluid particles and the dissipation of its momentum density. The latter depends on the kinematic viscosity coefficient $\upsilon = \mu/\rho$. Thus, Ya.I. Frenkel defines an average characteristic of motion of fluid molecules as a mobility, which is determined by the self-diffusion coefficient. Further, since, as it is demonstrated by Ya.I. Frenkel, the self-diffusion coefficient is proportional to the expression $\exp(-U_a/kT)$, the viscosity coefficient is described by the relation

$$\mu = A\exp\left(\frac{U_a}{kT}\right). \tag{3.6}$$

This relation was previously derived from other considerations in [5, 40]. Relationship (3.6) qualitatively correctly predicts the viscosity coefficient depending on temperature. Unlike gases, liquids experience a decrease in viscosity as temperature increases due to the destruction of short-range order in the fluid. The accuracy of formula (3.6) is determined by coefficient A. To calculate it, Ya.I. Frenkel analyzes the possible motion of a molecule surrounded by other molecules. The interaction with each of these molecules creates a corresponding potential well. The motion of a fluid molecule is defined as a successive jump from one potential well to another. As a result, the coefficient A is defined by the following relation

$$A = \frac{kT\tau_0}{\pi\sigma\delta^2}. \tag{3.7}$$

Here, τ_0 is the lifetime of a molecule in a given potential well, which in fact is the interaction time of molecules, and δ is the characteristic action radius of the corresponding potential. The author (Ya.I. Frenkel) acknowledges that the constant (3.7) is at least two orders of magnitude lower than what is required for formula (3.6) to provide quantitatively correct values of the viscosity coefficient. The estimate (3.7) was obtained under the assumption that the fluid was a quasi-solid. It is in this case that the parameter δ has a characteristic scale of the order of the molecule size. Actually, the fluid exhibits short-range order, and each molecule is in a self-consistent field, formed by its immediate surroundings. The action of this field has a characteristic scale that is of the order of the scale on which the short-range order decays. Taking this into account, the estimate (3.7) becomes much more realistic.

As previously stated, due to the lack of a rigorous theory of fluid viscosity, semi-empirical or empirical formulas are commonly used in practice. These formulas typically provide a temperature dependence of the form (3.6). A comprehensive list of these formulas can be found in books [115, 156]. The viscosity of a fluid remains relatively constant at small changes in pressure, but at high pressures, it increases significantly. This increase is particularly pronounced for fluids with complex molecules.

3.2 Effective Viscosity Coefficient of Coarse Dispersed Fluids

A dispersed fluid is a complex liquid that cannot always be described using continuum mechanics approaches. However, even when it is possible, the levels of approaches and description can vary. This will be discussed in more detail in the next to last chapter (see also Rudyak [118], Rudyak and Minakov [135]). It is important to distinguish between two-fluid (multifluid) and single-fluid descriptions. In the first case, the dispersed fluid is treated as a binary system comprising a carrier fluid and pseudogas[1] (pseudofluid) particles. Each component is characterized by its density, velocity, temperature, viscosity, and thermal conductivity. In addition, there is an exchange of momentum and energy between the components, resulting in cross-transport coefficients and interphase source forces appearing in the transport equations for each component. In the second case, the entire dispersed liquid is considered as a homogeneous continuum, characterized by a single density, velocity and temperature.

To account for the presence of different phases, effective viscosity and thermal conductivity coefficients are introduced. This approach is commonly used when simulating the flow of dispersed fluids. At the same time, while the density and velocity of such a continuum are determined quite easily, finding the effective transport coefficients requires special consideration. In this section we analyze the most commonly used models for determining the effective viscosity coefficient. The corresponding thermal transport coefficient will be discussed in Chap. 4.

A. Einstein first formulated the theory of the effective viscosity coefficient of a rarefied dispersed fluid in 1906 [29]. The theory considered a two-component rarefied dispersed fluid, where the carrier fluid was assumed to be incompressible, and the pseudogas of particles was so rarefied that their interaction could be neglected. It was assumed that a spherical particle with a radius of R_0, moving in a fluid, creates perturbations in the velocity and pressure fields around itself. These perturbations cause additional dissipation of fluid momentum, which results in a certain contribution to the viscosity coefficient. The pseudogas of particles is assumed to be so rarefied that the hydrodynamic perturbations of the carrier medium generated by each particle do not affect the dynamics of neighboring particles. Due to the small size of the particle introduced into the flow, the resulting fluid motion represents a superposition of translational, rotational, and deformational motions. Assuming the effect of inertia forces on the particle is negligible, it can be concluded that the particle moves with the surrounding fluid. As a result, the perturbations introduced by the particle only affects the deformation component of the fluid motion. It is this component that should be defined.

Let the particle, moving together with the fluid be at the point $\boldsymbol{r}_0$. In the absence of the particle, the fluid velocity at the point $\boldsymbol{r}$ in the vicinity of $\boldsymbol{r}_0$ can be found by knowing the velocity at the point $\boldsymbol{r}_0$, decomposing it into a Taylor series and limiting ourselves to the first two terms of the expansion

$$\boldsymbol{u}(\boldsymbol{r}) = \boldsymbol{u}(\boldsymbol{r}_0) + (\boldsymbol{r} - \boldsymbol{r}_0) \cdot \nabla \boldsymbol{u}. \tag{3.8}$$

In the particle-associated coordinate system, the unperturbed velocity of the fluid near the particle is a linear function of the coordinates

[1] Here we speak about pseudogas of particles, since their numerical density is small at usual volume concentrations in nanofluids, and indeed, one can speak about particle gas. In case the particle sizes are small enough (on the order of 1–5 nm) their mean free path can be of the order of the particle size. In this case the gas is considered as dense, and direct interactions of particles can be important.

$$\boldsymbol{u}(\boldsymbol{r}) = \boldsymbol{Q} \cdot \boldsymbol{r}. \tag{3.8a}$$

Comparing expressions (3.8) and (3.8a), and taking into account the symmetry properties of the medium, we find that the tensor $\boldsymbol{Q}$ is a strain rate tensor. Taking this into account, the fluid velocity and pressure in the presence of the particle can be found in the following form

$$\boldsymbol{u}(\boldsymbol{r}) = \boldsymbol{u}' + \boldsymbol{D} \cdot \boldsymbol{r}, \quad p = p_0 + p', \tag{3.9}$$

where $\boldsymbol{u}'$, p' are the velocity and pressure perturbations caused by the presence of the particle, and p_0 is the unperturbed pressure in the flow.

The symmetry of the flow under study suggests that the particle will remain stationary relative to the fluid. Therefore, the following boundary condition applies: $\boldsymbol{u}' = 0$ at $r = R_0$. However, perturbations must naturally dampen away from the sphere: $\boldsymbol{u}' \to 0, p' \to 0$ at $r \to \infty$. This leads to the Stokes problem, which enables determining the motion velocity and its perturbation. According to calculations by Batchelor [10], the velocity and pressure perturbations are described by the following functions

$$\boldsymbol{u}' = \frac{A}{r^5}\boldsymbol{D} \cdot \boldsymbol{r} + \left(\frac{B}{2r^5} - \frac{A}{2r^7}\right)(\boldsymbol{D} : \boldsymbol{rr})\boldsymbol{r}, p' = \frac{B\mu}{r^5}\boldsymbol{D} : \boldsymbol{rr}, \tag{3.10}$$

where A and B are constants.

Despite the fact that the contribution of the perturbations, induced by the particle on the velocity and pressure fields, has been calculated, the determination of the effective viscosity coefficient of the suspension, associated with the additional dissipation processes, induced by these fields, still remains a nontrivial problem. Indeed, the total momentum flux produced in the fluid by the fields (3.10) is equal to

$$\boldsymbol{J}_2' = (p_0 + p')\boldsymbol{U} - \mu(\boldsymbol{D} + \boldsymbol{D}'). \tag{3.11}$$

However, the presence of an effective viscosity coefficient μ_e allows for the representation of the stress tensor (3.11) in the following form

$$\boldsymbol{J}_2' = p_0\boldsymbol{U} - \mu_e\boldsymbol{D}. \tag{3.12}$$

It is quite obvious that the representation (3.12) is generally impossible at all.

The second problem is that the fields (3.10) describe the perturbations caused by a single particle, while it is necessary to consider the presence of a large number of particles someway distributed throughout the fluid. A. Einstein assumed that firstly, the particles are distributed uniformly, and secondly, that the perturbations introduced by other particles can be neglected when calculating the corresponding velocity and pressure fields due to their large distance between each other. Considering these guidelines, the stress flux through surface Σ, which encloses volume V of the suspension being considered, is equal to

$$\int_{\Sigma} \boldsymbol{u} \cdot \boldsymbol{J}_2' \cdot d\Sigma = \int_{\Sigma} \boldsymbol{u} \cdot \left[(p_0 + p')\boldsymbol{U} - \mu(\boldsymbol{D} + \boldsymbol{D}') \right] \cdot d\Sigma = \int_{\Sigma} \boldsymbol{u} \cdot (p_0 \boldsymbol{U} - \mu_e \boldsymbol{D}) \cdot d\Sigma.$$

The latter equality gives the equation for computing the effective viscosity coefficient. Performing the necessary calculations [10], it can be shown that the effective viscosity coefficient is equal to.

$$\mu_e = \mu(1 + 2.5\varphi) \tag{3.13}$$

where φ is the volume concentration of particles.

As demonstrated by the experimental data, Einstein's formula (3.13) is effective for small volume concentrations of the dispersed phase ($\varphi \leq 10^{-2}$). When calculating perturbations of velocity and pressure fields for larger particle concentrations in the fluid, it is necessary to consider their mutual effect. Strictly speaking, the velocity distribution of the fluid near a particle depends on the position of all other particles, unless they are too far apart. This results in the appearance of additional forces and moments acting from the fluid on the particles. In other words, particles interact with each other through the medium. This interaction is known as hydrodynamic interaction. However, finding a comprehensive solution to this complex hydrodynamic problem yet remains a challenge. To date, the investigation of this problem is limited to a sufficiently rarefied dispersed system, where the hydrodynamic interaction between pairs of particles can be considered [11, 12]. It has been demonstrated that the effective viscosity of a suspension can only be introduced for certain classes of flows, where the average stresses have a Newtonian form due to hydrodynamic interaction between particles. The corresponding contribution to the stresses turns out to be of the order of φ^2, and thus determines a correction to the Einstein's coefficient (3.13). This correction can only be explicitly calculated for certain types of flows. In particular, for an expansion flow with zero angular velocity of solid-state rotation, the effective viscosity coefficient takes the following form

$$\mu_e = \mu\left(1 + 2.5\varphi + 7.6\varphi^2\right). \tag{3.14}$$

Later G.K. Batchelor considered the contribution of Brownian motion of particles to their interaction. As a result, the formula (3.14) was refined and took the final form [13, 14]

$$\mu_e = \mu\left(1 + 2.5\varphi + 6.2\varphi^2\right). \tag{3.15}$$

The experimental data on the measurement of the effective viscosity coefficient of coarse dispersed fluids, available to date, allow asserting that formula (3.15) is a quite satisfactory approximation up to concentrations $\varphi \sim 0.2$.

If the particles have a non-spherical shape, determining the effective viscosity coefficient becomes more complex. Jeffery [58] has solved this problem for ellipsoidal particles, assuming that the orientations of the ellipsoids in the flow are equally probable due to Brownian motion. The effective viscosity coefficient, in the first approximation by concentration, is determined by the following relation

$$\mu_e = \mu\left[1 + B(4/3)\pi abcn_p\right], \tag{3.16}$$

where a, b, c are the semi-axes of the ellipsoid, n_p is the numerical density of dispersed particles, and B is some constant depending on the ratio of the semi-axes, provided that $B > 2.5$.

The effective viscosity coefficient of a dispersed medium increases as the particle shape deviates further from spherical. This is expected because the characteristic time of Brownian rotation is much smaller than the hydrodynamic times considered. Therefore, particle rotation leads to an increase in the effective volume they occupy. For an ellipsoid of rotation where $b = c$, the formula is somewhat simplified. If in addition $a \gg b$, such an ellipsoid simulates a continuous solid CNT quite well.

Einstein's (3.13) and Batchelor's (3.16) formulae are applicable for relatively small concentrations. Typically, nanofluid applications involve particle volume concentrations of no more than ten percent. For nanofluids containing CNTs, these concentrations are usually about one order of magnitude smaller. However, it is interesting and important to comprehend how Einstein's and Batchelor's approaches can be extended to higher concentrations. Batchelor followed Einstein's paradigm, focusing on the pairwise interaction of dispersed particles. However, further exploration of this path is not promising, since considering possible triple and quadruple interactions would lead to extremely difficult problems. In Nagatani's work [97], an interesting approach similar to Einstein's is presented. The effect of Brownian motion of particles was not taken into account, but it was assumed that the dispersed medium is described by stochastic fields of velocity $\boldsymbol{u}(\boldsymbol{r}, t)$ and density $\rho(\boldsymbol{r}, t)$, which satisfy the transport equations in the Stokes approximation

$$\nabla \cdot \boldsymbol{u} = 0, \quad \frac{\partial}{\partial t}\rho(\boldsymbol{r}, t)\boldsymbol{u}(\boldsymbol{r}, t) + \nabla p = \nabla \cdot [\mu(\boldsymbol{r}, t)\boldsymbol{D}(\boldsymbol{r}, t)].$$

The viscosity coefficient $\mu(\boldsymbol{r}, t)$ in this case is a random function, differing from the viscosity coefficient of the base fluid due to the presence of dispersed particles and their interactions. The equations are then solved using Fourier representation, and the low-frequency asymptotics of the resulting solutions are identified with the corresponding macroscopic fields. By performing the inverse Fourier transform, Nagatani [97] has obtained the macroscopic equations of motion of the dispersed medium

$$\nabla \cdot \boldsymbol{u} = 0, \quad \nabla p = \mu_e \nabla^2 \boldsymbol{u},$$

where the effective viscosity coefficient is equal to

$$\mu_e = \mu/(1 - 2.5\varphi). \tag{3.17}$$

It is evident that the Einstein coefficient (3.13) is derived from the first approximation by φ, , while the Batchelor coefficient (3.16) is obtained from the second approximation. However, a comparison with experimental data [155], conducted by Nagatani, has shown that formula (3.17) is applicable up to $\varphi \sim 0.4$.

Many studies have focused on deriving formulas that describe the effective viscosity coefficient of coarse dispersed suspensions [2, 33, 66, 93]. A fairly complete list of the formulas used can be found in reviews [50, 78, 109, 160]. However, it is important to emphasize that at quite small concentrations almost all formulas lead to a ratio of type (3.15), in which the second coefficient usually varies from 4.3 to 7.6. Formulas such as (3.15) result in a 25–30% increase in the viscosity coefficient at a 10% volume concentration of dispersed particles. In this book only volume concentrations of particles up to 10% are considered, making these ratios quite adequate. It should be noted that all of these formulas have been repeatedly verified using experimental data for dispersed particles with characteristic sizes ranging from 2–5 to 105 μm [28, 65, 70, 155].

3.3 Effective Viscosity Coefficient of Nanofluids

The viscosity study of nanofluids was started simultaneously with the study of their other thermophysical properties. As nanofluids are suspensions, it was initially assumed that their effective viscosity coefficient would be described by the theories summarized in the previous section. Surprisingly, this turned out not to be the case. First of all, it was found that even at sufficiently low particle concentrations, the viscosity coefficient of nanofluids was not described by Einstein's theory (3.13). The viscosity of nanofluids turned out to be much higher than predicted by this theory. At low nanoparticle concentrations, viscosity was described by a linear relationship of the form $\mu_e = \mu(1 + a\varphi)$, but the coefficient a varied from 4.3 to almost 40 [21, 23, 25, 36]. The excess viscosity increased rapidly with increasing concentration, however, this increase was not described by the relationship (3.15) either. Moreover, the heterogeneity of the results obtained increased further (see reviews by [50, 78, 63, 95, 109, 142]. This was in sharp contrast to the classical theories (3.13) and (3.15) as well as other approaches, applicable to coarse dispersed fluids. The significant difference in viscosity between nanofluids and coarse dispersed fluids suggests that the viscosity mechanisms in nanofluids are determined by some parameters other than concentration alone. Referring to Fig. 1.2, it is highly likely that the thermophysical characteristics of nanofluids depend on the size of nanoparticles. The second possible reason for such differences could be the material of the nanoparticles. Both possible reasons for the observed differences will be consistently considered in the following sections.

3.3.1 Experimental Measurements of Viscosity of Nanofluids

To date, there are various methods available for measuring fluid viscosity. A comprehensive overview of these methods and the equipment used can be found in the book [156]. It is important to note that the rheology of nanofluids, particularly those containing CNTs, may be non-Newtonian. Therefore, it is necessary to study the rheology of a nanofluid before measuring its viscosity. Various rotational viscometers are commonly used for this purpose. This book presents data on the rheology and viscosity of nanofluids obtained using such viscometers.

When studying the viscosity of nanofluids, the first issue that arises is determining its dependence on particle concentration. A typical dependence of the relative viscosity coefficient $\mu_r = \mu/\mu_b$ (μ_b is the viscosity coefficient of the base fluid) of ethylene glycol-based nanofluid with SiO_2 particles on their volume concentration is presented in Fig. 3.3a [129]. The particle concentration ranged from 0.25 to 8%, while particle sizes were 18.1 nm (rhombus symbols), 28.3 nm (triangles), and 45.6 nm (squares). The straight line corresponds to Einstein's formula (3.13). These measurements were taken at a temperature of 25 °C. The viscosity of all three fluids shown in Fig. 3.3 is different and much higher than that predicted by formulas (3.13) or (3.15).

In all cases, the dependence of the effective viscosity coefficient of nanofluid on the volume concentration of particles turns out to be quadratic

$$\mu = \mu_b\left(1 + a\varphi + b\varphi^2\right). \tag{3.18}$$

However, like coefficient *a*, coefficient *b* is not universal and is significantly larger than the corresponding value for ordinary suspensions (3.15). Several correlations obtained at different times are presented below. One of the earliest correlations was derived for a water-based nanofluid with TiO_2 particles [106]

$$\mu = \mu_b\left(1 + 5.45\varphi + 108.2\varphi^2\right). \tag{3.19a}$$

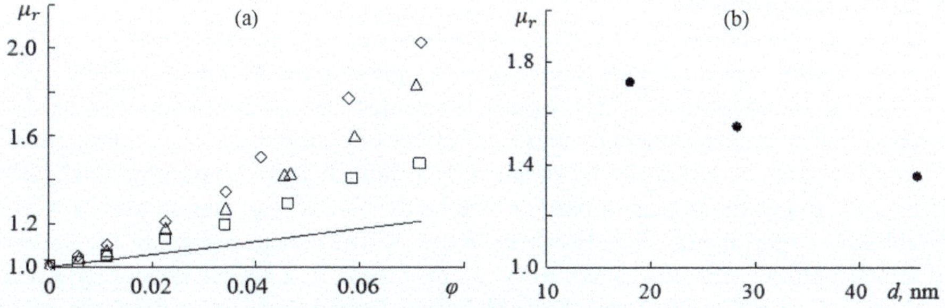

Fig. 3.3 Relative viscosity coefficient of ethylene glycol-based nanofluid with SiO_2 nanoparticles versus the nanoparticle volume concentration (**a**) and size (**b**)

Experimental data for water-based nanofluid with Al_2O_3 particles [159] were processed in [79], and the following correlation was obtained

$$\mu = \mu_b\left(1 + 7.3\varphi + 123\varphi^2\right). \tag{3.19b}$$

On the other hand, for water-based nanofluid with Al_2O_3 particles Buongiorno [18] presented the following formula

$$\mu = \mu_b\left(1 + 39.11\varphi + 533.9\varphi^2\right). \tag{3.19c}$$

For ethylene glycol-based nanofluid with TiO_2 particles Chen et al. [21] proposed the following correlation

$$\mu = \mu_b\left(1 + 10.6\varphi + 112.36\varphi^2\right). \tag{3.19d}$$

Oueslati and Bennacer [104], having processed a set of experimental data [99–101] for nanofluids with Fe_2O_3 particles obtained the following correlation

$$\mu = \mu_b\left(1 + 18.64\varphi + 248.3\varphi^2\right). \tag{3.19e}$$

The following formulas were obtained in [92] for water-based nanofluids with Al_2O_3 (150 nm) and TiO_2 (150 nm) particles, respectively

$$\mu = \mu_b\left(1 + 6\varphi + 72\varphi^2\right), \tag{3.19f}$$

$$\mu = \mu_b\left(1 + 7.1\varphi + 148\varphi^2\right). \tag{3.19g}$$

To summarize this brief review, it is important to note the extreme heterogeneity and observed differences in correlations that describe seemingly identical nanofluids [see (3.19b), (3.19c), and (3.19f)). Figure 3.3b provides a possible explanation for these discrepancies. The effective viscosity coefficient of nanofluids is dependent on the size of nanoparticles.

The viscosity coefficient of nanofluids exhibits size dependence as established for a simple model nanofluid with solid spheres employing the MD method [120, 123, 124]. It was found to increase as the particle size decreases. Almost simultaneously, researchers began experimental studies on effect of nanoparticle size, although its degree of influence could not be definitively established for some time [44, 100, 101, 151]. Today, the impact of particle size on the viscosity of nanofluids is a well-established experimental fact. However, there is still debate regarding the extent to which particle size affects viscosity, as discussed in a recent review by Koca et al. [63]. Figure 3.3b illustrates the particle size dependence of nanofluid viscosity, which increases as particle size decreases. In Fig. 3.3b, d represents the diameter of the nanoparticles. Later, this relationship was confirmed through experiments by multiple research teams. Indeed, the viscosity of nanofluids increases as the particle

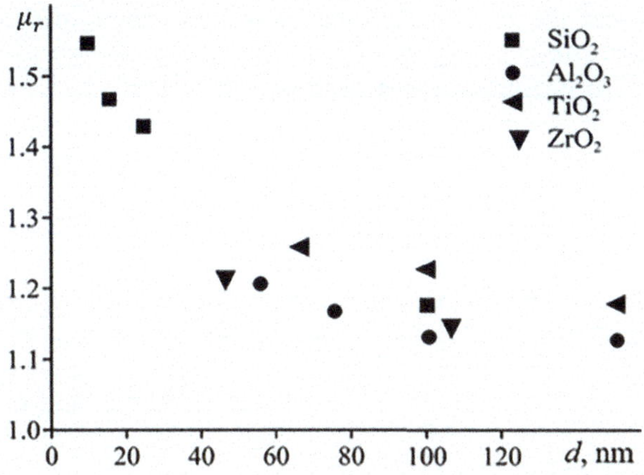

Fig. 3.4 Relative viscosity coefficient of nanofluids depending on nanoparticle size

size decreases, as illustrated in Fig. 3.4. The figure shows the dependence of the relative viscosity coefficient $\mu_r = \mu/\mu_b$ of water-based nanofluids on the size of SiO_2, Al_2O_3, TiO_2, and ZrO_2 nanoparticles at their constant volume concentration of 2% [92].

The temperature dependence of the viscosity coefficient is a crucial thermophysical characteristic of nanofluids. It is well-established that viscosity decreases with increasing temperature in fluids, but increases in gases due to different momentum transfer mechanisms. In gases, momentum equalization occurs through kinetic mechanism, molecules transfer the momentum during their movement and collisions. In fluids, momentum transfer is more complex. Here, due to the dense packing of molecules, their collective interaction makes it impossible to determine the mean free path of molecules, which is smaller than the characteristic size of the molecule. Therefore, the main mechanism of momentum transfer in fluids is related to the interaction of molecules. Molecule interactions redistribute momentum locally near the selected molecule, but this is insufficient to smooth out large-scale momentum fluctuations in the fluid. As previously mentioned, a fluid viscosity is largely dependent on its structure, specifically the nature of its short-range order. Short-range order exists at scales ranging from one to a few nanometers. Thus, while in gases viscosity is formed on scales of the order of the mean free path, in liquids it occurs on mesoscales much larger than the characteristic size of the molecules.

Attempts to take into account the fluid structure when calculating its viscosity can be found in the literature. This approach has been successfully applied to simulate the viscosity of non-Newtonian fluids [80]. Similar approaches are known for simulating the viscosity of simple fluids. For example, van Velzen and colleagues proposed a correlation [115] that defines the dependence of the viscosity coefficient of a fluid on temperature

$$\ln \mu = B^*\left(T^{-1} - T_0^{-1}\right), \tag{3.20}$$

where the B^* and T_0 parameters were determined by the structure of the fluid molecules.

To determine the dependence of a fluid viscosity coefficient on temperature, various empirical or semi-empirical formulas, such as (3.20), are commonly used. One typical formula is proposed by Andrade [5, 6]

$$\mu = Ae^{B/T}, \tag{3.21}$$

where A and B are some constants. In practice, the correlation proposed by Orrick and Erbar [115] is widely used to calculate the viscosity coefficient of a fluid

$$\ln \mu = A + BT^{-1} + \ln(\rho M), \tag{3.22}$$

where M is its molecular weight, the constants A and B for a number of fluids can be found in [115]. Some other semi-empirical correlations are also applied [115, 156]. However, qualitatively the nature of the dependence of the viscosity coefficient on temperature remains the same as in (3.21). Formula (3.22) also considers the physical properties of the fluid, namely, its molecular weight and density.

All of the formulas mentioned above indicate that the viscosity of the fluid decreases as the temperature increases. This is in line with the fact that the short-range order is destroyed as the temperature increases, first locally and then globally with the onset of the fluid–gas phase transition. Based on physical considerations, a decrease in viscosity with increasing temperature should be expected for nanofluids as well. There are many dozens of papers that have studied this topic, some of which are cited in the reviews mentioned above. Most studies on the dependence of viscosity on temperature in nanofluids indicate a decrease in viscosity with increasing temperature. The resulting viscosity–temperature relationships in all of these papers are quite typical. As an example, in Fig. 3.5a, the data displays the relationship between the viscosity coefficient and temperature for an ethylene glycol-based nanofluid with silicon dioxide nanoparticles [130]. It is seen that the viscosity of the fluids decreases as the temperature increases, regardless of the nanoparticle concentration.

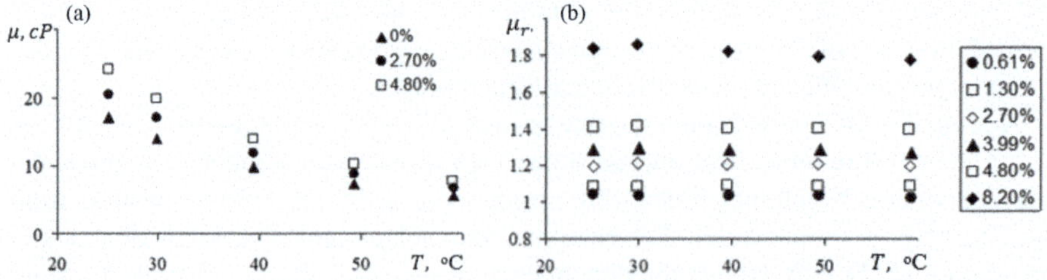

Fig. 3.5 Temperature dependence of viscosity coefficient of ethylene glycol-based nanofluid with SiO_2 particles of 28.3 nm (**a**), and relative viscosity coefficient of the same nanofluid (**b**) for different volume concentrations (**b**)

Quite a few correlations have been proposed to describe the temperature dependence of the nanofluid viscosity coefficient. However, these correlations are not universal and vary significantly with the concentration of nanoparticles, their material and size, as well as the viscosity of the base fluid. Therefore, it is important to understand how the relative viscosity coefficient of a nanofluid varies with temperature. The temperature dependence of μ_r for nanofluids of different concentrations, presented in Fig. 3.5a, is shown in Fig. 3.5 b. The relative viscosity coefficient remains constant with increasing temperature at low and moderate nanoparticle concentrations. Only at a concentration of 8.2% does it slightly decrease (by approximately 3%). A similar dependence was observed in [21, 100, 101], where nanofluids with TiO_2 particles in an ethylene glycol-based solution, SiO_2 particles in a water and ethylene glycol solution, and Al_2O_3 and CuO particles in a water-based solution were studied. Thus, at low concentrations of nanoparticles, the temperature dependence of the viscosity coefficient of nanofluids is primarily determined by that of the base fluid.

It was mentioned above that the nonuniversality of the formulas describing the viscosity of nanofluids can be explained by the effect of particle material in addition to the particle size. This dependence is difficult to establish experimentally for obvious reasons. Therefore, first it was shown on model nanofluids using the MD simulation method [131, 132], and will be discussed further in this chapter. This was only demonstrated experimentally two years later [137]. Conducting such an experiment is challenging because it requires studying the viscosity of at least two nanofluids based on the same basic fluid, but with nanoparticles of different materials and the same average size. It is important that the particle size distribution of both materials is similar.

Figure 3.6a compares the relative viscosity coefficients of water-based nanofluids containing SiO_2 (square symbols) and Al_2O_3 (triangle symbols) particles with a size of 100 nm [137]. The difference in viscosity coefficients between these two nanofluids is significant and increases with increasing particle volume concentration. These viscosity coefficients surpass the values predicted by formulas (3.13) (dashed line) and (3.15) (solid line). Figure 3.6 b displays the relative viscosity coefficients of two other nanofluids, both based on water with 150 nm Al_2O_3 and TiO_2 particles. The differences between the two nanofluids are significant. The comparison of Figs. 3.6a, b reveals that the viscosity of the nanofluid with Al_2O_3 particles slightly decreases with increasing particle size.

Finally, another important point needs to be discussed. In all classical theories of coarse dispersed fluids, the viscosity excess is determined solely by the particle concentration and is independent of the base fluid. This means that the relative viscosity coefficient in all dispersed fluids is universal and independent of the viscosity of the base fluid. However, this is not the case for nanofluids. A nanofluid with a higher viscosity of the base fluid will also have a higher relative viscosity coefficient. This is illustrated by Fig. 3.7, which represents the dependence of the relative viscosity coefficient on particle size for two water-based nanofluids (bottom symbols) and ethylene glycol (top symbols) with SiO_2 and Al_2O_3 nanoparticles. The volume concentration of particles was the same in all cases and equal to 1%.

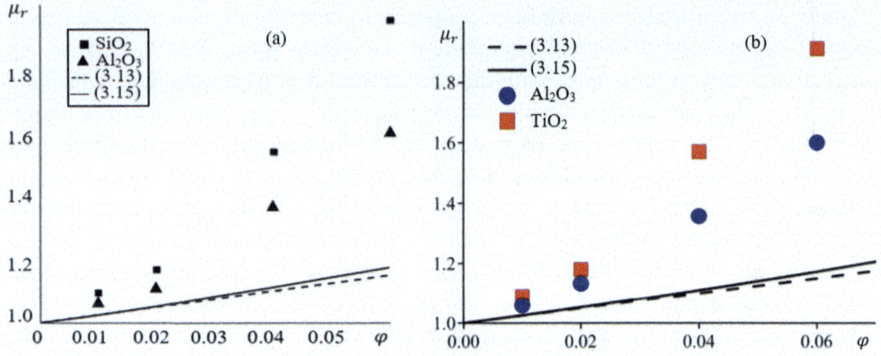

Fig. 3.6 Dependence of the relative viscosity coefficient of water-based nanofluids on the volume concentration of nanoparticles: **a** SiO_2 and Al_2O_3 of 100 nm size; **b** Al_2O_3 and TiO_2 of 150 nm size. The dashed line corresponds to formula (3.13), and the solid line to (3.15)

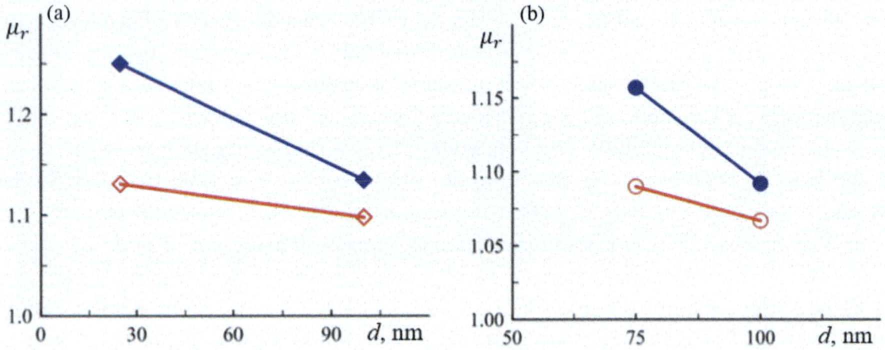

Fig. 3.7 Dependence of relative viscosity coefficient on particle size for nanofluids based on water (bottom symbols) and ethylene glycol (top symbols) with SiO_2 (**a**) and Al_2O_3 (**b**) nanoparticles

The viscosity of the nanofluids decreased with increasing particle size in all cases. However, the degree of viscosity decrease depends on the type of base fluid. The decrease is greater as the viscosity of the base fluid increases.

3.3.2 *MD Simulation of Nanofluid Viscosity*

The experimental data presented above indicate that nanofluids are non-standard dispersed fluids. Their viscosity cannot be described by classical theories proposed by A. Einstein, G.K. Batchelor, and others. It has been demonstrated that the viscosity of nanofluids depends not only on the concentration of nanoparticles but also on (i) their size, (ii) material, and (iii) the properties of the base fluid. This, in particular, indicates that the Einstein–Batchelor hydrodynamic mechanism of increasing the viscosity does not apply to nanofluid. Nanoparticles have dimensions of the order of a physically infinitesimal size for the base fluid, that is, in the metric of the base fluid they are material points. Material point does not generate the hydrodynamical macroscopic disturbance. Of course, earlier in Chap. 2 it was noted that microfluctuations of hydrodynamic fields of the base liquid were formed around the nanoparticles. They have a significant impact on the transport processes in the nanofluid, but it is a different type of influence.

Today, there are several known attempts to explain the non-standard behavior of the viscosity of nanofluids. One of the first was undertaken by Masoumi et al. [85]. They proposed to represent the effective viscosity coefficient of a nanofluid as the sum of the viscosity coefficient of the base fluid and the so-called apparent viscosity coefficient μ_{app}, presenting the effect of Brownian motion, so that

$$\mu = \mu_b + \mu_{app} = \mu_b + \frac{\rho_p V_p d_p^2}{72 C l_{mp}}, \tag{3.23}$$

where l_{mp} is the distance between the centers of nanoparticles determined by their volume concentration, and C is some correction factor, which the authors relate to the total shear stress acting on the nanoparticle. Here it is worth recalling again that the nanoparticle is a material point in the metric of the carrier nanofluid.

The model (3.23) has not received much attention in the literature, despite repeated attempts to relate various transport processes to Brownian motion (see Chap. 4). Additionally, simple estimates suggest that the appearance of the second term in (3.23) cannot account for the significant enhancement of viscosity, observed in the experiments with nanofluids. Therefore, the only method that can aid in understanding additional mechanisms of viscosity in nanofluids is the MD method. The use of the MD method to study transport processes in nanofluids has been recognized for quite a long time. However, early works primarily focused on simulating thermal conductivity. A review by Jabbari et al. [54] outlines the main stages of applying the MD method for viscosity simulation. The objective of most studies on nanofluid viscosity is to examine the relationship between the effective viscosity coefficient and nanoparticle concentration and size. Studies have shown that the enhancement viscosity coefficient of nanofluids is significantly higher than that predicted by theories for coarse dispersed fluids [46, 47, 60, 72–77, 112, 131–133]. However, there has not been a sufficiently systematic study of the relationship between viscosity and particle size. Most studies indicate that the viscosity coefficient of nanofluids increases as the particle size decreases.

Most authors attribute the significant increase in nanofluid viscosity to the formation of an adsorbed dense layer of base fluid molecules around the nanoparticles. Thus, Lu et al. [76] noted that the formation of an adsorbed dense layer of base fluid molecules around nanoparticles increases their equivalent radius and decreases their mobility, resulting in increased nanofluid viscosity. Ju [60] has demonstrated that the thickness of the adsorbed layer is dependent on the size of the nanoparticles. It is maximal for small nanoparticles up to 3 nm and then decreases slightly remaining almost the same for nanoparticles larger than 6 nm. The adsorption constant in water-based nanofluids with gold particles remains almost unchanged for particle sizes larger than 10 nm. Heyhat et al. [46] note that the formation of adsorbed layers significantly alters the density of the base fluid, which can affect the effective volume concentration of nanoparticles.

On the other hand, Lu et al. [76] demonstrated that the density of the adsorbed layer of the base fluid increases with the energy parameter ε_{ij} of the Lennard–Jones potential (2.24), resulting in an increase in the viscosity coefficient of the nanofluid. This suggests indirectly that the viscosity of the nanofluid should dependent on the intensity of the interaction between the molecules of the base fluid and the nanoparticles. At the same time, on the one hand, this indicates that the viscosity of the nanofluid should depend on the type of base fluid, and, on the other hand, it should depend on the material of the nanoparticles.

In [73], the significant excess of nanofluid viscosity is also attributed to the interaction between nanoparticles and molecules of the base fluid. The dependence on the nanoparticle material is also noted in the works [47, 131, 132]. Finally, it is worth noting that several studies have investigated the change in nanofluid viscosity with increasing temperature. It has been demonstrated that the viscosity decreases with increasing temperature, with the increase in the effect of particle size being particularly noticeable at lower temperature.

Nevertheless, in most studies, the mechanisms of momentum transfer processes in nanofluids, which determine their viscosity, have not been analyzed. To address this issue, it is necessary to study the damping character of the corresponding correlation functions for different nanofluids with different base fluids and nanoparticles. This is done below, following the works [124, 131–133]. The fluctuation–dissipation theorem (2.3) for the effective viscosity coefficient of nanofluid has the following form

$$\mu = \frac{V}{3kT}\int_0^{\tau_p} \langle \mathbf{J}(0) : \mathbf{J}(t)\rangle dt = \frac{V}{\mathrm{k}T}\int_0^{\tau_p} \chi_\mu(0,t)\mathrm{d}t, \tag{3.24}$$

where the microscopic stress tensor for the considered binary mixture has the form

$$\mathbf{J} = \frac{1}{V}\sum_{\alpha=1}^{2}\sum_{i=1}^{N_\alpha}\left(m_\alpha \mathbf{v}_i\mathbf{v}_j + \frac{1}{2}\sum_{\alpha=1,2 i\neq j}^{N_\alpha} \mathbf{F}_{ij}\mathbf{r}_{ij}\right). \tag{3.25}$$

Here the index $\alpha = 1$ refers to the carrier fluid molecules, and 2 refers to the nanoparticles, m_i is the mass of the corresponding particle (molecule or nanoparticle), V is the volume of the system, τ_p is the time to reach the plateau value, $\mathbf{F}_{ij}$ is the force acting on the particle (molecule or nanoparticle), $\mathbf{r}_{ij}$ is the radius-vector between the centers of the interacting particles. N_1 and N_2 are the number of molecules of the carrier fluid and nanoparticles, respectively. The angular brackets in (1) denote averaging over the equilibrium ensemble.

In [124, 131–133], the viscosity of argon-based nanofluids with lithium and aluminum particles and water-based nanofluids with copper particles were studied. In all cases, the interaction between the base fluid molecules was described using the Lennard–Jones potential (2.24). The interaction between the nanoparticles and the base fluid molecules was described using the RK potential (2.27), while the interaction between the nanoparticles with each other was described using the RKI potential (2.28).

The dependence of the relative viscosity coefficient of argon-based nanofluids with lithium and aluminum particles of 2 nm diameter on the particle volume concentration is shown in Fig. 3.8. Circle symbols represent the lithium particles and triangle symbols refer to the aluminum particles. The dashed line corresponds to Batchelor's formula (3.15). The viscosity of both nanofluids is significantly higher than predicted by classical theories which do not describe the viscosity of nanofluids even at low nanoparticle concentrations. Furthermore, Fig. 3.8 illustrates a significant difference in the viscosity coefficients of these two nanofluids indicating a particle material dependency. In all cases, the dependence of the effective viscosity coefficient on the particle concentration is determined by a formula similar to (3.18). For nanofluid with lithium particles the corresponding coefficients are: $a = 3.20$ and $b = 25.38$. The calculations by formula (3.18) with these coefficients are shown in Fig. 3.8 by the dash-dotted line. Coefficients $a = 3.25$ and $b = 13.06$ (as shown by the dashed line in Fig. 3.8) correspond to nanofluids with aluminum particles. This indicates that the viscosity of nanofluids is dependent on the material of the nanoparticles, and this difference becomes more pronounced as the concentration of nanoparticles increases.

To answer the question about the dependence of the viscosity coefficient on the size of nanoparticles, calculations were carried out for nanofluids with lithium nanoparticles with a diameter of 1, 2, 4 nm at volume concentrations of nanoparticles 2 and

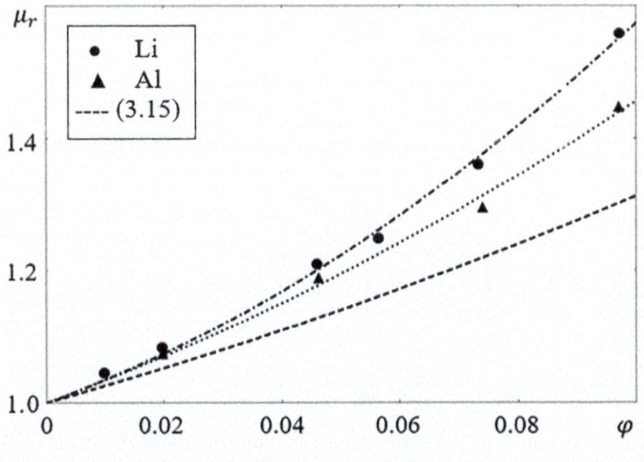

Fig. 3.8 Dependence of the relative viscosity coefficient of argon-based nanofluids with lithium and aluminum particles of 2 nm diameter on their volume concentration

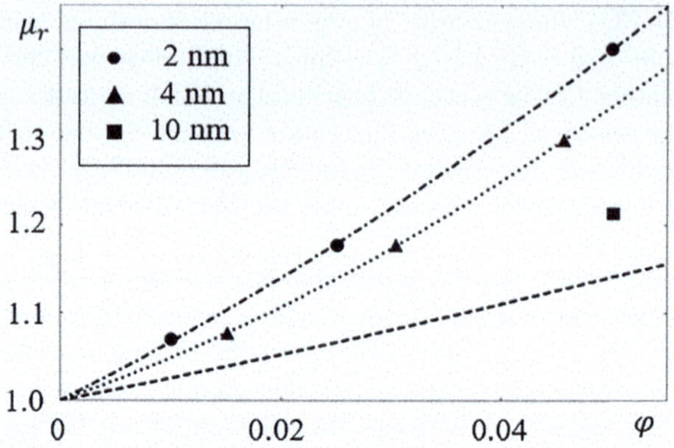

Fig. 3.9 Dependence of the relative viscosity coefficient of the water-based nanofluids on the volume concentration of the Cu nanoparticles of different size

4%. It was found [132] that, just as in experiments with a decrease in the size of nanoparticles, the viscosity coefficient increases.

A similar dependence was obtained for the viscosity coefficient of water-based nanofluid with copper nanoparticles [124]. The corresponding data, illustrating the dependence of the relative viscosity coefficient on the volume concentration of particles, are presented in Fig. 3.9. Here, round, triangular, and square symbols correspond to nanoparticles with sizes of 2, 4, and 10 nm, respectively, the dashed line corresponds to the formula (3.15). The viscosity coefficient decreases with increasing the nanoparticle size.

To understand the reason for the high viscosity of the nanofluids, it is necessary to systematically analyze the different contributions in (3.24). According to (3.24) and (3.25), the viscosity coefficient of the nanofluid is comprised of several different components

$$\mu = \mu_k + \mu_p + \mu_{kp}, \tag{3.26}$$

where the first term describes the kinetic contribution that determines the momentum transport

$$\mu_k = \frac{1}{kTV} \sum_{\alpha,\beta=1,2} \int_0^{\tau_p} \left\langle \sum_{i=1}^{N_\alpha} m_\alpha \dot{x}_i \dot{y}_i(t) \sum_{j=1}^{N_\beta} m_\beta \dot{x}_j \dot{y}_j(t+t_1) \right\rangle dt_1, \tag{3.27a}$$

the second term defines the potential contribution

$$\begin{aligned}\mu_p &= \mu_{p1} + \mu_{p2} + \mu_{p12} + \left(\mu_{p1p2} + \mu_{p2p1}\right) \\ &\quad + \left(\mu_{p1p12} + \mu_{p12p1}\right) + \left(\mu_{p2p12} + \mu_{p12p2}\right) \\ &= \frac{1}{4kTV} \sum_{\alpha,\beta=1,2} \int_0^{\tau_p} \langle \sum_{i\neq j}^{N_\alpha} F_{x,ij} y_{ij}(t) \sum_{k\neq l}^{N_\beta} F_{x,kl} y_{kl}(t+t_1) \rangle \mathrm{d}t_1 \\ &\quad + \frac{1}{kTV} \int_0^{\tau_p} \langle \sum_{i=1}^{N_1} \sum_{j=1}^{N_2} F_{x,ij} y_{ij}(t) \sum_{k=1}^{N_1} \sum_{l=1}^{N_2} F_{x,kl} y_{kl}(t+t_1) \rangle \mathrm{d}t_1 \\ &\quad + \frac{1}{2kTV} \sum_{\alpha} \int_0^{\tau_p} \langle \sum_{i\neq j}^{N_\alpha} F_{x,ij} y_{ij}(t) \sum_{k=1}^{N_1} \sum_{l=1}^{N_2} F_{x,kl} y_{kl}(t+t_1) \\ &\quad + \sum_{k=1}^{N_1} \sum_{l=1}^{N_2} F_{x,kl} y_{kl}(t) \sum_{i\neq j}^{N_\alpha} F_{x,ij} y_{ij}(t+t_1) \rangle \mathrm{d}t_1, \end{aligned} \tag{3.27b}$$

and the third one describes the cross effects

$$\begin{aligned}\mu_{kp} &= \frac{1}{kTV} \left[\sum_{\alpha,\beta=1,2} \int_0^{\tau_p} \left\langle \sum_{i=1}^{N_\alpha} m_\alpha \dot{x}_i \dot{y}_i(t) \frac{1}{2} \sum_{k\neq l}^{N_\beta} F_{x,kl} y_{kl}(t+t_1) \right\rangle \mathrm{d}t_1 \right. \\ &\quad \left. + \sum_{\alpha,\beta=1,2} \int_0^{\tau_p} \left\langle \sum_{i=1}^{N_\alpha} m_\alpha \dot{x}_i \dot{y}_i(t) \sum_{k=1}^{N_1} \sum_{l=1}^{N_2} F_{x,kl} y_{kl}(t+t_1) \right\rangle \mathrm{d}t_1 \right]. \end{aligned} \tag{3.27c}$$

A comparison of the contributions of each of these components to the viscosity coefficient (3.24) of water and a water-based nanofluid containing 2 nm diameter copper particles is presented in Table 3.1. Here the volume concentration of nanoparticles is 5%, pressure and temperature of water and nanofluid are identical. The sixth row shows the potential contribution to the viscosity coefficient of water molecules interacting with each other (μ_{p1}), the seventh row shows the contribution of water molecules interacting with nanoparticles (μ_{p12}), and the eighth row shows the contribution related to the correlation of intermolecular forces and interaction forces of water molecules with nanoparticles $(\mu_{p1p12} + \mu_{p12p1})$. The contributions of nanoparticles interacting with each other and the cross-correlation of intermolecular forces and nanoparticle interaction forces are negligible.

The analysis of Table 3.1 reveals that the total viscosity coefficient of the nanofluid is significantly higher than that of the carrier fluid. However, the kinetic contributions μ_k are lower. This decrease is due to the reduced number density of molecules in the nanofluid compared to the carrier fluid. However, the contribution of the nanoparticles to the kinetic energy is small due to their larger volume compared to that of the molecules provided that the average kinetic energy of both molecules

Table 3.1 Contributions of different components to the viscosity of nanofluids (μPa s)

	Water	Water–Cu
μ	863.1	1193.5
μ_k	7.08	6.90
μ_p	850.1	1181.7
μ_{kp}	5.94	4.89
μ_{p1}	850.1	967.9
μ_{p12}		128.4
$(\mu_{p1,p12} + \mu_{p12,p1})$		85.3

and nanoparticles remains conserved. It is important to note that due to the high fluid density, the contribution of kinetic processes to momentum transfer (viscosity) is generally small. For water, this contribution is approximately 0.8%, while for nanofluids, it is about 0.6%. The potential contribution of molecules in nanofluids (sixth row of the Table 3.1) slightly increases. The role of cross terms (fifth row of the Table) is also small, with these contributions being even smaller than the kinetic contributions for both water and nanofluids. The increase in viscosity of the nanofluid compared to the base fluid (water) is primarily due to the interaction between nanoparticles and molecules, as well as correlations between molecule–molecule and molecule–nanoparticle interactions. Figure 3.10 provides a qualitative illustration of typical molecule–nanoparticle interactions in nanofluids. Figure 3.10a displays the direct interactions between molecules and nanoparticle, while Fig. 3.10b shows typical pair correlations of interactions between molecules and nanoparticle. Figure 3.10c illustrates collective correlations of interactions between molecules of the adsorbed layer and nanoparticle, and finally, Fig. 3.10d demonstrates exchange interaction between nanoparticles through the base fluid molecules.

Thus, the increase in viscosity coefficient is mainly due to interactions between nanoparticles and molecules, as well as correlations between molecule–molecule and nanoparticle–molecule interactions, as shown schematically in Figs. 3.10a–c. How do these contributions depend on the nanoparticle size? It is easy to evaluate this qualitatively. Let consider particles with radii R and r. Assuming $R > r$, the number of particles in a unit volume at a given concentration is inversely proportional to the cubes of their radii: $N/n = r^3/R^3$. On the other hand, the total scattering cross-sections of the carrier fluid molecules with both sort of particles are equal to $\Omega = N\pi R^2$ and $\omega = n\pi r^2$. Therefore, the ratio of these cross-sections, which

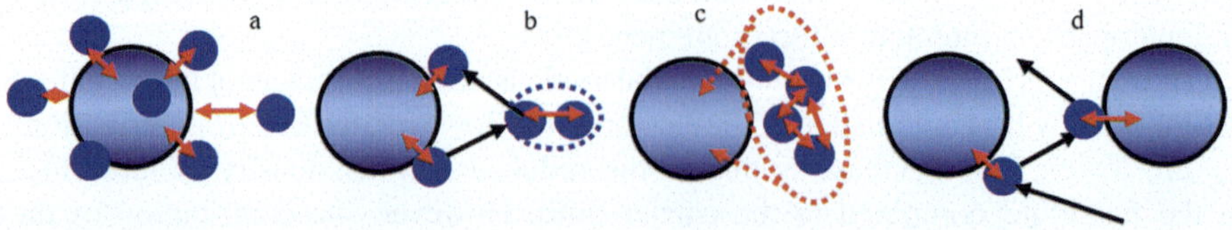

Fig. 3.10 Schematics of typical interactions between molecules and nanoparticles in nanofluid

exactly determines the magnitude of the potential contributions discussed, is inversely proportional to their radii

$$\frac{\Omega}{\omega} = \frac{r}{R}. \tag{3.28}$$

According to this estimate, the contribution to momentum transfer in nanofluids increases as the size of the nanoparticles decreases. Consequently, the viscosity of the nanofluid also increases with decreasing nanoparticle size, given a constant volume concentration. However, this is not the sole factor affecting nanofluid viscosity with respect to particle size. In fluids, exhibiting short-range order and near-environment molecules being in quasi-correlated states, one of the primary mechanisms of momentum transfer, as already mentioned, is the local destruction of short-range order (and the creation of new correlations). The presence of nanoparticles significantly alters the structure of the base fluid. An increase in fluid viscosity means an increase in fluid structuring. The fluid's structure can be accurately described by the radial distribution function $g_2(r)$. The radial distribution function of water exhibits a quasi-periodic damped structure that is typical of fluids (see Fig. 3.2). The introduction of nanoparticles further structures the fluid. Short-range order in nanofluids has been extensively studied by the author with colleagues. The studies conducted in [119, 122] examined a model argon-based nanofluid containing aluminum nanoparticles with diameters of 2 and 4 nm at a volume concentration of up to 5%. Another study [133] investigated a water-based nanofluid with 2% copper particles with diameter of 2 nm.

Figure 3.11 compares the radial distribution functions of water molecules and water molecules around a nanoparticle for a water-based nanofluid with copper particles (where r is measured in nm). The radial distribution functions show significant differences, indicating that the addition of nanoparticles further structures the fluid. The first peak of the radial distribution function in nanofluid increases by a factor of 2.5. Subsequent peaks also increase, and the length of the short-range order region in the nanofluid increases several times. At the same time, the volume of this region increases by approximately thirty times compared to pure water.

The structuring of carrier fluid molecules around nanoparticles increases the viscosity of the nanofluid. Moreover, as it was demonstrated in paper [122], the character of structuring depends significantly on the interparticle interaction parameters. Specifically, short-range order intensifies with an increase in the parameter ε of the intermolecular potential (2.24). As this parameter increases, the magnitude and extent of all maxima in the radial distribution function also increase. Structuring of the base fluid in nanofluid depends on the parameters of the interparticle interaction potential, which explains the effect of nanoparticle material on the effective viscosity coefficient of nanofluid.

It is clear that structuring of the base fluid takes place around particles of any size. However, the thickness of the structured fluid film is practically independent of the particle radius. Then a simple estimation shows that at a constant concentration of particles the nanofluid will be the more structured the smaller the size of the

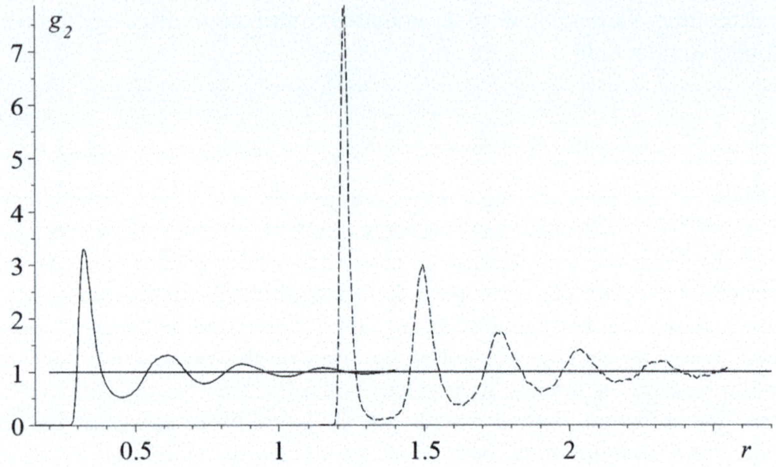

Fig. 3.11 Radial distribution functions of water molecules in pure water (solid line) and in nanofluids (dashed line)

particles dispersed in it. For example, in a nanofluid with a concentration of 5% and a particle size of 2 nm, the average distance between particles is approximately 2.5 nm. Consequently, the short-range order regions of molecules for neighboring particles overlap, resulting in a structured base fluid. As a result, the viscosity of the nanofluid increases as the particle size decreases.

3.4 Rheology of Nanofluids

Until now, the viscosity of Newtonian nanofluids which describe the stress tensor using relation (3.1) have been analyzed. However, an important property of nanofluids is that they can have non-Newtonian rheology. A defining feature of such fluids is the non-linear relationship between the stress tensor and the shear rate. When attempting to give this dependence the usual Newtonian form, the resulting effective viscosity coefficient becomes a function of shear rate. Additionally, it has been discovered that many fluids have a yield stress, meaning that their flow only begins once a certain shear rate is reached.

Many rheological relations have been derived experimentally for non-Newtonian fluids, relating the stress tensor to the shear rate [16, 22, 80, 86, 147]. The most commonly studied nanofluids are based on water, ethylene glycol, machine oils, etc. These are all ordinary Newtonian fluids. In general, their rheology does not change when different surfactants are used, since the concentrations of the latter are relatively low. However, in some cases, nanofluids have non-Newtonian rheology even at relatively low particle concentrations. In principle, the study of the rheology of nanofluids is an obligatory attribute accompanying a measure of their viscosity, since viscosity itself can only be discussed with respect to a Newtonian fluid. In fairly comprehensive reviews [50, 78, 109, 160] devoted to the study of the viscosity of nanofluids, relatively few works are noted in which their non-Newtonian rheology has been revealed. This is due to the fact that specialized rheometers were rarely used for this purpose, and in many works, when viscosity was studied using rotational viscometers the shear rate was not varied over a wide enough range, which is absolutely necessary. Nevertheless, to date there are already many works where the non-Newtonian character of the rheology of certain nanofluids has been reliably established (e.g. the review by Sharma et al. [143]), although the base fluids in these studies were Newtonian. For example, in [153], while studying a water-based nanofluid with TiO_2 particles, the nanofluid was found to be Newtonian at low particle volume concentrations. However, at particle concentrations above 0.1%, the nanofluid became thixotropic. Similar results were obtained later in [44].

Non-Newtonian behavior of ethylene glycol-based nanofluids was observed in [108], when increasing the concentration of Fe_2O_3 particles, as well as in [67], when increasing the concentration of CuO particles. The rheology of water-based nanofluids was also found to be non-Newtonian with increasing concentration of Fe_3O_4 nanoparticles [48].

The change in the rheology of nanofluids with increasing nanoparticle concentration in general should not be surprising. Similar behavior has been observed previously in studies of the rheology of conventional coarse dispersed fluids. The non-Newtonian behavior of coarse dispersed fluids with increasing concentration of dispersed particles is common [94, 107]. This behavior is due to several reasons, in particular, structuring of the dispersed fluid. However, the viscosity of nanofluids is highly dependent on particle size. Nevertheless, in most works where the change in the rheology of nanofluids was studied with an increase in the concentration of particles, their sizes practically did not vary. Furthermore, in a number of papers it is simply difficult to understand the size of the particles used in the nanofluid. Thus, there are virtually no works that have studied the effect of nanoparticle size on the rheology of nanofluids.

Since the viscosity of nanofluids depends in addition to particle concentration and size, on temperature and particle material, it is obvious that the same characteristics can also influence nanofluid rheology. Finally, it should be noted that in the preparation of stable nanofluids, various surfactants are often used to stabilize them. Surfactants usually have a rather complex structure, and their addition to the base fluid can change by itself structure and consequently rheology of base fluid. Thus, it is obvious that the possible change in the rheology of nanofluids prepared using Newtonian base fluids can be determined by a large number of parameters. Therefore, a systematic study of the effect of each parameter is necessary to determine the possible change in the rheology of nanofluids prepared using Newtonian base fluids. Experimental study of each parameter is challenging because it requires studying the rheology of a large number of nanofluids while varying only one parameter at a time. Today, however, it is already possible to identify key factors that can affect the rheology of nanofluids. This will be demonstrated further using the example of nanofluids based on water, ethylene glycol, and machine oil containing particles of various metal oxides (CuO, SiO_2, TiO_2, Al_2O_3, ZrO_2, Fe_2O_3, Fe_3O_4) (CuO, SiO_2, TiO_2, Al_2O_3, ZrO_2, Fe_2O_3, Fe_3O_4) and nanodiamonds [90, 91, 120]. Surfactants were not used in these examples. The particle volume concentrations ranged from 0.1 to 8%, and their sizes changed from 5 to 150 nm. The measurement data are shown in Tables 3.2, 3.3 and 3.4. Here, the first column indicates the base fluid (BF), the second column indicates the nanoparticle material (Np), the third column indicates the characteristic particle size (*d*), the fourth column indicates their volume concentration in percent (φ), and the fifth column indicates the rheology (Rh)—either Newtonian (N) or non-Newtonian (n–N). The base fluid in Table 3.2 was water (W), in Table 3.3, ethylene glycol (EG), and in Table 3.4, machine oil (O).

Table 3.2 Rheology of water-based nanofluids

BF	Np	*d* (nm)	φ (%)	Rh	BF	Np	*d* (nm)	φ (%)	Rh
W	SiO_2	10	0.25–2	N	W	TiO_2	71	2	N
W	SiO_2	16	0.25–2	N	W	TiO_2	100	2	N
W	SiO_2	25	0.25–6	N	W	TiO_2	150	1–6	N
W	SiO_2	100	0.25–2	N	W	ZrO_2	44	2–8	N
W	Al_2O_3	50	1–2	N	W	ZrO_2	105	2–8	N
W	Al_2O_3	75	1–2	N	W	C	5	0.25–2	n–N
W	Al_2O_3	100	1–2	N					
W	Al_2O_3	150	1–6	N					

Table 3.3 Rheology of ethylene glycol-based nanofluids

BF	Np	*d*, nm	φ (%)	Rh	BF	Np	*d* (nm)	φ (%)	Rh
EG	Al_2O_3	11	1–2	n–N	EG	SiO_2	25	1	N
EG	Al_2O_3	50	1–2	n–N	EG	SiO_2	100	1	N
EG	Al_2O_3	75	1	N	EG	TiO_2	150	0.25	N
EG	Al_2O_3	75	2	n–N	EG	TiO_2	150	2–6	n–N
EG	Al_2O_3	100	1	N	EG	CuO	100	0.1–1	N
EG	Al_2O_3	100	2	n–N	EG	C	5	0.25–2	N
EG	Al_2O_3	150	0.25	N					
EG	Al_2O_3	150	2–6	n–N					

Table 3.4 Rheology of machine oil-based nanofluids

BF	Np	*d* (nm)	φ (%)	Rh	BF	Np	*d* (nm)	φ (%)	Rh
O	Fe_3O_4	50	1	N	O	Fe_2O_3	18	1	N
O	Fe_3O_4	100	1	N	O	Fe_2O_3	50	1	N
O	C	5	0.25–2	n–N					

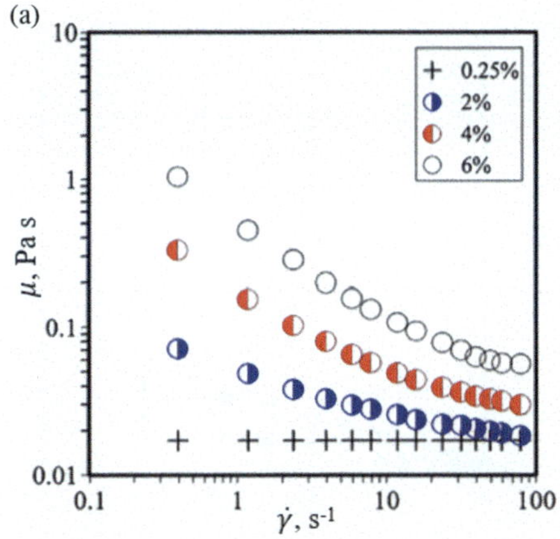

Fig. 3.12 Dependence of viscosity coefficient on shear rate for ethylene glycol-based nanofluids with Al_2O_3 (**a**) and TiO_2 (**b**) particles 150 nm in size

Tables 3.2, 3.3 and 3.4 present over three dozen nanofluids with particle concentrations and sizes varying within a rather broad range. Non-Newtonian behavior was observed in approximately 25% of all measurements. As an example, let consider data for ethylene glycol-based nanofluids with aluminum and titanium oxide particles of identical size (150 $\pm$ 5 nm) at four nanoparticle volume concentrations, equal to 0.25, 2, 4, and 6% at shear rates $\dot{\gamma}$ varying from 0.4 to 80 s^{-1}. The dependence of viscosity coefficient on shear rate is shown in Fig. 3.12. Pure ethylene glycol is a Newtonian fluid. The above nanofluids have exhibited also Newtonian behavior at low nanoparticle concentrations, $\varphi \leq 0.25$. However, as the particle concentration increases, nanofluids exhibit a pronounced non-Newtonian behavior in both cases, which further increases with higher nanoparticle concentration. Therefore, it can be concluded that an observable transition from Newtonian to non-Newtonian rheology of nanofluids (if such transition is a case) occurs with increasing nanoparticle concentration.

Similar behavior was observed in works [39, 88, 89] when studying the rheology of water-based nanofluids with CuO particles. In this study, a xanthan gum biopolymer was used as a surfactant, but its weight concentration did not exceed 0.03%. The nanofluid exhibited Newtonian behavior at a particle concentration of 0.25%, but became non-Newtonian as the concentration increased. Thus, the same conclusion was drawn that the transition from Newtonian to non-Newtonian rheological behavior of nanofluids occurs (again, if such transition is a case) with increasing nanoparticle concentration.

To determine the character of the nanofluid rheology, it is necessary to analyze the relationship between its stress tensor τ and the shear rate. Figure 3.13 illustrates this relationship for the ethylene glycol-based nanofluids under discussion. The character of these relationships varies qualitatively. At large particle concentrations (4 and 6%), a yield stress was observed, while at the lower concentration (2%), no such stress was detected within the measurement accuracy.

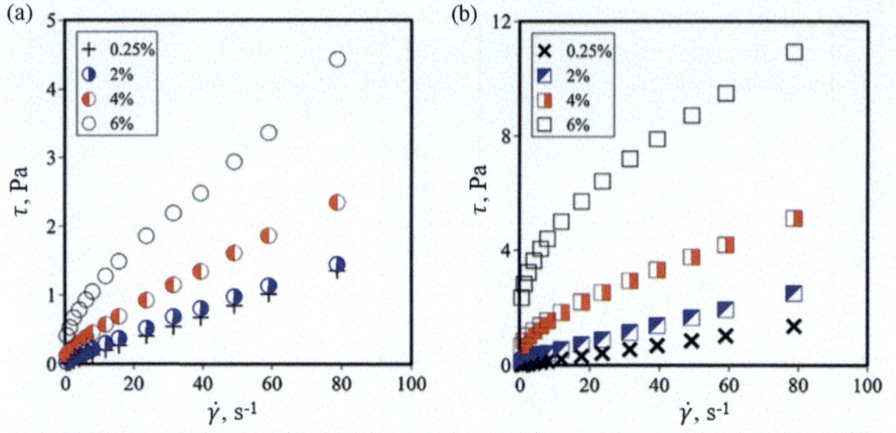

Fig. 3.13 Dependence of stress tensor on shear rate for ethylene glycol-based nanofluids with Al_2O_3 (**a**) and TiO_2 (**b**) particles 150 nm in size

The dependencies, presented in Figs. 3.12 and 3.13 show that nanofluids exhibit pseudoplastic or viscoplastic behavior. This means that their viscosity decreases with increasing shear rate (in viscoplastic fluids, starting from some stress limit value) [22, 59]. The power-law fluid model is typically used to describe pseudoplastic fluids

$$\mu = k_v \dot{\gamma}^{n-1}, \tag{3.29}$$

while viscoplastic fluids are described by the Hershel–Bulkly model

$$\mu_f = \left(\tau_y + k_v \dot{\gamma}^n\right)\dot{\gamma}^{-1}. \tag{3.30}$$

Here τ_y is the yield stress of the viscoplastic fluid, n is the fluid index, and k_v is the consistency parameter.

In paper [90], the authors constructed correlations for two nanofluids using experimental data. The results have shown that the nanofluids with a particle concentration of 2% were accurately described by model (3.29). For higher concentrations, model (3.30) was found to be more appropriate. The data obtained are summarized in Tables 3.5 and 3.6. As the particle concentration increases, the fluid index decreases, while the consistency factor and yield stress increase.

Table 3.5 Consistency factor and fluid index of nanofluids with Al_2O_3 and TiO_2 nanoparticles at a particle concentration $\varphi = 2\%$ (power-law)

Al_2O_3		TiO_2	
k_v, mPa s^n	n	k_v (mPa)·s^n	n
52.0	0.745	256	0.542

Table 3.6 Yield stress and rheological parameters of nanofluids with Al_2O_3 and TiO_2 nanoparticles at particle concentration $\varphi = 2\%$

φ (%)	Al_2O_3			TiO_2		
	τ_y (mPa)	k_ν (mPa) s^n	n	τ_y, mPa	k_ν(mPa) s^n	n
4	121.8	57.67	0.835	615.5	212.4	0.696
6	4382.3	140.0	0.746	2044	695.0	0.582

It has already been noted above that the viscosity of nanofluids depends not only on the concentration and size of the particles, but also on their material (see Sect. 3.3). It turned out that the rheology of nanofluids is also significantly influenced by the particle material. In the described experiments the particles of titanium and aluminum oxides with an average size of 150 nm were used. Their rheological characteristics at the same particle concentration turned out to be significantly different. These differences are illustrated in Fig. 3.14, where the dependences of viscosity and shear stress on shear rate are compared for a particle concentration of 6%. The viscosity, as well as the shear stress of the nanofluid with TiO_2 particles is higher than that of the nanofluid with Al_2O_3 particles.

As previously demonstrated, the viscosity of nanofluids is dependent on both nanoparticle concentration and size. Section 3.3.2 provides an explanation for this phenomenon, noting that the fluid surrounding the nanoparticles is significantly more structured than the fluid in the bulk. Therefore, if the nanoparticles are sufficiently small, the entire base fluid will become structured. Thus, it is reasonable to anticipate that alterations in the rheology of the nanofluid will be influenced by the size of the nanoparticles. To verify this assumption, a series of measurements were conducted on ethylene glycol-based nanofluids with a volume concentration of 1%, containing Al_2O_3 particles of varying sizes equal to 11, 50, 75, and 100 nm (see

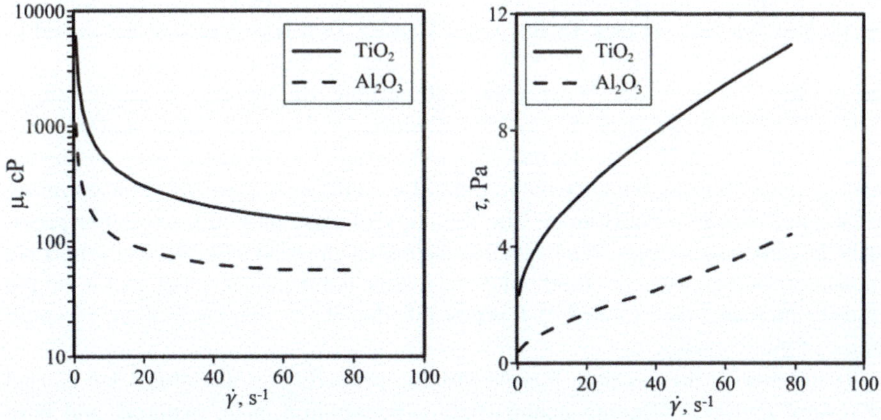

Fig. 3.14 Dependence of viscosity coefficient and shear stress on shear rate for ethylene glycol-based nanofluids with Al_2O_3 and TiO_2 particles 150 nm in size at $\varphi = 6\%$

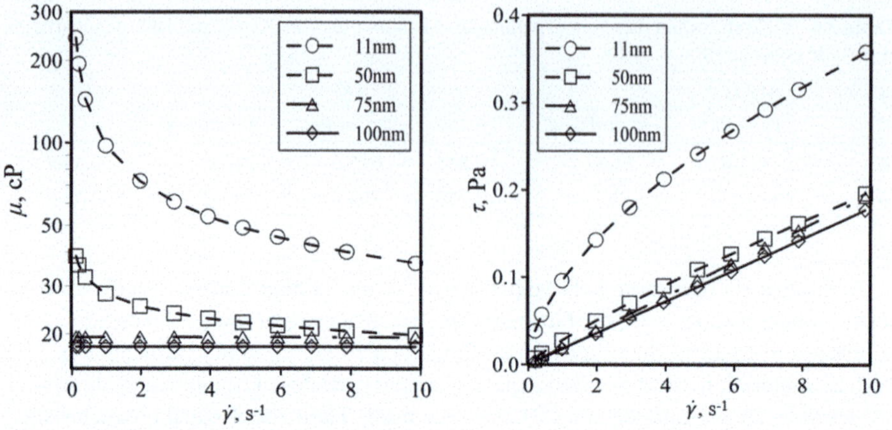

Fig. 3.15 Dependence of viscosity and shear stress of ethylene glycol-based nanofluids containing Al_2O_3 particles on shear rate

Table 3.3). The resulting dependence of the viscosity coefficient on the shear rate is presented in Fig. 3.15. It was observed that nanofluids with larger particles (75 and 100 nm) exhibit Newtonian behavior, while those with smaller particles exhibit non-Newtonian behavior. The rheology character is determined by the dependence of stresses on shear rate. The relevant data are also shown in Fig. 3.15. The rheology of the considered nanofluids with small particle sizes is well described by the power-law fluid model (3.29), where $k_\nu = 96.8$ mPa·s^n, $n = 0.572$ for nanofluids with 11 nm particles and $k_\nu = 28.1$ mPa·s^n; $n = 0.850$ for nanofluids with 50 nm particles.

It is evident from general considerations that the rheology of the nanofluid is affected by the base fluid. The reasons for this influence have yet to be understood. The experiments presented in this book investigate the impact of the base fluid on nanofluids containing diamond particles. The base fluids used were water, ethylene glycol, and machine oil (see Tables 3.3, 3.4 and 3.5). It was revealed that the ethylene glycol-based nanofluid was Newtonian, whereas the nanofluids based on water and machine oil were non-Newtonian. The particle volume concentration ranged from 0.25 to 2% in all cases. The results obtained for the water-based nanofluid are presented in Fig. 3.16. The degree of non-Newtonicity increased with increasing particle concentration, but even the nanofluid with the lowest particle concentration investigated was non-Newtonian. The rheology of this nanofluid, like other previously described nanofluids, can be well described by the model (3.29). As the particle concentration increases, the fluid index decreases and the consistency factor, on the contrary, increases (see Table 3.7). It is possible that at low shear rates there are small values of yield stress.

To summarize this section, non-Newtonian properties in nanofluids are caused by several factors, including nanoparticle concentration, size, material, and base fluid properties. As nanoparticle concentration increases, the rheology changes from Newtonian to non-Newtonian. All investigated non-Newtonian nanofluids were

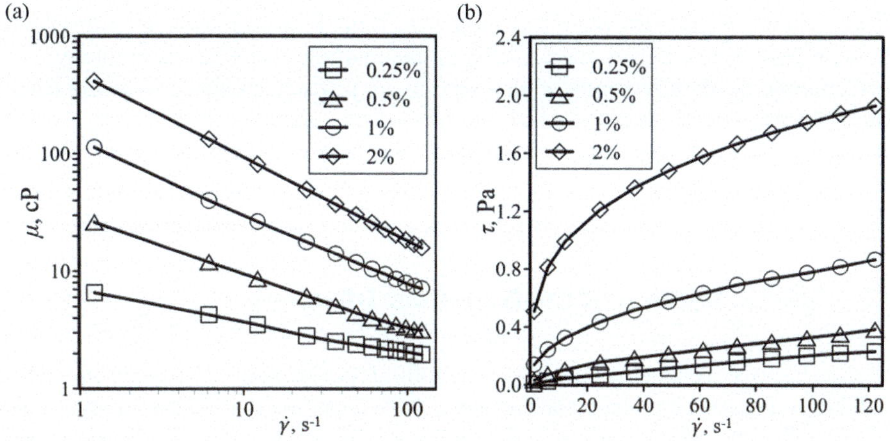

Fig. 3.16 Dependence of the viscosity coefficient (**a**) and shear stress (**b**) of water-based nanofluid with nanodiamond particles on shear rate

Table 3.7 Parameters of model (3.29) for water-based nanofluids with nanodiamond particles

$\varphi(\%)$	n	$k_\nu(\text{mPa})\cdot s^{-n}$
0.25	0.84	6.37
0.50	0.55	24.0
1.00	0.42	101
2.00	0.29	479

either pseudoplastic or viscoplastic. In the first case the rheology of nanofluids is accurately described by the power-law fluid model (3.29). In this case, with increasing particle concentration, the fluid index decreases while the consistency parameter increases. Viscoplastic fluids have a yield stress, and their rheology is described by the Herschel–Bulkley model (3.30). Here again, as the particle concentration increases, the fluid index decreases and the consistency parameter increases. It is important to note that the rheological model of a non-Newtonian nanofluid may change as the particle concentration increases. Specifically, a pseudoplastic fluid may become viscoplastic. This is generally consistent with existing literature data.

The rheology of nanofluid depends significantly on the particle size which is an important parameter. It has been revealed that in some cases, the transition from Newtonian to non-Newtonian behavior occurs at a specific volume concentration when decreasing the nanoparticle size. Apparently, it can be argued that nanofluids with small enough particles will always have a non-Newtonian rheology unless their concentrations are too low. Finally, nanofluids containing particles of the same size but related to different materials exhibit different rheological properties.

The rheology of nanofluids is significantly influenced by the properties of the base fluid. It is important to note that many nanofluids are prepared using surfactants, which should be considered as an additional component of the base fluid. The addition of surfactants can alter the rheology of the base fluid. However, even in cases where the base fluid is Newtonian, the addition of nanoparticles can alter its rheology.

Certainly, the data presented in this section are just statements of observed facts. The experimental data discussed did not reveal the reasons for the change in the rheology of nanofluids. Some of the questions that arise in this regard will be answered in the next section.

3.5 MD Simulation of the Nanofluid Rheology

As shown above, a change in rheology in nanofluids has been recorded experimentally and the factors contributing to such a change have been revealed. However, the cause of these changes in rheology remains unknown. In a typical molecular fluid, the viscosity coefficient is determined by the FDT and is a result of thermal (equilibrium) molecular fluctuations in the system's dynamic characteristics. In principle, it is not related to the presence of gradients of hydrodynamic quantities in the system. The nature of the stress tensor change in the fluid is related to its structure, which is characterized by the presence of short-range order. In the presence of sufficiently high shear rates, the character of the short-range order must change. Therefore, a change in the rheology of a fluid implies a change in its structure.

The change in the rheology of the nanofluid occurs as the shear rate increases. It is important to distinguish between the processes of forming the viscosity coefficient in equilibrium and changing the stress tensor in the presence of shear rate. The viscosity coefficient in a closed system is formed due to momentum equalization through thermal molecular fluctuations, while in an open system, the process is different. Two simultaneous processes occur within the open system: on the one hand, continuous momentum pumping and, on the other hand, equalization by molecular mechanisms. Therefore, a change in rheology due to shear rate occurs in an open system with a constant exchange of momentum between the subsystem and its surroundings. Rheology cannot change in a closed system.

Transport coefficients, which were discussed in Chap. 2, are a characteristic of a closed molecular system. However, there is one very important and challenging nuance here. It is important to note that transport coefficients describe the alignment of macroscopic inhomogeneities of density, momentum, and energy that arise in such a system due to the presence of macroscopic inhomogeneities of these characteristics formed for some reason. Such inhomogeneities may be small, and in this case the linear constitutive relations of the form (2.1) or (3.1) take place. However, in strongly nonequilibrium systems, these inhomogeneities can be significant, causing the constitutive relations to become nonlinear [20, 35, 116, 117, 118]. The emergence of such nonlinear transport processes is only possible when the system deviates significantly from local equilibrium, and they are generally characterized by large gradients of macroscopic observables. The observed change in rheology of nanofluids with increasing shear rate is different in nature. It occurs at relatively small velocity gradients, but with a constant influx of momentum into the subsystem under consideration, which forms the corresponding constitutive relations. In a certain sense, the system can be considered as a continuous succession of weakly nonequilibrium states (gradients may be small). Changes in states are related to the action of the surrounding, which forms the corresponding influx of momentum (and, certainly, energy) at the boundary.

The MD method described in Sect. 2.2 cannot be used to calculate transport coefficients in this situation. It is essential to consider the non-closed nature of the simulated system explicitly. The dynamics of the system's molecules or particles are no longer described by Newton's Eqs. (2.20) because the influence of the surrounding must be taken into account (for example, transfer of momentum at increasing shear rate). In principle, this problem can be solved in two simple ways. Either the surrounding can also be defined by a system of molecules, or the law of mutual interaction between the molecules of the simulated system and the surrounding can be defined. However, defining the interaction between the simulated system and the surrounding will considerably increase computational efforts and the conservation laws will not be fulfilled in the simulated subsystem, making it open. Additionally, Newton's equations will contain source terms that describe this interaction. In mechanics, possible positions and velocities of particles in the system are determined by introducing bonds, which can be either holonomic or non-holonomic. The corresponding system of equations describing the motion of molecules can be represented as [31]

$$m_i \ddot{\boldsymbol{r}}_i = \boldsymbol{F}_i - \alpha_i \boldsymbol{n}, \tag{3.31}$$

where m_i, $\boldsymbol{r}_i$ are the mass and coordinate of molecule i, $\boldsymbol{F}_i$ is the force acting on the molecule, $\boldsymbol{n}$ is the normal to the hypersurface bounded by the superimposed bonds, α_i are Lagrange multipliers.

The explicit form of Eqs. (3.31) is determined by the bonds imposed on the system. In particular, for simulating dynamics of a system of molecules in an external field of shear rate $\dot{\gamma} = \partial u/\partial y$ the system of equations takes the form.

$$\dot{\boldsymbol{r}}_i = \boldsymbol{v}_i + \dot{\boldsymbol{r}}_i \cdot \nabla \boldsymbol{u}, \quad \dot{\boldsymbol{v}}_i = \frac{\boldsymbol{F}_i}{m} - \boldsymbol{v}_i \cdot \nabla \boldsymbol{u} - \alpha_i \boldsymbol{v}_i. \tag{3.32}$$

These are the so-called SLLOD equations [31] or their analogs. However, SLLOD equations appear to be more critical in terms of the effect of the thermostat on the simulation results at high shear rates [152].

An alternative to the approach (3.32) in the MD method is the simulation of Couette flow between two walls (sometimes Poiseuille flow is used, [57], usually planar [8, 31, 57, 113], called the non-equilibrium MD (NEMD) method. The disadvantage of this approach for obtaining fluid rheology data in the bulk is obvious, the walls may affect on the system state. To obtain data, characteristic of fluid bulk, the channel height must be large enough. At the same time, it should be noted that the use of a cylindrical Couette flow corresponds exactly to the method of measuring the viscosity coefficient in a rotational viscometer. In this case, the viscosity coefficient is determined at a constant rotational velocity of the cylinder (or cylinders), which corresponds to stationary conditions and simulates a closed system with a created velocity gradient. This approach has been frequently used to study fluid rheology. A similar method involves using sliding periodic boundaries [68] at the boundary of the simulation cell.

The NEMD method has been used to study the rheology of nanofluids, although there are few publications on this topic. One of the first studies was conducted by McPhie et al. [87]. They considered a model nanocolloid system with a maximum particle size of 4.03 relative to the size of molecules of the carrier fluid, and a particle mass of 50. The interaction between molecules was described by the Lennard–Jones potential, while that of the particles was described by the shifted Lennard–Jones potential with a solid core [61]. The volume concentration of particles was 4, 10, 20, and 50%. An increase in particle concentration resulted in non-Newtonian behavior. It was discovered that as the shear rate increases, the relationship between viscosity and shear rate no longer depends on particle mass and becomes somewhat universal. Additionally, the authors attempted to describe the non-Newtonian behavior of the colloidal fluid. They state that the relative viscosity coefficient can be described by the equation $\mu_r = 1 - a\dot{\gamma}^2$, where the viscosity coefficient is normalized to the limiting value obtained at zero shear rate. The shear rate is generally the second invariant of the strain rate tensor, $\dot{\gamma} = \sqrt{\boldsymbol{D} : \boldsymbol{D}}$.

Very interesting data were obtained in a series of studies conducted by Grest et al. [37] and In 't Veld [51, 52]. The particle size ranged from 5σ to 20σ, where σ represents the characteristic size of the base fluid molecules. The authors considered very dense suspensions with volume concentrations ranging from 7 to 48%. They believed that such a system could simulate colloidal gels and glasses by changing the interaction force between nanoparticles and molecules of the base fluid. The interaction of molecules with nanoparticles was described by a potential [26, 32] At the same time, the strength of the interparticle interaction varied, in particular, the Hamacher constant [42]. It was found that the minimum size of nanoparticles at which the base liquid can be considered as a continuous medium is determined not only by their size, but also by the force of interaction of nanoparticles with the molecules of the carrier liquid. This, in particular, indirectly indicates the dependence of rheology on both the material of the nanoparticles and the type of base liquid. Finally, the rheological nature of the suspensions is demonstrated by the dependence of the radial distribution function of nanoparticles on their concentration. The change in the structure of the pseudofluid of nanoparticles with an increase in their concentration is clearly visible. Certainly, it can be challenging to interpret these data for typical nanofluids due to their extremely high particle concentrations. Additionally, it is worth noting that the authors were surprised by another conclusion. The NEMD simulation results have indicated that suspension rheology was solely dependent on nanoparticle concentration but was independent of their size.

The influence of nanofluid structure, in particular, translational order on the rheology of model nanofluids has been studied in [64]. The particles in these suspensions are spherical and their interaction is described by a potential [110, 111]. This potential depends significantly on the particle concentration, and the depth of the potential well increases with increasing concentration. The volume concentration of particles was varied from 0 to 40%. It was found that at low shear rates, Newtonian behavior is observed at all particle concentrations except the maximum concentration. The onset of pseudoplastic behavior is threshold in nature. The shear rates at which it begins to exhibit increase with increasing particle concentration. It is further shown that $\mu_r \sim a\dot{\gamma}^{-\alpha}$, where $\alpha = 0.8$. This is consistent with previously known data where α ranged from 0.5 to 1 [15, 158]. Additionally, it is noted that this relationship exhibits a certain self-similarity and is independent of particle concentration.

The effect of particle shape on the rheology of nanofluid was studied by Heine et al. [45]. The authors investigated nanofluids with particles shaped as spheres, rods, plates, and jacks, based on the Lennard–Jones potential. The volume concentration of nanoparticles was 0.075%. The characteristic size of the base fluid molecules was σ, and in nanoparticles it varied from 5.6σ to 22.5σ for spheres, and to 12σ for rods and plates, respectively. The study utilized the NEMD method with the Müller-Plathe approach [96]. The results have shown that the viscosity of the nanofluids was not significantly affected by the shape of the nanoparticles.

Wu et al. [163] conducted a study on the rheology of a machine oil-based nanofluid (PAO_4) with 1 nm graphene particles. The weight concentration of graphene was 3.06, 5.94, and 11.21%. The authors aimed to investigate the viscosity of the nanofluid using the NEMD method, which they found to converge much faster. The study revealed that the nanofluid was non-Newtonian and pseudoplastic. As the shear rate increased, the viscosity of the nanofluid decreased. The RAO 4 base fluid also exhibited non-Newtonian behavior.

Finally, it is worth mentioning the recent work by Devarajan et al. [27]. The size of the nanoparticles in this study was five times larger than that of the molecules, and their volume concentration ranged from 10 to 49%. The authors systematically studied the change in the pseudofluid structure of the nanoparticles by means of a radial distribution function. The results have shown that short-range order is formed with increasing concentration of nanoparticles. Moreover, there is a transition from a system in which there is virtually no short-range order (as in an ideal gas) to a densely packed system. At high particle concentrations, Newtonian behavior of nanofluids is not observed, but at lower concentrations it is observed at very low shear rates. The non-Newtonian behavior of nanofluids suggests the presence of viscoelastic properties in the studied fluids, which were also systematically analyzed.

The above brief review allows summarizing the main results obtained by the NEMD method for studying the rheology of nanofluids. The suspensions studied so far cannot be considered ordinary nanofluids due to their high volume concentrations of dispersed particles (up to 50%) and even higher weight concentrations. It has been reliably found that the suspensions exhibit pseudoplastic rheology with increasing shear rate. Additionally, it has been indicated that particle size does not affect the rheology. Thus, a number of questions remained unanswered. Let us formulate these questions. (i) How does the rheology of nanofluids changes with increasing particle volume concentration (the concentration should range from 1 to 5–9%), particle size, and material? (ii) Does the dependence of viscosity coefficient on shear rate change with increasing particle concentration, size, and material? (iii) How the base fluid affects the rheology of the nanofluid? (iv) How does the structure of nanofluid and nanoparticle pseudogas change with increasing shear rate? Some of these questions will be answered below based on our papers [121, 125, 126].

The rheology of benzene-based nanofluids containing aluminum and copper particles is considered. The size of the nanoparticles ranged from 3 to 6 nm. The maximum volume concentration of the particles was 6.11%. The NEMD method was utilized to simulate Couette flow between two plates. The simulation cell was a rectangular channel with molecules and nanoparticles of the simulated nanofluid contained between two walls. The corresponding screenshots are presented in Fig. 3.17. Periodic boundary conditions were imposed in the directions not bounded by walls. The cell length was chosen so that the results did not change when it was increased. Thus, correlations that may have been caused by the finiteness of the cell size, including possible self-interaction of nanoparticles were excluded. The walls of the channel were simulated using a set of copper atoms placed at the lattice nodes, with the atoms of the walls being stationary with respect to each other. The walls were given constant velocities in different directions. The required calculations were performed upon establishing a linear profile of the average velocity of molecules in the channel. The number of base fluid molecules varied from 10,000 to 250,000 while number of particles ranged from 10 to 60.

Since the aim of this work was to study the rheology of nanofluids in bulk, it was necessary to exclude the effect of walls on the calculation results. To do this, calculations were performed at different channel heights ranging from 5.9 to 27.5 nm. The results have shown that in latter case the effect of walls on the rheology data was practically absent. Therefore, all the data presented below correspond to this channel height.

The interaction of fluid molecules with each other and molecules with wall atoms was described by the truncated Lennard–Jones potential (2.24). The interaction between nanoparticles and base fluid molecules was described by the RK potential (2.27), while the interaction between nanoparticles was described by the RKI potential (2.28). Table 3.8 summarizes the parameters of the potential (2.24) for all molecules and atoms.

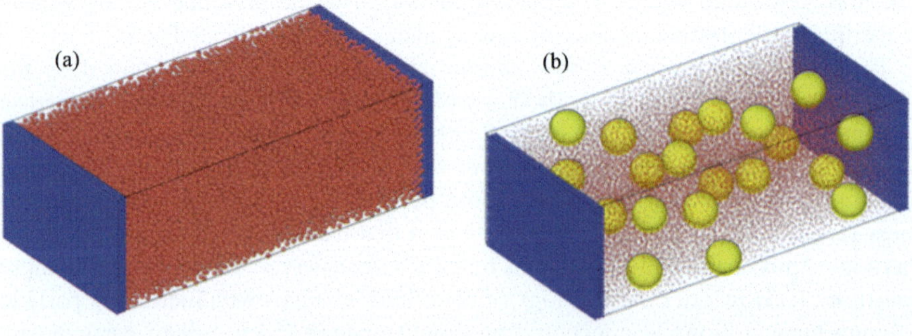

Fig. 3.17 The screenshot of the simulation cell with the base fluid (**a**) and with nanofluid (**b**)

Table 3.8 Lennard–Jones potential constants (2.24)

	ε/k (K)	σ (Å)		ε/k (K)	σ (Å)
C_6H_6	498.0	5.012	Al	857.6	2.56
Cu	1247.0	2.28	C	27.7	3.416

The simulations were conducted using the LAMMPS software package [150]. The Langevin thermostat was utilized, and its parameters were varied without affecting the simulation results. Benzene was simulated at atmospheric pressure and temperatures of 25 and 50 °C. The pressure in the channel was determined in its central zone, which occupies approximately one-third of the channel volume, using the virial theorem

$$p = \frac{2}{3V}\sum_{i=1}^{N}\left[\frac{\langle p_i^2\rangle}{2m} + \frac{1}{6V}\sum_{j\neq i}^{N}\langle \boldsymbol{r}_{ij}\cdot \boldsymbol{F}_{ij}\rangle\right]. \tag{3.33}$$

The viscosity coefficient was determined from the relation

$$\mu = \tau/\dot{\gamma}, \tag{3.34}$$

where the shear rate was determined from the velocity gradient using a pre-calculated velocity profile. By varying the velocity of the wall, one can gather information on the relationship between the viscosity coefficient and the shear rate.

The first question to be answered is simple: how does the rheology of nanofluid change with nanoparticle concentration? As an example, Fig. 3.18a shows the dependence of the viscosity coefficient on the shear rate of a benzene-based nanofluid with 3 nm diameter copper particles. The curves correspond to volume concentrations φ = 0, 0.02, 0.004, and 0.06 from bottom to top. Thus, the lower curve corresponds to pure benzene, and its viscosity coefficient at the minimum shear rate agrees well with the experimental value [157]. Benzene behaves as a Newtonian fluid up to relatively high shear rates, and its viscosity coefficient remains constant. However, when some critical shear rate $\dot{\gamma}_{cr}$ is reached, the viscosity coefficient begins to systematically decrease. Benzene becomes a pseudoplastic liquid. The dependence of the benzene viscosity coefficient on shear rate can be well approximated by the following correlation in the entire range of studied shear rates

$$\mu_{bf}(\dot{\gamma}) = \mu_0 \exp\left(-4.4\frac{\dot{\gamma}}{\dot{\gamma}_{cr}}\right), \tag{3.35}$$

where μ_0 is the viscosity coefficient of benzene in the Newtonian flow regime, and the critical shear rate in this case is $2 \times 10^9 s^{-1}$. Certainly, it should be realized that the critical shear rate is always determined with some accuracy, which depends on the magnitude of the deviation of the viscosity coefficient from the Newtonian value. The dotted line for benzene in Fig. 3.18 corresponds to the correlation (3.35).

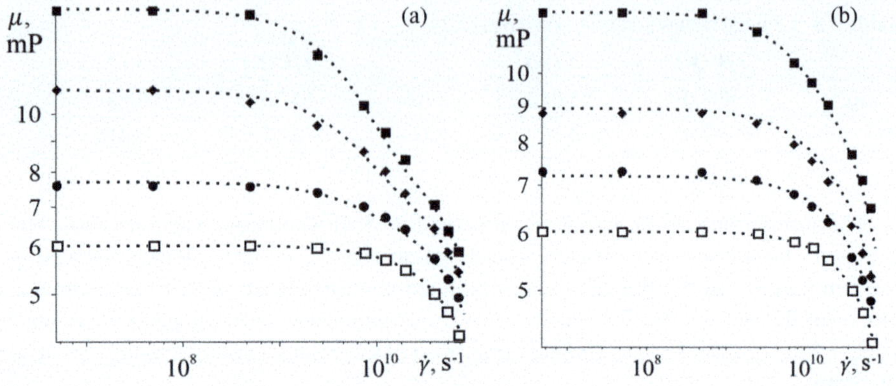

Fig. 3.18 Dependence of viscosity coefficient on shear rate of benzene-based nanofluid with 3 nm (**a**) and 6 nm (**b**) copper particles

As the shear rate increases, the viscosity of all nanofluids decreases, i.e. all liquids behave as pseudoplastic. However, as it is seen from the graphs, there is a Newtonian branch in the curves at all nanoparticle concentrations. In the Newtonian flow regime, the viscosity coefficient of the nanofluid is described by its dependence on the particle concentration (3.18), where $a = 8.38$ and $b = 300$. This value is considerably higher than the values predicted by classical theories for coarse dispersed fluids [14, 29]. The character of dependence (3.18) for nanofluids remains unchanged with increasing shear rate. However, it is important to note that the viscosity of nanofluids decreases with increasing shear rate, making it pseudoplastic. As a result, the coefficients a and b should decrease compared to the Newtonian flow regime, which is indeed the case. For instance, at $\dot{\gamma} = 8 \times 10^{10}\ \text{s}^{-1}$, these coefficients are $a = 4.7$ and $b = 10$, respectively.

A characteristic feature of the viscosity of nanofluids is that it is dependent on particle size. The previous section has experimentally shown how nanoparticle size affects the rheology of nanofluids with increasing shear rate. Figure 3.18b illustrates the viscosity dependence of a benzene-based nanofluid with 6 nm diameter copper particles on shear rate. Similar to Fig. 3.18a, the symbols represent volume concentrations of $\varphi = 0$, 0.022, 0.043, and 0.061 from bottom to top. The bottom symbols correspond to benzene. The corresponding exact viscosity values of the nanofluid with 3 and 6 nm particles (they are given in brackets) at different shear rates are given in Table 3.9. Comparison of the data in Fig. 3.18 and Table 3.9 shows that the viscosity of Newtonian nanofluids depends on the particle size and decreases with its increase. This statement is consistent with all available experimental data and MD simulations, as shown in the previous section. The viscosity coefficient is described by relation (3.18) with $a = 7.13$ and $b = 107$ in the Newtonian regime and $a = 2.87$ and $b = 25$ when $\dot{\gamma} = 8 \times 10^{10}\ \text{s}^{-1}$.

Analysis of Fig. 3.18 (also see Table 3.9) reveals that the change in rheology of the nanofluid is of a threshold nature in all cases. The corresponding critical shear

Table 3.9 Viscosity coefficients of benzene-based nanofluids with copper and aluminum particles at different shear rates and volume concentrations of nanoparticles

φ (%)	$\dot{\gamma} \times 10^{10}$ (s^{-1})									
	0.0005	0.005	0.05	0.25	0.75	1.25	2	4	5.5	7
Benzene	6.03	6.03	6.03	6.0	5.86	5.72	5.50	5.0	4.67	4.38
2.16 (Al, 3 nm)	6.62	6.62	6.55	6.4	6.22	6.01	5.66	5.05	4.71	4.38
2.16 (Cu, 3 nm)	7.58	7.58	7.55	7.38	7.10	6.81	6.51	5.72	5.29	4.93
4.3 (Cu, 3 nm)	10.94	10.94	10.43	9.53	8.63	8.11	7.52	6.43	5.86	5.43
6.11 (Al, 3 nm)	10.13	10.13	9.79	9.31	8.61	8.14	7.50	6.40	5.82	5.42
6.11 (Cu, 3 nm)	14.8	14.8	14.6	12.5	10.3	9.28	8.37	7.05	6.32	5.81
1.88 (Cu, 6 nm)	7.40	7.40	7.39	7.17	6.79	6.53	6.22	5.55	5.16	4.83
3.90 (Cu, 6 nm)	8.72	8.72	8.72	8.44	7.95	7.54	7.06	6.13	5.62	5.22
6.07 (Cu, 6 nm)	12.11	12.11	12.09	11.5	9.8	9.1	8.12	6.87	6.1	5.66

rate decreases with increasing particle concentration and decreasing particle size. Using the data provided in Table 3.9, a correlation can be constructed to determine the critical shear rate for the change in nanofluid rheology. This correlation has the following form

$$\dot{\gamma}_{cr}^* = 2 + 16.4(d^* - 3) + [60 + 15.6(6 - d^*)] \exp[-30(d^* - 1)\varphi]. \quad (3.36)$$

Here $\dot{\gamma}_{cr}^* = \dot{\gamma}/10^7$ is the normalized critical shear rate, and d^* is the dimensionless diameter of the nanoparticle ($d^* \geq 3$).

The viscosity coefficient of all nanofluids with spherical particles can be accurately described by a formula in the form of (3.18), except when their concentrations exceed 10%. As previously mentioned, this formula also applies to the viscosity of nanofluids in the pseudoplastic region. It is important to note that the coefficients in the formula are dependent on particle size and shear rate. Considering this fact, it is possible to construct a correlation that describes the viscosity of a benzene-based nanofluid with copper particles of any size. The correlation has the following form

$$\mu = \mu_{bf}(\dot{\gamma})[1 + a(\dot{\gamma}, d)\varphi + b(\dot{\gamma}, d)\varphi^2]. \quad (3.37)$$

Here $\mu_{bf}(\dot{\gamma})$ is the viscosity coefficient of the base fluid, determined by relation (3.35) and

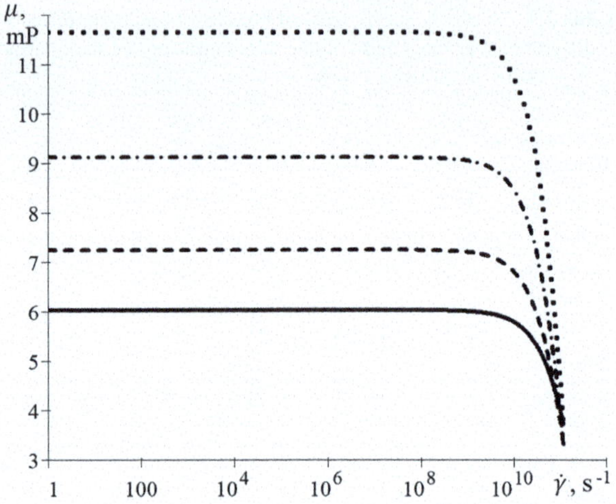

Fig. 3.19 Dependence of the viscosity coefficient of benzene-based nanofluid with 20 nm copper particles at different concentrations

$$a(\dot{\gamma}, d) = \left(8.35 - 4.647 \cdot 10^{-11}\dot{\gamma}\right) \exp\left[0.063\left(d^* - 3\right)\right] - \left(d^* - 3\right),$$

$$b(\dot{\gamma}, d) = \left[290 \exp\left(-1.5 \cdot 10^{-10}\dot{\gamma}\right) + 10\right] \exp\left[-0.413\left(d^* - 3\right)\right] + 8\left(d^* - 3\right).$$

The curves in Figs. 3.18a, b represent the approximations of (3.37), which are quite accurate. Another advantage of the correlation (3.37) is also that it enables the prediction of nanofluid rheology for particles of different sizes, generally speaking, of any size. For instance, Fig. 3.19 shows the rheology data of benzene-based nanofluids with 20 nm particles. The curves correspond to particle concentrations of 0, 2, 4, and 6% from bottom to top.

The data presented in Fig. 3.18 indicate that as the shear rate increases, the nanofluid becomes pseudoplastic. To describe the rheology of such a nanofluid most of the known experiments (see the previous section) use the power-law fluid model (3.29). In conventional nanofluids, the parameters of the model (3.29) change with increasing particle concentration, the fluid index decreases and the consistency factor usually increases. Additionally, experiments are typically conducted within a narrow range of shear rates, resulting in a significant decrease in viscosity at low shear rates. However, once a certain shear rate is reached, the viscosity of nanofluids stabilizes and no longer changes. The above described scenario (see Fig. 3.18 and Table 3.9) differs from this pattern, as the viscosity consistently decreases as the shear rate increases.

The rheology of the considered nanofluids can also be described by the model (3.29). However, the fluid index and consistency factor depend not only on the particle concentration but also on the shear rate. Figure 3.20 shows their evolution for nanofluids with 3 nm particles. As the particle concentration increases, the fluid index decreases and the consistency parameter increases. It is important to note that the consistency parameter is a dimensional quantity, and its dimensionality depends

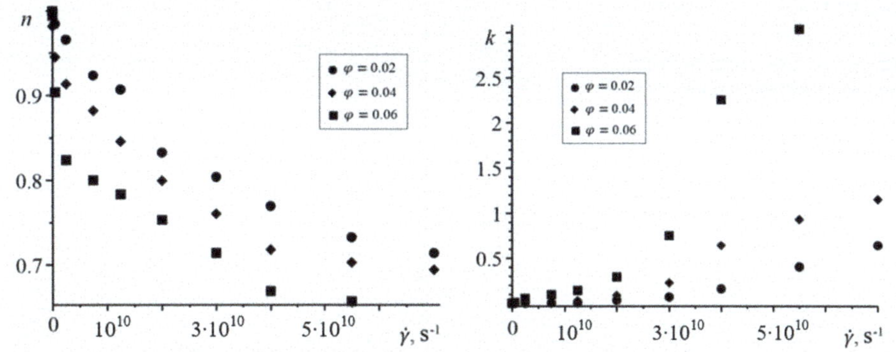

Fig. 3.20 Dependence of fluid index (left) and consistency parameter (right) on shear rate for nanofluids containing 3 nm copper nanoparticles at varying volume concentrations

on the current value of the fluid index. Therefore, the shape of the curves on the graphs will depend on the choice of measurement units. The values in Fig. 3.20 are given in the SI system.

As previously stated, the viscosity coefficient of nanofluids is dependent on the particle material. Figure 3.21 compares the viscosity–shear rate relationship of benzene-based nanofluids containing 3 nm diameter copper and aluminum particles. Copper particles are represented by dark symbols, while aluminum particles are represented by light symbols. Data are presented for two concentrations equal to 2 and 6% (from bottom to top). Indeed, both the viscosity and rheology of nanofluids are dependent on the nanoparticle material. Nanofluids with copper particles have a higher viscosity compared to those with aluminum particles. Additionally, the rheology of nanofluids with aluminum particles changes later with increasing shear rate. Physically, this is because copper and aluminum particles have significantly different interatomic interaction parameters. The size of atoms in copper and aluminum differs insignificantly, by about 10%. However, the interaction potential parameter (2.24) ε/k in copper is almost one and a half times higher than that in aluminum. This difference determines the character of the short-range order in the nanofluid and the structuring of the base fluid molecules in the vicinity of the nanoparticles.

The behavior of viscosity is primarily determined by the local change in the short-range order of the fluid in the presence of shear rate. The effect of shear rate on the radial distribution function of the base fluid needs to be studied first. This radial distribution function is presented in Fig. 3.22. A fragment of the first maximum of the radial distribution function ($\dot{\gamma} = 5 \times 10^6\ \text{s}^{-1}$) corresponds to the Newtonian flow regime, while the other two ($\dot{\gamma} = 8 \times 10^{10}$ and $2 \times 10^{11}\ \text{s}^{-1}$) correspond to the non-Newtonian regime (see Table 3.9 and Fig. 3.18). The inset shows the radial distribution function of benzene for the equilibrium system. Analysis of Fig. 3.22 leads to two conclusions. Firstly, the maxima of the radial distribution function systematically decrease with increasing shear rate. The exact values of the first and

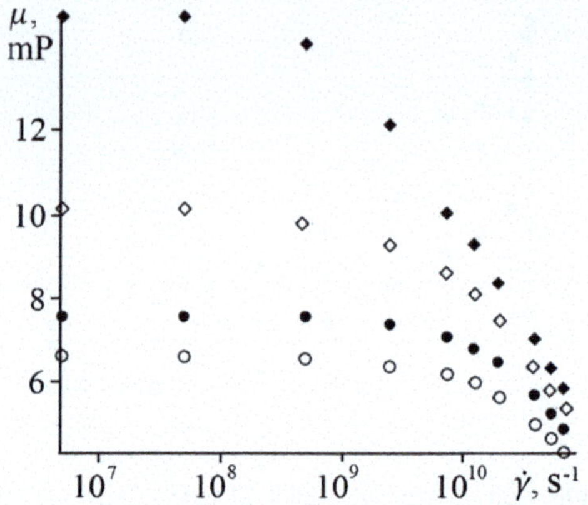

Fig. 3.21 Comparison of viscosity coefficients of nanofluids with copper (dark symbols) and aluminum (light symbols) nanoparticles with diameters of 3 nm

second maxima can be found in Table 3.10. Secondly, the distances between the maxima also decrease, indicating a weakening of the short-range order. It turned out that even a slight alteration in the topology of the radial distribution function of benzene can result in a significant decrease of almost 10% in its viscosity coefficient, as indicated in Table 3.10. Here it is shown how the maxima of the radial distribution function change with a small change in temperature.

The nanofluid is a binary system composed of base fluid molecules and nanoparticle pseudogas. The structure of the nanofluid is determined by the structure of

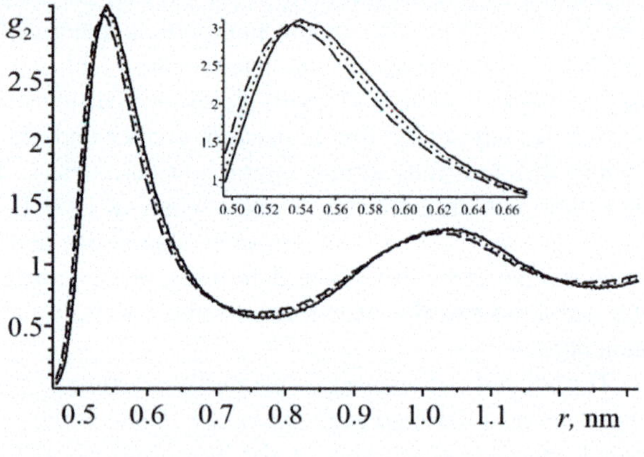

Fig. 3.22 Radial distribution functions of benzene molecules for varying shear rates. The inset shows the first maximum of this function

Table 3.10 RDF maxima of benzene for various shear rates at two temperatures

10^{10} s^{-1}	1st maximum, 298 K	2nd maximum, 298 K	1st maximum, 323 K	2nd maximum, 323 K
0.0005	3.11	1.293	2.93	1.269
8	3.06	1.276	2.88	1.261
20	3.03	1.259	2.83	1.234

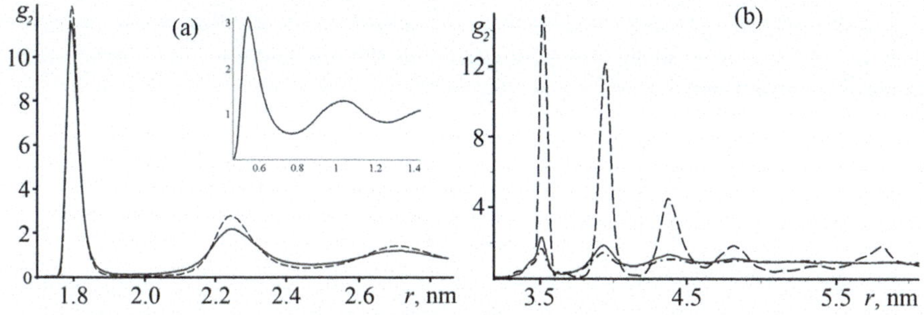

Fig. 3.23 Radial distribution functions of benzene molecules surrounding nanoparticles (**a**), and nanoparticles in benzene (**b**) at different shear rates

these components. Figure 3.23a shows the radial distribution function of the base fluid molecules around 3 nm copper nanoparticles. The volume concentration of the nanoparticles is 6.11%. The figure displays a dashed line for a Newtonian nanofluid and a solid line for a non-Newtonian nanofluid at $\dot{\gamma} = 1.7 \times 10^{11}\ \text{s}^{-1}$. The inset illustrates the g_2 function of benzene molecules in equilibrium. The first and second maxima of the nanofluid increase by several times compared to benzene in the Newtonian regime (see the inset), indicating a strong structuring of the base fluid molecules by nanoparticles. This is the reason for the sharp increase in the viscosity of the nanofluid compared to the base fluid. As the shear rate increases, the structuring of the base fluid molecules weakens, resulting in particular in a decrease in the viscosity of the nanofluid. The decrease in the maxima of the radial distribution function of benzene molecules appears to be minimal. From this point of view, it is useful to compare the maxima of the radial distribution function of benzene at temperatures of 25 and 50 °C, shown in Table 3.10. Here, the first maximum at the higher temperature decreases by almost 6%, and the second maximum decreases by about 5%. However, the viscosity coefficient of benzene decreases by nearly 30% when the temperature increases from 25 to 50 °C [157].

As noted above, the nanofluid structure is also determined by the structuring of the nanoparticle pseudogas. The corresponding radial nanoparticle–nanoparticle distribution function is shown in Fig. 3.23b. The dashed line represents the shear rate $\dot{\gamma} = 10^6 s^{-1}$, the solid line relates to $\dot{\gamma} = 4 \cdot 10^{10} s^{-1}$, and the dash-dotted line corresponds to $\dot{\gamma} = 8 \cdot 10^{10} s^{-1}$.. In the Newtonian flow regime ($\dot{\gamma} = 10^6 s^{-1}$) the first maximum is about 15 times the mean value. The subsequent maxima are due to the formation of layers of base fluid molecules around the nanoparticles. Of particular interest is a small maximum near $r = 6$ nm, which indicates the formation of a short-range order ensemble of nanoparticles. As the shear rate increases, the first maximum decreases by almost an order of magnitude. The maxima, caused by the structuring of the base fluid molecules around the nanoparticles, also decrease significantly. Finally, the radial distribution function of the nanoparticles is determined by their mean density, similar to an ideal gas.

The fluctuation–dissipation theorem (3.24) and (3.25) is generally valid for closed systems. In the case of shear flow (Couette flow), the microscopic stress tensor takes on the following form

$$\tilde{\boldsymbol{J}}_2 = \sum_{i}^{N} \left\{ m[\boldsymbol{v}_i - \boldsymbol{u}(\boldsymbol{r})][\boldsymbol{v}_i - \boldsymbol{u}(\boldsymbol{r})] + \frac{1}{2} \sum_{j \neq i}^{N} \boldsymbol{F}_{ij} \boldsymbol{r}_{ij} \right\}.$$

In case macroscopic flow velocity satisfies the condition $\boldsymbol{u}(\boldsymbol{r}) \ll \boldsymbol{v}_i$ the momentum equalization mechanism in the system that determines its viscosity does not change. The momentum equalization in the system is determined by thermal microfluctuations of its dynamic variables. In the opposite limiting case, when $\boldsymbol{u}(\boldsymbol{r}) \gg \boldsymbol{v}_i$, the system no longer perceives the impact of thermal fluctuations of particles on the momentum equalization in the system. The momentum equalization occurs due to an external force generating the shear velocity of the flow. As a result, the structure of fluid and nanofluid becomes more homogeneous, which leads to a decrease in viscosity. In addition, the difference between the viscosity of the base fluid and nanofluid decreases with increase in the shear rate. For example, according to Table 3.9, the viscosity of nanofluid containing 3 nm diameter copper particles at maximum concentration is 2.45 times higher than that of benzene in the Newtonian flow regime. In the pseudoplastic regime at $\dot{\gamma} = 2.5 \times 10^9\ \mathrm{s}^{-1}$ this excess is 2.08, while at $\dot{\gamma} = 7 \times 10^{10}\ \mathrm{s}^{-1}$ it is 1.32. As the concentration of nanoparticles decreases, the differences become even smaller. Additionally, at the minimum concentration of aluminum nanoparticles of the same size, the viscosity of the nanofluid becomes equal to that of benzene at a shear rate of $\dot{\gamma} = 7 \cdot 10^{10} s^{-1}$. Therefore, the system's ability to perceive the nanoparticle material and size decreases as the shear rate increases. In a sense, such a significant rise in the shear rate of the system is comparable to a sharp decrease in the system's temperature up to the onset of the superfluid regime.

It is possible to predict that with a further increase in shear rate, fluids, and nanofluids in particular, will reach a universal state with minimal viscosity. However, it is impossible to verify this assumption through direct calculation within the framework of the NEMD method used. The maximum shear velocity that can be achieved is about $5 \times 10^{11}\ \mathrm{s}^{-1}$. A further increase in wall velocity results in the slip effect when the velocity of walls is no longer transmitted to the fluid molecules. As a result, the actual shear rate in the liquid becomes significantly lower than shear rate determined by the velocity of the walls.

In the calculations presented in this book, nanofluids exhibited pseudoplastic behavior with increasing shear rate in all cases. However, in real experiments, viscoplastic nanofluids are also detected along with pseudoplastic nanofluids (as discussed in the previous section). Viscoplastic nanofluids begin to flow only upon reaching the yield stress, when the kinetic energy of the particle pseudogas exceeds the work of internal friction forces: $\left(\rho_p u_p^2/2\right) > \mu \dot{\gamma} S \delta x$, where ρ_p and u_p are, the density of the particle pseudogas and their macroscopic velocity, respectively, and S is the cross-sectional area of the channel. From here it is easy to find the corresponding critical concentration of particles.

3.6 Viscosity and Rheology of Nanofluids with CNT

The rheology and viscosity of nanofluids containing CNTs have been extensively studied for approximately two decades. Although hundreds of papers have been published on this topic, their comprehensive review is beyond the scope of this chapter. However, it is possible to summarize the current state of research and highlight the existing challenges and issues. To achieve this goal, it is reasonable to focus on papers published within the last decade and distinguish between those that studied nanofluids with MWCNTs and SWCNTs. The former category is more extensive, so we will begin with it.

Ponmozhi et al. [105] conducted a study on the rheology of nanofluids based on water with MWCNTs. The nanotubes had an outer diameter of 50–80 nm and a length of 10–20 μm, respectively. No surfactants were used, but the nanotubes were pretreated with sulfuric and nitric acids according to the method described in [164]. The MWCNTs were then grinded in a mill, so that their final length was unknown. The volume concentration ranged from 0.25 to 2.5%. The nanofluids were found to have non-Newtonian rheology. At low shear rates they were pseudoplastic fluids, however, they exhibited dilettante behavior starting from shear rates of 100 s^{-1}.

Halelfadl et al. [41] studied the viscosity of a water-based nanofluid with MWCNTs. SDBS was used as a surfactant. The nanotubes had an average diameter of 9.2 nm and a length of 1.5 μm. The volume concentration of MWCNTs ranged from 0.0055 to 0.55%. At the minimum CNT volume concentration, the nanofluid exhibited Newtonian behavior at all temperatures. However, at a volume concentration of MWCNTs equal to or greater than 0.278%, the nanofluids exhibited pseudoplastic behavior. The dependence of relative viscosity on shear rate in this case was well described by the Maron-Pierce model, which may indicate the effect of agglomeration. The relative viscosity coefficient was independent of temperature.

R. Sadri et al. [141] studied the viscosity and rheology of water-based nanofluids with MWCNTs and gum Arabic as a surfactant. The weight concentration of MWCNTs was 0.5%, and the surfactant concentration was 0.25%. The nanofluids exhibited viscoplastic behavior in all cases. At a constant shear rate, the viscosity increased with decreasing UST time. On the other hand, it decreased with increasing shear rate and the more so, the lower the temperature.

Dalkilic et al. [24] studied the viscosity of nanofluids containing MWCNTs with different base fluids, including water and a water–antifreeze mixture. In addition, CTAB and GA were used as surfactants. The volume concentrations of MWCNTs ranged from 0.5 to 2%. The measurements were performed using a capillary viscometer, but the rheology of these nanofluids was not analyzed. Meanwhile, it is clear that at such high concentrations of MWCNTs, nanofluids should have non-Newtonian rheology.

Ethelene glycol is commonly used as an initial base fluid in experiments, in addition to water. An example of a study on the rheology and viscosity of an EG-based nanofluid with MWCNTs of 60–100 nm diameter and 5–15 μm length can be found in [144]. The authors did not use surfactants, but pretreated the CNTs with nitric and sulfuric acids [30]. The weight concentration of CNTs ranged from 0.12 to 0.4%. At a CNT concentration of 0.2%, the nanofluid exhibited Newtonian behavior. Unfortunately, no data are given at higher concentrations. It is also indicated that with increasing temperature, the bonds of functional groups attached to the surface of CNTs, as a result of acid treatment, break down, leading to the destruction of the nanofluid.

The following example [3] is quite typical. The study examines the viscosity and rheology of a water-based nanofluid with MWCNTs. The nanofluid was provided by Nanocyl (Belgium) but the surfactant used is unknown. Moreover, the viscosity and rheology of the base fluid are not specified. The dispersed fluid used can hardly be considered a true nanofluid due to the large diameter and length of the MWCNTs (9 and 200 μm, respectively). The weight concentration of CNTs was 1%.

Leong et al. [69] studied several nanofluids with MWCNTs and different base fluids. In all cases, the authors used a mixture of 40% EG and 60% W as the initial base fluid with varying surfactants, including PVP, GA, and CTAB. The weight concentration of CNTs ranged from 0.01 to 0.5%. It is important to note that the MWCNTs had a special size, namely, diameters smaller than 30 nm and lengths of 10–30 nm. At low concentrations, the nanofluids exhibited Newtonian behavior, while at concentrations of 0.3% and above, they became non-Newtonian. However, due to the high shear rates, it was impossible to determine whether the nanofluid rheology was pseudoplastic or viscoelastic. The nanofluids that contained GA surfactant had the highest viscosity.

An interesting study was presented in the paper [103]. Here, the viscosity and rheology of water-based nanofluids with functionalized carboxyl groups of MWCNTs were studied. Two different surfactants, SDBS and Triton X-100, were used. The study examined the effect of aspect ratio for six different types, namely, 1250–3750, 500–3000, 200–400, 70–250, 20–200, and 10–40. The volume concentration of nanoparticles in all cases was the same and equal to 0.05%. At high shear rates, the nanofluids displayed Newtonian behavior. Additionally, the viscosity decreased with increasing aspect ratio.

In a recent paper [7] various fluids with both MWCNTs and SWCNTs were examined using different nanofluid preparation methods (functionalizing CNTs or using SDBS or Solsperset 27,000 surfactants). It is indicated that the rheology is described by the Maron and Pierce model [81] and the dependence of viscosity on temperature is discussed. At the same time, it should be noted that viscosity measurements were performed using a capillary viscometer and subsequent interpretation of the obtained data suggests Newtonian rheology of nanofluids. However, the nanofluid rheology has not actually been studied.

In general, when studying the viscosity of nanofluids, it is common to use capillary viscometers. It should be noted that, unfortunately, by now quite a lot of articles have been published where the rheology of nanofluids is not studied. Moreover, this is the case even when the authors use not only capillary but also rotational viscometers to measure viscosity, which allow studying rheology [1].

Yu et al. [166] attempted to predict the possible transition of nanofluid from Newtonian to non-Newtonian rheology by studying several water-based nanofluids with MWCNTs at volume concentrations ranging from 0.0047 to 0.2381%. The authors also studied the temperature dependence of viscosity, albeit in a narrow range (275−283 K). The critical concentration of MWCNTs was found to be 0.0571%. At lower concentrations, nanofluids exhibited Newtonian behavior. However, at higher concentrations, they displayed non-Newtonian behavior. It is further argued that this critical concentration is independent of temperature. Nonetheless, this conclusion should be treated with caution, since the rheology of a nanofluid may generally be affected by temperature.

Finally, one of the latest works can be mentioned, where the viscosity of several nanofluids based on water and a water-propylene glycol solution with MWCNTs is studied in the temperature range of 10–80 °C [109]. Triton X-100 was used as a surfactant, and the weight concentration of MWCNTs was varied from 0.005 to 0.1%. The results obtained were quite predictable: the nanofluids were Newtonian at low concentrations of MWCNTs and non-Newtonian at high concentrations (starting from 0.05%). At high shear rates, the viscosity was no longer dependent on the shear rate at all concentrations.

There are fewer publications on the rheology of nanofluids with SWCNTs compared to those with MWCNTs. Let us start with a fairly typical article [154], which studied the temperature dependence of the viscosity of nanofluids based on lubricating oil with SWCNTs. The CNTs had a diameter of 2 nm and a length of 10−15 μm. The measurements were conducted at temperatures ranging from 25 to 100 °C. No surfactants were used in the experiment, and the weight concentration of SWCNTs was varied from 0.01 to 0.2%. The study aimed to be practical, but the rheology of the nanofluids was not examined, despite calculating the viscosity index, which was found to be 32.94% at the highest concentration of SWCNTs. It is important to bear in mind that in non-Newtonian nanofluids, particularly at high CNT concentrations, the viscosity and therefore the viscosity index may vary with the shear rate.

The viscosity of W + CTAB-based nanofluids with SWCNTs at different temperatures (10–60 °C) was investigated in [165]. The weight concentration of nanotubes varied from 0.1 to 1%, with diameters of 1−2 nm and lengths of 5−30 μm. The presented data has shown that the viscosity increased almost linearly with increasing nanoparticle concentration and decreased with increasing temperature. However, the rheology of the nanofluids was not studied.

Ethylene glycol-based nanofluid was studied by M. Baratpour and colleagues [9]. The volume concentration of SWCNTs was varied from 0.0125 to 0.1%. It was found that at all concentrations the nanofluids were Newtonian, and their viscosity significantly exceeded the values predicted by the classical theories. At room temperature and maximum SWCNT concentration, the viscosity enhancement was found to be more than threefold.

The rheology and viscosity of W + SDS-based nanofluid with SWCNTs (500 nm tube length and 1−2 nm diameter) was systematically investigated by Sabiha et al. [140]. The volume concentrations of SWCNTs ranged from 0.05 to 0.25%. The nanofluids exhibited pseudoplastic behavior, but it is worth noting that the accuracy of viscosity measurements may have been insufficient at higher concentrations. In another paper mentioned above [7], the rheology of nanofluids based on W + SDBS or W + Solsperset 2700 was studied using SWCNTs from two different manufacturers. However, the measurements were conducted using a capillary viscometer, and therefore it is unclear whether the nanofluids studied exhibited Newtonian behavior or not. The weight concentrations of nanotubes ranged from 0.04 to 0.4%. No measurements were taken to determine the viscosity of the base fluids, although in some cases the surfactant concentration was as high as 20%.

A rather systematic study of the rheology of nanofluids based on W, EG, and isopropyl alcohol (IPA) with SWCNTs using SDBS, PVP, and SDS as surfactants was performed in [136, 139]. The nanotubes had a diameter of 1.6 nm and a length of approximately 4–5 μm. The weight concentration of SWCNTs ranged from 0.05 to 0.5%. In all cases the nanofluids were non-Newtonian and exhibited either pseudoplastic or viscoplastic behavior, with the rheology changing with temperature. The first cited work analyzed the micro-rheology of nanofluids and demonstrated their viscoelastic properties. The degree of elasticity was found to increase with higher SWCNT concentration.

Hence, even this brief review highlights the inconsistency of existing data due to the use of actually different nanofluids in various works. The base fluids differ, as different surfactants are often used, and CNTs can be both single-walled and multi-walled. However, predicting the rheology of specific CNT-containing nanofluids is crucial from a practical perspective. It is important to understand the potential differences when using nanofluids with SWCNTs and MWCNTs. The following sections aim to systematically study the viscosity and rheology of SWCNT and MWCNT nanofluids and compare the obtained data. The base fluids used, the range of weight concentrations of CNTs, and surfactants, if any, were consistent across all cases studied. Data on over two dozen different nanofluids are presented [127, 128, 136, 139].

3.6.1 Rheology of Nanofluids with SWCNTs

The rheology of nanofluids requires a preliminary study of the corresponding properties of the base fluids used. This is particularly important since surfactants are often used when preparing nanofluids with CNTs. The experiments presented below used the following base fluids: W + SDS, W + SDBS, W + PVP, EG + SDBS, and EG + PVP. When preparing nanofluid, the weight concentration of the surfactant was varied. It was either equal to the weight concentration of SWCNTs or twice that amount. At these concentrations, the base fluid, which consisted of water and SDBS or SDS, remained Newtonian, although its viscosity increased slightly. The increase in viscosity was proportional to the weight concentration of SDBS, but it was found to be small. For example, when the maximum weight concentration of W + SDBS solution was 1%, its viscosity only exceeded that of water by 2%. A similar pattern was observed in the W + PVP solution, which is also a Newtonian fluid. Its viscosity increased proportionally to the weight concentration of the surfactant, with an increase of about 3–4% at a surfactant weight concentration of 2%. The situation with EG is more complicated. In all cases, EG + PVP and EG + SDBS solutions remained Newtonian fluids, but their viscosity increased significantly with increasing surfactant weight concentration. For example, at a weight concentration of PVP equal to 1%, the viscosity of EG + PVP solution increased by 21%, and EG + SDBS solution increased by 5%.

During the second step of nanofluid preparation, the primary fluid underwent UST to destroy nanoparticle agglomerates. The mechanisms used in this process have already been discussed earlier. To objectively evaluate the impact of UST on the resulting nanofluid properties, it is important to first examine the effect of this treatment on the base solution. This is all the more important since the use of high molecular weight compounds as surfactants can result in changes to their structure under the impact of UST [84], which may cause a reduction in solution viscosity. To investigate this effect, the viscosity of W + PVP, W + SDBS, EG + PVP, and EG + SDBS solutions was measured before and after UST. The weight concentrations of surfactant in all cases did not exceed 2%. The first two solutions, W + SDBS and W + PVP, were low viscosity fluids. As previously stated, the addition of surfactants to water causes a slight change in viscosity for both solutions compared to that of water. The solutions remained almost unchanged after UST. However, the viscosity of the EG + PVP solution, with a weight concentration of surfactant equal to 1%, decreased by approximately 2.5%, and that of the EG + SDBS solution decreased by approximately 1%.

To study the effect of temperature on the viscosity of the base fluids, the corresponding measurements were taken within the temperature range from 10 to 50 °C. In all cases, the studied solutions had Newtonian rheology. The viscosity of both base fluids decreased with increasing temperature. However, at the surfactant concentrations considered, the relative viscosity coefficients of the solutions remain constant and do not depend on temperature. This indicates that the temperature dependence of the viscosity of these solutions is determined solely by the corresponding dependence of ethylene glycol or water, at the given concentrations of surfactant. This finding is consistent with previous research [38, 146].

The UST is an important step in preparing nanofluids. However, the duration of UST alone, usually noted in the studies, is not a meaningful indicator of the nanofluid since the effect obtained depends on the equipment used for UST. Instead, it is necessary to indicate the power consumed per unit volume of fluid. The effect of UST on the viscosity of the resulting W + SDBS-based nanofluid with SWCNTs is illustrated in Fig. 3.24. The plot illustrates the relationship between the viscosity coefficient of the nanofluid and the transferred power (U) during the UST. The curves represent weight concentrations of SWCNTs equal to 0.1, 0.2, and 0.3% from bottom to top. The weight concentration of SDBS is twice that of SWCNTs in all cases. The

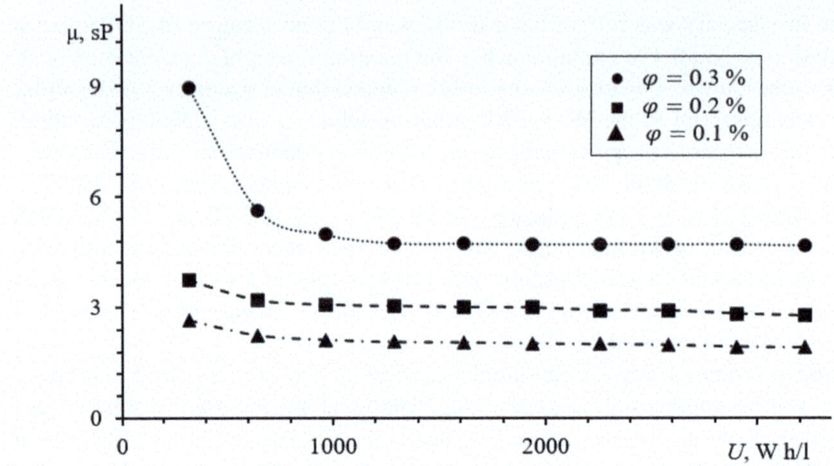

Fig. 3.24 Effect of UST on the viscosity of W + SDBS-based nanofluids at different concentrations of SWCNTs

nanofluid is considered ready for measurement when its viscosity no longer depends on the power pumped during UST. However, it is important to exercise caution as the use of UST can not only destroy CNT conglomerates but also break up nanotubes.

Systematic measurements, presented in papers [127, 128, 136, 139] have demonstrated that all investigated nanofluids (with EG, EG + PVP, EG + SDBS, W + PVP, W + SDBS, W + SDS, and EG + B as a base fluid), exhibit non-Newtonian behavior at weight concentrations of SWCNTs $w \geq 0.05\%$. The rheology of these nanofluids can be well described by either the power-law fluid model (3.29) or the Herschel–Bulkley model (3.30). As an example, Fig. 3.25 illustrates the relationship between viscosity, shear stress, and shear rate for an EG-based nanofluid at 25 °C. The weight concentrations of SWCNTs are 0, 0.05, 0.1, and 0.2% from bottom to top. The viscosity coefficient of EG at 25 °C is 17.2 cP. The indicated nanofluids exhibit viscoplastic behavior at all concentrations and their rheology is described by model (3.30). As the weight concentration of SWCNTs increases, the fluid index decreases, while the consistency parameter increases. Table 3.11 provides correlations that describe the dependence of the model parameters (3.30) on the weight concentration of SWCNTs. These correlations are applicable at $w \geq 0.05\%$ and are shown by dashed lines in Fig. 3.25. Additionally, the yield strength increases with increasing weight concentration, as clearly seen in Fig. 3.25.

The measurements, presented in Fig. 3.25, were performed at a temperature of 25 °C. With increasing temperature, the viscosity of this (as well as other) nanofluids decreases, however, all nanofluids remained non-Newtonian. This is exemplified in Fig. 3.26, which displays data for three different temperatures: 25, 35, and 45 °C (from top to bottom). The same EG-based nanofluids with SWCNTs were considered, where the weight concentration of nanotubes was 0.1%. The viscosity of pure EG at temperatures of 25, 35, and 45 °C is 17.2, 11.26, and 8.13 cP, respectively. The

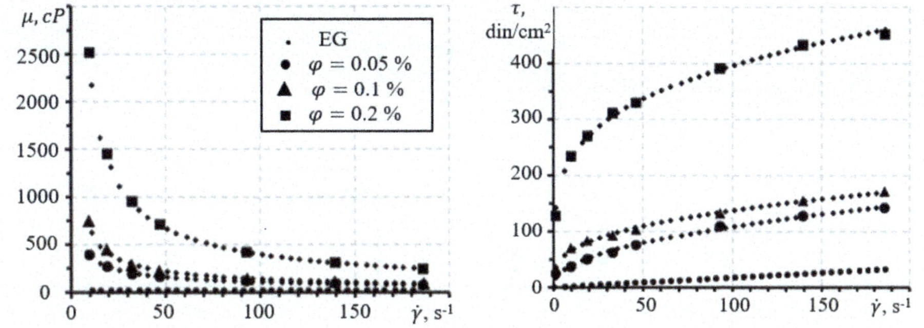

Fig. 3.25 Dependence of the viscosity coefficient and shear stress on shear rate for EG-based nanofluids with SWCNTs

Table 3.11 Correlations describing the rheology of SWCNT-containing nanofluids

BF	k_v dyn/(cm^{2}s^n)	n	τ_0 (dyn/cm)^2
EG	$0.173 + 199w + 200000w^5$	$1 - 10.2w + 40w^2 - 36.25w^3$	$105w + 420w^2$
EG + PVP	$0.173 + 13.4w + 204w^2 + 45900w^3$	$1 - 6.8w + 6w^2 - 12.5w^3$	$40w + 840w^2$
EG + SDBS	$0.173 + 91.5w + 2748w^2$	$1 - 9.6w + 31w^2 - 8.75w^3$	$71.8w + 917w^2$
W + PVP	$0.009 + 136w$	$1 - 10.4w + 14.5w^2 + 86.25w^3$	$27w$
W + SDBS	$0.009 + 0.136w + 1.95w^2$	$1 - 1.7w + 1.5w^2$	$0.4w$

corresponding values of the fluid parameters for model (3.30) are given in formulae (3.38), where all the coefficients have their respective dimensions and the temperature is given in °C

$$k_v(T) = 27.85 - 0.253T,\ n(T) = 0.401 - 0.001T,$$
$$\tau_0 = 5.34 + 0.85T - 0.18T^2. \tag{3.38}$$

It is important to note that as temperature increases, the degree of non-Newtonicity of the nanofluid decreases as well. This results in a decrease of almost 40% in the yield stress and consistency parameter. However, the fluid index increases by 11%. Additionally, the viscosity of the nanofluid decreases with increasing temperature at all shear rates. However, the viscosity of the base fluid (EG) decreases by almost half when the temperature increases from 25 to 45 °C. In contrast, the viscosity of the nanofluid decreases by about one and a half times on average.

Using surfactants does not qualitatively change the rheology of the nanofluid. As an example, Fig. 3.27 illustrates the dependence of the viscosity coefficient and shear stress of EG + SDBS-based nanofluid with SWCNTs on the shear rate. The weight concentrations were 0.2, 0.1, and 0.05% from top to bottom, while the SDBS concentration was 0.2% in all experiments. Comparison of the data presented in

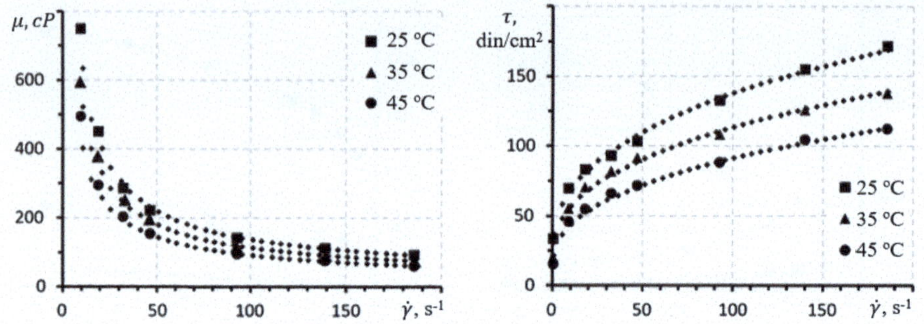

Fig. 3.26 Dependence of viscosity coefficient and shear stress on shear rate for EG-based nanofluid with SWCNTs at different temperatures, $w = 0.1\%$

Figs. 3.25 and 3.27 show that the presence of SDBS resulted in a marked increase in the viscosity of the nanofluid at all shear rates. The rheology of this nanofluid is also described by the Herschel–Bulkley model (3.30), and the corresponding correlation is summarized in Table 3.11. These correlations in Fig. 3.27 are depicted by the dashed lines. A similar pattern is observed for the EG + PVP-based nanofluid.

Ethylene glycol has a viscosity nearly twenty times that of water. In addition, due to the hydrophobicity of SWCNTs, the creation of water-based nanofluids requires the use of a surfactant. The resulting nanofluids are non-Newtonian in all cases. As an example, Fig. 3.28 shows the dependence of the viscosity coefficient and the shear stress on the shear rate for W + PVP-based nanofluids with SWCNTs. In all cases, the rheology of these three nanofluids is described by the Herschel–Bulkley model (3.30). Table 3.11 summarizes the corresponding correlation, depicted in Fig. 3.28 by the dashed line. As the concentration of SWCNT increases, the fluid index decreases while the consistency parameter and yield stress increase. It is important to note that the use of different surfactants can significantly affect the viscosity of nanofluids. Specifically, W + PVP-based nanofluids

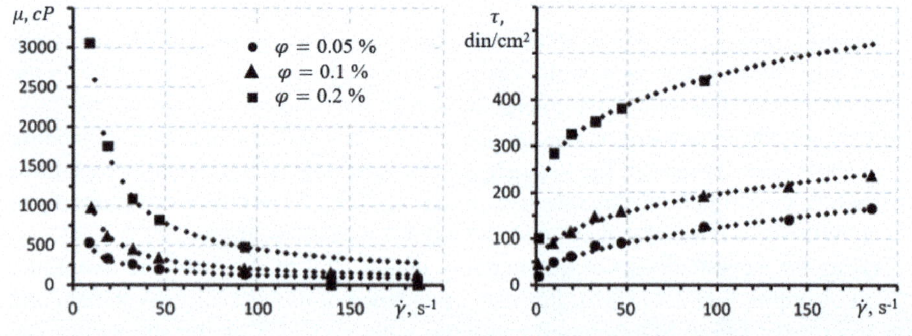

Fig. 3.27 Dependence of viscosity coefficient and shear stress on shear rate for EG + SDBS-based nanofluid with SWCNTs

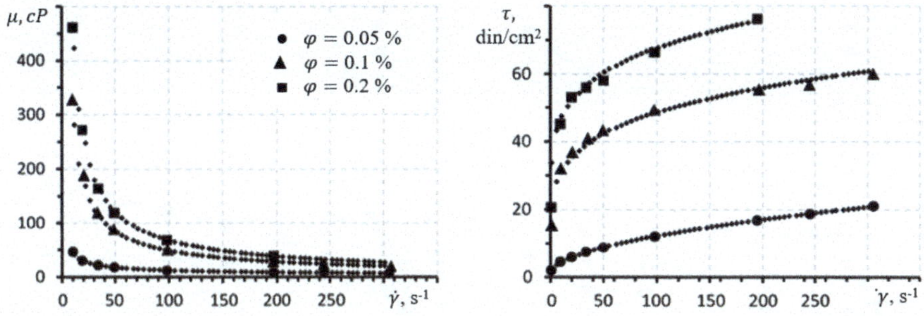

Fig. 3.28 Dependence of viscosity coefficient and shear stress on shear rate for W + PVP-based nanofluid with SWCNTs

Table 3.12 Viscosity coefficient of the W + EG mixture at different temperatures

	Viscosity coefficient, cP		
T (°C)	EG	W10% + EG90%	W20% + EG80%
25	17.,2	11.9	8.4
35	11.26	7.94	3.33
45	8.13	5.85	2.65

containing SWCNTs have viscosities approximately ten times higher than those of W + SDBS-based nanofluids. This is evident from the data presented in Table 3.11.

In practice, a mixture of water and ethylene glycol is one of the most common heat carriers. The data summarized below were obtained in nanofluids based on W10% + EG90% and W20% + EG80% without the use of surfactants [128]. These mixtures have significantly lower viscosities than ethylene glycol. The corresponding data obtained at three different temperatures are presented in Table 3.12.

The rheology of these based-fluids with SWCNTs at all studied concentrations is described by model (3.30). The fluid index decreases as the SWCNT concentration increases, while the consistency parameter and yield stress increase. The viscosity of nanofluid decreases with increasing water content at a constant shear rate. Additionally, increasing the temperature also changes the rheology of the nanofluid. The formula describing the temperature dependence of the Herschel–Bulkley model parameters for a nanofluid based on W10% + EG90% at 0.1% weight concentration of SWCNTs is as follows

$$k_v(T) = 28 - 0.3075T, n(T) = 0.366 - 0.00002T, \tau_0 = 25.9 - 0.42T,$$

and for a nanofluid based on W20% + EG80% it is described by the formula

$$k_v(T) = 23.9 - 0.236T, n(T) = 0.366 - 0.00004T, \tau_0 = 19.6 - 0.41T. \quad (3.39)$$

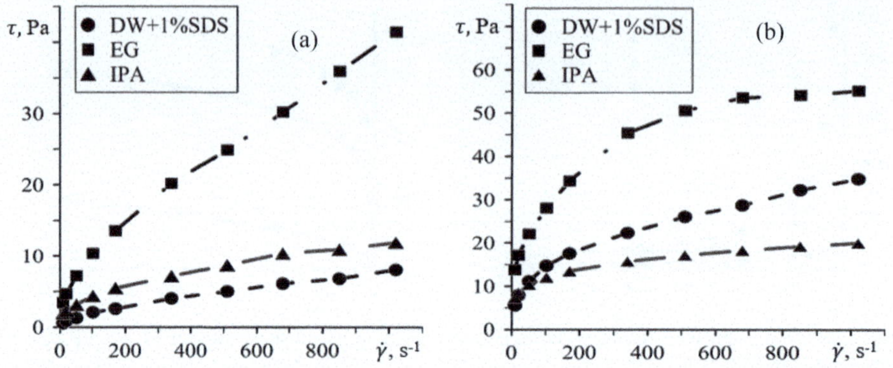

Fig. 3.29 Shear stress of nanofluids depending on the shear rate at different weight concentrations of SWCNTs: **a** 0.05%; **b** 0.25%

Thus, the consistency parameter decreases as the water content increases. However, for the nanofluid with lower water content, it decreases at a faster rate with increasing temperature than for the nanofluid with higher water content. The fluid index of both nanofluids remains relatively constant with increasing temperature.

To finish this section, it is worth noting that the rheology of EG + PVP and W + PVP-based nanofluids can be well described by the power-law fluid model (3.29), except when the shear rates are too low. However, the rheology of nanofluids may be affected by the concentration of SWCNT. Figure 3.29 illustrates the relationship between shear stress and shear rate for three different nanofluids: W + SDS, EG, and IPA. In Fig. 3.29a, the weight concentration of SWCNTs was 0.05%, while the SDS concentration was 1%. In Fig. 3.29b, the concentration of SWCNTs is five times higher. At low SWCNT concentration (see Fig. 3.29a), the rheology of water and IPA-based nanofluids conforms well to the power-law fluid model (3.29). However, the ethylene glycol-based nanofluid already exhibits viscoplastic behavior. Its rheology is described by the Herschel–Bulkley model (3.30). At high SWCNT concentrations (Fig. 3.29b), the rheology of all three nanofluids is described by Eq. (3.30). Therefore, nanofluids can (and usually do) change their rheology with increasing SWCNT concentration. In all cases, as the SWCNT concentration increases, the fluid index monotonically decreases while the consistency parameter increases (see Fig. 3.30).

3.6.2 *Rheology of Nanofluids with MWCNTs*

SWCNTs and MWCNTs are distinct objects, and it is expected that the properties of corresponding nanofluids may also differ. To ensure a fair comparison of viscosity and rheology between nanofluids with SWCNTs and MWCNTs, only nanofluids containing MWCNTs and the same base fluids as in the previous section are analyzed below. Let us start with the EG-based Taunit-M nanofluid with MWCNTs. The

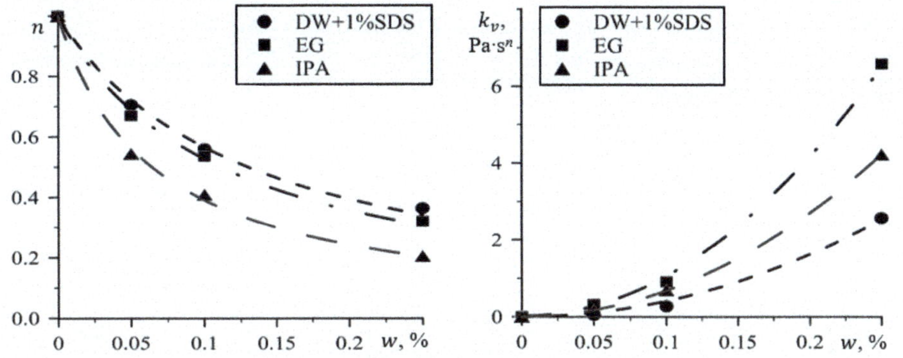

Fig. 3.30 The consistency parameter and the fluid index of nanofluids based on W + SDS, EG, and IPA depending on the weight concentration of CNTs

dependences of viscosity coefficient and shear stress on shear rate for these nanofluids are shown in Fig. 3.31. Strictly speaking, these fluids are considered viscoplastic and their rheology is described by Eq. (3.30). However, at the lowest weight concentration of MWCNTs, the yield stress is very small and falls within the measurement accuracy. The correlations presented in Table 3.13 support the observation that as concentration increases, k_v and τ_0 increase, while n decreases. The plot of shear stress versus shear rate in Fig. 3.31 indicates that the nanofluid at a concentration of 0.05% is close to Newtonian, with a fluid index of 0.965 and a yield stress of 0.018 dyn/cm^2.

The study of the rheology of EG + PVP-based nanofluids with MWCNTs resulted in an unexpected conclusion. These nanofluids were found to be Newtonian at all concentrations studied, up to 0.2%. The viscosity has shown an almost linear dependence on the weight concentration of MWCNTs (see Table 3.13). However, the increase in the viscosity coefficient could not be predicted by the classical Einstein formula (3.13). The viscosity coefficient of the nanofluid increases at a much faster rate with increasing MWCNTs concentration than predicted by formula (3.13).

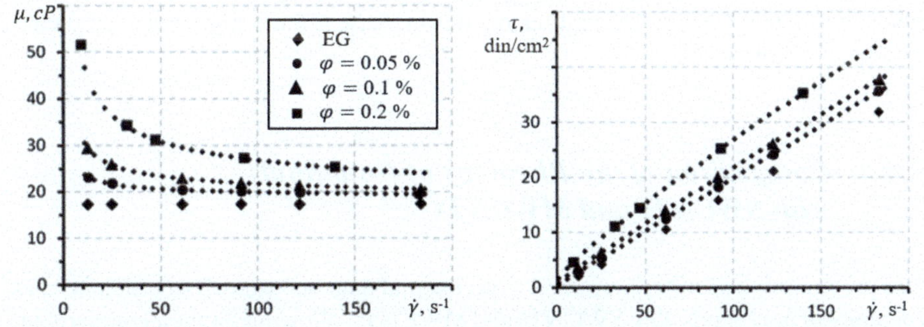

Fig. 3.31 Dependence of viscosity coefficient and shear stress on shear rate of EG-based nanofluids with MWCNTs

Table 3.13 Correlations describing the rheology of nanofluids with MWCNTs

BF	k_v, dyn/(cm^{2}s^n)	n	τ_0, dyn/cm^2
EG	$0.173 + w + 7.9w^2$	$1 - 0.707w$	$5.8w + 7.7w^2$
EG + PVP	$0.173 + 0.1w$	1	
EG + SDBS	$0.173 + 2.3w + 9.7\,w^2$	$1 - 1.05w$	$4,1w + 380w^3$
W + PVP	$\mu = 0.009 + 0.0093w$	1	
W + SDBS	$\mu = 0.009 + 0.0075w$	1	

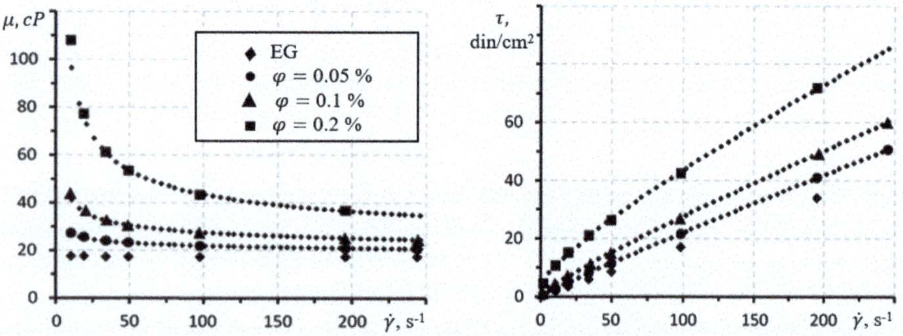

Fig. 3.32 Dependence of viscosity coefficient and shear stress on shear rate of EG + SDBS-based nanofluid with MWCNTs

On the other hand, EG + SDBS-based nanofluids with MWCNTs display non-Newtonian behavior, as shown in Fig. 3.32. The rheology of the nanofluid is accurately described by the power-law fluid model (3.29) at MWCNT concentrations of 0.05 and 0.1%. However, at a concentration of 0.2%, the nanofluid exhibits viscoplastic behavior, which is characterized by the emergence of a well-defined yield stress (Fig. 3.32). Finally, it is noted that the nanofluids based on W + PVP and W + SDBS with MWCNTs exhibit Newtonian behavior up to weight concentrations of 0.2% (see Table 3.13). In both cases, the viscosity coefficient increases linearly with MWCNT concentration, but this increase is much higher than predicted by Einstein's theory.

3.6.3 Comparison of the Rheology of Nanofluids with SWCNTs and MWCNTs

The previous two sections presented experimental data on viscosity coefficients and rheology of nanofluids with SWCNTs and MWCNTs, allowing for their comparison. The comparison of the data for EG-based nanofluids (Figs. 3.24 and 3.31) shows that both SWCNTs and MWCNTs nanofluids are viscoplastic. Additionally, the viscosity of nanofluids with MWCNTs is significantly lower than that of nanofluids with

SWCNTs. At high shear rates (185 s^{-1}), the viscosity of nanofluids with MWCNTs at maximum concentration is more than an order of magnitude lower than that of nanofluids containing SWCNTs. As the concentration decreases, this difference also decreases, and at the minimum concentration, the viscosity of nanofluids with MWCNTs becomes about five times lower. Conversely, this difference increases as the shear rate decreases. When the shear rate is changed from 9.3 to 186 s^{-1}, the viscosity of nanofluids containing SWCNTs varies between 3000 and 180 cP, while that of nanofluids containing MWCNTs it varies between 50 and 19 cP. Additionally, it was observed that nanofluids with MWCNTs are less non-Newtonian than those with SWCNTs. At the same weight concentrations of CNTs, the fluid index of EG-based nanofluids with SWCNTs varies between 0.52 and 0.24, and the consistency parameter varies between 9.84 and 140, whereas in nanofluids with MWCNTs, n varies from 0.92 to 0.7 and k_ν varies from 0.28 to 0.7.

Using surfactants slightly changed the viscosity of the base fluid (EG). Maximum surfactant weight concentration was 0.5%. As a result, the viscosity of the base fluids increased by about 10%, however, they remained Newtonian. The addition of both surfactants resulted in non-Newtonian behavior in the corresponding SWCNT-containing nanofluids. The degree of non-Newtonicity was slightly higher in the nanofluids based on EG + PVP and EG + SDBS compared to the EG-based nanofluids. In nanofluids containing MWCNTs, the situation differs. The EG + PVP-based nanofluids with MWCNTs are Newtonian, and their viscosity increases linearly with increasing MWCNT concentration in the investigated range. However, at all shear rates, the viscosity remains about an order of magnitude lower than in nanofluids with SWCNTs. On the other hand, EG + SDBS-based nanofluids with MWCNTs appear to be non-Newtonian, pseudoplastic at low concentrations (up to 0.1%), and viscoplastic at maximum concentrations. At all shear rates the viscosity of nanofluids with MWCNTs is several times lower than that of nanofluids with SWCNTs. However, the situation with W + PVP and W + SDBS nanofluids containing MWCNTs is different. These nanofluids are Newtonian and their viscosity increases linearly with increasing MWCNT concentration. It is worth noting that nanofluids created on these base fluids have significantly different viscosities.

Now, let's discuss why the viscosity of nanofluids with CNTs is significantly higher than that of the base fluid and conventional coarse dispersed fluids. Section 3.2.2 revealed that the structuring of the base fluid molecules near the nanoparticles is one reason. This structuring is also observed in the nanofluid with CNTs. This aspect will be discussed in Sect. 3.8. However, there is another reason for the increase in viscosity of the nanofluid with CNTs, and, perhaps, even more important. This reason is related to the change in the rheology of nanofluids with CNTs. Above, it has been experimentally established that nanofluids with a weight concentration of SWCNTs equal to or greater than 0.05% are non-Newtonian and have significantly higher viscosity than nanofluids with MWCNTs at the same weight concentration. The change in the rheology of nanofluids with CNTs is largely due to the formation of a percolation lattice. The appearance of this percolation structure was revealed quite a long time ago using atomic force microscopy [43]. The formation of a percolation lattice of CNTs is a threshold phenomenon. Percolation occurs at critical concentrations of CNTs, when the average distance between them is comparable to the strength of van der Waals forces. These forces are well described theoretically for objects significantly larger than molecules. The Hamaker theory [42] was the first in this series. Later theories [53, 71] enable more precise estimation of the characteristic scale of the attractive forces in such cases. Comparing this size with the average distance between CNTs (which is determined by their volume concentration), one can then estimate the critical concentration of CNTs at which the percolation lattice begins to form. It turns out that this critical concentration is quite low. It can be estimated as follows. The volume concentrations of SWCNTs φ_S and MWCNTs φ_M are determined respectively by the relations

$$\varphi_S = \pi R_S^2 L_S n_S, \varphi_M = \pi R_M^2 L_M n_M. \tag{3.40}$$

Here, the indices S and M refer to SWCNTs and MWCNTs, respectively, and n_S and n_M are their numerical densities. While considering the volume concentrations per unit volume, it is simply the number of SWCNTs and MWCNTs in this volume. At the same volume concentration, their ratio is determined as

$$\frac{n_S}{n_M} \sim \frac{R_M^2 L_M}{R_S^2 L_S} \gg 1. \tag{3.41}$$

The average distance between CNTs can be estimated as $(n)^{-1/3}$. Then, the critical percolation concentration of SWCNTs is approximately 0.0044%, while that of MWCNTs is 0.097%. Therefore, percolation structures appear at much lower concentrations in nanofluids with SWCNTs than in those with MWCNTs, which significantly affects their viscosity and rheology. It should be noted that the use of surfactants can significantly increase the critical percolation concentration of CNTs.

3.7 Microrheology of Nanofluids with CNTs

The rheology of nanofluids containing CNTs is highly dependent on both CNT concentration and temperature. This sensitivity is attributed to changes in the nanofluid structure. One method of studying this structure is through microrheology. For this purpose, the so-called passive technique of microrheology study, when only the thermal motion of CNTs is studied, was applied in the papers [127, 128, 136] using Rheolaser™ Master. The research technology is based on one of the methods of dynamic light scattering, called Diffusive Wave Spectroscopy (DWS) [82, 83, 162]. The samples of the suspension are placed in 20 ml glass vials and illuminated with a laser beam. The particles in the sample scatter the radiation, creating a speckle field due to photon interference. A CCD camera records the speckle field, which is constantly changing due to Brownian motion of the particles. The time-dependent mean square displacement (MSD) of the particles is then calculated.

In a viscous Newtonian fluid, particles diffuse into the fluid and their MSD increases linearly with time. The slope of the corresponding line is directly proportional to the viscosity of the sample. When particles move in a perfectly elastic medium, the MSD reaches a plateau corresponding to the equilibrium between the thermal energy and the elastic interaction energy [82]. The time dependence of MSD for EG-based nanofluids at two weight concentrations of SWCNTs (0.05 and 0.2%) is shown in Figs. 3.33a, b. The MSD(t) curves exhibit both inclined and gently sloping sections, indicating viscoelastic behavior. The section corresponding to the elastic behavior of the nanofluid increases as the concentration of SWCNTs increases (compare Figs. 3.33a, b). In contrast, the section on viscous behavior becomes gently sloping, leading to a decrease in the diffusion coefficient of SWCNTs. Figure 3.33c–e illustrate the evolution of MSD for EG-based nanofluids with MWCNTs at weight concentrations of 0.05, 0.2, and 1%. A viscous behavior is observed at the lowest concentration, while pronounced viscoelastic properties are exhibited only at a concentration of 1% (Fig. 3.33e). Comparing Fig. 3.33b, e, it is evident that the MSD curves are close to each other, however, the weight concentration of MWCNTs is five times higher in the latter. The plateau height on the MSD(t) curve indicates the fluid sample's elastic properties and is described by the elasticity index (EI). A decrease in the height of the plateau corresponds to an increase in the sample's elasticity. The decrease in the EI index with increasing SWCNT concentration for EG-based nanofluid is clearly visible when comparing Figs. 3.33a, b.

The elasticity index (EI) is directly proportional to the elastic modulus G' on the plateau and allows for the observation of changes in sample elasticity over time or for comparison of similar samples based on their elasticity. Figure 3.34

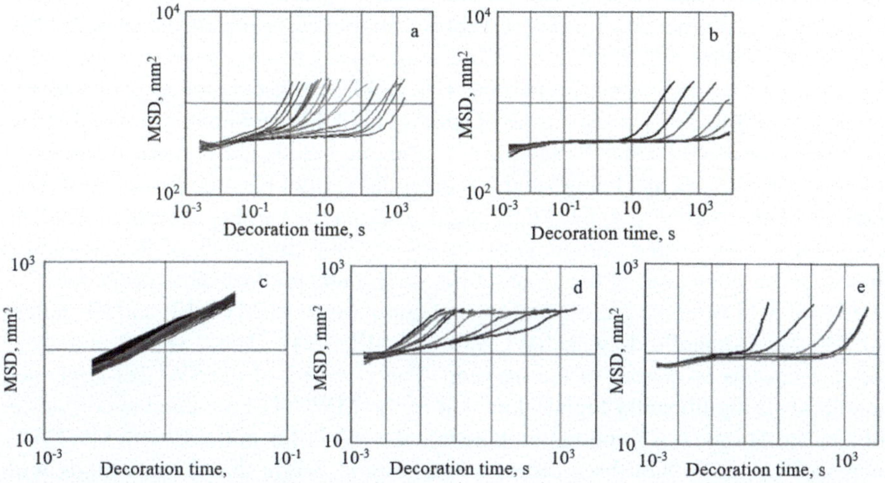

Fig. 3.33 Mean square displacement of SWCNTs (**a**, **b**) and MWCNTs (**c**–**e**) depending on the decorrelation time in EG-based nanofluids at different weight concentrations: 0.05% (**a**, **c**); 0.2% (**b**, **d**); 1.0% (**e**)

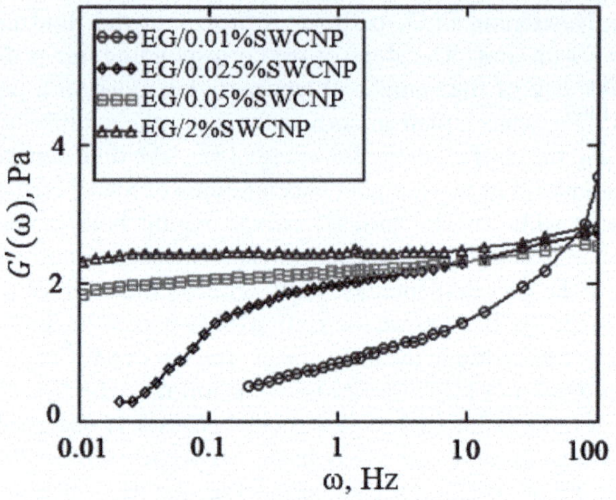

Fig. 3.34 Dependence of the elastic modulus of the EG-based nanofluid at 0.2% SWCNT concentration for different time points

provides an example of the elastic modulus dependence for an EG-based nanofluid with a SWCNT weight concentration of 0.2% at different instances on the inverse time of speckle field decorrelation (measured in Hz). The data obtained enable the determination of the elastic modulus of nanofluids and its dependence on CNT concentration.

The viscosity of the sample is determined by the macroscopic viscosity index (MVI), which is calculated as the inverse of the slope of the MSD curve in linear coordinates. A higher MVI index indicates a higher viscosity of the sample. This index enables observation of changes in sample viscosity over time. Another index that characterizes the viscous properties of the fluid is the fluidity index (FI), which is the inverse of the characteristic time of speckle field decorrelation, measured in Hz. It is generally accepted that the sample is considered mainly elastic/hard at low FI (of the order of 10^{-2} Hz and below) and viscous/fluid at high FI (of the order of 10 Hz). Table 3.14 summarizes the EI, MVI, and FI indices for various nanofluids studied. The table allows for a comparison of the viscoelastic properties of the mentioned fluids and an evaluation of the effect of the surfactants used on these properties.

Figure 3.35 presents a clear and typical comparison of MVI, FI, and EI indices for IPA-based nanofluids with SWCNTs and MWCNTs. The first plot characterizes the viscous properties of the medium. The viscosity of the SWCNT-contained nanofluids is significantly higher than that of the MWCNT-contained nanofluids at all concentrations. It is important to note that the MVI of nanofluids with MWCNTs increases by more than three orders of magnitude, while that of nanofluids with SWCNTs increases by about five times. The MVI characterizes the viscosity of the medium under investigation, while the FI characterizes its fluidity. The analysis of the change in this index with increasing CNT concentration indicates that the nanofluid with SWCNTs becomes primarily elastic starting from a concentration of 0.1%.

Table 3.14 EI, MVI and FI indexes for nanofluids at different CNT concentrations

Nanofluid	EI, nm^{-2}	MVI (nm^{-2} s)	FI (Hz)
0.2%SWCNT/EG + PVP	0.011	227	0.000348
0.2%SWCNT/EG + SDBS	0.0032	7	0.0004
0.2%SWCNT/EG	0.0035	78	0.000117
0.1%SWCNT/EG	0.003	50	0.00025
0.05%SWCNT/EG	0.0026	30	0.0015
0.2%SWCNT/DW + SDBS	0.006	3	0.009
0.2%SWCNT/DW + PVP	0.015	200	0.0008
1%MWCNT/EG	0.012	23	0.0007
0.2% MWCN/EG	0.0085	9	0.0018
0.1%MWCNT/EG	0.009	0.32	0.05
0.05%MWCNT/EG	0.006	0.0001	87
0.2%MWCNT/EG + PVP	0.0047	4	0.015
0.2%MWCNT/EG + SDBS	0.0115	0.011	1

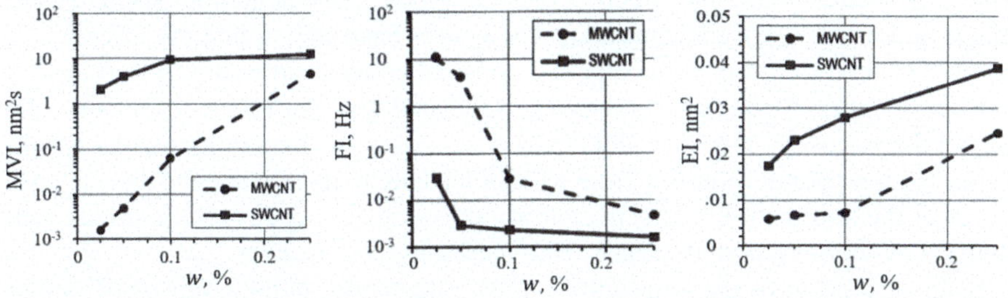

Fig. 3.35 Comparison of MVI, FI and EI indices for IPA-based nanofluids with SWCNTs and MWCNTs at various CNT concentrations

On the other hand, the nanofluid with MWCNTs is mainly elastic only at concentrations higher than 0.2%. Lastly, the EI index characterizes the elastic modulus of nanofluids. As the concentration of SWCNTs increases, the elastic modulus of the corresponding nanofluid increases by approximately two times, while that of nanofluids with MWCNTs increases by approximately four times (Fig. 3.35).

3.8 MD Simulation of the Viscosity of Nanofluids with CNTs

The use of MD method was extremely useful in studying the viscosity of nanofluids with spherical nanoparticles and elucidating the corresponding mechanism. However, directly modeling the viscosity of nanofluids with real CNTs is still not possible, since it requires the use of a huge number of both nanotubes and molecules. Only a few papers have been published on this problem so far. In one of the first studies, a viscosity simulation was performed on a water-based nanofluid containing SWCNTs with a diameter of 0.95 nm [55]. The interaction of carbon atoms in CNTs was described using the Tersoff potential, while their interaction with water molecules was described by the Lennard–Jones potential. Only a single CNT was used in the simulation cell to exclude any possible effects of CNT interactions on the nanofluid viscosity. Another drawback of this paper is that the increase in CNT concentration from 0.125 to 1% was achieved solely by elongating the nanotubes, resulting in varying their lengths from 0.22 to 1.78 nm. However, it is clear that in this case the viscosities of completely different nanofluids were actually compared, and it is simply impossible to draw meaningful conclusions about the dependence of viscosity of nanofluids with given CNTs on concentration. Additionally, in fact, a nanofluid containing nonspherical nanoparticles of very small size was considered. In another set of calculations, the length of the nanotubes remained constant at 2.5 nm (with a volume concentration of 0.734%), while the diameter varied from 1.11 to 1.59 nm. The results have shown that the variation in diameter had little to no effect on the viscosity of the corresponding nanofluids. The same authors also studied the temperature dependence of the viscosity of these nanofluids. The calculations seem reasonable at a constant CNT concentration. The results have shown that the viscosity of nanofluids decreases monotonically at a constant concentration of CNTs (i.e., at a constant length). At the same time, it is indicated that the relative viscosity (the ratio of the viscosity of the nanofluid to the viscosity of the base fluid) increases with increasing temperature, starting from a nanotube length of 0.88 nm, and this increase is greater the longer the CNT is. In contrast, the relative viscosity of nanofluids with particles of shorter length decreases with increasing temperature.

In the next paper by the same authors [56], calculations of the viscosity of similar nanofluids were performed. The CNTs had a diameter of 0.951 nm and lengths ranging from 0.22 to 1.3 nm, allowing for CNT concentrations to vary between 0.125 and 0.734%. The authors obtained qualitatively similar results to their previous work. It is important to mention that, according to the authors, the viscosity of nanofluids was significantly higher than that predicted by classical theories for coarse dispersed fluids.

In the series of studies [114, 98], nanofluid simulations were conducted using water with armchair type (6, 6) CNTs 3 nm long at temperatures within the range of 298–313 K. The volume concentration of CNTs ranged from 0.557 to 3%. The simulation cell contained a single CNT. The results have shown that the viscosity of the nanofluids was significantly higher than that of the base fluid and increased with increasing CNT concentration. Additionally, the viscosity of the nanofluid decreased with increasing temperature. It is noted that the viscosity of the nanofluid is significantly higher than that predicted by classical theories. In this case a comparison was conducted with Brinkman's theory [19]. Additionally, this work proposes a correlation for the viscosity coefficient depending on temperature and CNT concentration based on numerical analysis of MD data. However, caution should be exercised when considering the proposed correlation for two reasons. Firstly, the CNTs under consideration cannot be accurately described as nanotubes due to their length being comparable to their diameter. Secondly, the viscosity coefficient, calculated using Green–Kubo relations, did not reach the plateau values.

To conclude this forcedly very brief review of MD simulation of viscosity of nanofluids with CNTs it should be noted that still a number of fundamental issues remain unexplained. This concerns the following issues. (i) How does the viscosity of nanofluids vary with the aspect ratio (length-to-diameter ratio) of CNTs? (ii) How does the viscosity of nanofluids with CNTs depend on their concentration? (iii) How does the viscosity of nanofluids with CNTs compare with that with conventional spherical particles? The answers to these questions can be found below. This section discusses the MD simulation of the viscosity coefficient of benzene-based nanofluids containing SWCNTs, carbon, and copper nanoparticles [134]. Benzene was chosen as the base fluid due to the hydrophobic nature of CNTs, making it impossible to create a water-based nanofluid without surfactants. Therefore, the computation data quoted above, obtained using water as a base fluid in the calculations, are model data and cannot simulate a real nanofluid from this point of view. On the other hand, CNTs exhibit lyophilic properties towards benzene, similar to various alcohols, such as isopropanol and ethylene glycol. In this study, armchair-type CNTs with chirality (8, 8) and diameters of 1.1 nm were used, with lengths ranging from 1.1 to 14.6 nm. The weight concentrations of CNTs were 3, 9, and 15%, which approximately correspond to volume concentrations of 1, 3, and 5%, respectively. The number of benzene molecules in the simulation cell ranged from 6700 to 38,000. The number of CNTs in the calculations varied from 4 to 48, and that of nanoparticles ranged from 6 to 24.

The viscosity coefficients were calculated using the FDT (3.24). The Verlet scheme was used to integrate Newton's equations. The NPT thermostat of Nosé-Hoover [49, 102] was used for all calculations, performed at atmospheric pressure and a temperature of 25 °C. The interaction between benzene and carbon molecules was determined by the Lennard–Jones potential with parameters of $\sigma_1 = 0.5034$ nm and $\varepsilon_1/k_B = 544.3$ K. For carbon atoms $\sigma_2 = 3.4157$ Å and $\varepsilon_2/k_B = 27.70$ K.. To simulate CNTs, a set of carbon atoms interacting using the AIREBO potential [145] was used. The interaction of benzene molecules with graphite or copper nanoparticles was described by the RK potential (2.27), and the RKI potential (2.28) is used as the interaction potential of graphite or copper nanoparticles.

The viscosity coefficient (3.24) was considered calculated when it reached the plateau value. An example of this is shown in Fig. 3.36, which displays the evolution of the viscosity coefficient of a nanofluid with CNTs with length of 1.1 nm at a nanotube weight concentration of 3% [134]. The graph displays the benzene viscosity coefficient as a dashed line and the plateau value of the nanofluid viscosity coefficient as a solid line. Time is measured in femtoseconds, and the plateau is reached in approximately 30 picoseconds. The calculated viscosity coefficient of the nanofluid is 9.5% higher than that of the base fluid (benzene). The CNT, employed to exemplify the viscosity of the nanofluid in Fig. 3.37, had an aspect ratio of unity. As the aspect ratio increased, the time required reaching the plateau also slightly increased; it also increased with the concentration of CNTs in the nanofluid. Nevertheless, the

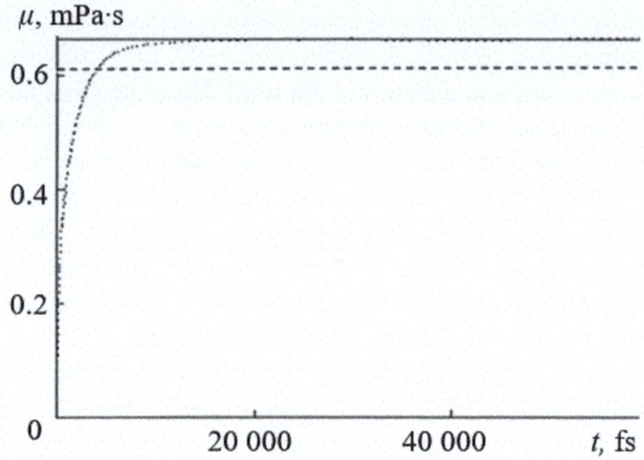

Fig. 3.36 Evolution of the viscosity coefficient of benzene-based nanofluid with CNTs

autocorrelation functions of viscosity (ACFV) $\chi_\mu(t)$ decayed relatively rapidly in all cases. For example, in the nanofluid comprising CNTs of maximum length (14.6 nm) at a weight concentration of 5%, the plateau value was reached in approximately 35 ps.

The dependence of the relative viscosity coefficient of nanofluids with CNTs of different lengths on their volume concentration is shown in Fig. 3.37. The nanotube lengths range from 1.1 to 14.6 nm from bottom to top, and the dashed line corresponds to Batchelor's theory [14]. Based on the analysis of the data, several conclusions can be drawn. The first conclusion is that the viscosity coefficient of all nanofluids is significantly higher than that of coarse dispersed fluids [14]. The second conclusion is that the viscosity coefficient of both coarse dispersed fluids and nanofluids depends on the volume concentration of CNTs, and this dependence is described by the square law (3.18). The coefficients involved are functions of CNT length and are presented in Table 3.15. The presented data analysis indicates a significant increase in the viscosity coefficient particularly at low CNT concentrations (see the coefficient a_1 in Table 3.15). The solid lines in Fig. 3.37 correspond to the dependences (3.18).

The third conclusion is that at a given concentration the viscosity coefficient increases with the length of CNTs. More precisely, viscosity depends on the aspect ratio of CNTs, since in this case the diameter of all CNTs was the same. Figure 3.38 shows the dependence of the relative viscosity coefficient of nanofluids on the CNT length at different CNT concentrations. The volume concentrations of CNTs are 1, 3, and 5% from the bottom to the top. The observed increase in the viscosity of the nanofluids with increasing CNT length is well described by the following logarithmic function

$$\eta_r(L) = a(\varphi)\ln(L) + b(\varphi). \tag{3.42}$$

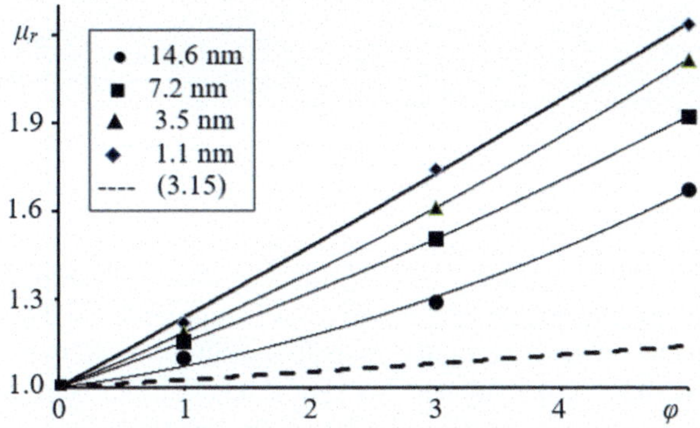

Fig. 3.37 Dependence of the relative viscosity coefficient of nanofluids on the CNT volume concentration

Table 3.15 Parameters of formula (3.18) for benzene-based nanofluids with CNTs

L, nm	1.1	3.6	7.4	14.6
a_1	7.38	16.36	19.59	23.44
a_2	144.23	38.39	45.14	26.52

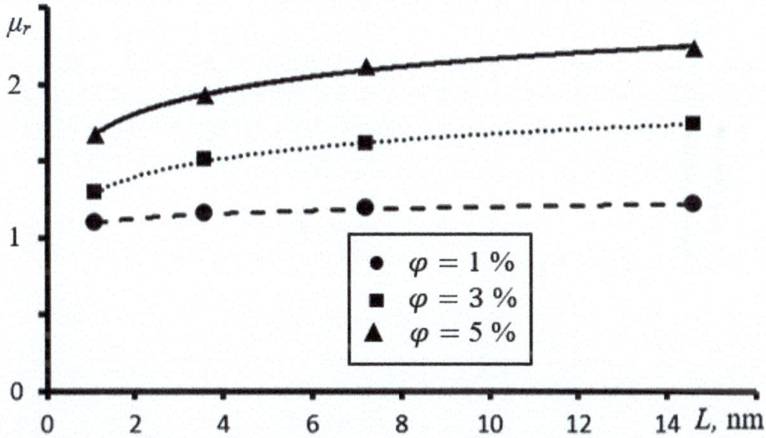

Fig. 3.38 Dependence of the relative viscosity coefficient of the nanofluid on the CNT length

The parameters $a(\varphi)$ and $b(\phi)$ in Eq. (3.42) are given in Table 3.16 These correlations are illustrated in Fig. 3.38 by means of various lines.

The increase in viscosity of nanofluids with CNTs, as in nanofluids with conventional nanoparticles, should be associated with a change in the structure of the base fluid and the possible formation of a short-range order of the CNT pseudogas. The structuring of base fluid molecules around spherical nanoparticles is early discussed

Table 3.16 Parameter values in formula (3.42)

$\varphi(\%)$	$a(\varphi)$	$b(\varphi)$
1	0.222	1.651
3	0.173	1.276
5	0.046	1.094

on Sect. 3.3.2. Similarly, the structuring of the base fluid molecules near the CNTs also occurs in the nanofluid. Figure 3.39 presents a comparison of the radial distribution functions (RDF) of benzene fluid molecules (solid line) and benzene molecules of the nanofluid with 1.1-nm diameter CNTs and 1.45, 3.56, 7.25, and 14.61 nm long. In all cases, the volume concentration of CNTs is 2%. As the length of the nanotubes increases, the RDF maxima of benzene molecules in the nanofluid and the short-range order extent also increase. At the maximum length of the CNTs, the first maximum of the RDF exceeds the corresponding value for pure benzene molecules more than twice. The second, third, and subsequent maxima increase even more. In addition, the extension of the short-range order in nanofluids with CNTs nearly doubles in comparison with that in pure benzene. The viscosity of the fluid also increases with the growth of the magnitude of the RDF maxima and their number.

Structuring of the base fluid molecules around CNTs is exactly one of the main factors influencing the growth of the viscosity of nanofluids. As previously mentioned, the structuring of the molecules depends on the surface area of nanoparticles or CNTs. The surface area of a CNT is directly proportional to its length. Consequently, the viscosity of nanofluids with CNTs increases with increasing their length. The ratio of the surface area of a spherical nanoparticle S_{sp} of radius R to the surface area of a CNT S_{cnt} of radius R_{cnt} and length L is determined as follows

$$\frac{S_{sp}}{S_{cnt}} = \frac{2R^2}{R_{cnt}^2 + R_{cnt}L}. \tag{3.42}$$

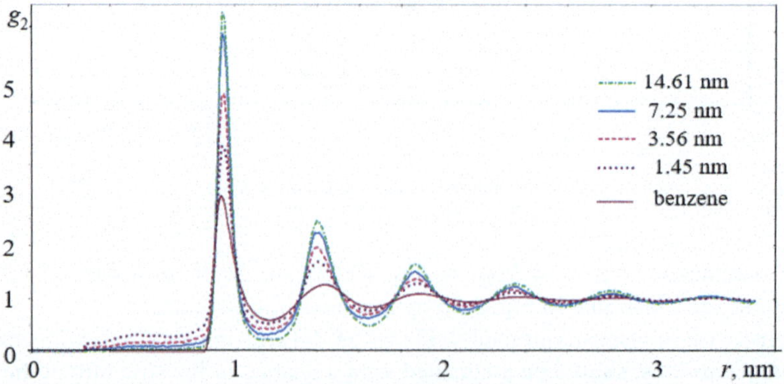

Fig. 3.39 Radial distribution functions of pure benzene molecules and benzene-based nanofluid with different size CNTs

Here, the following situations may be encountered. Let us $R \sim R_{cnt}$. This relationship for SWCNTs is satisfied if the diameter of the nanoparticles is less than 1–2 nm and the spherical particles are small. In this case $S_{sp}/S_{cnt} \sim R/L \ll 1$. The viscosity of nanofluids with spherical particles is less than that of nanofluids with CNTs, and only for the smallest CNTs ($L \sim R$) it will be almost the same in both cases. This is confirmed by the data presented in [134]. For large particles, if considering SWCNTs, $R > R_{cnt} \ll L$. In this case $S_{sp}/S_{cnt} \sim R^2/(R_{cnt}L) < 1$. Again, the viscosity of the nanofluids with CNTs will be greater than that of the nanofluids with spherical particles. It is evident that these two cases exhaust also all possible situations for the NFs with MWCNTs.

Thus the viscosity of nanofluids with CNTs will be larger than that of nanofluids with spherical particles (see data presented in paper [134]). In this regard, it is worth noting that viscosity of fluid with nonspherical particles is typically larger than that of the fluid containing spherical particles. The effective viscosity coefficient of a dispersed medium increases with increasing deviation of the particle shape from spherical. This is expected due to the shorter Brownian rotation time compared to the hydrodynamic times considered. Therefore, particle rotation leads to an increase in the effective volume they occupy.

The CNTs have the large length. Therefore they may active interact each other if their concentration is not very small. This, in particular, can lead to the structuring of the CNT pseudogas. Such structuring can also be studied by analyzing the corresponding RDFs, which are shown in Fig. 3.40. Here, in all cases, the volume concentration of CNTs is equal to 2%. The behavior of the RDF strongly depends on the CNT length. At the maximum CNT length, a short-range order from CNTs starts to form in the nanofluid.

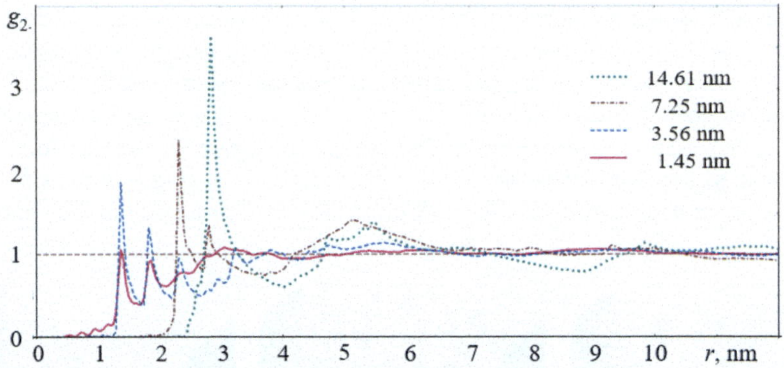

Fig. 3.40 Radial distribution functions of CNTs of different sizes in benzene-based nanofluid

3.9 MD Simulation of Rheology of Nanofluids with CNTs

In Sect. 3.5 the rheology modeling of conventional nanofluids was discussed in detail. The most interesting data from the MD simulation of the rheology of nanofluids with CNTs are presented below. The simulation was again performed using the NEMD method. Due to the anisotropy of CNTs, one of the main tasks was to study the effect of the CNT length on the viscosity and rheology of nanofluids. Their diameter was 1.16 nm, and their length varied from 1.45 to 14.96 nm. In all cases in this section, the volume concentration of CNTs was 2% [138]. The problem statement and the method of its solution are similar to those described in Sects. 3.5 and 3.8. The simulation cell was a rectangular parallelepiped. Typical screenshots are shown in Fig. 3.41. The viscosity coefficient was calculated from the shear stress of the nanofluid.

Since the goal was to study the rheology of these nanofluids in the bulk, it was necessary to ensure that the presence of walls did not affect the simulation results. It was shown in [138] that walls do not affect the viscosity and rheology of nanofluids if the distance between them is equal to (or greater than) 30 nm. In this work, therefore, all calculations were performed precisely at this distance between the walls. The viscosity of benzene is 6.04 mP, which is in high accuracy agreement with the experimental value [157].

It has been established that at relatively high shear rates (up to 10^8 s^{-1}) nanofluids are Newtonian. This is illustrated in Fig. 3.42, here in all cases the data from molecular dynamics simulations are marked with points. The lower curve corresponds to benzene, and the subsequent curves correspond from bottom to top to nanofluids with CNTs with lengths of 1.45, 5.38, 7.59 and 14.96 nm. The viscosity of nanofluids is consistent with the data presented above, where the viscosity of nanofluids with the same CNTs was modeled by the method of equilibrium molecular dynamics using the fluctuation–dissipation theorem. The viscosity of nanofluids increases with increasing CNT length, and it significantly exceeds the viscosity of the base fluid.

With increasing shear rate, the rheology of both benzene and all nanofluids changes, they become pseudoplastic. The critical shear rate for the transition from Newtonian to non-Newtonian rheology $\dot{\gamma}_{cr}$ depends on the length of the CNT and decreases with their growth. Figure 3.43 illustrates the dependence of the nanofluid

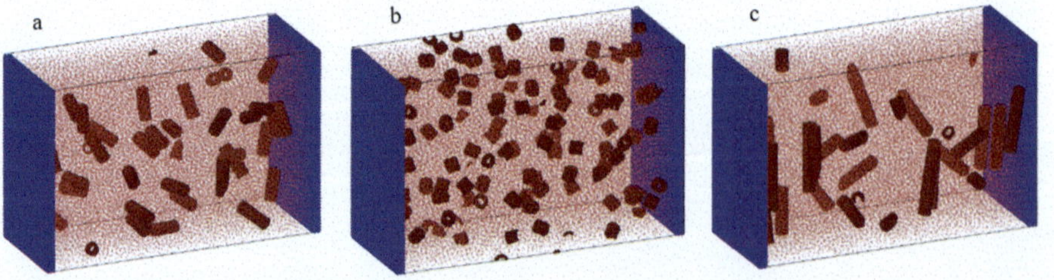

Fig. 3.41 Screenshots of calculated cells of benzene-based (red dots) nanofluids and CNTs with a diameter of 1.1 nm and a length of **a** 3.90 nm, **b** 1.45 nm, and **c** 7.59 nm

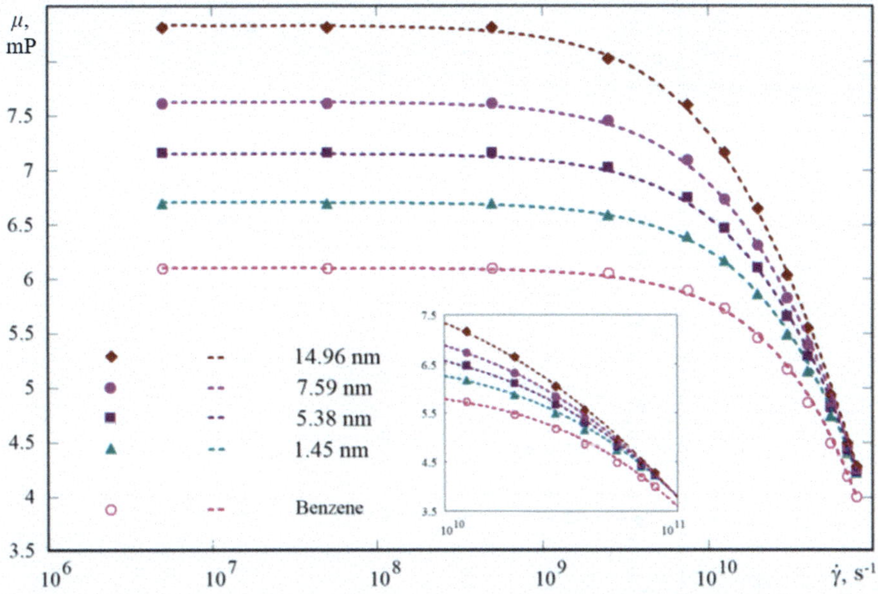

Fig. 3.42 Dependence of the nanofluid viscosity on shear rate for the nanotubes of various lengthes

viscosity on the length of the nanotubes at different shear rates. Here again, the solid marks correspond to the MD calculation data, and the shear rate increases from top to bottom from $5 \times 10^6\ \text{s}^{-1}$ to $8 \times 10^{10}\ \text{s}^{-1}$. Thus, the upper curve corresponds to the viscosity coefficient of nanofluids with CNTs of different lengths in the Newtonian flow regime (compare with Fig. 3.38). All subsequent curves describe the dependence of viscosity on the CNT length already in the pseudoplastic flow regime. As the length of the CNT increases, the viscosity also increases, but this increase decreases monotonically with increasing shear rate. Finally, at the maximum shear rate, the viscosity coefficients of all the nanofluids considered here are very close (practically, with calculation accuracy, they are the same, see insert in Fig. 3.40).

The obtained data from molecular dynamics simulation of the dependence of nanofluid viscosity on CNT length at different shear rates are well described by the following approximating correlation

$$\mu(\dot{\gamma}, L) = \mu_{bf}(\dot{\gamma}) + k_1(\dot{\gamma})\, L^{k_2(\dot{\gamma})}, \tag{3.43}$$

where μ_{bf} is described by the formula (3.35), and coefficients k_1 and k_1 depend on the shear rate and equal to

$$k_1(\dot{\gamma}) = 2.1 \cdot 10^{-5} + 2.8 \cdot 10^{-5} \cdot \exp\bigl(-0.45 \cdot 10^{-10} \cdot \dot{\gamma}\bigr),$$

$$k_2(\dot{\gamma}) = 1.32 - 0.76 \cdot \exp\bigl(0.06 \cdot 10^{-10} \cdot \dot{\gamma}\bigr).$$

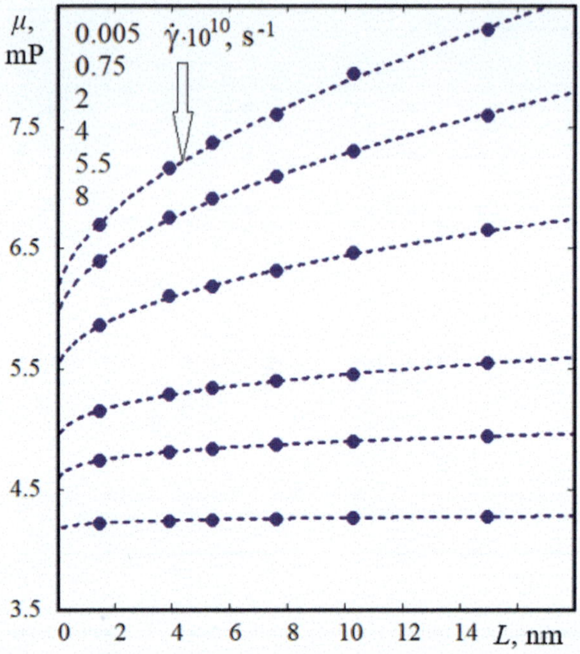

Fig. 3.43 Dependence of nanofluid viscosity on the length of nanotubes at different shear rates

The shorter the length of the nanotubes, the closer the viscosity of the nanofluid is to the viscosity of the base fluid at the current shear rate. As the length of the nanotubes increases, the increment in the viscosity of the nanofluid will increase according to a power law. However, with increasing shear rate, the difference between the viscosity of nanofluids with CNTs of different lengths decreases, and the viscosity of all nanofluids tends to the viscosity of the base fluid for any nanotube length. If in the Newtonian regime the viscosity of a nanofluid with the longest CNTs is approximately 40% higher than that of the base one, then at a shear rate of 10^{11} s^{-1} this difference is only 6% (see Figs. 3.42 and 3.43). In this sense, the behavior of nanofluids with CNTs is similar to the behavior of nanofluids with spherical particles. In both cases, the viscosity of all nanofluids converges. At high shear rates, the system ceases to sense the size of dispersed particles.

At high shear rates, the system ceases to distinguish between the differences in the sizes of its constituent particles. However, the asymmetry of sufficiently long tubes significantly affects the process of momentum transfer in the corresponding nanofluid. With increasing shear rate, CNTs turn around and orient along the flow. This orientation is well looked in the Fig. 3.44.

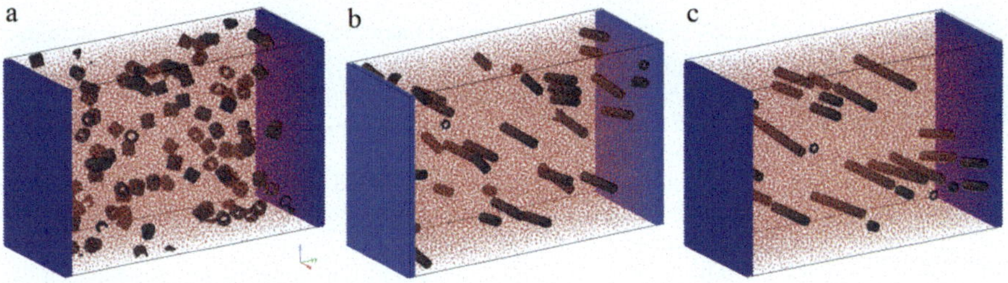

Fig. 3.44 Screenshots of calculated cells of nanofluids with the CNT diameter of 1.1 nm and a length of **a** 1.45 nm, **b** 3.90 nm and **c** 7.59 nm

References

1. Abdullah A, Mohamad IS, Hashim AYB et al (2016) Thermal conductivity and viscosity of deionized water and ethylene glycol-based nanofluids. J Mech Eng Sci (JMES) 10(3):2249–2261
2. Acrivos A, Chang EY (1986) A model for estimating transport quantities in two-phase materials. Phys Fluids 29(3–4):459–464
3. Aladag B, Halelfadl S, Doner N et al (2012) Experimental investigations of the viscosity of nanofluids at low temperatures. Appl Energy 97:876–880
4. Allen MP, Tildesley DJ (2017) Computer simulation of liquids. Oxford University Press, Oxford
5. Andrade da ENC (1930) Nature. Viscosity of liquids. Phil Magazine 125:309–324
6. Andrade da ENC (1934) A theory of the viscosity of liquids. Phil Magazine 17(112):497–514
7. Ansón-Casaos A, Ciria JC, Sanahuja-Parejo O et al (2020) The viscosity of dilute carbon nanotube (1D) and graphene oxide (2D) nanofluids. Phys Chem Chem Phys 22:11474–11484
8. Ashurst WT, Hoover WG (1975) Dense-fluid shear viscosity via nonequilibrium molecular dynamics. Phys Rev A 11(2):658–678
9. Baratpour M, Karimipour A, Afrand M et al (2016) Effects of temperature and concentration on the viscosity of nanofluids made of single-wall carbon nanotubes in ethylene glycol. Int Commun Heat Mass 74:108–113
10. Batchelor GK (2012) An introduction to fluid dynamics. Cambridge University Press, Cambridge
11. Batchelor GK, Green JT (1972) The hydrodynamic interaction of two-small freely-moving spheres in a linear flow field. J Fluid Mech 56(2):375–400
12. Batchelor GK, Green JT (1972) The determination of the bulk stress in a suspension of spherical particles to order c^2. J Fluid Mech 56(2):401–427
13. Batchelor GK (1976) Brownian diffusion of particles with hydrodynamic interaction. J Fluid Mech 74(1):1–29
14. Batchelor GK (1977) The effect of Brownian motion on the bulk stress in a suspension of spherical particles. J Fluid Mech 83(1):97–117
15. Berthier L, Barrat J-L (2002) Nonequilibrium dynamics and fluctuation-dissipation relation in a sheared fluid. J Chem Phys 116:6228
16. Bird RB, Armstrong RC, Hassager O (1987) Dynamics of polymeric liquids, vol 1, Fluid mechanics. Wiley, New York
17. Born M, Green HS (1947) A general kinetic theory of liquids III. Dynamical properties. Proc Royal Soc Lond Ser A Math Phys Sci 190:455–474
18. Boungiorno J (2006) Convective transport in nanofluids. J Heat Transfer 128:240–250
19. Brinkman HC (1952) The viscosity of concentrated suspensions and solutions. J Chem Phys 20:571

20. Chapman S, Cowling TG (1990) The mathematical theory of non-uniform gases. Cambridge University Press, Cambridge
21. Chen H, Ding Y, Tan C (2007) Rheological behavior of nanofluids. New J Phys 9:367
22. Chhabra RP, Richardson JF (2008) Non-Newtonian flow and applied rheology. Butterworth-Heinemann, Oxford
23. Colla L, Fedele L, Scattolini M et al (2012) Water-based Fe_2O_3 nanofluid characterization: thermal conductivity and viscosity measurements and correlation. Adv Mech Eng 2012:674947
24. Dalkilic AS, Küçükyıldırım BO, Eker AA et al (2017) Experimental investigation on the viscosity of water-CNT and antifreeze-CNT nanofluids. Int Com Heat Mass Transfer 80:47–59
25. Das SK, Putra PP, Thiesen P et al (2003) Temperature dependence of thermal conductivity enhancement for nanofluids. J Heat Transfer 125:567–574
26. Desai T, Keblinski P, Kumar SK (2005) Molecular dynamics simulations of polymer transport in nanocomposites. J Chem Phys 122(13):134910
27. Devarajan DS, Nourian P, McKenna GB et al (2020) Molecular simulation of nanocolloid rheology: Viscosity, viscoelasticity, and time concentration superposition. J Rheology 64:529
28. Eilers H (1941) Die viskosität von emulsionen hochviskoser stoffe als funktion der konzentration. Kolloid-Zeitschrift 97(3):313–321
29. Einstein A (1906) A new determination of molecular dimensions. Ann Phys 19:289–306
30. Esumi K, Ishigami M, Nakajima A et al (1996) Chemical treatment of carbon nanotubes. Carbon 34:279–281
31. Evans DJ, Morriss GP (1990) Statistical mechanics of nonequilibrium liquids. Academic Press, London
32. Everaers R, Ejtehadi MR (2003) Interaction potentials for soft and hard ellipsoids. Phys Rev E 67:041710
33. Frankel NA, Acrivos A (1967) On the viscosity of a concentrated suspension of solid spheres. Chem Eng Sci 22(6):847–853
34. YaI F (1955) Kinetic theory of liquids. Oxford University Press, Oxford
35. Garcia-Colin LS (1980) Non-linear transport theory. In: Garrdo L (ed) Lecture notes in physics. System far from equilibrium, vol 132. Springer, Berlin, pp 226–243
36. Garg J, Poudel B, Chiesa MG et al (2008) Enhanced thermal conductivity and viscosity of copper nanoparticles in ethylene glycol nanofluid. J Appl Phys 103:074301
37. Grest GS, in 't Veld PJ, Lechman JB, Tokuyama M, Oppenheim I, Nishiyama H (2008) Molecular simulations of nanoparticles in an explicit solvent. AIP Conf Proc 982:304–311
38. Grönroos A, Pentti P, Hanna K (2008) Ultrasonic degradation of aqueous carboxymethylcellulose: effect of viscosity, molecular mass and concentration. Ultrason Sonochem 15:644–648
39. Guzey DV, Minakov AV, Rudyak VYa et al (2014) Measuring the heat-transfer coefficient of nanofluid based on copper oxide in a cylindrical channel. Tech Phys Lett 40(3):203–206
40. de Guzman J (1913) Relation between fluidity and heat of fusion. Annales Soc Espan Fia Y Quim 11:353–362
41. Halefadl S, Estellé P, Aladag B et al (2013) Viscosity of carbon nanotubes based nanofluids: Influence of concentration and temperature. Int J Thermal Sci 71:111–117
42. Hamaker HC (1937) The London—Van der Waals attraction between spherical particles. Physica 4(10):1058–1072
43. Harish S (2014) Thermal conductivity enhancement of liquid and solid with single-walled carbon nanotubes. Doctoral thesis. University of Tokyo
44. He Y, Jin Y, Chen H et al (2007) Heat transfer and flow behavior of aqueous suspensions of TiO_2 nanoparticles (nanofluids) flowing upward through a vertical pipe. Int J Heat Mass Transfer 50(11–12):2272–2281
45. Heine DR, Petersen MK, Grest GS (2010) Effect of particle shape and charge on bulk rheology of nanoparticle suspensions. J Chem Phys 132(18):184509
46. Heyhat MM, Rajabpour A, Abbasi M et al (2018) Importance of nanolayer formation in nanofluid properties: Equilibrium molecular dynamic simulations for Ag-water nanofluid. J Mol Liquids 264:699–7052

47. Honarkhah R, Bakhshan Y, Rahmati M et al (2018) Investigation into the effects of nanoparticle size and channel depth on the thermophysical properties of water nanofluids in the nano-channel using molecular dynamics simulation. Environmental Energy Economic Res 2(1):21–36
48. Hong RY, Ren ZQ, Han YP et al (2007) Rheological properties of water-based Fe3O4 ferrofluids. Chem Eng Sci 62:5912
49. Hoover WG (1985) Canonical dynamics: equilibrium phase-space distributions. Phys Rev A 31(3):1695–1697
50. Hosseini SSH, Shahrjerdi A, Vazifeshenas Y (2011) A review of relations for physical properties of nanofluids. Australian J Basic App Sci 5(10):417–435
51. In't Veld PJ, Plimpton SJ, Grest GS (2008) Accurate and efficient methods for modeling colloidal mixtures in an explicit solvent using molecular dynamics. Computer Phys Com 179(5):320–329
52. In't Veld PJ, Petersen MK, Grest GS (2009) Shear thinning of nanoparticle suspensions. Phys Rev E 79:021401
53. Israelachvili J (2011). Intermolecular and surface forces. Elsevier Academic Press Inc.
54. Jabbari F, Rajabpour A, Saedodin S (2017) Thermal conductivity and viscosity of nanofluids: a review of recent molecular dynamics studies. Chem Eng Sci 174:67–81
55. Jabbari F, Saedodin S, Rajabpour A (2019) Experimental investigation and molecular dynamics simulations of viscosity of CNT-water nanofluid at different temperatures and volume fractions of nanoparticles. J Chem Eng Data 64(1):262–272
56. Jabbari F, Rajabpour A, Saedodin S (2019) Viscosity of carbon nanotube/water nanofluid. J Thermal Anal Calorimetry 135(3):1787–1796
57. Jabbarzadeh A, Atkinson JD, Tanner RI (2003) Effect of moleculars shape on rheological properties in molecular dynamics simulation of star, H, comb, and linear polymer melts. Macromol 36(13):5020–5031
58. Jeffery GB (1922) The motion of ellipsoidal particles immersed a viscous fluid. Proc R Soc London A 102:161–179
59. Johnson RW (1998) Handbook of fluid dynamics. CRC Press LLC, Boca Raton
60. Ju S-P (2005) A molecular dynamics simulation of the adsorption of water molecules surrounding an Au nanoparticle. J Chem Phys 122:094718
61. Kihara T (1976) Intermolecular forces. Wiley, New York
62. Kalmanovich Y (1982) Kinetic theory of non-ideal gases and non-ideal plasmas. Pergamon Press, Oxford
63. Koca DH, Doganay S, Turgut A et al (2018) Effect of particle size on the viscosity of nanofluids: a review. Renew Sustain Energy Rev 82:1664–1674
64. Krekelberg WP, Truskett TM, Ganesan V (2009) Relationship between shear viscosity and structure of model colloidal suspension. Chem Eng Com 197(1):63–75
65. Kriger IM (1972) Rheology of monodispersed latices. Adv Colloid Interface Sci 3(1):111–132
66. Krieger IM, Dougherty TJ (1959) A mechanism for non-Newtonian flow in suspensions of rigid spheres. J. Rheology 3(1):137–145
67. Kwak K, Kim C (2005) Viscosity and thermal conductivity of copper oxide nanofluid dispersed in ethylene glycol. Korea-Australia Rheology J 17(2):35–40
68. Lees AW, Edwards SF (1972) The computer study of transport process under extreme conditions. J Phys C 5:1921–1929
69. Leong KY, Fadhillahanafi NM, Chew SP (2016) Viscosity characteristic of carbon nanotube based nanofluids at room temperature. ARPN J Eng App Sci 11(10):6584–6588
70. Lewis TB, Nielsen LE (1968) Viscosity of dispersed and aggregated suspensions of spheres. Trans Soc Rheol 12(3):421–443
71. Lifshits EM, Dzyaloshinsky IE, Pitaevsky LP (1961) General theory of Van der Waals forces. Adv Phys Sci 73:381–422
72. Lou Z, Chen C, Cui YC et al (2022) Interfacial interaction–driven rheological properties of quartz nanofluids from molecular dynamics simulations and density functional theory calculations. J Mol Modeling 28:189

73. Lou Z, Yang M (2015) Molecular dynamics simulations on the shear viscosity of Al_2O_3 nanofluids. Comput Fluids 117:17–23
74. Loya A, Stair JL, Jafri AR et al (2015) A molecular dynamic investigation of viscosity and diffusion coefficient of nanoclusters in hydrocarbon fluids. Comp Mat Sci 99:242–246
75. Loya A, Stair JL, Ren G (2015) Simulation and experimental study of rheological properties of CeO_2–water nanofluid. Int Nano Lett 5:1–7
76. Lu G, Duan Y-Y, Wang X-D (2014) Surface tension, viscosity, and rheology of water-based nanofluids: a microscopic interpretation on the molecular level. J Nanoparticle Res 16(9):2564
77. Lu W-Q, Fan Q-M (2008) Study for the particle's scale effect on some thermophysical properties of nanofluids by a simplified molecular dynamics method. Eng Anal Boundary Elements 32(4):282–289
78. Mahbubul IM, Saidur R, Amalina MA (2012) Latest developments on the viscosity of nanofluids. Int J Heat Mass Transfer 55:874–885
79. Maïga SEB, Nguyen CT, Galanis N et al (2004) Heat transfer behaviors of nanofluids in a uniformly heated tube. Superlatt Microstruct 35(3–6):543–557
80. Malkin AYa, Isaev AI (2006) Rheology. Concepts, methods, and applications. Chemical Publishing, Toronto
81. Maron HM, Pierce PI (1956) Application of Ree-Eyring generalized flow theory to suspensions of spherical particles. J Colloid Sci 11:80–95
82. Mason TG (2000) Estimating the viscoelastic moduli of complex fluids using the generalized Stokes-Einstein equation. Rheological Acta 39:371–378
83. Mason TG, Gang H, Weitz DA (1997) Diffusing-wave-spectroscopy measurements of viscoelasticity of complex fluids. J Optical Soc America A14:139–149
84. Mason TJ, Lorimer JP (2002) Applied sonochemistry, Wiley VCH
85. Masoumi N, Sohrabi N, Behzadmehr A (2009) A new model for calculating the effective viscosity of nanofluids. J Phys D: Applied Phys 42(5):055501
86. Mewis J, Wagner NJ (2012) Colloidal suspension rheology. Cambridge University Press, Cambridge
87. McPhie MG, Daivis IK, Snook PJ (2006) Viscosity of a binary mixture: Approach to the hydrodynamic limit. Phys Rev E 74:021301
88. Minakov AV, Rudyak VYa, Guzei DV et al (2015a) Measurement of the heat transfer coefficient of a nanofluid based on water and copper oxide particles in a cylindrical channel. High Temperature 53(2):246–253
89. Minakov AV, Rudyak VYa, Guzei DV et al (2015) Measurements of the thermal-conductivity coefficient of nanofluids by the hot-wire method. J Eng Phys Thermophys 88(1):149–162
90. Minakov AV, Rudyak VYa, Pryazhnikov MI (2018) Rheological behavior of water and ethylene glycol nanofluids with oxide nanoparticles. Colloids Surfaces A: Physicochem Eng Aspects 554: 279–285
91. Minakov AV, Rudyak VYa, Pryazhnikov MI (2018a) About rheology of nanofluids. AIP Conf Proc 2027:030141
92. Minakov AV, Rudyak VYa, Pryazhnikov MI (2020) Systematic experimental study of the viscosity of nanofluids. Heat Transfer Eng 41(5):457–460
93. Mooney M (1951) The viscosity of a concentrated suspension of spherical particles. J Colloid Sci 6–8(1):162–170
94. Morrison ID, Ross S (2002) Colloidal dispersions. Wiley, New York
95. Murshed SMS, Estellé PA (2017) State of the art, review on viscosity of nanofluids. Renew Sustain Energy Rev 76:1134–1152
96. Müller-Plathe F (1999) Reversing the perturbation in nonequilibrium molecular dynamics: An easy way to calculate the shear viscosity of fluids. Phys Rev E 59(5):4894–4898
97. Nagatani T (1979) Statistical theory of effective viscosity in a random suspension. J Phys Soc Jpn 47(1):320–326
98. Namarvari H, Razmara N, Miranda CR et al (2021) Effect of SWCNT volume fraction on the viscosity of water based nanofluids. J Mol Modeling 27:253

99. Namburu PK, Kulkarni DP, Dandekar A et al (2007) Experimental investigation of viscosity and specific heat of silicon dioxide nanofluids. Micro Nano Lett 2(3):67–71
100. Namburu PK, Kulkarni DP, Misra D et al (2007) Viscosity of copper oxide nanoparticles dispersed in ethylene glycol and water mixture. Exp Thermal Fluid Sci 32:397–402
101. Nguyen CT, Desgranges F, Galanis N et al (2008) Viscosity data for Al_2O_3–water nanofluid—hysteresis: is heat transfer enhancement using nanofluids reliable? Int J Thermal Sci 47(2):103–111
102. Nosé S (1984) A unified formulation of the constant temperature molecular-dynamics methods. J Chem Phys 81(1):511–519
103. Omrani AN, Esmaeilzadeh E, Jafari M et al (2019) Effects of multi walled carbon nanotubes shape and size on thermal conductivity and viscosity of nanofluids. Diamond Rel Mat 93:96–104
104. Oueslati FS, Bennace R (2011) Heterogeneous nanofluids: natural convection heat transfer enhancement. Nanoscale Res Lett 6:222
105. Ponmozhi J, Gonçalves FAMM, Ferreira AGM et al (2010) Thermodynamic and transport properties of CNT-water based nanofluids. J Nano Research 11:101–106
106. Pak BC, Cho YI (1998) Hydrodynamic and heat transfer study of dispersed fluids with submicron metallic oxide particles. Exp Heat Transfer 11:151–170
107. Pal R (2016) Fundamental rheology of disperse systems based on single-particle mechanics. Fluids 1:40
108. Pastoriza-Gallego MJ, Lugo L, Soyo JLL et al (2011) Rheological non-Newtonian behavior of ethylene glycol-based Fe_2O_3 nanofluids. Nanoscale Res Lett 6:560
109. Patra AK, Nayak MK, Misra A (2020) Viscosity of nanofluids—A review. Int J Thermofluid Sci Technol 7(2):070202
110. Puertas AM, Fuchs M, Cates ME (2002) Comparative simulation study of colloidal gels and glass. Phys Rev Lett 88:098301
111. Puertas AM, Fuchs M, Cates ME (2003) Simulation study of nonergodicity transitions: gelation in colloidal systems with short-range attractions. Phys Rev E 67:031406
112. Puliti G, Paolucci S, Sen M (2012) Thermodynamic properties of gold–water nanofluids using molecular dynamics. J Nanopart Res 14:1296
113. Rapaport DC (2005) The art of molecular dynamics simulation. Cambridge University Press, Cambridge
114. Razmara N, Namarvari H, Meneghini JR (2019) A new correlation for viscosity of model water-carbon nanotube nanofluids: molecular dynamics simulation. J Mol Liquids 293:111438
115. Reid RC, Prausnitz JM, Sherwood TK (2004) The properties of gases and liquids. McGraw-Hill, New York
116. Rudyak VYa (1987) Statistical theory of dissipative processes in gases and liquids. Science, Novosibirsk
117. Rudyak VYa (2004) Statistical aerohydromechanics of homogeneous and heterogeneous media, vol 1, kinetic theory. NSUACE, Novosibirsk
118. Rudyak VYa (2005) Statistical aerohydromechanics of homogeneous and heterogeneous media, vol 2, hydromechanics. NSUACE, Novosibirsk
119. Rudyak VYa (2013) Viscosity of nanofluids. Why it is not described by the classical theories. Adv Nanoparticles 2:266–279
120. Rudyak VYa (2019) Thermophysical characteristics of nanofluids and transport process mechanisms. J Nanofluids 8:1–16
121. Rudyak VYa (2023) Rheology of nanofluids. Experiments and molecular dynamics modeling. J Phys Conf Series 2585:012009
122. Rudyak VYa, Belkin AA (2019) On the effect of nanoparticles on fluid structure. Colloid J 81(4):487–490
123. Rudyak VYa, Belkin AA, Egorov VV et al (2006) Modeling of transport processes by molecular dynamics method. II. The resistance force and effective viscosity in nanosuspension. Preprint 1(18), NSUACE, Novosibirsk

124. Rudyak VYa, Belkin AA, Krasnolutskii SL (2021) Molecular dynamics modelling transport processes of fluids and nanofluids in bulk and nanochannels. In: Köhler S (ed) Advances in molecular dynamics simulations research. Nova science publisher, New York, pp 1–86
125. Rudyak VYa, Belkin AA, Rafalskaya TA (2023) Molecular dynamics modeling rheology of nanofluids. Tech Phys Lett 49(10):1–4
126. Rudyak VYa, Belkin AA, Rafalskaya TA (2024) Molecular dynamics study of the rheology of benzene-based nanofluids with metal particles. J Mol Liquids 403:124805
127. Rudyak VYa, Dashapilov GR, Minakov AV et al (2022) Viscoelastic properties of nanofluids with carbon tubes. Tech Phys Lett 48:184–187
128. Rudyak VYa, Dashapilov GR, Minakov AV et al (2023) Comparative characteristics of viscosity and rheology of nanofluids with multi-walled and single-walled carbon nanotubes. Diamond Rel Mat 132:109616
129. Rudyak VYa, Dimov SV, Kuznetsov VV et al (2013) Measurement of the viscosity coefficient of an ethylene glycol–based nanofluid with silicon dioxide particles. Doklady Phys 58(5):173–176
130. Rudyak VYa, Dimov SV, Kuznetsov VV (2013a) On the dependence of the viscosity coefficient of nanofluids on particle size and temperature. Tech Phys Lett 39(9):779–782
131. Rudyak VYa, Krasnolutskii SL (2014) Dependence of the viscosity of nanofluids on nanoparticle size and material. Phys Lett A 378:1845–1849
132. Rudyak VYa, Krasnolutskii SL (2015) Simulation of the nanofluid viscosity coefficient by the molecular dynamics method. Technical Physics. 60(6):798–804
133. Rudyak VYa, Krasnolutskii SL, Belkin AA et al (2021a) Molecular dynamics simulation of water based nanofluids viscosity. J Thermal Anal Calorimetry 145(6):2983–2990
134. Rudyak VYa, Krasnolutskii SL, Lezhnev EV (2024) Molecular dynamics study of nanofluids viscosity with carbon tubes. Nanosystems Phys Chem Math 15(1):37–45
135. Rudyak VYa, Minakov AV (2016) Modern problems of micro and nanofluidics. Science, Novosibirsk
136. Rudyak V, Minakov A, Pryazhnikov M (2021) Preparation, characterization, and viscosity studding the single-walled carbon nanotube nanofluid. J Mol Liquids 329(1):115517
137. Rudyak VYa, Minakov AV, Smetanina MS et al (2016) Experimental data on the dependence of the viscosity of water—and ethylene glycol–based nanofluids on the size and material of particles. Doklady Phys 61(3):152–154
138. Rudyak V, Rafalskaya T (2024) Molecular dynamics simulation of viscosity and rheology of nanofluids with single wall carbon nanotubes. E3S Web of Conf 578:01032
139. Rudyak VYa, Tret′yakov DS (2021) Rheological properties of water- and ethylene glycol-based nanofluids with single wall carbon nanotubes. J Eng Phys Thermophys 94(5):1208–1216
140. Sabiha MA, Mostafizur RM, Saidur R et al (2016) Experimental investigation on thermo-physical properties of single walled carbon nanotube nanofluids. Int J Heat Mass Transfer 93:862–871
141. Sadri R, Ahmadi G, Togun H et al (2014) An experimental study on thermal conductivity and viscosity of nanofluids containing carbon nanotubes. Nanoscale Res Lett 9:151
142. Said Z, Sundar LS, Tiwari AK et al (2022) Recent advances on the fundamental physical phenomena behind stability, dynamic motion, thermophysical properties, heat transport, applications, and challenges of nanofluids. Phys Reports 946(2):194
143. Sharma AK, Tiwari AK, Dixit AR (2016) Rheological behavior of nanofluids: A review. Renew Sustain Energy Rev 53:779–791
144. Singh N, Chand G, Kanagaraj S (2012) Investigation of thermal conductivity and viscosity of carbon nanotubes–ethylene glycol nanofluids. Heat Transfer Eng 33(9):821–827
145. Stuart SJ, Tutein AB, Harrison JA (2000) A reactive potential for hydrocarbons with intermolecular interactions. J Chem Phys 112(14):6472–6486
146. Taghizadeh MT, Abdollahi R (2015) Ultrasonic degradation of polyvinyl pyrrolidone (PVP): effect of power of ultrasound, temperature and concentration. American Chem Sci J 9(3):16508
147. Tanner RI, Walters K (1998) Rheology: an historical perspective. Elsevier, Amsterdam

148. Tang W, Advani SG (2006) Drag on a nanotube in uniform liquid argon flow. J Chem Phys 125:174706
149. Temperley HN, Rowlinson JS, Rushbrooke GS (1968) Physics of simple liquids. North-Holland Publishing Company, Amsterdam
150. Thompson AP et al (2022) LAMMPS—A flexible simulation tool for particle-based materials modelling at the atomic, meso, and continuum scales. Comp Phys Com 271:108171
151. Timofeeva EV, Smith DS, Yu W et al (2010) Particle size and interfacial effects on thermo-physical and heat transfer characteristics of water-based α-SiC nanofluids. Nanotechnology 21:215703
152. Travis KP, Daivis PJ, Evans DJ (1995) Computer simulation algorithms for molecules undergoing planar Couette flow: A nonequilibrium molecular dynamics study. J Chem Phys 103:1109
153. Tseng WJ, Lin K-C (2003) Rheology and colloidal structure of aqueous TiO2 nanoparticle suspensions. Mater Sci Eng A 355:186–192
154. Vakili-Nezhaad G, Dorany A (2012) Effect of single-walled carbon nanotube on the viscosity of lubricants. Energy Procedia 14:512–517
155. Vand V (1948) Viscosity of solutions and suspensions I. Theory. J Phys Chem 52(2):277–299
156. Viswanath DS, Ghosh TK, Prasad DHL et al (2007) Viscosity of liquids: theory, estimation, experiment, and data. Springer
157. Volkov AI, Zharsky IM (2005) The great chemical handbook. Modern School, Moscow
158. Wang L, Chen H, Witharana S (2013) Rheology of nanofluids: a review. Recent Pat Nanotechnol 7:232–246
159. Wang X, Xu X, Choi SUS (1999) Thermal conductivity of nanoparticle–fluid mixture. J Thermophys Heat Transfer 13(4):474–480
160. Wang X-Q, Mujumdar AS (2008) A review on nanofluids—Part I: theoretical and numerical investigations. Brazilian J Chem Eng 25(4):613–630
161. Warren BE (1969) X-ray diffraction. Addison-Wesley Pub Co., Reading, Mass
162. Weitz DA, Pine DJ (1993) Diffusing-wave spectroscopy. In: Brown W (ed) Dynamic light scattering. Clarendon Press, New York
163. Wu L, Keer LM, Lu J et al (2018) Molecular dynamics simulations of the rheological properties of graphene–PAO nanofluids. J Mat Sci 53(23):15969–15976
164. Xie H, Lee H, Youn W et al (2003) Nanofluids containing multiwalled carbon nanotubes and their enhanced thermal conductivities. J App Phys 94(8):4967–4971
165. Xing M, Yu J, Wang R (2015) Thermo-physical properties of water-based single walled carbon nanotube nanofluid as advanced coolant. App Thermal Eng 87:344–351
166. Yu L, Bian Y, Liu Y et al (2019) Experimental investigations on rheological properties of water based nanofluids with low MWCNT concentrations. Int J Heat Mass Transfer 135:175–185

Chapter 4
Thermal Conductivity of Nanofluids

4.1 Thermal Conductivity of Simple Fluids

This book primarily discusses weakly nonequilibrium processes, where the fluxes are proportional to the gradients of relevant macroscopic variables. The relationship between the heat flux $\boldsymbol{J}_3$ and the temperature gradient has a simple form

$$\boldsymbol{J}_3 = \lambda \nabla T, \tag{4.1}$$

where λ is the thermal conductivity coefficient, which, like the viscosity coefficient, is a function of state (density, temperature, etc.).

The interpretation of thermal conductivity is well comprehended and transparent in the case of gases, particularly rarefied ones. Here heat transfer is carried out by molecules during their movement and collisions. Therefore, in rarefied gases, the thermal conductivity coefficient is proportional to the mean free path of molecules and is independent of density [25]. Similar to momentum transfer, there is only one transfer mechanism for thermal conductivity, which has been previously called as kinetic. The presence of one mechanism, in particular, means that the coefficients of viscosity and thermal conductivity are related to each other. It is easy to see that they are simply proportional to each other, $\lambda \sim \mu$. However, this is not the case in fluids. A fluid, unlike a gas, is characterized by the presence of a short-range order. This means that fluid molecules are quasi-bonded, and thermal conductivity can be interpreted as heat (energy) transfer by means of these quasi-bonds. In this sense, thermal conductivity mechanism is some collective process taking place in the fluid.

Although the kinetic theory for fluids cannot be constructed in principle, the resonance caused by the derivation of transport coefficients based on the solution of the Boltzmann equation was quite stunning. It is not surprising that Debye's first attempt to derive the thermal conductivity coefficient of a liquid included as a starting

V. Rudyak, *Thermophysical Properties of Nanofluids*, Fluid Mechanics and Its
Applications 140, https://doi.org/10.1007/978-3-031-97995-8_4

point a formula of the type obtained for the thermal conductivity of a rarefied gas. P. Debye assumed that heat transfer was accomplished by phonons, similar to a solid. Therefore, the thermal conductivity coefficient was expressed as [40]

$$\lambda = \frac{1}{3}\rho a C_v l_{ph}, \tag{4.2}$$

where a is the speed of sound, C_v is the specific heat capacity, and l_{ph} is the mean free path of phonons.

In fact, this formula was also the key one in the theory proposed by Bridgman [22]. He believed that the mean free path of photons is determined by the average distance between molecules. As a result, the formula for the thermal conductivity coefficient was obtained in the following form

$$\lambda = 3kan^{2/3}, \tag{4.3}$$

where n is the numerical density of the fluid molecules.

As indicated in the monograph [21] formula (4.3) fits the experimental data quite well. However, setting the numerical coefficient to 2.8 improves this compliance even further. Later, numerous semi-empirical and empirical formulas have been proposed for calculating the thermal conductivity coefficient, some of which are available in Reid et al. [27]. It is important to note that the thermal conductivity of fluids depends also on pressure and temperature. The thermal conductivity of simple fluids, which are the majority, decreases as temperature increases. This decrease is almost linear below and near the boiling point T_0 [27]

$$\lambda(T) = \lambda(T_0)[1 + \alpha(T - T_0)], \tag{4.4}$$

where α is some constant. There are also other experimental correlations, for example [107].

Relation (4.4) does not describe the temperature dependence of thermal conductivity in certain polar fluids and fluids with multiple hydroxyl groups, such as water, phenol, glycerol, and others. These fluids are known as anomalous fluids, and their thermal conductivity actually increases with increasing temperature. At pressures up to approximately 30 atm, the thermal conductivity of fluids remains relatively constant.

4.2 Effective Thermal Conductivity Coefficient of Coarse Dispersed Fluids

Maxwell [102] first calculated the effective thermal conductivity coefficient of a coarse dispersed fluid. The idea of the method he used was to first find the changes in the temperature field of the carrier fluid caused by the dispersed particles. Then, using the disturbances found in this way and the Fourier law (4.1), he determined the effective thermal conductivity coefficient. In fact, it was in this paradigm that A. Einstein acted, calculating the effective viscosity coefficient of a dispersed fluid. Of course, he knew Maxwell's work, and in that sense A. Einstein was following in his footsteps.

Let's examine Maxwell's derivation more closely. So, at the first stage, the objective is to identify disturbances in the temperature field of an incompressible resting fluid with thermal conductivity coefficient λ_f, caused by an isolated particle of radius R with thermal conductivity coefficient λ_p. It is assumed that a constant temperature gradient is maintained in the fluid. Thus, the formulated problem is reduced to the solution of the following equation

$$\nabla^2 T = 0, \tag{4.5}$$

with boundary conditions

$$T_p = T_f, \lambda_p \frac{\partial T_p}{\partial r} = \lambda_f \frac{\partial T_f}{\partial r} \tag{4.5a}$$

at $r = R$. In addition, the temperature T and its gradient are bounded $\nabla T = \boldsymbol{A}$ at infinity. The solution of the problem (4.5) and (4.5a) has the following form [86, 102]

$$T = \begin{cases} \frac{3\lambda_f}{\lambda_p + 2\lambda_f}(\boldsymbol{A} \cdot \boldsymbol{r}), & |\boldsymbol{r}| < R, \\ (\boldsymbol{A} \cdot \boldsymbol{r}) + \frac{\lambda_f - \lambda_p}{\lambda_p + 2\lambda_f}\left(\frac{R}{r}\right)^3 (\boldsymbol{A} \cdot \boldsymbol{r}). & |\boldsymbol{r}| > R. \end{cases} \tag{4.6}$$

Assuming that the particles in the suspension are well-separated and that the temperature fields around each particle are independent and determined by the solution (4.6), it can consider the system as consisting of a carrier fluid and a fluid ball of radius Ω with a dispersed particle placed in the center. The temperature distribution on the surface of this ball is then equal to

$$T = \boldsymbol{A} \cdot \boldsymbol{r} + \frac{\lambda_f - \lambda_p}{\lambda_p + 2\lambda_f}\left(\frac{R}{\Omega}\right)^3 (\boldsymbol{A} \cdot \boldsymbol{r}).$$

The $(R/\Omega)^3$ ratio determines the volume concentration φ of particles. This ball has a thermal conductivity defined by the corresponding effective coefficient λ_e. Then, the temperature distribution on its surface is defined by the relation

$$T = \boldsymbol{A} \cdot \boldsymbol{r} + \frac{\lambda_f - \lambda_e}{\lambda_e + 2\lambda_f}\left(\frac{R}{\Omega}\right)^3 (\boldsymbol{A} \cdot \boldsymbol{r}).$$

From the comparison of the last two relations one can obtain the well-known Maxwell's formula for the thermal conductivity coefficient of a coarse dispersed fluid

$$\lambda_e = \lambda_f \left[1 + \frac{3(1-\kappa)\varphi}{1+2\kappa-(1-\kappa)\varphi} \right], \tag{4.7}$$

where $\kappa = \lambda_f / \lambda_p$.

Formula (4.7) is derived for non-interacting spherical particles. Later, Bruggeman [23] proposed a model that accounts for the interaction of randomly distributed particles. In this case, the effective thermal conductivity coefficient is determined implicitly by the following equation

$$\frac{(\lambda_p - \lambda_e)\varphi}{\lambda_p + 2\lambda_e} + \frac{(\lambda_f - 2\lambda_e)(1-\varphi)}{\lambda_f + 2\lambda_e} = 0 \tag{4.8}$$

Formally, this model has no limitations on the particle concentration, but in the limit of small concentrations, it leads to the same results as Maxwell's model. Furthermore, Bruggeman's model for the thermal conductivity coefficient of a dispersed fluid results in a rather complicated quadratic equation. Since we will only be discussing the thermal conductivity of nanofluids with small particle volume concentrations, we can limit ourselves to using model (4.7).

Models (4.7) and (4.8) were derived for spherical particles. Hamilton and Grosser [58] extended Maxwell's model to nonspherical particles, resulting in the following equation

$$\lambda_e = \lambda_f \left[\frac{1+(n-1)\kappa+(n-1)(1-\kappa)\varphi}{1+(n-1)\kappa-(1-\kappa)\varphi} \right]. \tag{4.9}$$

Here, $n = 3/\psi$ is the shape parameter, and ψ is the so-called sphericity, which is defined as the ratio of the surface area of a sphere with a volume equal to the particle volume to the particle surface area. The literature reports different values for the sphericity parameter [169, 179, 182], but the most widely accepted values are $\psi = 1$ for a spherical particle and $\psi = 0.5$ for a cylindrical particle.

It is evident that for spherical particles, formula (4.9) can be simplified to (4.7). Typically, the thermal conductivity of dispersed particles is much higher than that of the base fluid, especially for metal particles. For example, while the thermal conductivity of water is 0.6 W/m K, that of aluminum and copper is 209.3 and 389.6 W/m K, respectively. For metallic particles with $\kappa \ll 1$, Eqs. (4.7) and (4.9) are reduced to the following forms

$$\lambda_e = \lambda_f \frac{1+2\varphi}{1-\varphi}, \tag{4.7a}$$

$$\lambda_e = \lambda_f \frac{1 + (n-1)\varphi}{1 - \varphi}. \tag{4.9a}$$

The model (4.9) for cylindrical particles gives the following formula

$$\lambda_e = \lambda_f \left[\frac{1 + 5\kappa + 5(1-\kappa)\varphi}{1 + 5\kappa - (1-\kappa)\varphi} \right]. \tag{4.9b}$$

A comparison between this formula and (4.7) for spherical particles reveals that the thermal conductivity of fluids with dispersed cylindrical particles is higher than that of fluids with spherical particles at the same volume concentration.

The thermal conductivity of dispersed fluids has been studied by many researchers since the works mentioned above were published [23, 58, 102]. Relevant data can be found in reviews [37, 38, 63, 189]. For the purposes of this book the aforementioned models are quite adequate, although additional models may be mentioned later.

4.3 Thermal Conductivity of Nanofluids

As already mentioned, the study of thermophysical properties of nanofluids was initiated due to the expectation of their possible thermophysical applications. Over the past thirty years, hundreds of papers have been published with thermal conductivity measurements of nanofluids. This large number of papers is due, in part, to the fact that for a long time the data obtained seemed (and were) very contradictory. At the same time, it became evident that the thermal conductivity of nanofluids, as well as their viscosity, is not described by classical theories (see Sect. 4.2). Reviews published in different years provide a comprehensive background on the studies of thermal conductivity of nanofluids [9, 12, 37, 54, 55, 57, 117, 189, 190]. This section summarizes experimental data obtained by the author and their research team in different years [104, 121, 123, 130, 142] that are consistent with data available in literature.

The thermal conductivity coefficient was measured using the hot-wire method. A review of commonly used methods for measuring the thermal conductivity of nanofluids can be found in papers [37, 54]. A detailed description of the setup used and its testing is provided in our paper [104]. The measuring setup is based on the Wheatstone Bridge circuit, which is used to measure the electrical resistance of a heated wire [128, 129]. The measurement principle of thermal conductivity by the hot wire method is based on the linear relationship between the increase in wire temperature and the logarithm of the heating time. The thermal conductivity coefficient was determined by averaging five to ten measurements. The nanofluid temperature was kept constant during the measurements. The test bench was repeatedly tested for thermal conductivity measurements of various known fluids, including water, ethylene glycol, machine oil, and others. The final relative error in measuring the thermal conductivity coefficient of the liquid using this method did not exceed 3%.

4.3.1 Dependence of Thermal Conductivity on Particle Concentration

The dependence of the thermal conductivity coefficient on the nanoparticle concentration triggered the subsequent appearance of a huge number of publications, since it became evident that the obtained data were not universal and could not be described by Maxwell's theory (4.7) even at low particle concentrations. When studying the viscosity of nanofluids, it became evident that it depends on various parameters. To understand the reasons for the similar behavior of thermal conductivity, it is necessary to sequentially consider the relevant experimental data and study the effect of different parameters of the nanofluid on its thermal conductivity.

Data on the measured relative thermal conductivity coefficients $\lambda_r = \lambda/\lambda_f$ of several water-based nanofluids with oxide particles at 25 °C are presented in Table 4.1. In all cases, the thermal conductivity of nanofluids increases significantly with increasing particle concentration. The excess thermal conductivity of nanofluids over the corresponding value for water at 6% particle volume concentration ranges from 17 to 24%, which is considerably higher than the values determined by formula (4.7). For illustration purposes, Fig. 4.1 compares the thermal conductivity coefficients of the nanofluids presented in the table with the values calculated using formula (4.7).

One characteristic feature of nanofluid thermal conductivity is that its growth rate slows down as particle concentration increases. This is clearly demonstrated in Fig. 4.1b for the nanofluid containing 44 nm zirconium dioxide particles. This behavior of the nanofluid thermal conductivity has been observed in numerous experiments [1, 76, 77, 79, 92, 197], as well as in molecular dynamic simulations [131, 136] (see also Sect. 4.5). For small particle concentrations this dependence can be approximated by a simple formula.

$$\lambda_r = 1 + b_1\varphi - b_2\varphi^2. \tag{4.10}$$

However, the coefficients in this formula are not universal. So for nanofluid with Al_2O_3 particles of size 150 nm.

Table 4.1 Dependence of relative thermal conductivity coefficient of water-based nanofluids on particle concentration

Al_2O_3 (150 nm)		TiO_2 (150 nm)		ZrO_2 (44 nm)		ZrO_2 (105 nm)	
φ	λ_r	φ	λ_r	φ	λ_r	φ	λ_r
0.01	1.059	0.01	1.048	0.02	1.077	0.02	1.087
0.02	1.131	0.02	1.100	0.04	1.143	0.04	1.169
0.04	1.178	0.04	1.146	0.06	1.172	0.06	1.216
0.06	1.240	0.06	1.206	0.08	1.185	0.08	1.283

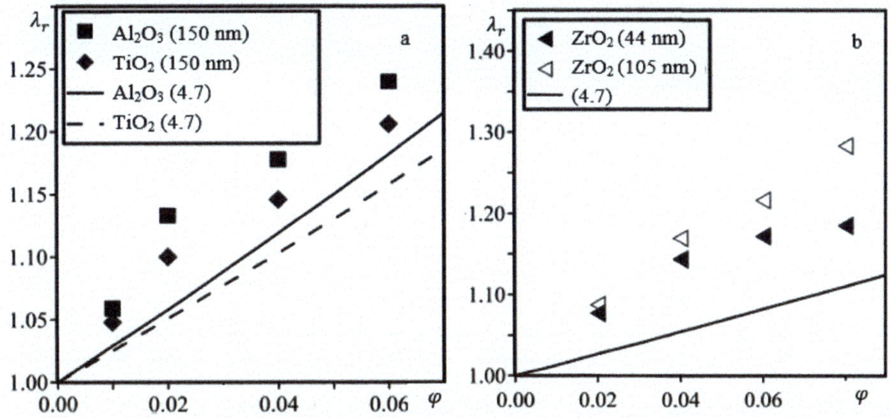

Fig. 4.1 The relative thermal conductivity coefficients of water-based nanofluids with Al_2O_3, TiO_2 (**a**), and ZrO_2 (**b**) particles depending on their concentration

$$\lambda_r = 1 + 6.42\varphi - 39.5\varphi^2, \tag{4.10a}$$

for nanofluid with TiO_2 particles of size 150 nm

$$\lambda_r = 1 + 4.82\varphi - 23.1\varphi^2, \tag{4.10b}$$

for nanofluid with ZrO_2 particles of size 44 nm

$$\lambda_r = 1 + 4.68\varphi - 29.6\varphi^2, \tag{4.10c}$$

and for nanofluid with ZrO_2 particles of size 105 nm

$$\lambda_r = 1 + 4.55\varphi - 12.7\varphi^2. \tag{4.10c}$$

Nanofluids containing metal particles exhibit qualitatively similar dependencies of the thermal conductivity coefficient on concentration. As an example, Figs. 4.2a, b show the dependence of the relative thermal conductivity coefficients of nanofluids based on water and ethylene glycol with metal aluminum particles (average particle size of 90.7 nm) and nanofluids based on ethylene glycol with copper particles of two different sizes (51.8 and 98.5 nm), which further are denoted as Cu I and Cu II, respectively (see Sects. 1.4 and 5.3). Here, the thermal conductivity coefficients of the nanofluids are presented depending on their weight concentrations *w*. This is due to the high densities of the metals considered; copper particles having density $\rho_p = 8.93$ g/cm^3, and aluminum particles having $\rho_p = 2.70$ g/cm^3. The weight concentration of copper particles in ethylene glycol-based nanofluids ranged from 2.5 to 20%, corresponding to a volume concentration variation from 0.32 to 3%.

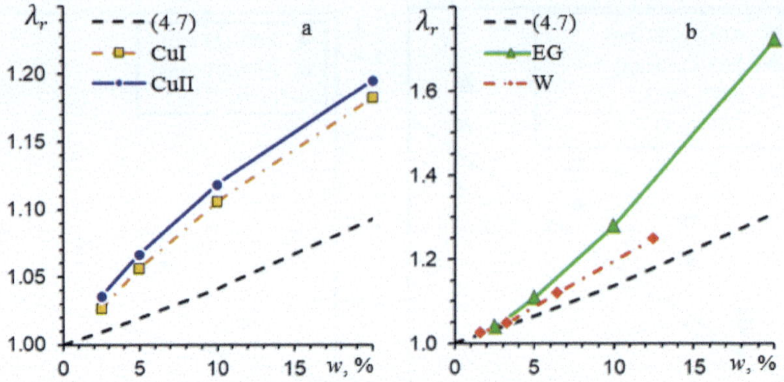

Fig. 4.2 The relative thermal conductivity coefficient of EG-based nanofluid with Cu I and Cu II particles (**a**), and EG- and W-based nanofluids with aluminum particles (**b**) depending on the nanoparticle weight concentration

Similarly, in water-based nanofluids with aluminum particles, the weight concentrations of particles varied from 1.67 to 12.5%, corresponding to volume concentration variations ranged from 0.5 to 5%. Weight concentrations of aluminum particles in ethylene glycol-based nanofluids varied from 2.5 to 20%, corresponding to volume concentration variations from ranged 1.05–9.35%.

In Fig. 4.2 the dashed line corresponds to Maxwell's formula (4.7). In all cases, the enhancement of the thermal conductivity coefficient of these nanofluids is significantly higher than that predicted by Maxwell's theory, and this enhancement increases rapidly with increasing particle concentration. At maximum particle concentration, the excess thermal conductivity coefficient of ethylene glycol-based nanofluid with aluminum particles reaches 72%, whereas Maxwell's theory predicts only about 30%. Comparison of Fig. 4.2a, b shows that nanofluids with more massive copper particles have slightly higher thermal conductivity than nanofluids with aluminum particles at the same volume concentration (recall that the maximum volume concentration of ethylene glycol-based nanofluid with copper particles is 3%, while the same volume concentration of nanofluid with aluminum particles corresponds to a weight concentration of 7.3%).

Dependences (4.10) are determined at a constant temperature. A change in temperature can significantly alter the thermal conductivity of any fluid. Figure 4.3 illustrates the nature of this change for two nanofluids: one based on water with Al_2O_3 particles and the other based on ethylene glycol with TiO_2 particles. The dependence of the relative coefficient of thermal conductivity is shown here. The volume concentrations of particles increase from bottom to top and are successively equal to 2, 4 and 6% in Fig. 4.3a, and 1, 2, 4 and 6% in Fig. 4.3b. In both cases, the particle diameter was 150 nm, and the temperature ranged from 20 to 50 °C. It is important to note that the thermal conductivity of the base fluids, both water and ethylene glycol increases almost linearly in the considered temperature range.

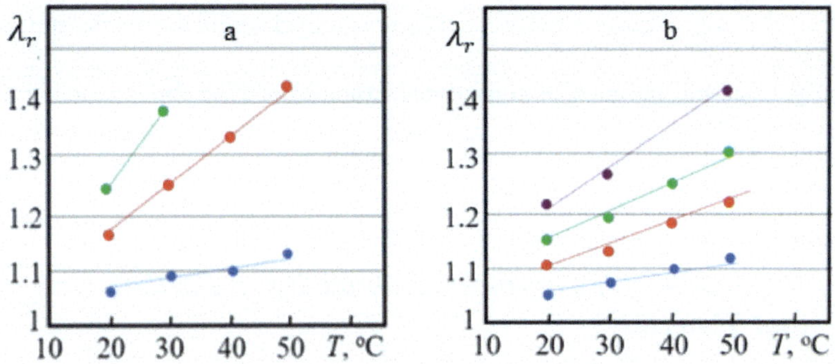

Fig. 4.3 Temperature dependence of relative thermal conductivity coefficient of nanofluids with different volume concentrations of nanoparticles: **a** based on water with Al_2O_3 particles, and **b** based on ethylene glycol with TiO_2 particles

In all cases, the thermal conductivity of nanofluids based on water and ethylene glycol increased with temperature. This increase was greater than the increase in thermal conductivity of the base fluid. For example, the thermal conductivity of pure water increases by approximately 6.6% as the temperature rises from 20 to 50 °C. Meanwhile, the thermal conductivity of the 2% water-based nanofluid with Al_2O_3 particles increases by 9.7%, and that of the 4% nanofluid increases by almost 25% (Fig. 4.3a). The thermal conductivity of the 6% nanofluid increases even more, as seen in Fig. 4.3a. Similar behavior is observed in ethylene glycol-based nanofluids (Fig. 4.3b). It is worth noting that the growth of thermal conductivity with increasing temperature in the above range was linear to the accuracy of measurement, meaning that the thermal conductivity increased in proportion to temperature. The data obtained are consistent with those previously published in literature [30, 81, 177, 189, 193]. The degree of increase in thermal conductivity is directly proportional to the concentration of nanoparticles and temperature. Considering that the thermal conductivity coefficient of base fluids also increases with temperature, while the viscosity coefficient decreases, it is clear that using nanofluids as a heat carrier is more effective at higher temperatures.

In the examples presented in Fig. 4.3, water or ethylene glycol was utilized as base fluids. As previously noted, these fluids are anomalous since their thermal conductivity increases with temperature. However, normal fluids, which exhibit a decrease in thermal conductivity with increasing temperature, can also serve as base fluids. In such cases, the resulting nanofluids will also exhibit an excess thermal conductivity. For example, Li et al. [92] studied the thermal conductivity of a nanofluid composed of diathermic oil and 30 nm SiC particles. In this case the thermal conductivity coefficient of the base fluid decreased with temperature, but adding nanoparticles increased it. Starting from a volume concentration of 0.6%, this growth becomes linear and at a concentration of 0.8% it reaches almost 13%. However, according to Yu et al. [191], the relative thermal conductivity coefficient of kerosene-based nanofluid with Fe_3O_4

particles of 155 nm size remained practically unchanged within the temperature range of 10–60 °C.

Finally, it should be noted that a typical example of base fluids whose thermal conductivity decreases with increasing temperature are those involving surfactants. This issue will be discussed in detail in Sect. 4.6.3.

4.3.2 Dependence of Thermal Conductivity on Particle Size and Material

The non-universality of the thermal conductivity coefficient of nanofluids [as seen by relations (4.10)] is primarily due to the dependence on particle size. Analysis of Table 4.1 and Figs. 4.1b and 4.2a shows that the coefficients b_i in formula (4.10) are non-universal because they are functions of the nanoparticle size. To answer the question about the nature of this dependence unambiguously, it was necessary to measure the thermal conductivity of nanofluids with particles of different sizes, while maintaining the same volume concentration. Such measurements were conducted using several dozen nanofluids containing copper, silicon, aluminum, zirconium, and titanium oxide particles. The particle sizes ranged from 10 to 150 nm, and the base fluid in all cases was water with a nanoparticle volume concentration of 2%. The measurements were taken at a temperature of 25 °C. Table 4.2 presents some of the obtained data, which are also illustrated in Fig. 4.4 [123].

The data obtained unambiguously show that the thermal conductivity coefficient of nanofluids depends on the particle size and the larger their size, the higher it is. The increase in thermal conductivity coefficient with increasing nanoparticle size is monotonic. However, the work [123] revealed that for nanofluids containing small nanoparticles of SiO_2, Al_2O_3, and TiO_2, the relative thermal conductivity values obtained were lower than those calculated using formula (4.7). This phenomenon was observed for SiO_2 particles with a diameter $d \leq 30$ nm, for TiO_2 particles with a diameter $d \leq 40$ nm, and for Al_2O_3 particles with a diameter $d \leq 50$ nm. Such a low excess of the thermal conductivity coefficient for these nanofluids has been recorded in previous studies (see, for example, papers [16, 36, 39, 41, 43, 73, 100, 118, 171] although this was usually not related to the size of the nanoparticles. However,

Table 4.2 Dependence of the relative thermal conductivity coefficient of water-based nanofluids on the particle size

SiO_2		Al_2O_3		TiO_2		ZrO_2	
d	λ_r	d	λ_r	d	λ_r	d	λ_r
10	1.015	50	1.061	71	1.066	44	1.077
16	1.027	75	1.082	100	1.084	105	1.087
25	1.040	100	1.100	150	1.100	–	–
100	1.072	150	1.133	–	–	–	–

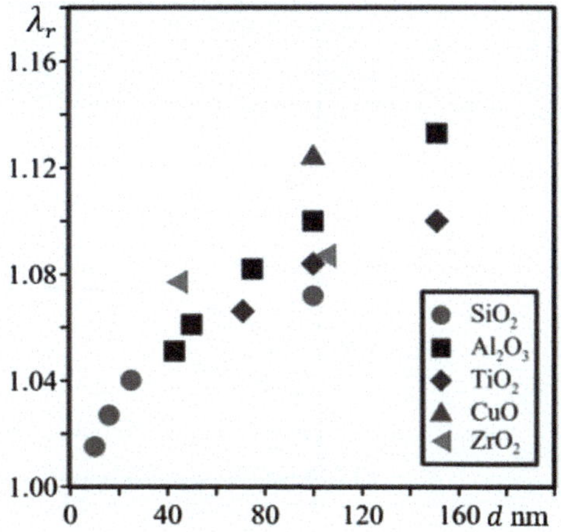

Fig. 4.4 Dependence of the relative thermal conductivity of water-based nanofluids on particle size

in all cases, the thermal conductivity coefficient of nanofluids then increases with increasing particle size. Starting from certain particle size, the measured thermal conductivity coefficients of nanofluids exceed those predicted by classical theories. At the concentrations studied, this excess ranges from a few percent to about 10–15%. At the same time, the relative excess thermal conductivity coefficient can be as high as 25–30%.

All classical theories of thermal conductivity of dispersed fluids assume that it depends on the particle material, as indicated by formulas (4.7)–(4.9). To account for this dependence, the thermal conductivity coefficient of the particle material is introduced. The data of nanofluids with particles of different materials are summarized in Fig. 4.4 and Table 4.2. Since presented nanofluids contain particles of the same size but different materials, it is clear that the thermal conductivity of nanofluid depends on the particle material. The question arises, what is the reason for this dependence?

To address this issue, the first step is to determine whether there is a systematic dependence of the thermal conductivity coefficient of nanofluids on the thermal conductivity of nanoparticles. This dependence is crucial in deriving formula (4.7) (see Sect. 4.1). The answer to this question was provided in the paper [123], and the corresponding illustration can be found in Fig. 4.5, which presents the dependence of relative thermal conductivity coefficients of five water-based nanofluids containing ZrO_2, TiO_2, SiO_2, CuO, and Al_2O_3 particles, on the thermal conductivity of the particle material. In all cases, the average particle size was close to 100 nm. The volume concentration of particles was 2%. The dashed line corresponds to the dependence determined by Maxwell's formula (4.7). The particles should be arranged in the following order in terms of increasing thermal conductivity coefficients of the material: ZrO_2, TiO_2, SiO_2, CuO, Al_2O_3. On the other hand, in terms of the ascending thermal conductivity coefficients of the nanofluid, the data in Fig. 4.5 should be arranged as follows: SiO_2, TiO_2, ZrO_2, Al_2O_3, and CuO. It is evident that there is no direct correlation between the thermal conductivity of the nanofluid and the material of the nanoparticles.

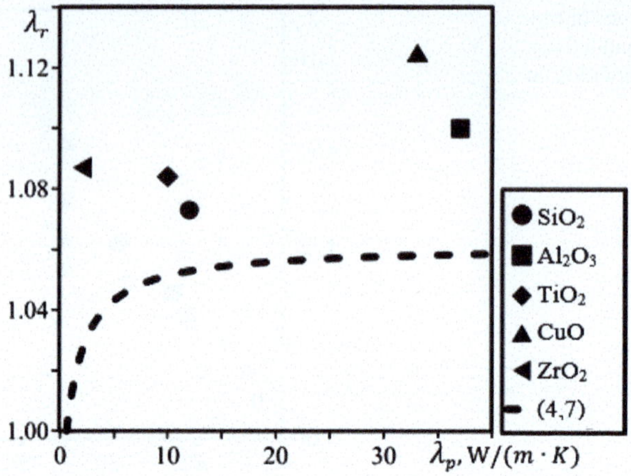

Fig. 4.5 Dependence of relative thermal conductivity coefficient of nanofluid on thermal conductivity of particle material

The thermal conductivity coefficient of the nanofluid does not correlate with other important thermophysical characteristics of the nanoparticle material either. This is illustrated by Fig. 4.6a, b, which shows the dependences of the relative thermal conductivity coefficient of nanofluid on the thermal diffusivity α and heat capacity C_p of the particle material. The data presented is for the same nanofluids as depicted in Fig. 4.5.

At the same time, experiments (Fig. 4.4) demonstrate the dependence of thermal conductivity of nanofluids on the particle material. On the other hand, employing the molecular dynamics simulation [133, 136] it was found that the thermal conductivity of nanofluids is greater the higher the density of nanoparticles. This relationship

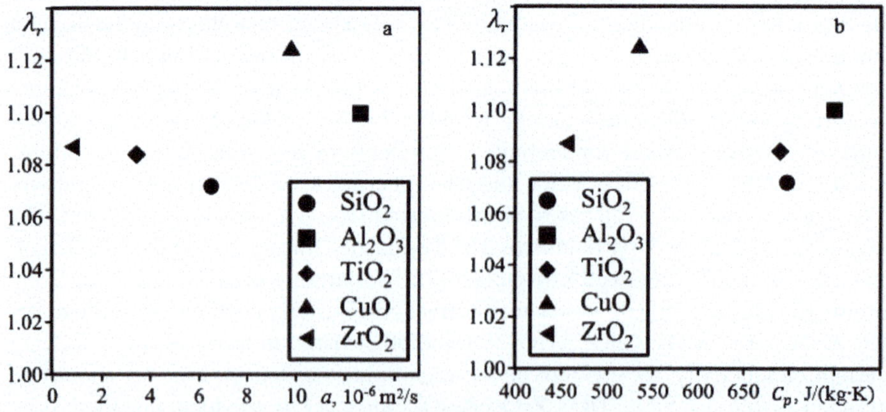

Fig. 4.6 Dependence of the relative thermal conductivity of nanofluids on thermal diffusivity (**a**) and heat capacity (**b**) of the particle material

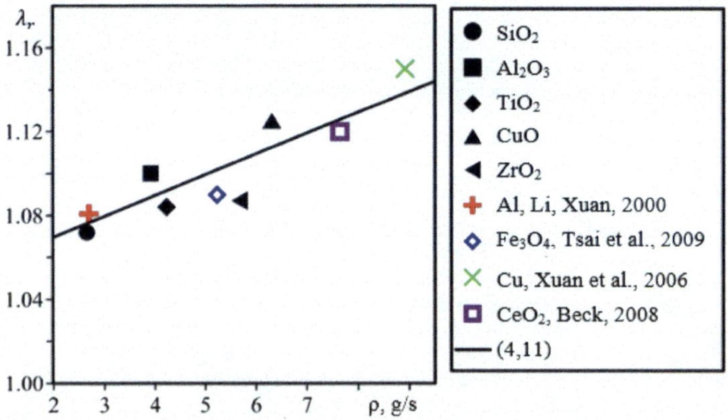

Fig. 4.7 Relative thermal conductivity coefficient of water-based nanofluids with volume concentration of 2% versus density of particle material

is confirmed by the experimental data presented in Fig. 4.7 [123], which shows a dependence close to linear. Figure 4.7 presents also the data obtained by other authors [15, 91, 170, 181], where the particle sizes were 100 nm.

Thus, it can be stated that indeed the thermal conductivity of nanofluids depends on the particle material, but this dependence is determined by their density, rather than by the thermal conductivity coefficient of the material. From this standpoint, nanofluids with metal particles, whose density significantly exceeds the density of other materials, should give the greatest excess thermal conductivity coefficient. Almost all known experimental data support this statement. As an example, the relative thermal conductivity coefficients of nanofluids based on water and ethylene glycol are compared in Fig. 4.8a, b, respectively. The nanofluids contain particles of Al_2O_3 (150 nm), ZrO_2 (105 nm), ZrO_2 (44 nm) TiO_2 (150 nm) [123], as well as nanodiamonds (5 nm) [122]. The results show that nanofluids with metal particles have a higher thermal conductivity coefficient than those with oxide particles. To achieve equal excess thermal conductivity coefficients in nanofluids with oxide particles, their volume concentrations should be several times higher than those for nanofluids with metal particles. Nanofluids with nanodiamonds exhibit the lowest excess thermal conductivity, but their size is more than an order of magnitude lower.

The second factor, most influencing the thermal conductivity coefficient of nanofluids is nanoparticle size. What should this dependence be? Numerous correlations are known in the literature (see [9, 37, 54, 55, 57] and the literature cited there). Ceotto and Rudyak [24] demonstrated that the relative thermal conductivity coefficient of water-based nanofluids with Al_2O_3 particles can be approximated by the formula $\lambda_r = 1 + A\sqrt{\varphi \tilde{d}}$, where $\tilde{d} = d/\sigma$, σ is the effective size of the base fluid molecules, and the coefficient $A = 0.48$. This correlation has achieved an accuracy of approximately 3%. However, it was noted by the authors that the coefficient A for

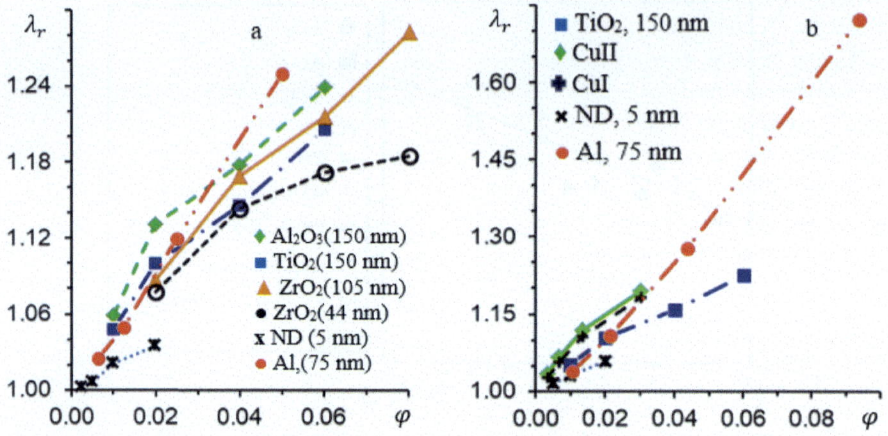

Fig. 4.8 Comparison of the thermal conductivity of water-based (**a**) and ethylene glycol-based (**b**) nanofluids with metal and oxide particles

other materials in this formula should be dependent on the nanoparticle material. In a later work [123], this was taken into account and the formula was revised to the following form

$$\lambda_r = 1 + (0.0193 + 0.00383\tilde{\rho})\sqrt{\varphi\tilde{d}}, \tag{4.11}$$

where $\tilde{\rho} = \rho_p/\rho_f$, ρ_p, ρ_f are the material densities of the nanoparticles and the fluid, respectively.

Formula (4.11) is quite universal and its accuracy was demonstrated in [123] when approximating experimental data of water-based nanofluids with SiO_2, TiO_2 и Al_2O_3 particles.

4.3.3 Dependence of Thermal Conductivity on the Base Fluid

The thermal conductivity of the base fluids can vary significantly, which in turn should affect the thermal conductivity of the corresponding nanofluids. However, there are relatively few studies that have systematically investigated this relationship. The first comprehensive study was conducted in [173, 176]. These studies examined nanofluids based on water, ethylene glycol, vacuum pump oil, and machine oil. It was noted that the nanofluid based on ethylene glycol had the highest thermal conductivity. Later, Xie et al. [175] investigated the effect of the base fluid. They developed a method to create a stable nanofluid using deionized water, ethylene glycol, and decene with MWCNTs. The thermal conductivity enhancement was found to decrease as the thermal conductivity of the base fluid increased. This conclusion contradicts the

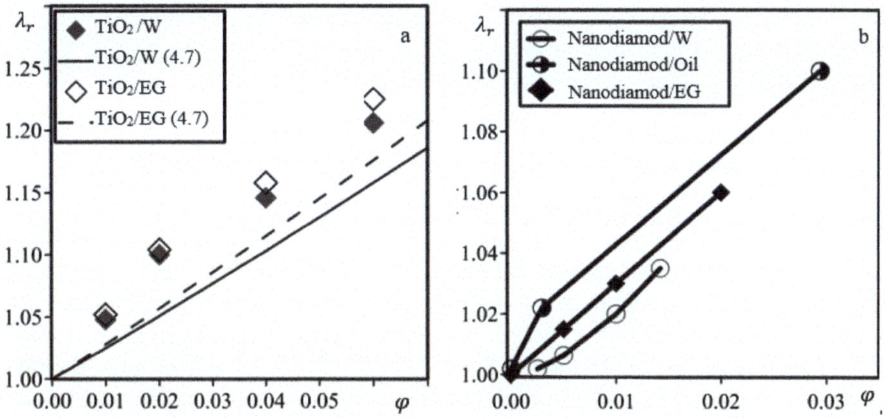

Fig. 4.9 Dependence of relative thermal conductivity coefficient on particle concentration in nanofluids based on **a** water and ethylene glycol with TiO_2 particles; **b** water, ethylene glycol, and machine oil with nanodiamond particles

above cited paper. However, it is important to note that the nanofluids investigated were very specific. Therefore, it was crucial to test the applicability of this conclusion on ordinary nanofluids.

Studying the effect of base fluid properties on nanofluid thermal conductivity can be challenging due to the need to compare data for at least two nanofluids with the same particle concentration and size but different base fluids. This comparison was carried out in the paper [123]. First, nanofluids based on water and ethylene glycol with titanium dioxide particles measuring 150 nm were considered. The obtained data are presented in Fig. 4.9a. The relative thermal conductivity coefficients exhibit systematic but not significant differences. In all cases, the ethylene glycol-based nanofluid has a higher thermal conductivity enhancement than the water-based nanofluid. Additionally, water has a thermal conductivity coefficient almost two and a half times higher than that of ethylene glycol.

In the second series of measurements, nanofluids containing diamond particles with an average size of 5 nm were used. The nanofluids were based on water, ethylene glycol, and machine oil. The results have shown significant differences (see Fig. 4.9b). The relative thermal conductivity coefficient of the nanofluid based on machine oil was significantly higher than that of the nanofluids based on ethylene glycol and water. At the same time, the thermal conductivity coefficient of machine oil was nearly half that of ethylene glycol and almost five times lower than that of water.

Thus, the thermal conductivity of the nanofluid is significantly affected by the base fluid. The relative thermal conductivity coefficient of nanofluids decreases as the thermal conductivity coefficient of the base fluid increases. This is a natural phenomenon; in fluids with higher thermal conductivity, the thermal conductivity enhancement will be smaller, assuming all other factors are equal. The addition of nanoparticles to a fluid is most effective when the fluid has a low thermal conductivity.

However, the influence of the base fluid on the effective thermal conductivity of the nanofluid is not yet well understood.

4.4 Mechanisms of Thermal Conductivity of Nanofluids

The previous section presented experimental data that indicate a significant enhancement in the thermal conductivity coefficient of nanofluids. Nanofluids containing oxide particles with a volume concentration of up to 6–8% exhibit an enhancement of up to 30%. In nanofluids containing metal particles, this enhancement is even higher (see reviews [9, 12, 37, 54, 55, 57, 117, 187, 189, 190]. The measured thermal conductivity enhancement is considerably greater than the values predicted by classical theories, indicating the emergence of alternative heat transfer mechanisms in nanofluids Keblinski et al. [78] were the first who have conducted a thorough analysis of potential mechanisms. Four main mechanisms were proposed for the thermal conductivity of nanofluids: (1) Brownian motion, (2) layering of base fluid molecules near the nanoparticle surface, (3) specifics of energy transfer by nanoparticles, and (4) clustering of nanoparticles. These mechanisms are commonly discussed in majority of contemporary works and should be considered in detail.

4.4.1 Effect of Brownian Motion

Nanoparticles, like the molecules of the base fluid, are in thermal motion. Since their size usually significantly exceeds the size of the base fluid molecules, their interaction with these molecules is usually collective. Therefore, the nature of thermal motion is largely determined by this interaction, and one can indeed speak of Brownian motion of nanoparticles. Due to this, they can transfer energy during their movement and transfer it when they collide with each other. However, the velocities of Brownian motion of nanoparticles are small. Therefore already Keblinski et al. [78] and later Eastman et al. [44] have shown that such a heat transfer channel in nanofluids is inefficient even for the smallest particles. In addition, with an increase in the particle size, the influence of this effect should decrease, whereas experiments demonstrate an increase in the thermal conductivity of the nanofluid with increasing particle size. Using the MD method, the first cited study has found that the collision of nanoparticles also cannot significantly increase the thermal conductivity of nanofluids.

The enhancement of thermal conductivity due to Brownian motion of nanoparticles is often attributed to the resulting microconvection of the base fluid near the particles. However, Evans et al. [48] have demonstrated that this mechanism has a small effect on thermal conductivity. In this connection there is a certain point to be made. Thermal conductivity is an energy transfer property of a fluid (including nanofluid) due to thermal microfluctuations in the motion of its particles (see the fluctuation–dissipation theorem). Therefore, any convective motions, including microconvection, are irrelevant to the thermal conductivity of the fluid, as well as ordinary convective thermal conductivity occurring in flows. Moreover, thermal conduction should not be confused with the contribution to heat transfer associated with thermodiffusion.

Later, Shukla and Dhir [155] studied the effect of Brownian motion of nanoparticles on the thermal conductivity of nanofluids. They divided the total heat flux due to Brownian motion into a kinetic component and a component caused by interaction. The study has shown that the kinetic contribution had a negligible effect on a typical nanofluid. On the other hand, it has been observed that the contribution of interactions plays an important role in enhancing the thermal conductivity of nanofluids. Additionally, it is important to note that the Brownian diffusion of nanoparticles is inversely proportional to their radius (2.12). If this is the case, the contribution of Brownian motion should decrease rapidly as nanoparticle size increases. However, on the contrary, experiments show an increase in thermal conductivity with increasing particle size.

Despite the fact that the effect of Brownian motion of nanoparticles is really small, many papers have proposed models of the heat conductivity coefficient in the form of superposition of the coefficient (4.7) and some additional term λ_{br} caused by Brownian motion

$$\lambda_e = \lambda_f \left[1 + \frac{3(1-\kappa)\varphi}{1 + 2\kappa - (1-\kappa)\varphi} \right] + \lambda_{br}. \tag{4.12}$$

Various versions of such formulas can be found in Xuan et al. [180], Bhattacharya et al. [20], Kumar et al. [84], Koo and Kleinstreuer [82, 83], Jang and Choi [64], Prasher et al. [120], Yu-Hua et al. [192], Shukla and Dhir [155], Nie et al. [115], Jung and Yoo [72]. In many works, the thermal conductivity coefficient of nanofluid is considered as a combination of the thermal conductivity coefficient proper, called the static coefficient, and a dynamic coefficient attributed to the factors such as Brownian motion of nanoparticles, microconvection generated by them, etc. In this case, the second term also includes various similarity parameters such as Reynolds, Prandtl, Nusselt numbers, etc. Examples of this approach can be found in the papers [19, 32, 120, 178]. In this regard, it is worth recalling once again that this second part of the thermal conductivity coefficient defined in this way is not the thermal conductivity itself. Moreover, the similarity parameters for nanofluids are themselves complex functions of the concentration of nanoparticles and their size. Such formulas simply cannot be used for nanofluids (see Chap. 7).

4.4.2 Peculiarities of Energy Transfer by Nanoparticles

When analyzing the thermal conductivity of nanofluids, Keblinski et al. [78] focused on the thermal conductivity of the nanoparticles themselves. This suggests that the thermal conductivity of the nanofluid is a superposition of the thermal conductivity of the base fluid and the nanoparticles. In nonmetals, thermal conductivity is primarily due to phonons, whereas in metals, electron thermal conductivity is also present, and the latter is significantly greater than phonon thermal conductivity. On the other hand, the thermal conductivity mechanism of nonmetal nanoparticles is determined by the mean free path of phonons

$$l_{ph} \sim (10bT_m)/T, \tag{4.13}$$

where b and T_m are, respectively, the lattice constant and the fusion temperature of the material. For nanoparticles with a diameter of approximately 20 nm or less, a ballistic phonon propagation mechanism occurs due to the phonon mean free path being significantly greater than the particle size. This allows the authors [78] to consider that since in this case (for small nanoparticles) even at small volume concentrations (about 5%) the distance between the particles is relatively small, high thermal conductivity of nanofluids is due to the transfer of phonons between nanoparticles.

This concept has faced several objections. Firstly, it fails to explain the observed increase in thermal conductivity of nanofluids with increasing nanoparticle size. For particles that are large enough, the ballistic mechanism of phonon propagation ceases to work. Diffusion of phonons occurs in the nanoparticle material in such cases. Finally, the ballistic thermal conductivity does not explain the significantly higher thermal conductivity of nanofluids with metal particles, where electronic conduction mechanism is dominating. For this case, Avsec [11] demonstrated that conduction electrons in nanoparticles also move ballistically.

Secondly, with an increase in nanoparticle size at a given concentration, the average distance between them in the nanofluid also increases rapidly. The phonons in the base fluid are quickly damped. Additionally, the Kapitza resistance [75] occurs at the fluid–nanoparticle interface, which further reduces the efficiency of phonon heat transfer in nanofluids.

Kapitza resistance was first discovered in fluid helium, but it was later found to occur at ambient temperatures as well. When macroscopic bodies are in contact, it can be assumed with great accuracy that their contacting surfaces have the same temperature. In this case, the Kapitza resistance is much lower than the total thermal resistance. It is linked to the appearance of a temperature difference (ΔT) at the interface, which is related to the heat flux ($q = \sigma_K \Delta T$), where σ_K represents the Kapitza conductivity, whose inverse value is known as the Kapitza resistance $R_K = 1/\sigma_K$. The relationship between the thermal conductivity coefficient and the Kapitza conductivity is expressed as $\lambda = \sigma_K h_K$, where h_K is the Kapitza radius. This radius defines the distance at which the temperature drop is the same as at the interface. This radius is usually small if the contact between the phases is good, and decreases with increasing temperature [160]. The behavior of the fluid–solid interface is highly dependent on the properties of the adsorption layer of fluid molecules. In the case of nanoparticles, the adsorption layer can significantly affect energy transfer.

Benveniste [17] generalized Maxwell's formula taking into account the interfacial Kapitza resistance. Similar generalizations can also be found in Benveniste et al. [18], Chiew and Glandt [29], Ni et al. [114], Nan et al. [111], Ju and Li [71], Xue [184], where ellipsoidal and cylindrical particles were considered as well. The use of all formulas derived from these works necessitates an explicit calculation of the Kapitza resistance, which, however, remains an unresolved issue. According to Barrat and Chiaruttini's research [14], the Kapitza resistance is associated with the particle surface's wettability. When the base fluid molecules have a weak interaction with the particle surface (non-wetted surface), the energy transfer efficiency is low, resulting in relatively large values of R_K. The solid–fluid interface typically has a negative overall contribution to the macroscopic thermal conductivity of nanofluids, which is proportional to the total interface area and increases as the particle size decreases [166, 167]. In conclusion, it is important to note once again that taking into account the Kapitza resistance weakens the effect of the ballistic energy transfer in nanofluids. This circumstance does not contribute to explaining the significant excess of their thermal conductivity.

4.4.3 Nanoparticle Clustering

The clustering of nanoparticles is usually noted as one of the main reasons for significantly higher thermal conductivity of nanofluids compared to the corresponding value of the base fluid. The degree of particle clustering increases with increasing nanoparticle concentration and decreasing the size. This phenomenon becomes particularly pronounced when the average distance between particles is on the order of the van der Waals forces acting between them. The recorded increase in thermal conductivity with increasing nanoparticle size suggests that clustering has a positive effect on the thermal conductivity of nanofluids. During contact interactions between nanoparticles, Kapitza resistance is absent, and the entire particle agglomerate operates under the ballistic thermal conductivity mechanism. At the same time, several factors that reduce the importance of clustering when explaining the high thermal conductivity of nanofluids can be noted. First of all, adsorbed layers of the base fluid molecules are formed on the surface of the nanoparticles. This creates an interphase boundary, which on the one hand forms the Kapitsa resistance, and on the other hand, prevents the ballistic propagation of phonons. It is further obvious that with an increase in the characteristic size of aggregated particles, their sedimentation time decreases sharply. This leads to the fact that the concentration of nanoparticles in the nanofluid decreases, resulting in a decrease in thermal conductivity. Such a decrease was observed, for example, in the work of Jana et al. [67]. Indeed, the sedimentation rate can be estimated using the following relation

$$v_s \sim \frac{(\rho_p - \rho_f)gR^2}{\mu}, \tag{4.14}$$

where g is the acceleration of gravity. Thus, as the characteristic particle size increases, the sedimentation rate of the particles increases rapidly. It is important to note that clustering of particles can disrupt the homogeneity of their distribution in the nanofluid and its stability.

4.4.4 *Effect of Layering of Base Fluid Molecules Near the Nanoparticle*

The presence of the layering of base fluid molecules has already been noted in the previous chapter. This is an important factor in nanofluid transport processes, as it leads to a significant increase in viscosity. Many authors have pointed out that the significant enhancement of nanofluid thermal conductivity can be explained by layering [44, 78, 183, 185, 188]. The base fluid layering around the nanoparticle suggests the existence of some composite particle consisting of a solid core (the nanoparticle) and a spherical layer of ordered fluid with a thickness of δR. It is assumed that this layer has a thermal conductivity coefficient greater than that of the corresponding base fluid. This leads to an overall increase in the thermal conductivity of the nanofluid. The total particle volume concentration increases since the volume of the composite particle is larger than the volume of the nanoparticle, becoming equal to $\varphi_* = \varphi(1 + \delta R/R)^3$. Following Maxwell's theory, the effective thermal conductivity coefficient can be derived, which was done in Yu and Choi [188]. Several authors have developed a model for calculating the effective thermal conductivity coefficient of nanofluids, taking into account the presence of ordered fluid layers at the nanoparticle surface. In particular, Xue and Xu [186] calculated the thermal conductivity of nanofluids using water and ethylene glycol with aluminum oxide particles of varying sizes. It was determined that at a given concentration the thermal conductivity of the nanofluid increases as the size of the nanoparticles decreases.

Thus, summarizing this brief review of the enhancement mechanisms of thermal conductivity of nanofluids discussed in the literature, we can note the following. Brownian motion of nanoparticles is certainly an important factor in nanofluid dynamics. However, it alone cannot fully explain the high thermal conductivity excess of nanofluids. The high thermal conductivity of nanoparticles alone also cannot explain the high thermal conductivity of nanofluids. Furthermore, Sect. 4.3 experimentally demonstrated that there is no direct correlation between the thermal conductivity of nanofluids and the thermal conductivity of nanoparticles. Although clustering does contribute to the thermal conductivity of nanofluids, it is important to keep in mind several circumstances. First, clustering increases thermal conductivity only if the phonon mean free path is larger than the characteristic particle size. For sufficiently large particles this is not the case. Furthermore, as the size of the forming nanoparticle clusters increases, they will sediment, reducing the volume concentration of particles in the nanofluid. Finally, the presence of the base fluid layering near the nanoparticles is an important factor in the morphology of the nanofluid. At moderate concentrations of small nanoparticles, the base fluid becomes layered. This results in a higher thermal conductivity of nanofluid compared to the base fluid. However, as the nanoparticle size increases, the significance of this effect decreases. Therefore, it is impossible to explain the increase in thermal conductivity of the nanofluid by such layering. All of the above mechanisms do indeed occur, but they are not sufficient to explain the observed high thermal conductivity of nanofluids.

4.5 MD Simulation of Thermal Conductivity of Nanofluids

The idea of using the MD method to study the thermal conductivity of nanofluids was born almost immediately with the beginning of their active study. These studies aimed to determine the dependence of the thermal conductivity coefficient on nanoparticle concentration and to elucidate the relevant mechanisms. Several dozen publications have been published to date, making it impossible to provide a comprehensive review of all studies in this book. Therefore, only the main characteristic features of these publications will be considered. The complexity of MD simulation is well understood, therefore, only model nanofluids were considered. Liquid argon [33, 34, 42, 68, 108, 141, 152, 159, 163, 172, 194] xenon [42], water [96, 145] and ethylene glycol [80, 149] were very often used as the base fluid. The nanoparticles used in studies are typically metal particles, including copper [35, 80, 90], platinum [42, 147], gold [148], zinc, and aluminum [139]. Metal oxides have also been utilized [95, 108, 110, 164, 196] This list can be extended to include more exotic nanofluids that have been used as model base fluids. The nanoparticles used were typically less than 3 nm in size, and their volume concentrations did not exceed 5%. In most of the studies, nanofluids were created by adding a single particle, which did not allow for a proper understanding of the influence of nanoparticle interactions and dynamics on the transport processes. The main conclusions obtained earlier by the MD method will be summarized below.

First of all, practically all the works noted a significant excess of thermal conductivity of nanofluids compared to base ones. This excess was significantly higher than predicted by Maxwell's theory (4.7). Additionally, several studies reported a nonlinear increase in the thermal conductivity coefficient. At low particle concentrations, the thermal conductivity coefficient increases linearly with increasing particle concentration. However, this growth slows down at higher concentrations [68, 141, 150, 152]. This behavior of the thermal conductivity coefficient depending on the particle concentration was also observed experimentally (see Sect. 4.3).

A very interesting result was obtained in Loya et al. [95]. It was found that the thermal conductivity excess of the nanofluid depends significantly on the thermal conductivity of the base fluid. The study simulated the thermal conductivity of nanofluids based on water and eicosane, ($C_{20}H_{42}$) with 0.4 nm CuO nanoparticles at a weight concentration of 3%. In the first case, the thermal conductivity increased by 25% (from 0.6 to 0.75 $\mathrm{Wm^{-1}K^{-1}}$), and in the second case, it increased almost threefold (from 0.15 to 0.4 $\mathrm{Wm^{-1}K^{-1}}$). This is consistent with experimental data, which shows that the enhancement thermal conductivity of the nanofluid is the higher the lower thermal conductivity of the base fluid (see Sect. 4.3.3).

Most studies on MD simulation of nanofluid thermal conductivity have focused on verifying the thermal conductivity mechanisms discussed in Sect. 4.4. The first thing that was extensively discussed was the influence of Brownian motion of nanoparticles on nanofluid thermal conductivity. However, almost all papers have found that Brownian motion had no appreciable effect on nanofluid thermal conductivity, although the arguments in favor of this conclusion have varied [48, 78, 87, 152, 159].

The impact of nanoparticle aggregation has not been extensively studied because calculations typically involved only a small number of nanoparticles, often just one. Kang et al. [74] conducted a study on the effect of particle aggregation on the thermal conductivity of argon-based nanofluid with copper particles, considering several configurations of copper particles. First, the authors considered spherical particles of given radius, which unfortunately was not specified. These particles form bonded pairs, triplets, and quadruplets. In the latter cases, the particles were arranged in a line, triangle, or square. Thus, the study considers nanofluids with particles of significantly different morphologies. The thermal conductivity coefficient was calculated using FDT. The calculation results indicate that particle clustering significantly increases the thermal conductivity of the corresponding nanofluids. Unfortunately, it is unclear at what particle volume concentrations the calculations were performed. Later, the effect of particle aggregation on thermal conductivity was investigated in Topal and Servantie [168] using a Lennard–Jones model based nanofluid, fluid whose atoms interact with each other by the potential (2.24). Nanoparticles were created from such atoms, but in addition to the potential (2.24), the finitely extensible nonlinear elastic (FENE) potential [56] acted between them. The nanoparticles obtained in this way are actually clusters whose size was 4σ and whose concentration was 2, 3, 6, and 10%. However, the particles were not spherical and varied in shape. The ability of the particles to combine was adjusted by changing the van der Waals interactions between the atoms of the nanoparticles. The calculations performed have shown that the aggregation of nanoparticles has no significant effect on the thermal conductivity of the nanofluid. This conclusion is in agreement with the study [150], where a water-based nanofluid with SiO_2 particles was simulated.

The influence of the dense layers of the base fluid formed at the interphase boundary on the thermal conductivity has also been studied very extensively using the MD method [35, 90, 94, 108]. The layering of base fluid molecules near the surface of nanoparticles is a long-established phenomenon, which is explained by the fact that the potential of van der Waals forces acting on the molecule is significantly larger than the corresponding intermolecular forces (see Fig. 2.2). It is also worth noting that these forces are usually much greater in metal particles than, for example, in metal oxides. Additionally, liquid molecules are adsorbed onto the surface of the particles. The width of the ordered fluid layer near nanoparticles at small particle diameters depends on their size, but starting from a diameter of about 3–4 nm, the width of this layer practically does not change and is 1.5–2 nm. This width can be easily estimated, for example, from the radial distribution function. The thickness of the layer is determined by the characteristic size of the short-range order of the base fluid molecules near the surface of the particles. Several models have been constructed to quantify the influence of these layers on the excess thermal conductivity [50, 88, 174, 188]. However, reliable evidence on the magnitude of such enhancement has not actually been obtained. According to one of the recent papers [172], the presence of these layers has no significant effect on the thermal conductivity of the nanofluid. However, Chen et al. [26] express the opposite opinion. It is difficult to make a definitive diagnosis because this contribution cannot be determined experimentally. Additionally, the existence of Kapitza resistance must always be considered.

There were few studies that investigated the effect of nanoparticle size and material on the thermal conductivity of nanofluids. Some studies have even presented data that contradicts experimental findings [34, 35, 97]. In the following sections, we will discuss the thermal conductivity mechanisms using MD simulations of an argon-based nanofluid with aluminum and zinc particles ranging in size from 1 to 4 nm and a volume concentration of nanoparticles varying from 1 to 5% [141].

4.5.1 MD Simulation of Thermal Conductivity of Argon-Based Nanofluid

The simulation was carried out using the standard MD method and the original SibMD software package, which has been previously employed to solve various problems in nanofluid transport theory [131, 132, 133, 134, 136, 137, 138, 140, 141]. The cubic cell with periodic boundary conditions was considered when performing simulation. The interaction between atoms in the base medium was determined using the Lennard–Jones potential (2.24). The interaction between base medium molecules and nanoparticles was described by the RK potential (2.27), while the interaction between nanoparticles was defined by the potential (2.28). The parameters for the interaction potential of argon molecules were $\sigma = 3.405$ Å and $\varepsilon/k = 119.8$ K [62]. For zinc the corresponding values were $\sigma = 2.46$ Å, $\varepsilon/k = 1040$ [10] and for aluminum, $\sigma = 2.551\,1$ Å, $\varepsilon/k = 957.6.6$ K [61]. The parameters σ_{12} and ε_{12} were determined using the simplest combinatorial relations: $\sigma_{12} = \sqrt{\sigma\tilde{\sigma}}$ and $\varepsilon_{12} = \sqrt{\varepsilon\tilde{\varepsilon}}$. Newton's equations were integrated using the Verlet scheme.

The thermal conductivity coefficient of the nanofluid (binary mixture) was calculated using the FDT, which relates the thermal conductivity coefficient to the correlation function of the corresponding dynamic variables. Calculating the thermal conductivity coefficient for the binary system under consideration is challenging because it requires excluding the diffusive heat flux. If this is done, the thermal conductivity coefficient of the nanofluid can be written in the following form [198]

$$\lambda = \lambda_Q - \lambda_D \equiv \frac{L_Q}{T^2} - \frac{L_2^2}{L_{22}T^2}. \tag{4.15}$$

The coefficients in relation (4.15) are equal to the integrals of the equilibrium correlation functions

$$L_Q = \frac{V}{3}\int_0^{t_p} \langle \boldsymbol{j}_Q(0) \cdot \boldsymbol{j}_Q(t)\rangle \mathrm{d}t, \tag{4.16}$$

$$L_{22} = \frac{V}{3}\int_0^{t_p} \langle \boldsymbol{j}_{D2}(0) \cdot \boldsymbol{j}_{D2}(t)\rangle \mathrm{d}t, \quad L_2 = \frac{V}{6}\int_0^{t_p} \langle \boldsymbol{j}_Q(0) \cdot \boldsymbol{j}_{D2}(t)\rangle \mathrm{d}t + \frac{V}{6}\int_0^{t_p} \langle \boldsymbol{j}_{D2}(0) \cdot \boldsymbol{j}_Q(t)\rangle \mathrm{d}t. \tag{4.17}$$

where the diffusion flux of nanoparticles $\boldsymbol{j}_{D2}$ and the heat flux $\boldsymbol{j}_Q$ are defined as follows

$$\boldsymbol{j}_{D2}(t) = \frac{1}{V}\sum_{\alpha=2,i=1}^{N_2} \boldsymbol{v}_{2,i}(t), \boldsymbol{j}_Q(t) = \boldsymbol{j}_Q^K(t) + \boldsymbol{j}_Q^P(t) + \boldsymbol{j}_Q^C(t). \tag{4.18}$$

The latter formula includes three distinct terms, $\boldsymbol{j}_Q^K$, $\boldsymbol{j}_Q^P$, and $\boldsymbol{j}_Q^C$. These terms specify the heat flux resulting from kinetic and potential energy transfer, as well as the heat flux resulting from collisions

$$\boldsymbol{j}_Q^K(t) = \frac{1}{V}\left[\frac{m_1}{2}\sum_{\alpha=1,i=1}^{N_1} \boldsymbol{v}_{1,i}(t)v_{1,i}^2(t) + \frac{m_2}{2}\sum_{\alpha=2,i=1}^{N_2} \boldsymbol{v}_{2,i}(t)v_{2,i}^2(t)\right],$$

$$\boldsymbol{j}_Q^P(t) = \frac{1}{V}\left[\frac{1}{2}\sum_{\alpha=1,i=1}^{N_1}\sum_{\alpha=1,j\neq i} \boldsymbol{v}_{1,i}(t)\Phi_{11,ij}(r_{ij}(t)) + \frac{1}{2}\sum_{\alpha=1,i=1}^{N_1}\sum_{\alpha=2,j=1}^{N_2} \boldsymbol{v}_{1,i}(t)\Phi_{12,ij}(r_{ij}(t))\right.$$
$$\left. + \frac{1}{2}\sum_{\alpha=2,i=1}^{N_2}\sum_{\alpha=1j=1}^{N_1} \boldsymbol{v}_{2,i}(t)\Phi_{12,ij}(r_{ij}(t)) + \frac{1}{2}\sum_{\alpha=2,i=1}^{N_2}\sum_{\alpha=2,j=1}^{N_2} \boldsymbol{v}_{2,i}(t)\Phi_{12,ij}(r_{ij}(t))\right],$$

$$\boldsymbol{j}_Q^C(t) = \frac{1}{V}\left\{\frac{1}{2}\sum_{\alpha=1,i=1}^{N_1}\sum_{\alpha=1,j\neq i} \boldsymbol{r}_{ij}(t)\left[\boldsymbol{F}_{11,ij}(t)\cdot\boldsymbol{v}_{1,j}(t)\right] + \frac{1}{2}\sum_{\alpha=1,i=1}^{N_1}\sum_{\alpha=2,j=1}^{N_2} \boldsymbol{r}_{ij}(t)\left[\boldsymbol{F}_{12,ij}(t)\cdot\boldsymbol{v}_{2,j}(t)\right]\right.$$
$$\left. + \frac{1}{2}\sum_{\alpha=2,i=1}^{N_2}\sum_{\alpha=1,j\neq i} \boldsymbol{r}_{ij}(t)\left[\boldsymbol{F}_{11,ij}(t)\cdot\boldsymbol{v}_{1,j}(t)\right] + \frac{1}{2}\sum_{\alpha=2,i=1}^{N_2}\sum_{\alpha=2,j=1}^{N_2} \boldsymbol{r}_{ij}(t)\left[\boldsymbol{F}_{22,ij}(t)\cdot\boldsymbol{v}_{2,j}(t)\right]\right\} \tag{4.19}$$

The index $\alpha = 1$ refers to argon atoms and the index 2 refers to nanoparticles, m_i is the particle (atom or nanoparticle) mass, $\Phi_{11,ij}$ are the interaction potentials of the base fluid atoms, $\Phi_{12,ij}$ are the interaction potentials of its atoms and particles, $\Phi_{22,ij}$ are the interaction potentials of the nanoparticles, and $\boldsymbol{F}_{\alpha\beta,ij}$ are the corresponding forces, N_1 and N_2 are the number of argon atoms and nanoparticles, respectively.

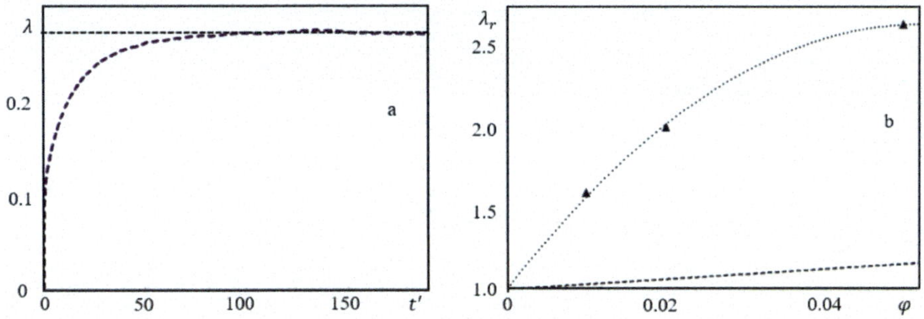

Fig. 4.10 **a** Evolution of the thermal conductivity coefficient W/(m·K); **b** dependence of the relative thermal conductivity coefficient on the volume concentration of zinc nanoparticles with a diameter of 2 nm for argon-based nanofluid

According to Eqs. (4.16) and (4.18), the simulated thermal conductivity coefficient varies with time. The actual value of the thermal conductivity coefficient is attained when it reaches the plateau corresponding to integration time t_p, beyond which it remains constant. Figure 4.10a illustrates the evolution of the function $\lambda(t)$ (4.15), which determines the thermal conductivity coefficient for a nanofluid containing 2 nm zinc particles with a volume concentration of 4.2%. The time t' is measured in $\tau_0 = \sigma/c$ units, where c represents the thermal velocity of the base molecules. The plateau is typically reached after approximately 100 τ_0.

The typical dependence of the relative thermal conductivity coefficient λ_r of the same nanofluid as shown in Fig. 4.10 a on the volume concentration of nanoparticles $((N\sigma^3)/V = 0.707$, a $T = 300$ K) is presented in Fig. 4.10b. The simulation data is represented by triangular symbols, while the dashed line 2 corresponds to Maxwell's formula (4.7). The thermal conductivity of the considered nanofluid is significantly higher than that of both the base fluid and the coarse dispersed fluids (4.7). Thus, the thermal conductivity of a 2% nanofluid is almost double the value determined by Maxwell's formula. The dependence of the thermal conductivity coefficient of the nanofluid on the concentration slows down with its growth and is well described by a formula (4.10). This formula is represented in Fig. 4.10b by the dashed line, and in this case $b_1 = 63.1$, $b_2 = -607.9$. As already indicated, this fact is also observed experimentally. On the other hand, at low concentrations of nanoparticles (when the second term in formula (4.10) can be neglected), the thermal conductivity increases linearly with concentration.

As can be seen from (4.15), the thermal conductivity coefficient of the nanofluid is a superposition of two terms. While the first term λ_Q is present in both pure fluid (single component medium) and nanofluid, the second term λ_D is present only in nanofluid. As shown in the paper [141], the contribution of the diffusion term is not large (about 10%), and starting from a certain concentration, this contribution decreases. The latter is quite natural, since the diffusion of nanoparticles decreases as their concentration increases.

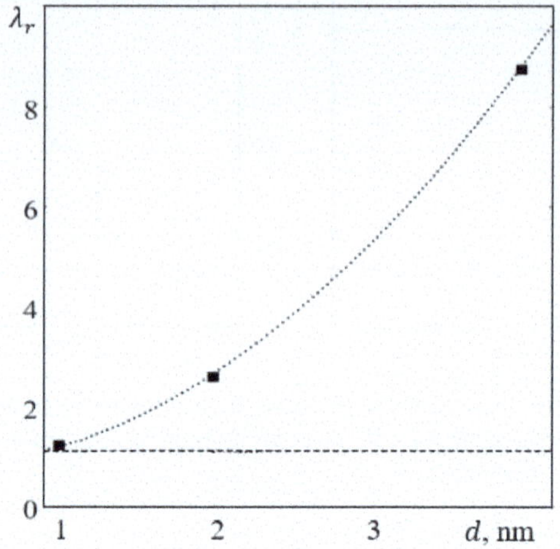

Fig. 4.11 Dependence of the relative thermal conductivity coefficient of argon-based nanofluids with zinc particles on the nanoparticle diameter

The dependence of the relative thermal conductivity coefficient on the diameter of zinc particles is presented in Fig. 4.11. The volume concentration of nanoparticles was 4.2%. Here the dotted line shows calculations by formula (4.7) for suspensions with the same volume concentration of particles. In all cases, the nanofluid with zinc particles has a thermal conductivity significantly higher than that of both the base fluid and the calculation by formula (4.7). In addition, the thermal conductivity of the nanofluid increases with increasing particle concentration.

In classical theories of the thermal conductivity of nanofluids, the material of the dispersed particles is taken into account by their thermal conductivity (4.7). However, this is not the case for nanofluids containing metal nanoparticles. Such particles have thermal conductivity orders of magnitude higher than that of the base fluid, and in this case Maxwell's equation no longer depends on the particle material and takes the form (4.7a). Thus, classical theories do not actually give the dependence of the thermal conductivity coefficient of dispersed fluids on the particle material. However, the data presented show that such a dependence exists. For example, the thermal conductivity coefficient for an argon-based nanofluid with zinc particles, shown in Fig. 4.11, is approximately 2.4 times greater than the corresponding value for a nanofluid with aluminum particles [141]. The density of zinc is also greater than the density of aluminum. It can therefore be argued that the thermal conductivity of the nanofluid depends on the material of the nanoparticles, and it increases with increasing density of the material of the nanoparticles. The density of zinc is 2.95 times greater than the density of aluminum. Thus, the dependence of the thermal conductivity of the nanofluid increases almost linearly with increasing density of the nanoparticles.

4.5.2 MD Analysis of Thermal Conductivity Mechanisms in Nanofluids

Analysis of formulas (4.15)–(4.19) shows that the thermal conductivity coefficient is determined by a number of different contributions, such as heat transfer due to the motion of fluid atoms and nanoparticles, their interaction, etc. To find the reason for the significant enhancement thermal conductivity of the nanofluid over the corresponding value for the base fluid, it is necessary to analyze all possible components. The contribution to the thermal conductivity coefficient (4.15) consists of six terms

$$\lambda_Q = \lambda_Q^K + \lambda_Q^P + \lambda_Q^C + \lambda_Q^{KP} + \lambda_Q^{KC} + \lambda_Q^{PC}, \tag{4.20}$$

$$\lambda_Q^K = \frac{V}{3kT^2}\int_0^{t_p} \chi_Q^K(t)\mathrm{d}t = \frac{V}{3kT^2}\int_0^{t_p} \langle \mathbf{j}_Q^K(0)\cdot \mathbf{j}_Q^K(t)\rangle \mathrm{d}t,$$

$$\lambda_Q^P = \frac{V}{3kT^2}\int_0^{t_p} \chi_Q^P(t)\mathrm{d}t = \frac{V}{3kT^2}\int_0^{t_p} \langle \mathbf{j}_Q^P(0)\cdot \mathbf{j}_Q^P(t)\rangle \mathrm{d}t,$$

$$\lambda_Q^C = \frac{V}{3kT^2}\int_0^{t_p} \chi_Q^C(t)\mathrm{d}t = \frac{V}{3kT^2}\int_0^{t_p} \langle \mathbf{j}_Q^C(0)\cdot \mathbf{j}_Q^C(t)\rangle \mathrm{d}t,$$

$$\lambda_Q^{KP} = \frac{2V}{3kT^2}\int_0^{t_p} \chi_Q^{KP}(t)\mathrm{d}t = \frac{2V}{3kT^2}\int_0^{t_p} \langle \mathbf{j}_Q^K(0)\cdot \mathbf{j}_Q^P(t)\rangle \mathrm{d}t,$$

$$\lambda_Q^{KC} = \frac{2V}{3kT^2}\int_0^{t_p} \chi_Q^{KC}(t)\mathrm{d}t = \frac{2V}{3kT^2}\int_0^{t_p} \langle \mathbf{j}_Q^K(0)\cdot \mathbf{j}_Q^C(t)\rangle \mathrm{d}t,$$

$$\lambda_Q^{PC} = \frac{2V}{3kT^2}\int_0^{t_p} \chi_Q^{PC}(t)\mathrm{d}t = \frac{2V}{3kT^2}\int_0^{t_p} \langle \mathbf{j}_Q^P(0)\cdot \mathbf{j}_Q^C(t)\rangle \mathrm{d}t \tag{4.21}$$

The diffusive contribution λ_D (4.15) can be represented similarly

$$\lambda_D = \lambda_D^K + \lambda_D^P + \lambda_D^C + \lambda_D^{KP} + \lambda_D^{KC} + \lambda_D^{PC}$$
$$= (T^2 L_{22})^{-1}\Big[(L_2^K)^2 + (L_2^P)^2 + (L_2^C)^2 + 2L_2^K L_2^P + 2L_2^K L_2^C + 2L_2^P L_2^C\Big],$$

$$L_2^K = \frac{V}{3}\int_0^{t_p} \langle \mathbf{j}_Q^K(0)\cdot \mathbf{j}_{D2}(t)\rangle \mathrm{d}t,$$

$$L_2^P = \frac{V}{3}\int_0^{t_p} \langle j_Q^P(0)\cdot j_{D2}(t)\rangle \mathrm{d}t,\quad L_2^C = \frac{V}{3}\int_0^{t_p} \langle j_Q^C(0)\cdot j_{D2}(t)\rangle \mathrm{d}t. \tag{4.22}$$

As a result, it is possible to calculate all six contributions to the thermal conductivity coefficient of nanofluid, namely, kinetic (K), potential (P), collisional (C), kinetic–potential (KP), kinetic–collisional (KC), and potential–collision (PC)

$$\begin{aligned}
\lambda^K &= \lambda_Q^K - \lambda_D^K = \frac{1}{T^2}\left[L_Q^K - \frac{\left(L_2^K\right)^2}{L_{22}}\right], \lambda^P = \lambda_Q^P - \lambda_D^P = \frac{1}{T^2}\left[L_Q^P - \frac{\left(L_2^P\right)^2}{L_{22}}\right],\\
\lambda^C &= \lambda_Q^C - \lambda_D^C = \frac{1}{T^2}\left[L_Q^C - \frac{\left(L_2^C\right)^2}{L_{22}}\right], \lambda^{KP} = \lambda_Q^{KP} - \lambda_D^{KP} = \frac{1}{T^2}\left[L_Q^{KP} - 2\frac{L_2^K L_2^P}{L_{22}}\right],\\
\lambda^{KC} &= \lambda_Q^{KC} - \lambda_D^{KC} = \frac{1}{T^2}\left[L_Q^{KC} - 2\frac{L_2^K L_2^C}{L_{22}}\right], \lambda^{PC} = \lambda_Q^{PC} - \lambda_D^{PC} = \frac{1}{T^2}\left[L_Q^P - 2\frac{L_2^C L_2^P}{L_{22}}\right],
\end{aligned} \tag{4.23}$$

The calculation data for argon and argon-based nanofluids with zinc particles of 2 and 4 nm diameter at their volume concentration of 4.2% are summarized in Table 4.3. The thermal conductivity coefficient of pure argon is mainly determined by the coefficients λ^C (62%), which represents the energy flux associated with collisions of argon atoms with each other, and λ^{KC} (22%), which results from the correlation of the kinetic energy flux and energy flux due to collisions. This is expected because of the high density of argon. The heat transfer resulting from molecular motion (λ^K) is minimal, accounting for only about 8%.

How does the thermal conductivity structure of nanofluid differ from that of pure argon? As can be seen from Table 4.3, along with positive contributions, negative contributions appear in the binary systems as well. The negative contributions include potential (P), potential–kinetic (PK), and potential–collisional (PC) terms. These negative contributions increase as the nanoparticle size increases. This means that in addition to the normal heat flux, there is also a flux consistent with the temperature gradient (negative thermal conductivity). Collisional contribution λ^C is important for both pure fluid and nanofluids, in nanofluids, it increases as the nanoparticle size increases. For example, in a nanofluid with a nanoparticle with diameter of 2 nm, the contribution increases by a factor of one and a half. In contrast, in a nanofluid with a nanoparticle diameter of 4 nm, this contribution increases by a factor of 3.6. However, with increasing particle size, the contribution of λ^C to the total thermal conductivity coefficient decreases slightly from 35% for nanofluid with 2 nm particles to 29.6% for 4 nm particles. The increase in the collisional contribution compared to the pure fluid is due to the interaction of the base fluid molecules with the nanoparticles, but as shown in Sect. 3.2.2 the effective scattering cross-section of such interactions decreases with increasing particle size [see formula (3.21)].

The largest contribution to thermal conductivity is due to the term λ^{KC}. It ranges from 50% for nanofluid with 2 nm particles to almost 80% for nanofluid with 4 nm

Table 4.3 Contributions of different components to the thermal conductivity coefficient of nanofluid, W/(m K)

	Ar	Ar–Zn, ($d = 2$ nm)	Ar–Zn, ($d = 4$ nm)
λ_Q		0.2822	0.8073
λ_D		0.0132	0.0117
λ	0.1068	0.2690	0.7955
λ_Q^K		0.1394	0.5557
λ_D^K		0.0479	0.1998
λ^K	0.0085	0.0915	0.3559
λ_Q^C		0.0965	0.2354
λ_D^C		0.0023	0.0001
λ^C	0.0650	0.0942	0.2353
λ_Q^P		0.0164	0.1006
λ_D^P		0.0231	0.1232
λ^P	0.0005	– 0.0068	– 0.0226
λ_Q^{KC}		0.1560	0.6324
λ_D^{KC}		0.0210	0.01109
λ^{KC}	0.0235	0.1350	0.6213
λ_Q^{KP}		– 0.0833	– 0.4644
λ_D^{KP}		– 0.0666	– 0.3138
λ^{KP}	0.0026	– 0.0221	– 0.1506
λ_Q^{PC}		– 0.0388	– 0.2525
λ_D^{PC}		– 0.0146	– 0.0087
λ^{PC}	0.0066	– 0.0242	– 0.2438

particles of the total value of the thermal conductivity coefficient. In contrast to pure argon, the kinetic contribution of λ^K is also significant reaching 34% for nanofluid with 2 nm nanoparticle diameter and almost 45% for nanofluid with 4 nm nanoparticle diameter of the total value of λ. Thus, the increase in thermal conductivity of nanofluid compared to pure fluid is mainly due to the significant increase in kinetic λ^K and kinetic–collisional λ^{KC} contributions.

To evaluate the various mechanisms of thermal conductivity enhancement in nanofluids, it is essential to understand not only the values of the different contributions but also the dynamics of the corresponding correlation functions. Figure 4.12 presents the evolution of the correlation functions of the heat flux fluctuations that contribute the largest portion to the thermal conductivity coefficient: $\chi_Q^K(t)$, $\chi_Q^C(t)$, $\chi_Q^{KC}(t)$, [see formulas (4.21)]. Figure 4.12 a displays data for argon, while the Fig. 4.12 b shows data for an argon-based nanofluid with 2 nm diameter zinc particles ($\varphi = 4.2\%$). The correlation functions $\chi(t)$ are normalized to the value of $\chi^K(0)$

for pure argon, and time is normalized to τ_0. The dotted line represents the correlation function $\chi_Q^K(t)$, the dash line depicts the correlation function $\chi_Q^C(t)$, and the dotted-dash line shows the correlation function $\chi_Q^{KC}(t)$. For pure argon, all considered correlation functions decay exponentially quite fast with a characteristic time of the order of the molecule interaction time. The evolution of correlation functions for nanofluids is more complex. According to the study [137, 138], the velocity autocorrelation function (VACF) of a nanoparticle is a superposition of two exponentials. The relaxation time of the first exponential is related to the interaction time of the nanoparticle with individual molecules of the base fluid, while the second exponential is related to the interaction time of the nanoparticle with microfluctuations of the hydrodynamic fields of the base fluid [see Sect. 2.3, formula (2.32)]. This latter time is at least an order of magnitude greater than the former. The correlation functions that determine the main contributions to the thermal conductivity of nanofluid behave similarly. Their evolution also occurs in two stages. At the first stage, the relaxation time is equal to that of the correlation functions of the base fluid thermal conductivity coefficient. Starting from a time approximately equal to $2\tau_0$ (see to Fig. 4.12b), it is described by a relaxation time τ_2.

Let us consider the relaxation mechanisms of the VACF of the nanoparticle in more detail. The effectiveness of the kinetic mechanism is directly proportional to the particle size and mass. Already for particles with a size of 1–2 nm, its contribution to the total relaxation of correlation functions is minimal. From this size and up, the relaxation of correlation functions of a nanoparticle will be mainly due to its interaction with microfluctuations of hydrodynamic fields of the carrier medium. As demonstrated in Chap. 2, the scale of these fluctuations is described by formula (2.32), which can be expressed as follows

$$\tau_2 \sim \frac{R}{c_p} \sim \frac{R}{c}\sqrt{\frac{M}{m}} \sim \tau_1 \left(\frac{d}{\sigma}\right)^{5/2} \sqrt{\frac{\rho_p}{\rho_f}}, \tag{4.24}$$

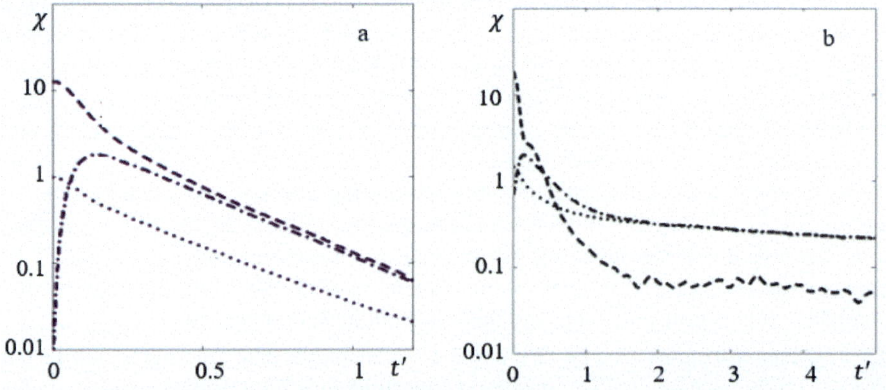

Fig. 4.12 Correlation functions of the heat flux fluctuations in pure argon (**a**) and in Ar–Zn nanofluids (**b**)

For the nanoparticles being considered, the time (4.24) exceeds the characteristic relaxation time of the base fluid molecules τ_1 by more than an order of magnitude. This implies that the characteristic damping of all correlation functions that determine the thermal conductivity coefficient of the nanofluid is at least an order of magnitude larger than that of the base fluid. This is confirmed by the calculations shown in Fig. 4.12. The correlation functions relax slowly, resulting exactly in a significant increase in the thermal conductivity coefficient of nanofluids compared to the base fluid. According to estimation (4.24), the relaxation time τ_2, and therefore the thermal conductivity coefficient, increases with increasing nanoparticle size and density. Note that formula (4.24) provides an upper estimate of the relationship between the time τ_2 and the nanoparticle size. This is because the characteristic size of the resulting microfluctuations stabilizes at a certain particle size.

4.6 Thermal Conductivity of Nanofluids with CNTs

Due to the significant amount of experimental data available on the thermal conductivity of nanofluids with CNTs, it is not feasible to provide a comprehensive review in this section. This topic has been extensively researched for approximately two decades. The majority of studies have focused on nanofluids with MWCNTs, with considerably fewer works dedicated to SWCNTs. However, it is possible to identify the most common characteristics of previously conducted measurements by limiting ourselves to works from the last decade. The data obtained during this time are typical and can help identify current challenges that require clear answers. A systematic review of works from the previous decade can be found in Younes et al. [187], Estellé et al. [46], Soltanimehr and Afrand [158], Tawfik [162], Akhilesh et al. [4].

Nanofluids are typically prepared using water and ethylene glycol as base fluids. Due to the hydrophobic nature of CNTs, it is often necessary to use surfactants or functionalize the CNTs during preparation. The weight concentration of MWCNTs typically ranges from 0.01 to 3%. The diameters of the nanotubes vary from 20 to 150 nm, and their length varies widely, ranging from 5 to 50 μm. Here are some typical examples. The study [124] investigated the thermal conductivity of a water-based nanofluid with xanthan gum (XG) as a surfactant. The weight concentration of MWCNTs ranged from 0.01 to 0.1%, with an outer diameter of 20–30 nm and a length ranging from 10 to 30 μm. The study has found that the thermal conductivity coefficient increased with increasing the concentration and temperature. The degree of increase was greater at higher temperatures, reaching 15% at $T = 70$ °C and a weight concentration of 0.1%.

In Sadri et al. [146] a water-based nanofluid with the addition of gum arabic (GA) surfactant at a MWCNT weight concentration of 0.5% was studied. The study has shown that at 45 °C, the thermal conductivity exceeded that of water by 14%. The thermal conductivity coefficient of W + GA-based nanofluid was also studied in Thakur et al. [165]. The volume concentration of MWCNTs varied from 0.2 to

0.8%, with a diameter of 18–20 nm and a length of 20 μm. At the maximum concentration, the enhancement of the thermal conductivity coefficient has reached 12% at $T = 30$ °C and 23% at $T = 70$ °C. The growth of the thermal conductivity coefficient with temperature was almost linear. Previously, the thermal conductivity coefficient of similar nanofluid was studied in Garg et al. [52], the diameter of MWCNTs was 10–20 nm and the average length was about 20 μm. It is noted that at a weight concentration of 1% and a temperature of 35 °C, the thermal conductivity enhancement was about 20%. Finally, we can mention the paper of Rashmi et al. [125], where the thermal conductivity coefficient of W + GA-based nanofluid with MWCNTs of 20 nm diameter and 30 mm length was measured. The thermal conductivity coefficient increased almost linearly with temperature and MWCNT concentration with the maximum enhancement observed at a maximum concentration of 0.1% and $T = 60$ °C.

Jiang et al. [69] conducted a study on the thermal conductivity of W + PVP based nanofluid. The volume concentration of MWCNTs ranged from 0.22 to 1%, and the temperature ranged from 30 to 90 °C. The CNTs had a diameter of 11 nm and a length of 10 μm. The study has found a maximum thermal conductivity enhancement of 13.5%, and the relative thermal conductivity was almost independent of temperature. Earlier, Fadhillahanafi et al. [49] revealed that nanofluids with a weight concentration of MWCNTs equal to 0.5% exhibited an enhancement of the thermal conductivity coefficient of up to 22.2%. The MWCNTs had a diameter of 10–30 nm and a length of 10–30 μm.

Phuoc et al. [119] used chitosan as a surfactant to disperse MWCNTs with a diameter of 20–30 nm and a length of 10–30 μm. The enhancement of the thermal conductivity coefficient of nanofluids with MWCNT weight concentrations ranging from 0.5 to 3% was 2.3 and 13%, respectively.

A very interesting work was presented by Estellé et al. [46]. The authors studied the dependence of the thermal conductivity coefficient of four different water-based nanofluids with MWCNTs and three different surfactants: SDBS, lignin, and sodium polycarboxylate. The volume concentration of MWCNTs varied from 0.005 to 0.55% and the temperature ranged from 20 to 40 °C. In addition, the effect of aspect ratio (AR), which was equal to 160 and 90 (CNT diameter was 9.2 and 11.4 nm) on the thermal conductivity coefficient was studied. The thermal conductivity of nanofluids was found to increase with increasing CNT concentration and temperature. However, it is noted that the aspect ratio and surfactant type had little effect on the thermal conductivity excess.

As mentioned above, CNT functionalization is commonly used when creating nanofluids. Various techniques of CNT surface functionalization are known, including fluid-phase or gas-phase oxidation of CNTs with reagents such as nitric acid, nitrogen dioxide, and hydrogen peroxide. These techniques are successful in polar media such as water and polar organic solvents. In non-polar organic solvents, advanced functionalization methods are necessary. For example, polystyrene and polyisoprene, as well as carboxyl groups, can be used. Below are some examples of thermal conductivity studies of nanofluids with functionalized MWCNTs.

Abdullah et al. [2] investigated the thermal conductivity of a W + PVP-based nanofluid containing hydroxyl functionalized MWCNTs (MWCNT-OH). The weight concentrations ranged from 0.1 to 1%, and the PVP concentrations were either 0.1 or 0.2%. The maximum enhancement of the thermal conductivity coefficient was 8.86% at a concentration of 0.8%. However, a further increase in concentration, the thermal conductivity coefficient did not increase or even decreased. This may indicate that the nanofluid becomes unstable at higher concentrations.

Omrani et al. [116] considered a water-based nanofluid with MWCNT-COOH at a volume concentration of 0.05%. The thermal conductivity coefficient was systematically measured and its dependence on the aspect ratio was analyzed. The volume concentration of the SDBS or Triton X-100 surfactants was also 0.05%. The aspect ratio ranged from 2500 to 25 on average. The temperature was varied from 10 to 45 °C. It was found that the thermal conductivity coefficient was highest at the highest aspect ratio and maximum temperature, reaching 36%. Conversely, this coefficient was lowest at the minimum aspect ratio and a temperature of 10 °C, measuring only 3%. Therefore, for nanofluids with MWCNTs of the same diameter, the thermal conductivity coefficient increased with longer CNTs.

In paper [5], the authors investigated the thermal conductivity of a water-based nanofluid containing two types of functionalized MWCNTs. However, the functionalization technique was not stated clearly. The weight concentration of the nanofluid was varied from 0.01 to 0.1%, and the experiments were conducted at a temperature of 40°°C. The results have shown a maximum increase in the thermal conductivity coefficient of 11.6 and 7.1%, respectively. The authors noted a non-monotonic relationship between concentration and thermal conductivity coefficient. In one type of nanofluid, the thermal conductivity coefficient decreased when the concentration reached 0.05%. This non-monotonicity seems unjustified at such low CNT concentrations. This work is mentioned here because many studies in the literature have used commercially produced nanofluids with unclear fabrication technology.

A rather unconventional functionalization method with polyethylene glycol (PEG) using Fischer esterification method was applied in paper [99]. The water-based nanofluids contained MWCNT-PEG, whose weight concentration was varied from 0.01 to 0.1%, and their diameter and length were 24 nm and 10–30 μm, respectively. The thermal conductivity excess was about 17.5% at 35 °C.

In addition to water, ethylene glycol (EG) or a W + EG mixture in varying proportions are widely used as base fluids. Surfactant or CNT functionalization and often both are usually employed when preparing the nanofluid. For example, Bai et al. [13] used functionalized MWCNTs to prepare the nanofluid, while W + EG with different weight concentrations of EG (ranging from zero to 100%) served as the base fluid. Additionally, PVP was added. The MWCNTs had an outer diameter of approximately 8 nm and a length of 0.5–2 μm. The enhancement of the thermal conductivity was highly dependent on the solution composition, with a maximum excess of 6% observed at 60 °C and EG volume concentration of 25%. A little earlier in the study [106], the thermal conductivity of the EG + W-based nanofluid with a fixed volume concentration of 20:80% and functionalized MWCNTs was investigated. The concentration of MWCNTs ranged from 0.025 to 0.8%. At the maximum concentration of MWCNTs, there was a 27.3% increase in thermal conductivity coefficient. Ganeshkumar et al. [51] also studied nanofluids based on EG + W (50:50). At MWCNT weight concentration of 0.9%, the enhancement of the thermal conductivity of 11% was recorded. However, a further increase in MWCNT concentration up to 1.5% resulted in a decrease in thermal conductivity. The authors associate this effect with an increase in the thermal resistance caused by the formation of a percolation lattice and refer to papers [28, 109], where such a reduction was also recorded. It also looks quite strange that the thermal conductivity coefficient practically did not change with increasing temperature.

Singh et al. [156] studied the thermal conductivity of EG-based nanofluid with MWCNTs (with diameter of 60–100 nm, and a length of 5–15 μm), functionalized with carboxyl groups using the technique described in Esumi et al. [47]. The weight concentration of CNTs varied from 0.12 to 0.4% and the temperatures ranged from 25 to 60 °C. The results have shown that the thermal conductivity coefficient increased linearly with increasing CNT concentration, with a maximum enhancement of 72% at 0.4% concentration.

EG-based nanofluids containing functionalized MWCNT-OH were used to measure the thermal conductivity coefficient in Abdullah et al. [2]. The CNT concentration varied from 0.1 to 1%, in addition, PVP was used. The maximum enhancement of the thermal conductivity coefficient was 5.37% at a concentration of 0.2% and a temperature of 25 °C. However, at higher concentrations, the thermal conductivity coefficient was lower. Additionally, the thermal conductivity of the nanofluid decreased with increasing temperature and became lower than that of ethylene glycol.

Shamaeil et al. [154] investigated the thermal conductivity coefficient of EG-based nanofluids with double-walled functionalized CNTs. The study examined nanofluids with a maximum volume concentration of 0.6%, an outer diameter of 2–4 nm, and a length of 50 μm. The results have shown that the thermal conductivity coefficient increased with increasing CNT concentration and temperature. The maximum enhancement was 24.9% at a concentration of 0.6% and a temperature of 52 °C. The authors of paper [103] used nitric acid treatment of MWCNTs to create an EG-based nanofluid. They observed an enhancement of 25.4% in the thermal conductivity coefficient at $T = 55$ °C and MWCNT weight concentration of 4%.

There are practically no works that study ethylene glycol-based nanofluids without the use of surfactants or CNT functionalization. There are also relatively few studies on the thermal conductivity of nanofluids with SWCNTs. Relevant data can be found in the reviews mentioned above. The following are typical works published in the past decade, where the weight concentration of SWCNTs ranged from 0.01 to approximately 1%. The CNTs had a diameter of 1.3–2 nm and a length of 2–5 μm. Therefore, the relevant data will only be mentioned in the review where it is necessary to draw specific attention to them. The nanofluids with CNTs were prepared using the same base fluids as those used in preparing nanofluids with MWCNTs. In general, the thermal conductivity of the nanofluid increases with temperature, but the presence of surfactants can significantly alter this dependence, as shown below.

Chen and Xie [27] studied the thermal conductivity of water-based nanofluid with functionalized hydroxyl groups of SWCNTs and MWCNTs. The enhancement of the thermal conductivity coefficient at $T = 55$ °C and 0.2% concentration was 15.6. Harish et al. [59], Harish [60] studied the thermal conductivity of nanofluid with SWCNTs based on water and ethylene glycol. Sodium deoxycholate (DOC) was used as a surfactant in the former case, while a bile salt was used in the latter. At a volume concentration of 0.3%, the enhancement of the thermal conductivity coefficient was approximately 11% in the former case and 14.8% in the latter at a volume concentration of 0.21%. As with the conventional spherical particles in nanofluid [123], the lower the thermal conductivity coefficient of the base fluid, the higher is the excess in nanofluid.

Nasiri et al. [113] conducted a very intriguing study on the thermal conductivity of nanofluids containing SWCNTs and MWCNTs of two types. The first variety had an outer diameter less than 10 nm (MWCNT1), while the second had a diameter of 10–20 nm (MWCNT2). The length of CNTs was 10 μm in all cases. The nanofluid was prepared using potassium persulfate (KPS) as an oxidizing agent of CNTs, with a weight concentration of 0.25% in all cases. The study has found that the nanofluid with SWCNTs exhibited the enhancement of the thermal conductivity coefficient ranging from 13 to 35.3% as the temperature varied from 10 to 50 °C. The nanofluids with MWCNT1 and MWCNT2 have shown corresponding values of 6–17 and 4–15.5%, respectively.

In Esfe et al. [45], the thermal conductivity of ethylene glycol-based nanofluids with functionalized SWCNTs, whose volume concentration varied from 0.02 to 0.75%, was investigated. The measurements were carried out in the temperature range of 30–50 °C. At the maximum temperature and concentration, the enhancement of the thermal conductivity coefficient was 45%.

In addition to water and ethylene glycol, other base fluids, including both common and quite rare ones, such as antifreeze, machine oil, jatropha seed oil, solar glycol, various alcohols, poly-alpha-olefin, and others, have been used when preparing the nanofluid. Relevant examples can be found in Tam et al. [161], Younes et al. [187], Estellé et al. [46], Soltanimehr and Afrand [158], Tawfik [162], Akhilesh et al. [4], Nanda et al. [112], Ahmadi et al. [3], Amani et al. [8], Ali et al. [6], Kumar et al. [85], Ghasemi et al. [53], Rehman et al. [126]. Finally, ionic nanofluids have recently received significant attention as base fluids [70, 98, 151, 105, 195]

To summarize this brief review, it is important to note that the obtained data are dissimilar due to already known circumstances. In fact, the works studied different nanofluids, making it difficult and sometimes impossible to compare their properties. However, in all cases, the use of a nanofluid with CNTs significantly increases the thermal conductivity coefficient by several percent or even tens of percent compared to that of the base fluid. Meanwhile, the weight concentrations of CNTs are typically only a fraction of a percent. Next, the majority of studies show that the thermal conductivity coefficient of nanofluids increases monotonically with increasing CNT concentration. However, some studies indicate a non-monotonic relationship starting from a certain CNT concentration. It is important to note that MWCNTs and SWCNTs have significantly different properties. Despite this, there are only a few studies that have simultaneously investigated the properties of nanofluids containing both SWCNTs and MWCNTs. However, such study is necessary to develop universal recommendations for the practical application of nanofluids. This will be further discussed in this section. The nanofluids based on W + S, EG, EG + S, IPA with SWCNTs and MWCNTs were considered. PVP, SDBS, and SDS were used as surfactants. In total, data from more than two dozen nanofluids were analyzed.

4.6.1 Thermal Conductivity of Base Fluids

The review presented above demonstrates that the thermal conductivity coefficient of nanofluids increases with rising CNT concentration and temperature, as established in the vast majority of studies. The main objective of this chapter is to determine the degree of growth of the thermal conductivity coefficient in nanofluids containing SWCNTs and MWCNTs. It is well-established that the degree of growth of the thermal conductivity coefficient of nanofluids with temperature at concentrations not exceeding 0.5% is almost always linear. Therefore, all measurements presented below were carried out only at a temperature of 25 °C. The data used below were mainly obtained from Rudyak et al. [143, 144]

The preparation technology of many nanofluids involves the use of surfactants. Therefore, it is necessary to obtain information on the thermal conductivity of the base fluid (BF) before measuring the thermal conductivity of nanofluids. In this case, the base fluid is a complex solution, BF = IBF + S. Table 4.4 presents data on the thermal conductivity coefficient of W + SDS and W + SDBS base fluids depending on the weight concentration of surfactant. Here, for convenience, the relative thermal conductivity coefficient of the base fluid $\lambda_{rbf} = \lambda_{bf}/\lambda_{ibf}$ is presented, where λ_{bf} and λ_{ibf} are thermal conductivity coefficients of base fluid and initial base fluid, respectively.

Thermal conductivity coefficients of W + SDS and W + SDBS base fluids are lower than that of water. The reduction in thermal conductivity is found to be quite significant; at a weight concentration of surfactant of 0.4%, the thermal conductivity of the base solution is reduced by more than 3%. This is in agreement with works [46, 146], which also examined the thermal conductivity of complex base fluids containing gum arabic and lignin as surfactant additives to water. Generally speaking, the indicated decrease in thermal conductivity of base fluids increases with increasing temperature. This effect is particularly pronounced in the case of the W + SDS

Table 4.4 Thermal conductivity coefficient of W + SDBS and W + SDS base fluids

w(%)	λ, W + SDBS W/(m K)	λ_{rbf}	w(%)	λ, W + SDS W/(m K)	λ_{rbf}
0	0.6037	1.000	0	0.6037	1.000
0.1	0.6053	0.995	0.1	0.6048	0.989
0.2	0.5935	0.983	0.2	0.5898	0.977
0.4	0.5862	0.971	0.4	0.5835	0.967

Table 4.5 Thermal conductivity coefficient of EG + PVP and W + PVP base fluids

$w(\%)$	λ, EG + PVP, BW/(m·K)	λ_{rbf}	$w(\%)$	λ W + PVP, BW/(m K)	λ_{rbf}
0	0.2600	1.0000	0	0.6021	1
0.1	0.2610	1.004	0.1	0.603	1.0015
0.2	0.2648	1.018	0.2	0.6039	1.003
0.4	0.2663	1.024	0.4	0.6057	1.006

solution. The base W + GA fluid also has a lower thermal conductivity than water, but this decrease is almost independent of temperature.

On the other hand, the thermal conductivities of the W + PVP and EG + PVP base fluids are higher than those of the original base fluids (see Table 4.5). This increase is particularly significant for the EG + PVP solution.

4.6.2 *Thermal Conductivity of Nanofluids Based on EG and IPA*

The thermal conductivity of the initial base fluid can be significantly altered, including a reduction, by using surfactants in the preparation of nanofluids. In additional, the use of surfactants complicates the nanofluid preparation process. Therefore, it is advisable to first analyze the thermal conductivity of nanofluids prepared without surfactants. In this section, we analyze the thermal conductivity of nanofluids based on ethylene glycol and isopropyl alcohol. The thermal conductivity coefficients of ethylene glycol and isopropyl alcohol are 0.258 and 0.137 W/(m K), respectively.

The preparation of nanofluids using the two-step technology requires the use of ultrasonic treatment (UST). Therefore, it is important to investigate the effect of UST time on the thermal conductivity of CNT-containing nanofluids. The influence of UST time on the relative thermal conductivity coefficient of the nanofluid can be demonstrated using the example of the IPA-based nanofluid with SWCNTs, the corresponding data are given in Table 4.6. The thermal conductivity coefficient of the nanofluid slightly decreased with increasing UST time. However, if the treatment lasted more than 90 min, the thermal conductivity coefficient remained almost unchanged. The UST time after which the thermal conductivity coefficient is practically unchanged indicates reaching the most stable nanofluid. These stable nanofluids were studied further.

Systematic data of the relative thermal conductivity coefficient of EG- and IPA-based nanofluids with SWCNTs and MWCNTs are summarized in Table 4.7 (in units of W/(m·K)). The nanofluids are denoted as follows: SWCNTs as S, MWCNT1 as M1, and MWCNT2 as M2. The analysis of the data presented in Table 4.7 reveals several conclusions. Firstly, it is noteworthy that the thermal conductivity coefficient of nanofluids with SWCNTs is significantly higher. Thus, at a weight concentration of

Table 4.6 Effect of UST time of the IPA-based nanofluid with SWCNTs on its thermal conductivity coefficient

t (min)	λ/λ_{bf}	t (min)	λ/λ_{bf}
60	1.136	10	1.278
90	1.125	90	1.246
200	1.123	200	1.243
w= 0.05%		w= 0.1%	

0.2%, the enhancement of the thermal conductivity coefficient of EG-based nanofluid is 19.4% and that of IPA-based nanofluid is 51.1%. Here, as well as in nanofluids with conventional spherical particles, the enhancement is higher the lower the thermal conductivity coefficient of the base fluid. The second important conclusion is that the enhancement of the thermal conductivity coefficient in nanofluids with MWCNTs is significantly lower than that of the nanofluids with SWCNTs. The maximum enhancement of the thermal conductivity coefficient is observed in the IPA-based nanofluid with M2 and is only 5.8% at equal weight concentration. In contrast, the ethylene glycol-based nanofluid achieves an enhancement of 6.2% at a weight concentration of just 1%. However, as shown in Fig. 4.13 the thermal conductivity coefficient of the nanofluid in all cases increases almost linearly with an increase in the weight concentration of CNTs.

Finally, it should be noted that the thermal conductivity coefficients of nanofluids containing both types of MWCNTs are practically identical (see Table 4.7). Now regarding the comparison with classical theories of thermal conductivity of dispersed fluids [58, 102], it is important to note that at the small weight concentrations of CNTs considered (while volume concentrations are even smaller), the enhancement of the thermal conductivity coefficient predicted by these theories is about two orders of magnitude lower.

Table 4.7 Dependence of the relative thermal conductivity coefficient of EG- and IPA-based nanofluids with SWCNTs, MWCNT1 and MWCNT2

w, %	EG/S	λ_r	EG/M1	λ_r	EG/M2	λ_r	IPA/S	λ_r	IPA/M2	λ_r
0.05	0.268	1.039	0.262	1.017	0.261	1.013	0.154	1.124	0.139	1.015
0.10	0.289	1.120	0.264	1.022	0.263	1.020	0.171	1.248	0.141	1.031
0.20	0.308	1.194	0.265	1.027			0.207	1.511	0.145	1.058
0.25					0.268	1.038				
1.00			0.274	1.062						

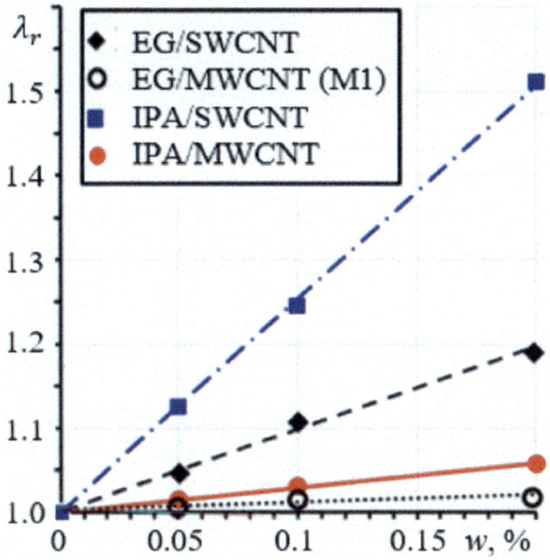

Fig. 4.13 Dependence of the relative thermal conductivity coefficient of EG- and IPA-based nanofluids with SWCNTs and MWCNTs on their weight concentration

4.6.3 Effect of Surfactants on Thermal Conductivity of Nanofluids

When CNTs are not functionalized, surfactants are often required in the preparation of nanofluids. This is particularly true when preparing water-based nanofluids since CNTs are hydrophobic. The following section presents thermal conductivity data for nanofluids prepared using PVP, SDBS, and SDS surfactants. First, we will consider EG + S based nanofluids. The corresponding data are summarized in Table 4.8. Comparison of the data in Tables 4.8 and 4.7 reveals that the thermal conductivity coefficient of EG + PVP-based nanofluid with SWCNTs exceeds that of EG-based nanofluid. At SWCNT weight concentration of 0.2%, the thermal conductivity coefficient of EG + PVP-based nanofluid is almost twice as high as that of EG-based nanofluid. Also, the thermal conductivity coefficient of the EG + PVP-based nanofluid with MWCNTs is nearly double that of the EG-based nanofluid.

The use of SDBS allows to further increase the enhancement of the thermal conductivity coefficient. At a weight concentration of 0.2% for SWCNTs, the

Table 4.8 Thermal conductivity of EG + PVP- and EG + SDBS-based nanofluids with SWCNTs and MWCNT1 depending on weight concentration

w (%)	EG + PVP/S	λ_r	EG + PVP/M1	λ_r	EG + SDBS/S	λ_r	EG + SDBS/M1	λ_r
0.05	0.293	1.136	0.265	1.027	0.284	1.101	0.265	1.027
0.10	0.303	1.174	0.270	1.043	0.318	1.233	0.275	1.066
0.20	0.357	1.384	0.272	1.054	0.372	1.465	0.292	1.132

Table 4.9 Thermal conductivity of nanofluids based on W + PVP, W + SDBS, and W + SDS with SWCNTs and MWCNT1 depending on weight concentration

w, %	W + PVP/S	λ_r	W + PVP/M1	λ_r	W + SDBS/S	λ_r	W + SDBS/M1	λ_r
0.05	0.626	1.038	0.618	1.025	0.623	1.033	0.618	1.025
0.10	0.642	1.065	0.629	1.043	0.645	1.070	0.628	1.057
0.20	0.676	1.121	0.637	1.056	0.676	1.121	0.652	1.081

Table 4.10 Thermal conductivity of nanofluid based on W + SDS with SWCNTs and MWCNT1 depending on weight concentration

w (%)	W + SDS/S	λ_r	W + SDS/M2	λ_r
0.05	0.623	1.033	0.611	1.013
0.10	0.630	1.045	0.626	1.038
0.20	0.664	1.101	0.649	1.076

enhancement is 46.5%, and for MWCNTs, it is 13.2%. Therefore, using SDBS as a surfactant is more effective in increasing thermal conductivity than using PVP. Here we should specify that in Table 4.8, the relative thermal conductivity coefficient was determined with respect to the original base fluid (EG). It should be noted that the use of SDBS leads to a decrease in thermal conductivity of the base fluid (see Table 4.4), whereas PVP, on the contrary, slightly increases it. Therefore, the observed enhancement of thermal conductivity coefficient of the nanofluid with SDBS will be even greater.

Table 4.9 summarizes the thermal conductivity coefficients of nanofluids based on W + PVP, W + SDBS, and W + SDS containing SWCNTs and MWCNT M1, and Table 4.10 summarizes those of nanofluids based on W + SDS. In all cases, the nanofluid with MWCNTs exhibits a significantly lower enhancement of thermal conductivity coefficient than the nanofluid with SWCNTs. The use of PVP and SDBS yields almost identical results, while the use of SDS results in slightly lower values of enhancement of thermal conductivity. The highest enhancement is observed at a weight concentration of 0.2%. The nanofluid with SWCNTs shows a 12.1% enhancement, while the nanofluid with MWCNTs shows an 8.1% enhancement. It is noteworthy that the effectiveness of PVP and SDBS is similar. The thermal conductivity coefficient increases almost linearly with the increase in weight concentration in all cases.

4.6.4 Discussion of the Presented Experimental Data

Before drawing any general conclusions, it is necessary to understand the extent to which the data presented in the previous two sections are consistent with those known in the literature. Despite the large number of publications, making an adequate comparison is quite challenging because most papers do not provide exact numerical data for thermal conductivity coefficient measurements. Furthermore, the thermal conductivity coefficient of nanofluids is often discussed in relation to both temperature and concentration at the same time, which makes it difficult to identify accurately the individual effects of these parameters. Finally, several papers provide data on the dependence of the thermal conductivity coefficient on the volume concentration of CNTs. However, comparing this data with corresponding data obtained by changing the weight concentration is usually impossible. However, in certain cases, such a comparison can be made. In all of the examples provided below, our data agrees with the literature data, with an accuracy that is practically within the range of their measurement error. For example, the thermal conductivity coefficients of nanofluids based on EG + PVP with SWCNTs, obtained in this study align with the findings of Leong [89], while those based on W + SDBS containing SWCNTs coincide with the results of Almanassra et al. [7]. Esfe et al. [45] obtained rather similar values of enhancement of thermal conductivity coefficient of EG-based nanofluids but with functionalized SWCNTs. In study [59] of water-based nanofluids, the enhancement of the thermal conductivity of approximately 15% was achieved at a SWCNT volume concentration of 0.2%. However, it is important to note that a different surfactant (sodium deoxycholate) was used, and the diameter of the tubes was slightly smaller. Additionally, the UST time during the nanofluid preparation was three times shorter than in the present study. As it was shown above in Table 4.6, this circumstance slightly reduces the thermal conductivity. Choi et al. [31] reported that at a weight concentration of 1.14%, the thermal conductivity of nanofluid with SWCNTs increased by 19.4%. The diameter of the nanotubes was 1.33 nm. However, the authors did not specify the type of surfactant used and the exact temperature at which the measurements were taken. These factors can significantly affect the measured value.

Comparing data for nanofluids containing MWCNTs can be challenging due to significant differences in their inner and outer diameters and lengths. However, in some cases, it is possible. For example, the data for W + PVP-based nanofluids (Table 4.9) are in reasonable agreement with those of Fadhillahanafi et al. [49]. Slightly higher values were obtained in Jiang et al. [69], however, the measurements were performed at a higher temperature. Similar thermal conductivity values for the W + SDBS based nanofluid (N1) were obtained in Estellé et al. [46]. Additionally, higher values were reported in Omrani et al. [116], but in this work MWCNTs were functionalized. Similar thermal conductivity coefficients were obtained in Meng et al. [103] for EG-based nanofluids with functionalized MWCNTs. However, the magnitude of the thermal conductivity excess strongly depends on the method of functionalization.

It has been already noted that in the study of nanofluids with conventional spherical particles it was found that the excess thermal conductivity is higher the lower the thermal conductivity of the base fluid. This trend was also observed in nanofluids with CNTs. Figure 4.14 illustrates the dependence of the relative thermal conductivity coefficient of nanofluids, based on three different base fluids, on the weight

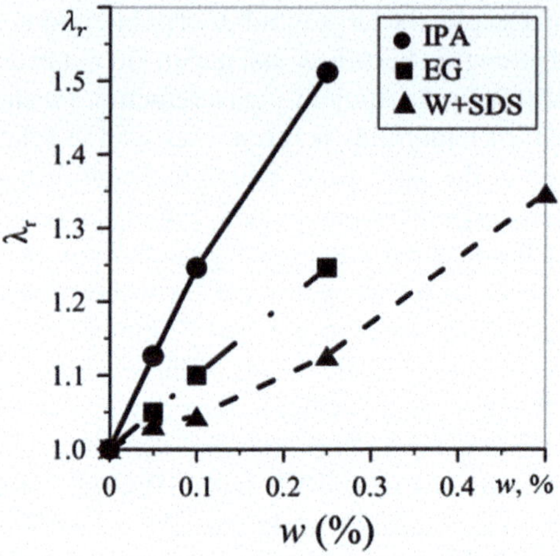

Fig. 4.14 Dependence of the relative thermal conductivity coefficient on the weight concentration of nanofluids with SWCNTs

concentration of SWCNTs. The thermal conductivity coefficient of the base fluids considered was highest in water, followed by ethylene glycol, and then isopropyl alcohol. Figure 4.14 demonstrates that, similar to nanofluids with spherical nanoparticles, the enhancement of the thermal conductivity of nanofluids with SWCNTs is greater when the thermal conductivity of the base fluid is lower. Therefore, the use of nanofluid as a refrigerant will be more efficient when the thermal conductivity of the base fluid is lower.

When using conventional spherical particles, nanofluids typically have volume concentrations ranging from 1 to 10%. However, the data provided for nanofluids with CNTs shows significantly lower concentrations of the latter. From a practical point of view, it is interesting to compare the values of the obtained enhancement of the thermal conductivity coefficient for nanofluids with spherical nanoparticles and with CNTs. Such a comparison is shown in Fig. 4.15. Here the data for water-based nanofluids with 95 nm aluminum, as well as 150 nm Al_2O_3, TiO_2, and ZrO_2 particles are compared [123]. Only the nanofluid with SWCNTs was prepared using water and SDS surfactant, while surfactants were not used in the other cases. The weight concentration of SWCNTs was 0.2%. The analysis of Fig. 4.15 clearly illustrates the advantages of using SWCNTs in nanofluids compared to conventional nanoparticles.

4.6.5 *Thermal Conductivity and Rheology*

The use of nanofluids in most thermophysical applications implies the presence of flow. The viscosity of nanofluids is a significant factor in such thermophysical applications. Conventional nanofluids have a much higher viscosity than their base

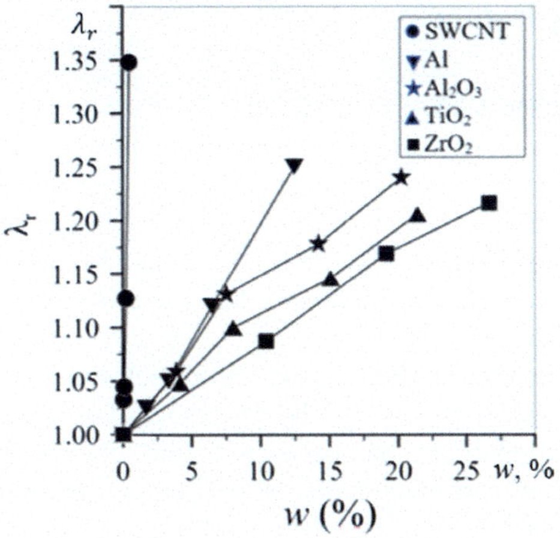

Fig. 4.15 Dependence of relative thermal conductivity coefficient of water-based nanofluids with different nanoparticles and SWCNTs on concentration

fluids (see Chap. 3). Additionally, the rheology of nanofluids can vary with changes in particle concentration or size. A similar situation occurs in several nanofluids containing CNTs. The viscosity and rheology measurements of nanofluids with CNTs, presented in Chap. 3, indicate that these fluids are generally pseudoplastic or viscoplastic. It was observed that the fluid index typically decreases, while the consistency parameter conversely increases with increasing CNT concentration. The rheology of the nanofluid can also be affected by increasing temperature. Changes in rheology have significant consequences for the thermal conductivity of real flows. The shear rate varies in different flow regions. For example, in the Poiseuille flow in a pipe it is maximum near the wall and minimum at the pipe axis. This means that the viscosity in such a flow is not uniform. Additionally, viscosity will also change with increasing CNT concentration. For example, the viscosity coefficient of EG-based nanofluids with SWCNTs is described by formula (3.30), the parameters of which are functions of w

$$k_\nu = 0.173 + 199w + 200000w^5, n = 1 - 10.2w + 40w^2, \tau_0 = 105w + 420w^2.$$

It follows that given value of weight concentration may correspond to different nanofluids. As shown in Fig. 4.13, the thermal conductivity coefficient of nanofluids increases almost linearly with concentration. However, concentration $w = 0.1$ corresponds to nanofluids with $k_\nu = 22.073$, $n = 0.38$, and concentration $w = 0.2$ corresponds to nanofluids with $k_\nu = 103.973$, $n = 0.22$. These are completely different nanofluids. Thus, as shown in Fig. 4.13, the thermal conductivity coefficient for EG-based nanofluids with SWCNTs varies for each point of w and hence corresponds to different nanofluids.

4.7 Concluding Remarks

In summarizing the discussion of thermal conductivity of nanofluids, it is important to reiterate some key findings. First of all, it should be noted that the thermal conductivity of almost all nanofluids is significantly higher than that of their base fluids. This enhancement of thermal conductivity is often much greater than predicted by classical theories for ordinary coarse dispersed fluids. Furthermore, the degree of thermal conductivity enhancement tends to increase as the thermal conductivity of the base fluid decreases.

Nanofluids containing conventional spherical particles and those containing CNTs are two distinct types of dispersed fluids. The thermal conductivity of nanofluids with conventional nanoparticles increases with the size of the particles. At the same weight concentration of nanoparticles and CNTs, the thermal conductivity of nanofluids with SWCNTs is significantly higher (at least an order of magnitude) than nanofluids with conventional particles. This statement indirectly suggests that the thermal conductivity mechanisms of these two types of nanofluids are significantly different.

The thermal conductivity coefficient of nanofluids with conventional particles is practically independent of their thermal conductivity. It increases with increasing particle concentration, until it reaches a certain limit value, after which it changes little. The effect of exceeding the thermal conductivity coefficient is influenced not only by the size but also by material density of the nanoparticles. In this respect, it is important to note that as the material density of the particles increases their weight concentration in the nanofluid also increases. Therefore, the higher the weight concentration of particles, the higher the thermal conductivity of the nanofluid over the corresponding value for the base liquid.

Currently, there is no commonly accepted model for the thermal conductivity of nanofluids, despite numerous attempts to construct such models. The four commonly stated mechanisms of the nanofluid thermal conductivity are discussed in detail in Sect. 4.4. It can be reasonably stated that clustering in a well-prepared nanofluid cannot explain the significant enhancement of nanofluid thermal conductivity. In a well-prepared nanofluid, the average size of nanoparticles should remain stable throughout its lifetime. If clustering were to occur, particles with macroscopic sizes would appear in the fluid. However, this would result in thermal conductivity no longer depending on the size of nanoparticles (as is the case in classical theories), and such particles would sediment quickly. However, neither of these phenomena is observed in practice.

The ballistic phonon mechanism of creating an additional excess thermal conductivity is rightly rejected by almost all authors. It acquires certain sense when clustering is taken into account. However, there should be no noticeable clustering in a well-prepared nanofluid. It is also worth noting that the MD simulations, described in Sect. 4.5, do not take this mechanism into account, since nanoparticles are considered to be solids, within which no energy transfer takes place. Nevertheless, there is a noticeable excess thermal conductivity of the nanofluid.

MD simulation provides a clear explanation of the enhancement mechanism in the thermal conductivity of nanofluids. The excess thermal conductivity of nanofluids is due to the interaction of nanoparticles with microfluctuations of hydrodynamic fields formed around them. These microfluctuations propagate at the speed of sound, and their scale is practically independent of the size of nanoparticles. The formation of microfluctuations should not be confused with the formation of a fluid layer around nanoparticles, which increases thermal conductivity. This is the so-called layering, which is also often used to explain the increase in thermal conductivity of nanofluids. Today, direct modeling has actually shown that this mechanism cannot explain the observed increase in thermal conductivity of nanofluids. At the same time, MD simulation points to another mechanism. The largest contribution to the thermal conductivity is brought by the λ^{KC} term due to the correlation of kinetic and collision energy fluxes. In addition, in nanofluid, the kinetic contribution λ^{K} is also significant due to the energy transfer by both the base fluid molecules and nanoparticles. The increase in thermal conductivity due to the increase in mobility of the base fluid molecules has been previously reported by Sarkar and Selvam [152].

Comprehending the thermal conductivity mechanisms of nanofluids with CNTs is a complex issue. On the one hand, some authors refer to the four mechanisms proposed for conventional nanofluids, as discussed in detail in Sect. 4.4. Their applicability for nanofluids with CNTs in general does not look more convincing than for ordinary nanofluids. On the other hand, the percolation structure formed in nanofluids with CNTs, even at low their concentrations, significantly contributes to the thermal conductivity of the nanofluid. Here the ballistic (or at least quasi-ballistic) heat transfer in such a structure is worth took into accounted. However, further studies in this direction are necessary. The MD method has been also used to study the thermal conductivity of nanofluids containing CNTs. Relevant literature can be found in Li and Cao [93], Sarode et al. [153], Jabbari et al. [65, 66], Sobczak et al. [157]. Simulating such nanofluids is challenging, and therefore, simulations are usually limited to relatively short CNTs, often considering just a single nanotube. Currently, the main focus is on studying the thermal conductivity coefficient in relation to CNT concentration and temperature. However, the energy transfer mechanisms in such fluids have not been thoroughly studied, nor has the formation of the percolation structure of CNTs and its influence on nanofluid thermal conductivity. These areas require further research in the future.

In conclusion, it is important to note once again that most nanofluids containing CNTs exhibit non-Newtonian behavior. Their rheology continuously changes with increasing CNT concentration and temperature. This suggests that with changes to the mentioned parameters, which is typical for many flows, it is appropriate to discuss the effective thermal conductivity of nanofluids. It is necessary to consider how to introduce this effective thermal conductivity in practical calculations of nanofluid flows.

References

1. Abareshi M, Goharshadi E, Zebarjad S et al (2010) Fabrication characterization and measurement of thermal conductivity of Fe_3O_4 nanofluids. J Magn Magn Mater 322(24):3895–3901
2. Abdullah A, Mohamad IS, Hashim AYB et al (2016) Thermal conductivity and viscosity of deionized water and ethylene glycol-based nanofluids. J Mech Eng Sci (JMES) 10(3):2249–2261
3. Ahmadi H, Rashidi A, Nouralishahi A et al (2013) Preparation and thermal properties of oil-based nanofluid from multi-walled carbon nanotubes and engine oil as lubricant. Int Com Heat Mass transfer 46:142–147
4. Akhilesh M, Santarao K, Babu MVS (2018) Thermal conductivity of CNT-watered nanofluids: a review. Mechanics Mech Eng 22(1):207–214
5. Ali AJ, Tugolukov EN (2019) An experimental study on the influence of functionalized carbon nanotubes CNT Taunit series on the thermal conductivity enhancement. IOP Conf Ser: Mat Sci Eng 693:012001
6. Ali N, Teixeira JA, Addali A (2018) A review on nanofluids: fabrication, stability, and thermophysical properties. J Nanomat 2018:6978130
7. Almanassra IW, Manasrah AD, Al-Mubaiyedh UA et al (2020) An experimental study on stability and thermal conductivity of water/CNTs nanofluids using different surfactants: a comparison study. J Mol Liquids 304:11025
8. Amani M, Amani P, Mahian O (2017) Multi-objective optimization of thermophysical properties of eco-friendly organic nanofluids. J Clean Prod 166:350–359
9. Ambreen T, Kim MH (2020) Influence of particle size on the effective thermal conductivity of nanofluids: a critical review. Appl Energy 264(16):114684
10. Arefyev KM (1983) Transport phenomena in gas and plasma. Energoatomizdat, Leningrad
11. Avsec J (2008) The combined analysis of phonon and electron heat transfer mechanism on thermal conductivity for nanofluids. Int J Heat Mass Transfer 51(19–20):4589–4598
12. Azmi W, Sharma K, Mamat R et al (2016) The enhancement of effective thermal conductivity and effective dynamic viscosity of nanofluids—A review. Renew Sustain Energy Rev 53:1046–1058
13. Bai M, Liu J, He J et al (2020) High efficiency heat transfer and antifriction characteristics of SMWCNTs nanofluids. Diamond Related Mat 105:107792
14. Barrat JL, Chiaruttini F (2003) Kapitza resistance at the liquid-solid interface. Mol Phys 101:1605–1610
15. Beck MP (2008) Thermal conductivity of metal oxide nanofluids. The Ph.D. thesis. Georgia Institute of Technology
16. Beck MP, Yuan Y, Warrier P et al (2009) The effect of particle size on the thermal conductivity of alumina nanofluids. J Nanopart Res 11:1129–1136
17. Benveniste Y (1987) Effective thermal conductivity of composites with a thermal contact resistance between the constituents: nondilute case. J Appl Phys 61:2840–2843
18. Benveniste Y, Chen T, Dvorak GJ (1990) The effective thermal conductivity of composites reinforced by coated cylindrically orthotropic fibers. J Appl Phys 67:2878–2884
19. Bhattacharya P, Phelan PE, Prasher R (2006) Brownian-motion-based convective-conductive model for the effective thermal conductivity of nanofluids. J Heat Transfer 128:588–595
20. Bhattacharya P, Saha SK, Yadav A et al (2004) Brownian dynamics simulation to determine the effective thermal conductivity of nanofluids. J Appl Phys 95:6492–6494
21. Bird RB, Steward WE, Lightfoot EN (1965) Transport phenomena. John Wiley, New York
22. Bridgman PW (1923) The thermal conductivity of liquids under pressure. Proc Am Acad 59(7):141–169
23. Bruggeman DAG (1935) Berechnung verschiedener physikalischer Konstanten von heterogenen Substanzen. I. Dielektrizitatskonstanten undleitfahigkeiten der Mischkorper aus isotropen Substanzen. Ann Phys 24:636–679
24. Ceotto D, Rudyak VYa, (2016) Phenomenological formula for thermal conductivity coefficient of water-based nanofluids. Colloid J 78(4):509–514

25. Chapman S, Cowling TG (1990) The mathematical theory of non-uniform gases. Cambridge University Press, Cambridge
26. Chen J, Han K, Wang S (2019) Investigation of enhanced thermal properties of Cu–Ar nanofluids by reverse non-equilibrium molecular dynamics method. Powder Technol 356:559–565
27. Chen L, Xie H (2010) Surfactant-free nanofluids containing double- and single-walled carbon nanotubes functionalized by a wet-mechanochemical reaction. Thermochim Acta 497:67–71
28. Chen L, Xie H, Li Y et al (2008) Nanofluids containing carbon nanotubes treated by mechanochemical reaction. Thermochim Acta 477:21–24
29. Chiew YC, Glandt ED (1987) Effective conductivity of dispersions: the effect of resistance at the particle surfaces. Chem Eng Sci 42:2677–2685
30. Colla L, Fedele L, Scattolini M et al (2012) Water-based Fe_2O_3 nanofluid characterization: thermal conductivity and viscosity measurements and correlation. Adv Mech Eng 2012:674947
31. Choi TY, Maneshian MH, Kang B et al (2009) Measurement of the thermal conductivity of a water-based single-walled carbon nanotube colloidal suspension with a modified 3-ω method. Nanotechnology 20:315706
32. Chon CH, Kihm KD, Lee SP et al (2005) Empirical correlation finding the role of temperature and particle size for nanofluid (Al_2O_3) thermal conductivity enhancement. Appl Phys Lett 87(15):153107
33. Cu W, Bai M, Lv J et al (2011) On the influencing factors and strengthening mechanism for thermal conductivity of nanofluids by molecular dynamics simulation. Ind Eng Chem Res 50(23):13568–13575
34. Cui W, Shen Z, Yang J et al (2014) Influence of nanoparticle properties on the thermal conductivity of nanofluids by molecular dynamics simulation. Roy Soc Chem 4(98):55580–55589
35. Cui W, Shen Z, Yang J et al (2016) Molecular dynamics simulation on the microstructure of absorption layer at the liquid–solid interface in nanofluids. Int Com Heat Mass Transfer 71:75–85
36. Daniel M-C, Astruc D (2004) Gold nanoparticles: assembly, supramolecular chemistry, quantum-size-related properties, and applications toward biology, catalysis, and nanotechnology. Chem Rev 104:293–346
37. Das PK (2017) A review based on the effect and mechanism of thermal conductivity of normal nanofluids and hybrid nanofluids. J Mol Liquids 240:420–446
38. Das SK, Choi SUS, Yu W et al (2008) Nanofluids: science and technology. Wiley, Hoboken, New Jersey
39. Das SK, Putra N, Thiesen P et al (2003) Temperature dependence of thermal conductivity enhancement for nanofluids. J Heat Transfer 125:567–574
40. Debye P (1914) Zustandsleihung und quantenhypothese mit cinem Anhang über Wärmeleitung, Vorträge über die Kinetische Theorie der Materie and der Elektrizität. Teubner, Berlin
41. Duangthongsuk W, Wongwises S (2009) Measurement of temperature-dependent thermal conductivity and viscosity of TiO_2–water nanofluids. Exp Therm Fluid Sci 33(4):706–714
42. Eapen J, Li J, Yip S (2007) Mechanism of thermal transport in dilute nanocolloids. Phys Rev Lett 98(2):028302
43. Eastman JA, Choi US, Li S et al (1997) Enhanced thermal conductivity through the development of nanofluids. Mater Res Soc 457:3–11
44. Eastman JA, Phillpot SR, Choi SUS et al (2004) Thermal transport in nanofluids. Annu Rev Mater Res 34:219–246
45. Esfe H, Firouzi M, Afrand M (2018) Experimental and theoretical investigation of thermal conductivity of ethylene glycol containing functionalized single walled carbon nanotubes. Physica E 95:71–77
46. Estellé P, Halelfadl S, Maré M (2015) Thermal conductivity of CNT water based nanofluids: experimental trends and models overview. J Thermal Eng 1(2):381–390

47. Esumi K, Ishigami M, Nakajima A et al (1996) Chemical treatment of carbon nanotubes. Carbon 34:279–281
48. Evans W, Fish J, Keblinski P (2006) Role of Brownian motion hydrodynamics on nanofluids thermal conductivity. Appl Phys Lett 88:93–116
49. Fadhillahanafi NM, Leong KY, Risby MS (2013) Stability and thermal conductivity characteristics of carbon nanotube based nanofluids. Int J Autom Mech Eng (IJAME) 8:1376–1384
50. Feng Y, Yu B, Xu P et al (2007) The effective thermal conductivity of nanofluids based on the nanolayer and the aggregation of nanoparticles. J Phys D Appl Phys 40(10):3164
51. Ganeshkumar J, Kathikaman D, Raja K et al (2017) Experimental study on density, thermal conductivity, specific heat, and viscosity of water-ethylene glycol mixture dispersed with carbon nanotubes. Thermal Sci 21:255–265
52. Garg J, Poudel B, Chiesa M et al (2008) Enhanced thermal conductivity and viscosity of copper nanoparticles in ethylene glycol nanofluid. J Appl Phys 103:074301
53. Ghasemi A, Hassani M, Goodarzi M et al (2019) Appraising influence of COOH-MWCNTs on thermal conductivity of antifreeze using curve fitting and neural network. Physica A: Statistical Mech Applications 514(15):36–45
54. Gonçalves I, Souza R, Coutinho G et al (2021) Thermal conductivity of nanofluids: a review on prediction models. Appl Sci 11(6):2525
55. Guo Z (2020) A review on heat transfer enhancement with nanofluids. J Enhanc Heat Transfer 27:1–70
56. Grest GS, Kremer K (1986) Molecular dynamics simulation for polymers in the presence of a heat bath. Phys Rev A 33:3628–3631
57. Gupta M, Singh V, Kumar R et al (2017) A review on thermophysical properties of nanofluids and heat transfer applications. Renewable Sustainable Energy Rev 74:638–670
58. Hamilton RL, Crosser OK (1962) Thermal conductivity of heterogeneous two-component systems. I&EC Fundam 1:182–191
59. Harish S, Ishikawa K, Einarsson E et al (2012) Enhanced thermal conductivity of ethylene glycol with single-walled carbon nanotube inclusions. Int J Heat Mass Transfer 55(13):3885–3890
60. Harish S (2014) Thermal conductivity enhancement of liquid and solid with single-walled carbon nanotubes. Doctoral thesis. The University of Tokyo
61. Heinz H, Vaia RA, Farmer BL et al (2008) Accurate simulation of surfaces and interfaces of face-centered cubic metals using 12–6 and 9–6 Lennard-Jones potentials. J Phys Chem C 112:17281–17290
62. Hirschfelder JO, Curtiss CF, Bird RB (1984) The molecular theory of gases and liquids. Wiley
63. Hosseini SSH, Shahrjerdi A, Vazifeshenas Y (2011) A review of relations for physical properties of nanofluids. Australian J Basic Appl Sci 5(10):417–435
64. Jang SP, Choi SUS (2004) Role of Brownian motion in the enhanced thermal conductivity of nanofluids. Appl Phys Lett 84:4316–4318
65. Jabbari F, Saedodin S, Rajabpour A (2018) Experimental investigation and molecular dynamics simulations of viscosity of CNT-water nanofluid at different temperatures and volume fractions of nanoparticles. J Chem Eng Data 64:1
66. Jabbari F, Rajabpour A, Saedodin S (2021) Thermal conductivity of CNT–water nanofluid at different temperatures, volume fractions, and diameters: experimental investigation and molecular dynamics simulations. Microfuidics Nanofuidics 25:102
67. Jana S, Salehi-Khojin A, Zhong W-H (2007) Enhancement of fluid thermal conductivity by the addition of single and hybrid nano-additives. Thermochim Acta 462:45–55
68. Jia T, Zhang Y, Ma H et al (2012) Investigation of the characteristics of heat current in a nanofluid based on molecular dynamics simulation. Appl Phys A 108(3):537–544
69. Jiang H, Zhang Q, Shi L (2015) Effective thermal conductivity of carbon nanotube-based nanofluid. J Taiwan Inst Chem Eng 55:76–81
70. Jóźwiak B, Boncel S (2020) Rheology of ion nanofluids—A review. J Mol Liquids 302:112568

71. Ju S, Li ZY (2006) Theory of thermal conductance in carbon nanotube composites. Phys Lett A 353:194–197
72. Jung J-Y, Yoo JY (2009) Thermal conductivity enhancement of nanofluids in conjunction with electrical double layer (EDL). Int J Heat Mass Transfer 52:525–528
73. Kang HU, Kim SH, Oh JM (2006) Estimation of thermal conductivity of nanofluid using experimental effective particle volume. Experimental Heat Transfer 19(3):181–191
74. Kang H, Zhang Y, Yang M et al (2012) Molecular dynamics simulation on effect of nanoparticle aggregation on transport properties of a nanofluid. J Nanotech Eng Med 3(2):021001
75. Kapitza PL (1941) The study of heat transfer in helium II. J Phys USSR 4(1–6):181–210
76. Karimi A, Sadatlu MAA, Saberi B et al (2015) Experimental investigation on thermal conductivity of water-based nickel ferrite nanofluids. Adv Powder Technology 26(6):1529–1536
77. Karimipour A, Malekahmadi O, Karimipour A et al (2020) Thermal conductivity enhancement via synthesis produces a new hybrid mixture composed of copper oxide and multi-walled carbon nanotube dispersed in water: experimental characterization and artificial neural network modeling. Int J Thermophys 41:116
78. Keblinski P, Phillpot SR, Choi SUS et al (2002) Mechanism of heat flow in suspensions of nano-sized particles (nanofluids). Int J Heat Mass Transfer 45:855–863
79. Keblinski P, Prasher R, Eapen J (2008) Thermal conductance of nanofluids: is the controversy over? J Nanoparticles Res V:1089–1097
80. Khamliche T, Khamlich S, Moodley M et al (2021) Laser fabrication of Cu nanoparticles based nanofluid with enhanced thermal conductivity: experimental and molecular dynamics studies. J Mol Liquids 323:114975
81. Kleinstreuer C, Yu F (2011) Experimental and theoretical studies of nanofluid thermal conductivity enhancement: a review. Nanoscale Res Lett 6:22–35
82. Koo J, Kleinstreuer C (2004) A new thermal conductivity model for nanofluids. J Nanoparticle Res 6(6):577–588
83. Koo J, Kleinstreuer C (2005) Impact analysis of nanoparticle motion mechanisms on the thermal conductivity of nanofluids. Int Com Heat Mass Transfer 32:1111–1118
84. Kumar DH, Patel HE, Kumar VRR et al (2004) Model for heat conduction in nanofluids. Phys Rev Lett 93(14):144301
85. Kumar PG, Sakthivadivel D, Meikandan M et al (2019) Experimental study on thermal properties and electrical conductivity of stabilized H_2O-solar glycol mixture based multi-walled carbon nanotube nanofluids: developing a new correlation. Heliyon 5:e02385
86. Landau LD, Lifshitz EM (2013) Course of theoretical physics. Vol 6. Fluid mechanics. Elsevier
87. Lee SL, Saidur R, Sabri MFM et al (2016) Molecular dynamic simulation: studying the effects of Brownian motion and induced micro-convection in nanofluids. Numer Heat Transfer A: Appl 69(6):643–658
88. Leong KC, Yang C, Murshed SMS (2006) A model for the thermal conductivity of nanofluids–the effect of interfacial layer. J Nanopart Res 8(2):245–254
89. Leong KY (2016) The effect of surfactant on stability and thermal conductivity of carbon nanotube based nanofluids. Thermal Sci 20(2):429–436
90. Li L, Zhang Y, Ma H et al (2009) Molecular dynamics simulation of effect of liquid layering around the nanoparticle on the enhanced thermal conductivity of nanofluids. J Nanoparticle Res 12(3):811–821
91. Li Q, Xuan YM (2000) Experimental investigation on transport properties of nanofluids. Heat Transfer Sci Techn 2000:757–762
92. Li X, Zou C, Zhou L et al (2016) Experimental study on the thermo-physical properties of diathermic oil based SiC nanofluids for high temperature applications. Int J Heat Mass Transfer 97:631–637
93. Li Y-W, Cao B-Y (2011) Thermal conductivity of single-walled carbon nanotube with internal heat source studied by molecular dynamics simulation. Int J Thermophys 34(12):2361–2370
94. Liang Z, Tsai H-L (2011) Thermal conductivity of interfacial layers in nanofluids. Phys Rev E 83:1–8

95. Loya A, Najib A, Aziz F et al (2022) Comparative molecular dynamics simulations of thermal conductivities of aqueous and hydrocarbon nanofluids. Beilstein J Nanotech 13:620–628
96. Loya A, Stair JL, Ren G (2014) The study of simulating metal oxide nanoparticles in aqueous fluid. Int J Eng Res Technol 4:1954–1960
97. Lu W-Q, Fan Q-M (2008) Study for the particle's scale effect on some thermophysical properties of nanofluids by a simplified molecular dynamics method. Eng Analysis Boundary Elements 32(4):282–289
98. Maghlany MEl, Minea A, (2019) Novel empirical correlation for ionanofluid PEC inside tube subjected to heat flux with application to solar energy. J Thermal Anal Calorim 135(2):1161–1170
99. Manasrah AD, Laoui T, Zaidi SJ et al (2017) Effect of PEG functionalized carbon nanotubes on the enhancement of thermal and physical properties of nanofluids. Exp Thermal Fluid Sci 84:231–241
100. Manna I, Chopkar M, Das PK (2006) Synthesis and characterization of nanofluid for advanced heat transfer applications. Scripta Mater 55:549–552
101. Masuda H, Ebata A, Teramae K et al (1993) Alteration of thermal conductivity and viscosity of liquid by dispersing ultra-fine particles. Netsu Bussei 4(7):227–239
102. Maxwell JC (1881) A treatise on electricity and magnetism. Clarendon Press, Oxford
103. Meng Z, Wu D, Wang L et al (2012) Carbon nanotube glycol nanofluids: photo-thermal properties, thermal conductivities and rheological behavior. Particuology 10(5):614–618
104. Minakov AV, Rudyak VYa, Guzei DV et al (2015) Measurements of the thermal-conductivity coefficient of nanofluids by the hot-wire method. J Eng Phys Thermophys 88:149–162
105. Minea AA, Murshed SMS (2018) A review on development of ionic liquid based nanofluids and their heat transfer behavior. Renew Sustain Energy Rev 91:584–599
106. Mirbagheri MH, Akbari M, Mehmandoust B (2018) Proposing a new experimental correlation for thermal conductivity of nanofluids containing of functionalized multiwalled carbon nanotubes suspended in a binary base fluid. Int Comm Heat Mass Transfer 98:216–222
107. Missenard A (1965) Conductivité thermique des solides, Liquids, gaz et de leurs Mélanges. Editions Eyrolles, Paris
108. Mohebbi A (2012) Prediction of specific heat and thermal conductivity of nanofluids by a combined equilibrium and non-equilibrium molecular dynamics simulation. J Mol Liquids 175:51–58
109. Munkhbayar B, Tanshen MR, Jeoun et al (2013) Surfactant-free dispersion of silver nanoparticles into MWCNT-aqueous nanofluids prepared by one-step technique and their thermal characteristics. Ceram Int 39:6415–6425
110. Muraleedharan MG, Sundaram DS, Henry A et al (2017) Thermal conductivity calculation of nano-suspensions using Green-Kubo relations with reduced artificial correlations. J Phys: Condensed Matter 29(15):155302
111. Nan C-W, Liu G, Lin Y et al (2004) Interface effect on thermal conductivity of carbon nanotube composites. Appl Phys Lett 85:3549–3551
112. Nanda J, Maranville C, Bollin SC et al (2008) Thermal conductivity of single-wall carbon nanotube dispersions: role of interfacial effects. J Phys Chem C 112:654–658
113. Nasiri A, Shariaty-Niasar M, Rashidi AM et al (2012) Effect of CNT structures on thermal conductivity and stability of nanofluid. Int J Heat Mass Transfer 55(5–6):1529–1535
114. Ni F, Gu GQ, Chen KM (1997) Effective thermal conductivity of nonlinear composite media with contact resistance. Int J Heat Mass Transfer 40:943–949
115. Nie C, Marlow WH, Hassan YA (2008) Discussion of proposed mechanisms of thermal conductivity enhancement in nanofluids. Int J Heat Mass Transfer 51:1342–1348
116. Omrani AN, Esmaeilzadeh E, Jafari M et al (2019) Effects of multi walled carbon nanotubes shape and size on thermal conductivity and viscosity of nanofluids. Diamond Related Mat 93:96–104
117. Ozerinc S, Kakac S, Yazicioglu AG (2010) Enhanced thermal conductivity of nanofluids: a state-of-the-art review. Microfluidics Nanofluidics 8(2):145–170

118. Pak BC, Cho YI (1998) Hydrodynamic and heat transfer study of dispersed fluids with submicron metallic oxide particles. Exp Heat Transfer 11:151–170
119. Phuoc TX, Massoudi M, Chen R-H (2011) Viscosity and thermal conductivity of nanofluids containing multi-walled carbon nanotubes stabilized by chitosan. Int J Thermal Sci 50(1):12–18
120. Prasher R, Bhattacharya P, Phelan PE (2005) Thermal conductivity of nanoscale colloidal solutions (nanofluids). Phys Rev Lett 94(2):025901
121. Pryazhnikov MI, Minakov AV, Guzeil DV et al (2016) Study of the dependence the thermal conductivity of nanofluids on different parameters. J Phys: Conf Ser 754:092003
122. Pryazhnikov MI, Minakov AV, Lyamkin AI et al (2020) Experimental study of transport coefficients of aqueous suspensions of nanodiamonds. Colloid J 82:705–712
123. Pryazhnikov MI, Minakov AV, Rudyak VYa et al (2017) Thermal conductivity measurements of nanofluids. Int J Heat Mass Transfer 104:1275–1282
124. Rashidi S, Rashimi W, Abdullan L et al (2017) Thermal stability and conductivity of carbon nanotube nanofluid using xanthan gum as surfactant. Sains Malaysiana 46(7):1017–1024
125. Rashmi W, Ismail AF, Sopyan I et al (2011) Stability and thermal conductivity enhancement of carbon nanotube nanofluid using gum arabic. J Exp Nanosci 6(6):567–579
126. Rehman WUr, Zulkifli Merican ZMA, Bhat AH et al (2019) Synthesis, characterization, stability and thermal conductivity of multi-walled carbon nanotubes (MWCNTs) and eco-friendly jatropha seed oil based nanofluid: an experimental investigation and modeling approach. J Mol Liquids 293:111534
127. Reid RC, Prausnitz JM, Sherwood TK (1977) The properties of gases and liquids. McGraw–Hill Book Comp, New York
128. Richard RG, Shankland IR (1989) A transient hot-wire method for measuring the thermal conductivity of gases and liquids. Int J Thermophys 10(3):673–686
129. Rodert HM (1981) A transient hot wire thermal conductivity apparatus for fluids. J Res Natl Bureau Standards 86(5):457–493
130. Rudyak VYa (2019) Thermophysical characteristics of nanofluids and transport process mechanisms, J Nanofluids 8:1–16
131. Rudyak VYa, Belkin AA (2003) Nanoparticle velocity relaxation in condensed carrying medium. Tech Phys Lett 29(7):560–562
132. Rudyak VYa, Belkin AA (2004) Mechanisms of collective nanoparticles interaction with condensed solvent. Thermophys Aeromech 11(2):54–63
133. Rudyak VYa, Belkin AA (2010) Simulation of transport coefficients. Nanosystems: Phys Chem Math 1:156–177
134. Rudyak VYa, Belkin AA (2013) Transport processes of nanoparticles in gases and liquids. Adv Struct Mater 4:135–168
135. Rudyak VYa, Belkin AA, Ivanov DA et al (2008a) The simulation of transport processes using the method of molecular dynamics. Self-diffusion coefficient. High Temp 46:30–39
136. Rudyak VYa, Belkin AA, Tomilina EA (2010) On the thermal conductivity of nanofluids. Tech Phys Lett 36:660–662
137. Rudyak VYa, Kharlamov GV, Belkin AA (2000) The velocity autocorrelation function of nanoparticles in a hard-sphere molecular system. Tech Phys Lett 26:553–556
138. Rudyak VYa, Kharlamov GV, Belkin AA (2001) Diffusion of nanoparticles and macromolecules in dense gases and liquids. High Temp 39:264–271
139. Rudyak VYa, Krasnolutskii SL (2014) Dependence of the viscosity of nanofluids on nanoparticle size and material. Phys Lett A 378:1845–1849
140. Rudyak VYa, Krasnolutskii SL (2015) Simulation of the nanofluid viscosity coefficient by the molecular dynamics method. Tech Phys 60(6):798–804
141. Rudyak VYa, Krasnolutskii SL (2017) Simulation of the thermal conductivity of a nanofluid with small particles by molecular dynamics methods. Tech Phys 62(10):1456–1465
142. Rudyak VYa, Minakov AV, Krasnolutskii SL (2016a) Physics and mechanics of heat exchange processes in nanofluid flows. Phys Mesomech 19(3)456–474

143. Rudyak VYa, Minakov AV, Pryazhnikov MI et al (2022) Measurement of the thermal conductivity and heat transfer coefficient of nanofluids with single-walled nanotubes. High Temp 60(5):631–638
144. Rudyak VYa, Pryazhnikov MI, Minakov AV et al (2023) Comparison of thermal conductivity of nanofluids with single-walled and multi-walled carbon nanotubes. Diamond Related Mat 139:110376
145. Sachdeva P, Kumar R (2009) Effect of hydration layer and surface wettability in enhancing thermal conductivity of nanofluids. Appl Phys Lett 95:223105
146. Sadri R, Ahmadi G, Togun H et al (2014) An experimental study on thermal conductivity and viscosity of nanofluids containing carbon nanotubes. Nanoscale Res Lett 9:151
147. Sankar N, Mathew N, Sobhan C (2008) Molecular dynamics modeling of thermal conductivity enhancement in metal nanoparticle suspensions. Int Commun Heat Mass Transfer 35:867–872
148. Shalkevich N, Escher W, Bürgi T et al (2010) On the thermal conductivity of gold nanoparticle colloids. Langmuir 26(2):663–670
149. Sarkar S, Ghosh NK (2022) Equilibrium molecular dynamics simulation for investigation on the enhancement of thermal conductivity in ethylene glycol-based ZnO nanofluid. IOP Conf Ser: Mat Sci Eng 1258:012004
150. Sedighi M, Mohebbi A (2014) Investigation of nanoparticle aggregation effect on thermal properties of nanofluid by a combined equilibrium and nonequilibrium molecular dynamics simulation. J Mol Liquids 197:14–22
151. Samira A, Hossein M, Branch A et al (2022) Investigation of thermophysical properties of Io nanofluids containing multi-walled carbon nanotubes and graphene. Iran J Chem Chem Eng 41(2):380–391
152. Sarkar S, Selvam RP (2007) Molecular dynamics simulation of effective thermal conductivity and study of enhanced thermal transport mechanism in nanofluids. J Appl Phys 102(7):1–7
153. Sarode A, Ahmed Z, Basarkar P et al (2017) A molecular dynamics approach of the role of carbon nanotube diameter on thermal interfacial resistance through vibrational mismatch analysis. Int J Thermal Sci 122:33–38
154. Shamaeil M, Firouzi M, Fakhar A (2016) The effects of temperature and volume fraction on the thermal conductivity of functionalized DWCNTs/ethylene glycol nanofluid. J Therm Anal Calorim 126:1455–1462
155. Shukla RK, Dhir VK (2008) Effect of Brownian motion on thermal conductivity of nanofluids. J Heat Transfer 130(4):042406
156. Singh N, Chand G, Kanagaraj S (2012) Investigation of thermal conductivity and viscosity of carbon nanotubes–ethylene glycol nanofluids. Heat Transfer Eng 33(9):821–827
157. Sobczak J, Vallejo JP, Traciak J et al (2021) Thermophysical profile of ethylene glycol based nanofluids containing two types of carbon black nanoparticles with different specific surface areas. J Mol Liquids 326:115255
158. Soltanimehr M, Afrand M (2016) Thermal conductivity enhancement of COOH functionalized MWCNTs/ethylene glycol–water nanofluid for application in heating and cooling systems. Appl Therm Eng 105:716–723
159. Sun C, Lu W-q, Lu J et al (2011) Molecular dynamics simulation of nanofluid's effective thermal conductivity in high-shear-rate Couette flow. Int J Heat Mass Transfer 54(11–12):2560–2567
160. Swartz ET, Pohl RO (1989) Thermal boundary resistance. Rev Mod Phys 61(3):605–668
161. Tam NT, Phuong NV, Khoi PH et al (2020) Carbon nanomaterial-based nanofluids for direct thermal solar absorption. Nanomaterials 10:1199
162. Tawfik MM (2017) Experimental studies of nanofluid thermal conductivity enhancement and applications: a review. J Renew Sustain Energy Rev 75:1239
163. Teng K-L, Hsiao P-Y, Hung S-W et al (2007) Enhanced thermal conductivity of nanofluids diagnosis by molecular dynamics simulations. J Nanosci Nanotechnol 8(7):1–9
164. Teng T-P, Hung Y-H, Teng T-C et al (2010) The effect of alumina/water nanofluid particle size on thermal conductivity. Appl Therm Eng 30(14–15):2213–2218

165. Thakurm M, Gangacharyulu D, Singh G (2017) Effect of temperature and multiwalled carbon nanotubes concentration on thermophysical properties of water based nanofluids. Int J Mech Prod Eng Res Dev (IJMPERD) 7(4):151–160
166. Timofeeva EV, Routbort LJ, Singh D (2009) Particle shape effects on thermophysical properties of alumina nanofluids. J Appl Phys 106:014304
167. Timofeeva EV, Smith DS, Yu W et al (2010) Particle size and interfacial effects on thermophysical and heat transfer characteristics of water-based α-SiC nanofluids. Nanotechnology 21:215703
168. Topal I, Servantie J (2019) Molecular dynamics study of the thermal conductivity in nanofluids. Chem Phys 516:147–151
169. Trisaksri V, Wongwises S (2007) Critical review of heat transfer characteristics of nanofluids. Renew Sustain Energy Rev 11:512–523
170. Tsai T-H, Kuo L-S, Chen P-H, Yang C-T (2009) Thermal conductivity of nanofluid with magnetic nanoparticles. PIERS Online 5(3):231–234
171. Turgut A, Tavman I, Chirtoc M et al (2009) Thermal conductivity and viscosity measurements of water-based TiO_2 nanofluids. Int J Thermophys 30:1213–1226
172. Wang X, Jing D (2019) Determination of thermal conductivity of interfacial layer in nanofluids by equilibrium molecular dynamics simulation. Int J Heat Mass Transfer 128:199–207
173. Wang X, Xu X, Choi SUS (1999) Thermal conductivity of nanoparticle–fluid mixture. J Thermophys Heat Transfer 13(4):474–480
174. Xie H, Fujii M, Zhang X (2005) Effect of interfacial nanolayer on the effective thermal conductivity of nanoparticle-fluid mixture. Int J Heat Mass Transfer 48(14):2926–2932
175. Xie H, Lee H, Youn W et al (2003) Nanofluids containing multiwalled carbon nanotubes and their enhanced thermal conductivities. J Appl Phys 94(8):4967–4971
176. Xie H, Wang J, Xi T et al (2002) Dependence of the thermal conductivity of nanoparticle-fluid mixture on the base fluid. J Mat Sci Lett 21:1469–1471
177. Xie H, Yu W, Li Y et al (2011) Discussion on the thermal conductivity enhancement of nanofluids. Nanoscale Res Lett 6:124–136
178. Xu J, Yu B, Zou M et al (2006) A new model for heat conduction of nanofluids based on fractal distributions of nanoparticles. J Appl Phys 39:4486–4490
179. Xuan Y, Li Q (2000) Heat transfer enhancement of nanofluids. Int J Heat Fluid Flow 21:58–64
180. Xuan Y, Li Q, Hu W (2003) Aggregation structure and thermal conductivity of nanofluids. J American Inst Chem Eng (AIChE) 49(4):1038–1043
181. Xuan Y, Li Q, Zhang X et al (2006) Stochastic thermal transport of nanoparticle suspensions. J Appl Phys 100:043507
182. Xuan Y, Roetzel W (2000) Conceptions for heat transfer correlation of nanofluids. Int J Heat Mass Transfer 43:3701–3707
183. Xue QZ (2003) Model for effective thermal conductivity of nanofluids. Phys Lett A 307:313–317
184. Xue QZ (2006) Model for the effective thermal conductivity of carbon nanotube composites. Nanotechnology 17:1655–1660
185. Xue QZ, Keblinski P, Philpot SR et al (2004) Effect of liquid layering at the liquid–solid interface on thermal transport. Int J Heat Mass Transfer 47:4277–4284
186. Xue Q, Xu W-M (2005) A model of thermal conductivity of nanofluids with interfacial shells Mat. Chem Phys 90:298–301
187. Younes H, Christensen G, Li D et al (2015) Thermal conductivity of nanofluids: review. J Nanofluids 4:107–132
188. Yu W, Choi SUS (2004) The role of interfacial layers in the enhanced thermal conductivity of nanofluids: a renovated Hamilton-Crosser model. J Nanoparticle Res 6:355–361
189. Yu W, France DM, Choi SUS et al (2007) Review and assessment of nanofluid technology for transportation and other applications. Report Argonne National Lab ANL/ESD/07-9
190. Yu W, France DM, Routbort JL et al (2008) Review and comparison of nanofluid thermal conductivity and heat transfer enhancements. Heat Transfer Eng 29(5):432–460

191. Yu W, Xie H, Chen L et al (2010) Enhancement of thermal conductivity of kerosene-based Fe_3O_4 nanofluids prepared via phase-transfer method. Colloids Surfaces A 355(1–3):109–113
192. Yu-Hua L, Wei Q, Jian-Chao F (2008) Temperature dependence of thermal conductivity of nanofluids. Chinese Phys Lett 25(9):3319
193. Zhang X, Gu H, Fujii M (2007) Effective thermal conductivity and thermal diffusivity of nanofluids containing of spherical cylindrical nanoparticles. Experimental Thermal Fluid Sci 31:593–599
194. Zhang L, Tian L, Zhang A et al (2021) Molecular dynamics simulations of the effects of a nanoparticle surface adsorption layer on the thermal conductivity of a Cu–Ar nanofluid. Int J Thermophys 42:44
195. Zheng L, Bu XX, Fan BH et al (2016) Study on thermodynamic property for ionic liquid [C_4mim][Lact](1-butyl-3-methylimidazolium lactic acid). J Thermal Anal Calorim 123(2):1619–1625
196. Zheng ZM, Wang B (2018) A prediction model for the effective thermal conductivity of nanofluids considering agglomeration and the radial distribution function of nanoparticles. Acta Mech Sin 34:507–514
197. Zhu HT, Zhang CY, Tang YM et al (2007) Novel synthesis and thermal conductivity of CuO nanofluid. J Phys Chem C 111(4):1646–1650
198. Zubarev D (1974) Nonequilibrium statistical thermodynamics. Consultants Bureau, New York

Chapter 5
Electrical Conductivity of Nanofluids

5.1 Electrical Conductivity of Simple and Coarse Dispersed Fluids

The active study of the physical characteristics of nanofluids, which has begun in the 90s of the last century, was given reason at first by their possible thermophysical applications. That is why, their thermal conductivity, viscosity, and rheology was measured first of all. At the same time, even in the first works on the creation of nanofluids, their ζ-potential was actively measured. It has become clear that the electrical properties of nanofluids are an important indicator of their stability, and their electrical conductivity also needs to be systematically measured. One of the first works to draw attention to this issue was a paper by. Cruz et al. [11]. Later, this problem was repeatedly discussed [22, 34, 63].

Recently, it has become clear that knowledge of the electrical conductivity of nanofluids is essential for solar energy conversion [44, 55]. Additionally, knowledge of this property is crucial for using nanofluids in cooling processors [43], since it plays a significant role in optimizing their performance. In the last decade, several papers have been published on the possibility of using nanofluids to increase the breakdown voltage of transformer oils [14]. Additionally, nanofluids can be potentially applied in proton exchange membrane fuel cells (PEMFC), which is rather exotic and quite interesting research field [25, 61, 62]. Certainly, electrical conductivity is an extremely important factor when studying nanofluid flows in electric and magnetic field. Finely, there is a fairly wide range of ionic nanofluids whose electrical conductivity is very important [27, 36, 51].

There is another important reason that should be considered. In the previous chapter, we discussed thermal conductivity mechanisms. For nanofluids, these mechanisms are not trivial. On the other hand, the thermal conductivity and electrical conductivity of a solid are linked by the Wiedemann–Franz–Lorenz law [26], which states that the ratio of these two coefficients is proportional to temperature

V. Rudyak, *Thermophysical Properties of Nanofluids*, Fluid Mechanics and Its Applications 140, https://doi.org/10.1007/978-3-031-97995-8_5

$$\lambda/\sigma = LT, \tag{5.1}$$

where L is the Lorenz number.

The ratio (5.1) is due to the fact that the mechanisms of thermal conductivity and electrical conductivity in metals are the same and involve the movement of free electrons. Although the ratio (5.1) does not hold true at all temperatures, it is still interesting to explore to what extent the electrical conductivity and thermal conductivity mechanisms of nanofluids are related, and whether they are related at all.

The electrical conductivity of a liquid is determined by its ability to conduct electric current under the influence of an electric field. Typically, liquid dielectrics are the base fluid for creation of nanofluids (ionic liquids will not be discussed here). The electrical conductivity of liquid dielectrics is determined by the structure of their molecules. In nonpolar liquids, the conductivity depends on the presence of dissociated impurities, and in polar liquids it is also determined by the dissociation of the molecules of the liquid itself. Thus, the current generated in a liquid under the action of an electric field is caused precisely by the movement of ions formed in it and is described by Ohm's law

$$\boldsymbol{J} = \sigma \boldsymbol{E}, \tag{5.2}$$

where $\boldsymbol{J}$ is the current density vector, $\boldsymbol{E}$ is the vector of the electric field strength, and σ is the specific electrical conductivity coefficient.

The electrical conductivity of a dielectric liquid is determined by the mobility of ions l_i, which in turn is proportional to the strength of the applied electric field and inversely proportional to the drag force acting on it from the liquid. The latter can be described by Stokes law

$$l_i \sim 1/(6\pi r_i \mu), \tag{5.3}$$

where r_i is the effective radius of the ion.

The mobility of the ion is thus inversely proportional to its radius and the viscosity coefficient of the liquid. In addition, the velocity of the ion depends on the electric field strength and its charge. In the liquids under consideration, mainly single-charged ions are formed. Finally, with increasing temperature, the ion velocity increases leading to an increase in the electrical conductivity of the liquid. This is mainly due to a reduction in the viscosity of the liquid with an increase in temperature.

Base fluids can be either non-polar or polar. Polar liquids have molecules with a dipole moment. A typical example of a polar liquid is water, whose molecule consists of two hydrogen atoms with a positive charge and an oxygen atom with a negative charge. The resulting dipole moment of the water molecule is determined by the topology of these atoms within the molecule. Electrical conductivity of polar liquids is always higher than that of non-polar ones. It is also worth noting that the electrical conductivity of a liquid is higher, the greater its dielectric constant. With an increase in the dielectric constant, the degree of dissociation of polar liquid molecules also increases. Thus, the dielectric constant of water at 20 °C is 80.4, and that of ethylene glycol is 37.

The electrical conductivity of coarse dispersed fluid was first studied by C.J. Maxwell, who derived the corresponding formula. The idea behind its derivation is actually the same as that used to derive the corresponding thermal conductivity coefficient (4.7). Therefore, here we will not discuss the derivation of the electrical conductivity coefficient (see also the paper [3]). Instead, only the final result is given below [32]

$$\sigma_e = \sigma_{bf}\left[1 + \frac{3(1-\omega)\varphi}{1 + 2\omega - (1-\omega)\varphi}\right]. \tag{5.4}$$

where $\omega = \sigma_{bf}/\sigma_p$, σ_{bf} and σ_p are the coefficients of specific electrical conductivity of the base fluid and dispersed particles, respectively.

If $\omega\ w \ll 1$ (low electrical conductivity of the base fluid compared to particles), formula (5.4) reduces to the following

$$\sigma_e = \sigma_{bf}\frac{1 + 2\varphi}{1 - \varphi}. \tag{5.5}$$

In the inverse limiting case, when $w \gg 1\ \omega \gg 1$, (non-conducting particles), we have

$$\sigma_e = \sigma_{bf}\frac{1 + 2\varphi}{1 + \varphi/2}. \tag{5.6}$$

Fricke [17] has developed a theory that generalizes Maxwell's theory for particles in the form of ellipsoids. The relations obtained for spherical particles reduce to (5.4), but in general, they are extremely difficult to apply in practice. Even for biaxial ellipsoids, the formulas are extremely cumbersome. However, it is possible to consider a special case of strongly elongated ellipsoids with a large aspect ratio. These particles can be thought of as some CNT models.

5.1.1 Electrical Conductivity of Nanofluids with Oxide and Nitride Particles

The electrical conductivity of nanofluids is a relatively understudied property. There have been comparatively few published papers on this topic. In the papers of Minea [34], Poongavanam et al. [41], and Bhanvase and Barai [5], a fairly detailed review of the electrical conductivity of conventional nanofluids with spherical nanoparticles is presented. This section provides a brief overview of the data obtained over the past 15 years.

Water, ethylene glycol, their mixtures, as well as bio-glycol and engine oils were commonly used as base fluids. The electrical conductivity of nanofluids containing MgO, CuO, ZnO, TiO_2, SiO_2, In_2O_3, Al_2O_3, Fe_3O_4, Si_3N_4, and β-SiC nanoparticles was studied. In some cases, surfactants were added when preparing nanofluids. Due to the variety of nanofluid types and their heterogeneity, the data obtained were often contradictory. However, recent studies have allowed for the formulation of quite objective results.

A comprehensive study on the electrical conductivity of various nanofluids was conducted by Shoghl et al. [57]. The authors examined water-based nanofluids containing CuO, TiO_2, MgO, Al_2O_2, and ZnO particles. The weight concentrations of these nanoparticles varied from 0.01 to 2%. The study has found that all the nanofluids exhibited significantly higher electrical conductivities compared to water. Additionally, it increased almost linearly with increasing the concentration of nanoparticles. It is noted that the electrical conductivity of nanofluids depends significantly on the material of the particles used. Nanofluids containing MgO particles exhibit the highest electrical conductivity. The authors also point out that an increase in conductivity may be linked to the formation of double electric layers around nanoparticles (see Sect. 5.3). This leads to the appearance of additional charges in the nanofluid, which in turn contributes to the increased conductivity. This, in particular, determines the increase in electrical conductivity with increasing particle concentration. It is shown that Maxwell's theory (5.4), however, does not describe real experimental data for any nanofluid. In some cases, a surfactant (SDS) was used when preparing nanofluids, which significantly increased the electrical conductivity of the base fluid (W + SDS, where W is water). At the same time, the electrical conductivity of a W + SDS-based nanofluid is significantly dependent on the material of the nanoparticles. The conductivity of these nanofluids increases compared to the base fluid when Al_2O_3, MgO, and ZnO nanoparticles are used, while the conductivity of the fluids on the same base containing TiO_2 and CuO nanoparticles is lower than that of the base fluid. The disadvantage of this certainly interesting work is that it lacks information about the size of the nanoparticles used. This makes it difficult to understand whether nanoparticle size effects electrical conductivity.

An extensive study on the electrical conductivity of nanofluids based on water and ethylene glycol containing Al_2O_3, CuO, and Cu nanoparticles was conducted by Sarojini et al. [52]. The Al_2O_3 nanoparticles had three different sizes, including 150, 20–30, and 80 nm, while the CuO and Cu nanoparticles were approximately 89 nm in diameter. The volume concentration of these nanoparticles varied from 0.05 to 1%. Nanofluids were produced using the standard two-step method. It has been revealed that the electrical conductivity of water-based nanofluids containing CuO and Al_2O_3 nanoparticles increases linearly with increasing nanoparticle concentration, while a nonlinear increase is observed in nanofluids containing copper nanoparticles. It is worth noting that the latter type of nanofluid demonstrates a significantly lower increase in electrical conductivity compared to nanofluids containing oxide nanoparticles. The authors note that this is counterintuitive, as oxide particles have much lower electrical conductivity compared to copper particles. They attribute this phenomenon to the fact that the formation of surface charges on both types of particles is influenced by the polarization of the nanoparticles, which in turn is determined by the effective dielectric constant and particle density. The dielectric constant of oxide particles is significantly higher than that of copper particles, while their density is lower. This results in an increase in the electrical conductivity of the nanofluid as the effective dielectric constant increases. Similar results have been obtained for ethylene glycol-based nanofluids, although the increase in electrical conductivity in this case is an order of magnitude lower than that for water-based nanofluids. An important finding is the established relationship between particle size and electrical conductivity. The electrophoretic mobility of particles has been shown to increase with decreasing particle size, which in turn leads to an increase in electrical conductivity of the nanofluid due to a decrease in particle size. In all cases, increasing temperature has been found to increase the electrical conductivity of the nanofluid. Finally, the effect of a surfactant (SDS) on the electrical conductivity of a nanofluid was studied. It was revealed that the addition of surfactant reduces the electrical conductivity of the nanofluid, and the higher the concentration, the greater the reduction.

The influence of water purification on the preparation of nanofluids was studied in Iglesias et al. [23]. The authors studied nanofluids containing Al_2O_3 particles with a size of 15 nm. The volume concentration of particles ranged from 0.25 to 2% and the temperature was between 298.15 and 348.15 K. Water was purified through double and triple distillation. As a result, the electrical conductivity of the nanofluid differed by a factor of about two and was higher for water that had been double distilled. The electrical conductivity of both nanofluids increased with increasing the concentration of aluminum oxide and temperature. The increase in electrical conductivity was greater in nanofluids prepared using higher-quality water. The presence of impurities in lower-quality water had a more significant effect on the conductivity of nanofluids with higher particle concentrations and at higher temperatures. Later, the electrical conductivity of nanofluids based on aluminum oxide particles of 40 nm in size was studied [24]. A comparison was made with previous work, but the results were quite contradictory. The authors note that in all cases, nanofluids with larger particles have higher electrical conductivity. However, the enhancement of the electrical conductivity depends on the concentration of particles and temperature, and decreases with increasing concentration.

The electrical conductivity of water-based nanofluids using the SDBS surfactant and Al_2O_3 nanoparticles of 50 nm in size was studied in Zawrah et al. [63]. A rather surprising result was obtained in this research. Initially, the electrical conductivity of the nanofluid increased with increasing volume concentration. However, starting from a concentration of 0.2 vol%, it decreased. The authors explain this decrease by the fact that as the number of nanoparticles per unit volume increases, not all of them can form a double electrical layer (DEL).

Finally, we can mention the works of Minea and Luciu [35] and Minea [33], which studied the electrical conductivity of water-based nanofluids with Al_2O_3 nanoparticles of 12 and 10 ± 5 nm size. It was found that the electrical conductivity of these nanofluids increased almost linearly with an increase in the volume concentration of nanoparticles, reaching a maximum value of 4%. At the maximum particle concentration, the electrical conductivity increased by 390.11% at a temperature of

60 °C and by 379.6% at 25 °C. It is noted that Maxwell's theory is not applicable to describing the electrical conductivity of nanofluids. This is primarily due to the fact that the formation of DEL around nanoparticles plays an important role in electrical conductivity of nanofluids. The paper by Zakaria et al. [62] presents data on the study of a water-based nanofluid containing same particles with a size of 13 nm. The excess electrical conductivity of this nanofluid has reached 516% at a particle volume concentration of 0.5%. The authors note that a value predicted by Maxwell's theory is an order of magnitude lower than the value observed experimentally, even at a low concentration of particles. This difference increases with increasing concentration. In this work, the nanofluid in question was considered as an alternative refrigerant for use in proton exchange membrane (PEM) fuel cells. The results have shown that using a nanofluid with a particle concentration of 0.5% was 187% more effective than using water.

The electrical conductivity of a water-based nanofluid containing CuO particles with a size of 12 nm was studied in a temperature range of 298–348 K in the paper [10]. The volume concentration of the nanoparticles varied from 0.3 to 2%. It was demonstrated that with increasing concentration and temperature, the electrical conductivity increased monotonically. It has been indicated that the change in electrical conductivity with increasing temperature was due to a corresponding increase in the electrical conductivity of the base fluid. The data obtained were compared with those from the paper [23], which studied water-based nanofluids containing Al_2O_3 nanoparticles of almost similar size (15 nm). It was noted that the electrical conductivity of nanofluids with copper oxide nanoparticles was higher than that of those with aluminum oxide.

The study of the electrical conductivity of water-based nanofluids containing magnetite (Fe_3O_4) and maghemite (γ-Fe_2O_3) nanoparticles was conducted by Bagheli et al. [2] and Nurdin [38], respectively. In the former study, tetramethyl ammonium hydroxide was used as a surfactant to prepare the nanofluids. The average size of the maghemite nanoparticles was 9.5 nm, while the size of magnetite nanoparticles was not specified. In both cases, the electrical conductivity of the nanofluids significantly exceeded the corresponding values of the base fluid and increased with increasing particle concentration and temperature. The greatest increase in electrical conductivity was observed for maghemite nanofluids, which was 160.49% at a particle volume concentration of 2.5% and a temperature of 60 °C. At the same time, the excess electrical conductivity of magnetite nanofluid was 360% at the same temperature and particle concentration of 0.5%.

The electrical conductivity of water-based nanofluids with Ag and SiC nanoparticles has also been studied in Heyhat et al. [21]. In the preparation of nanofluids containing silver nanoparticles with a size of 20 nm, a PVP surfactant was used. Silicon carbide nanoparticles with a diameter of 55 nm were used in the preparation of the nanofluid with CTAB surfactant. It has been found that the electrical conductivity of nanofluids containing Ag particles increased with increasing both their concentration and temperature. However, while the dependence on the temperature at small particle concentration was nearly linear, at higher concentrations of particles, it became non-linear. At the highest concentration of silver nanoparticles, the electrical conductivity of the nanofluid increased by more than 100 times compared to pure water. A similar trend was observed with another nanofluid, although the increase in conductivity with increasing nanoparticle concentration was almost always linear.

Along with water, ethylene glycol is the most common base fluid. The study of electrical conductivity in ethylene glycol-based nanofluids is both interesting and important because ethylene glycol has different properties compeering with water. Firstly, its electrical conductivity is several times lower, and secondly, its viscosity is about twenty times higher. Therefore, comparing the electrical conductivity of nanofluids based on water and ethylene glycol is extremely useful for understanding the mechanisms of electrical conduction in nanofluid systems. There are approximately two dozen papers that describe experiments on the electrical conductivity of ethylene glycol-based nanofluids containing various nanoparticles. A series of papers have been published by [64–66]. The first paper studied the electrical conductivity of a nanofluid containing aluminum nitride particles with a diameter of 20 nm. The second paper focused on nanofluids with silicon oxide particles ranging from 7 to 14 nm, while the third paper examined silicon nitride particles (20 nm). The weight concentration of the particles differed in each case, ranging from 5 to 20% in the first study, from 1 to 5% in the second, and from 1 to 10% in the third. In the paper [16], the electrical conductivity of the ethylene glycol-based nanofluid with indium oxide nanoparticles was studied. The volume concentration of nanoparticles varied in the range of 0.16–0.81%. In all cases, electrical conductivity increased almost linearly with increasing nanoparticle concentration. All measurements were conducted at a constant temperature of 298.15 K. In nanofluid with AlN particles, it increased by 573.9 times, with SiO_2 particles—by 155%, and with Si_3N_4 particles—by 2394.98 times. In a nanofluid containing In_2O_3 particles, the maximum increase in electrical conductivity was observed at the highest concentration and temperature of 333.15 K. The excess was more than 27,700 times. In all cases, the Maxwell model failed to describe the experimental data, significantly underestimating the results, particularly, by several orders of magnitude in the first and third cases.

In a recent study [15] the electrical conductivity of an ethylene glycol-based nanofluid with titanium dioxide particles from three different manufacturers was investigated. The nanoparticles had sizes of 4–8, 2, and 21 nm, respectively. The weight concentration of the nanoparticles ranged from 1 to 20% and the temperature was between 10 and 45 °C. In all cases, a significant excess in the electrical conductivity of nanofluids compared to that of the base fluid was observed. However, the degree of excess electrical conductivity varies depending on the size and phase composition of TiO_2 nanoparticles (anatase or rutile). The most significant excess in electrical conductivity was observed in nanofluids containing particles in the anatase phase with a declared average particle size of 4–8 nm. At a maximum concentration and temperature of 283.15 K, this excess was 4586%.

One can also mention the study [59], which investigated the electrical conductivity of ethylene glycol-based nanofluids containing ZrO_2 particles. The volume concentration of particles varied in the range of 0.2–10.2%, and the average particle size of the powder used for the preparation of nanofluid was 5–25 nm. All measurements were performed at a temperature of 298.15 K. It has been shown that the electrical conductivity of the nanofluid increased almost linearly with increasing particle concentration reaching a maximum approximately 4.5 times higher than the initial value.

As mentioned above, in addition to pure water and ethylene glycol, the base fluids were prepared also by mixing water and ethylene glycol [9, 52, 62], as well as using polyethylene-glycol PEG-400 [7, 8, 60], bio glycol [1], and transformer oil [12]. It is important to note that the main patterns observed above were also seen using these base fluids. Almost all studies report a significant increase in the electrical conductivity of nanofluids compared to the corresponding value for the base fluid. Generally, the electrical conductivity of a nanofluid increases linearly with increasing particle concentration and temperature, although some studies report a certain deviations from this trend. It is important to note that the classical Maxwell's theory (5.4) predicts values that are significantly lower than those observed experimentally. In most studies, this difference has been associated with the emergence of a new mechanism of electrical conductivity in nanofluids due to the formation of DEL around nanoparticles. Nevertheless, most studies have not taken into account the effect of nanoparticle size on electrical conductivity and no conclusions have been drawn about the effects caused by the base fluid. These factors will be addressed in the next section on the example of several nanofluids. Additionally, the basic mechanisms of electrical conductivity in nanofluids containing conventional nanoparticles will be discussed.

5.1.2 Electrical Conductivity of Nanofluids with Metal Particles

This section presents the findings of studies conducted by Rudyak et al. [47, 48]. The nanofluids, whose electrical conductivity will be discussed in detail below, were prepared using a two-step method. Ethylene glycol and distilled water with aluminum and copper particles were employed in the preparation of the nanofluids. The aluminum nanopowder was subjected to a treatment with palmitic acid, with a weight content varying in the range of 8–10%. The bulk density of aluminum powder was 0.32 g/cm^3. The particles of copper nanopowders were of two sizes (designated as CuI and CuII). In both cases, the powders comprised approximately 98% metallic copper. The bulk density of these powders was approximately 5 g/cm^3. The specific surface area of all powders was determined by the Brunauer–Emmett–Teller method [6], with values of 24.5 m^2/g for aluminum particles, 12 m^2/g for CuI, and 6.8 m^2/g for CuII. Histograms of particle size distribution (PSD) are shown in Fig. 1.16. The average size of aluminum nanoparticles was 75.2 nm. CuI had an average size of 51.5 nm (Fig. 1.16b), while Cu II had an average size of 79.4 nm (Fig. 1.16c).

The maximum weight concentration of nanoparticles was 20%. The corresponding volume concentration of particles was determined by the formula (1.5). The density of ethylene glycol was $\rho_f = 1.11$ g/cm^3, water—$\rho_f = 0.997$ g/cm^3, copper particles—$\rho_p = 8.93$ g/cm^3, and aluminum particles—$\rho_p = 2.70$ g/cm^3. The density of the nanofluid was determined employing the following formula

$$\rho_{nf} = \varphi \rho_p + (1 - \varphi)\rho_f. \quad (5.7)$$

Formula (5.7) indicates that the density of the nanofluid increases linearly with an increase in the volume concentration of particles. At the same time, the multiplier at φ, which determines the increase in the density of the nanofluid compared to the base fluid, is $\left(\rho_p/\rho_f - 1\right)$. It is large enough for copper nanoparticles (7.93), but significantly smaller for aluminum nanoparticles (1.7). The weight concentration of copper particles in ethylene glycol-based nanofluids ranged from 2.5 to 20%, corresponding to volume concentration variations from 0.32 to 3%. Nanofluids comprising aluminum particles were prepared using both water and ethylene glycol. In the first case, the weight concentration of particles ranged from 0.63 to 5%, which corresponded to volume concentrations from 0.23 to 1.91%. In the second case, the weight concentrations of aluminum particles varied from 2.5 to 20%, which corresponded to volume concentrations from 1.05 to 9.34%.

The quality of the resulting nanofluid was controlled by optical density (permeability) and stability. This was achieved by utilizing a UV-3600 spectrophotometer. In addition, the size distribution of nanoparticles in the nanofluid, the zeta potential, attenuation spectra, and the speed of sound were measured. This was accomplished by employing an acoustic and electroacoustic spectrometer (DT1202, Dispersion Technologies). This instrument permits the measurement of a wide range of particle volume concentrations (0.1–60%) of the studied nanofluids. The size distributions of nanoparticles obtained in this way are presented in Fig. 5.1, which displays data for an ethylene glycol-based nanofluid with copper particles (Fig. 5.1a) and aluminum particles (Fig. 5.1b). In all cases, the data are presented at a maximum weight concentration of 20%. A comparison of the data with the corresponding histograms in Fig. 1.16 reveals that the particle sizes in the nanofluid following ultrasonic treatment have undergone a significant change in comparison to the initial particle sizes in nanopowders. The average size of CuI copper particles is approximately 81 nm, CuII is 107 nm, and aluminum is 95 nm.

All stability studies conducted, including zeta potential measurements, have demonstrated that the nanofluids produced were highly stable. In addition to electrical conductivity, the thermal conductivity of these nanofluids and their rheology were also evaluated. The specific electrical conductivity was measured using an ANION 7025 conductometer, with the measurement interval spanning from 10^{-8} to 10 S/m. The permissible limit of the relative error when measuring the specific electrical conductivity was $\pm$ 2%. The tests conducted on ethylene glycol and water confirmed this accuracy.

The measurements have demonstrated that the electrical conductivity of all nanofluids exhibited a notable enhancement in comparison to the corresponding values of the base fluids. Figure 5.2 presents the relative electrical conductivity coefficient $\sigma_r = \sigma/\sigma_{bf}$ of an ethylene glycol-based nanofluid (Fig. 5.2a) and water (Fig. 5.2b) with aluminum nanoparticles, illustrating the variation in electrical

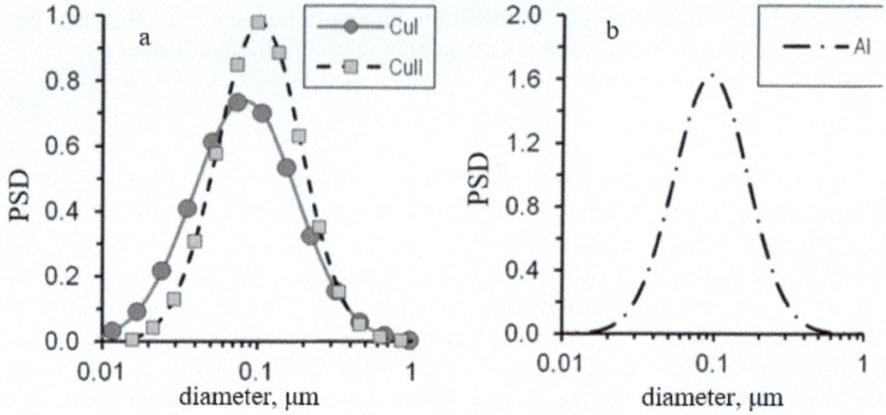

Fig. 5.1 Particle size distribution (PSD) of copper (**a**) and aluminum (**b**) particles in an ethylene glycol-based nanofluid, $w = 20\%$

conductivity as a function of weight concentration. Here, σ and σ_{bf} represent the electrical conductivity coefficients of the nanofluid and the base fluid, respectively. In both cases, the electrical conductivity of nanofluids increases with an increase in the weight concentration of nanoparticles, and this increase is almost linear; however, it significantly depends on the base fluid. The excess electrical conductivity of ethylene glycol-based nanofluids was approximately 23% (Fig. 5.2a), with the electrical conductivity of ethylene glycol being 0.91 μS/cm. Conversely, the electrical conductivity of water-based nanofluids increased by more than fifteen times (Fig. 5.2b), with the electrical conductivity of water being 3.1 μS/cm.

The dependence of the relative electrical conductivity coefficient of ethylene glycol-based nanofluids with CuI and CuII particles on their weight concentration is shown in Fig. 5.3. It can be observed that the dependence of electrical conductivity on particle concentration also increases almost linearly. It is important to note that the electrical conductivity of nanofluids, as well as other transport coefficients (see

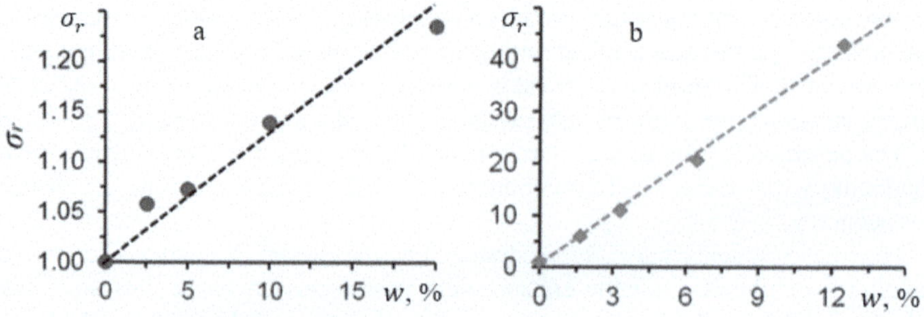

Fig. 5.2 The relative thermal conductivity coefficient of a nanofluid based on ethylene glycol (**a**) and water (**b**) with aluminum nanoparticles depending on their weight concentration

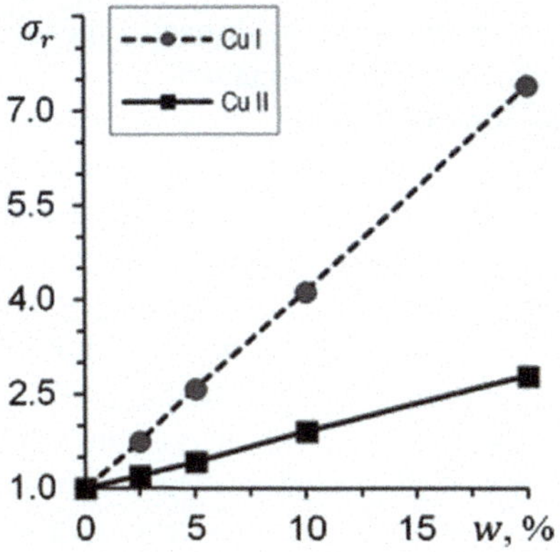

Fig. 5.3 Relative electrical conductivity coefficients of ethylene glycol-based nanofluids with copper particles depending on their weight concentration

Chaps. 3 and 4), depends on the size of the nanoparticles. In contrast to thermal conductivity, the electrical conductivity of nanoparticles is inversely proportional to their size. The excess electrical conductivity of the nanofluid containing CuII particles at a concentration of 20% is approximately two and a half times that of the base fluid, while the electrical conductivity of the nanofluid containing CuI particles is almost seven and a half times that of the base fluid.

The second significant factor is that the excess electrical conductivity of the nanofluid is dependent on the electrical conductivity of the nanoparticles (see Figs. 5.2a and 5.3). Thus, the electrical conductivity of copper is 1.57 times greater than that of pure aluminium. Furthermore, the excess electrical conductivity of nanofluids comprising CuII and Al particles (the sizes of these particles are similar) at a weight concentration of 20% differs by more than two times. It is important to note that the L-ALEXTM aluminum nanoparticles used in the experiments were treated with palmitic acid. Therefore, naturally, the electrical conductivity of these nanoparticles was lower than that of pure aluminum particles.

5.1.3 *Electrical Conductivity Mechanisms of Nanofluids with Spherical Particles*

Nanofluid is a complex dispersed fluid. Its electrical conductivity is dependent on the mobility of all particles present in the suspension. In liquid dielectrics, in the presence of weak electric fields, electrophoretic conductivity mainly occurs. In this case, for example, hydroxyl ($H_2O \Leftrightarrow OH^- + H^+$) and hydroxonium ($H_2O \Leftrightarrow HO^- + H_3O^+$) ions are formed in water. Furthermore, water contains a variety of ions, including sodium, calcium, chloride, and numerous other molecules of various gases. It is important to note that in addition to electrophoretic ionic conductivity in water, the presence of hydroxyl and hydroxonium also gives rise to weak proton conductivity due to the exchange of protons between them. In ethylene glycol, ions are formed according to the following scheme: $C_2H_4(OH)_2 \Leftrightarrow C_2H_4(O_2)^- + 2H^+$. It is not difficult to determine which ions appear in other base liquids. In the nanofluid with aluminum particles discussed in the previous section, palmitic acid was present, in which H^- and H^+ ions can form in the presence of an electric field. Using surfactants in the preparation of nanofluids results in the formation of additional ions associated with their ionization in an electric field. These ions contribute to the electrical conductivity of the base fluids. The electrical conductivity component due to these ions is proportional to their mobility (5.3), which is inversely proportional to the systematic drag force acting on them from the liquid.

Now we need to analyze the influence of nanoparticles on the electrical conductivity of the nanofluid. Upon contact of two condensed phases (base fluid and solid nanoparticles), the properties of the interfacial surface layer of each phase change. The thickness of this layer can be several nanometers and, in general, depends on the size of the nanoparticles. The properties of the interfacial layer formed around nanoparticles differ significantly in structure, physical, and chemical properties from the bulk properties of these phases. Typically, the interfacial layers are electroneutral. However, due to the uneven distribution of charged particles, DELs consisting of charges of the opposite sign are formed in them. The formation of DEL is associated with the tendency of heterogeneous systems to reduce the Gibbs energy of the surface layer. This leads to the orientation of polar molecules, ions, and electrons in the surface layer. As a result, in the interfacial layer, the contacting phases acquire charges of the opposite sign but of equal magnitude. The DEL forming near the nanoparticle is schematically shown in Fig. 5.4a. Roughly speaking, two rows of ions of the opposite sign are formed at the interface of the contacting phases. Since ions in the liquid phase are in thermal motion, in reality this scheme is dynamic and blurred. The boundary of the layer, at which the molecules of the base fluid practically lose their mobility plays an important role. The plane separating the mobile external environment from the stationary layer is called the sliding plane. In this plane the potential has a value of ζ and is called the zeta potential (ζ-potential).

In the absence of an external electric field, the ion atmosphere is symmetrical and has a spherical shape (Fig. 5.4a). A multidirectional movement of ions and

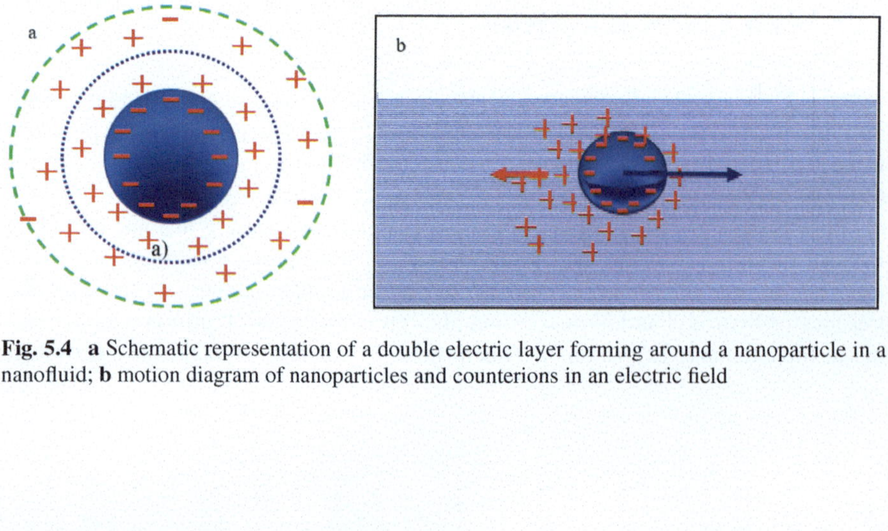

Fig. 5.4 **a** Schematic representation of a double electric layer forming around a nanoparticle in a nanofluid; **b** motion diagram of nanoparticles and counterions in an electric field

nanoparticles occurs in an external electric field, as schematically shown in Fig. 5.4b. This movement results in a violation of the spherical symmetry of the distribution of counterions and their charges. Consequently, the electrical conductivity of nanofluids also depends on the mobility of nanoparticles l_{np}. This mobility is also inversely proportional to the drag force acting on the nanoparticle in the liquid. For not too small particles, which exactly are considered in this paper, this force can also be described by the Stokes formula [45]

$$l_{np} \sim 1/(6\pi R\mu). \tag{5.8}$$

Estimates (5.3) and (5.8) explain the observed experimental data. Electrical conductivity increases inversely with the size of the nanoparticles, while it is inversely proportional to the viscosity of the base fluid (recall that the viscosity of ethylene glycol is almost twenty times higher than that of water). Here it should be borne in mind, however, that at sufficiently high concentrations of nanoparticles, they no longer move in a stationary liquid, but in a flow of counterions moving towards it. This phenomenon will slow down the movement of nanoparticles, thereby somewhat reducing the electrical conductivity of the nanofluid. Additional deceleration of nanoparticles also occurs due to the violation of symmetry in the distribution of counterions, since it generates the interaction of nanoparticles with counterions formed around them. This deceleration force also increases with increasing nanoparticle concentration.

The aforementioned estimates are applicable exclusively to Newtonian fluids. Nevertheless, the data presented in Chap. 3 indicate that the nanofluids under consideration in certain cases exhibit non-Newtonian rheological properties. In non-Newtonian liquids, the situation is considerably more intricate, primarily due to the nonlinear dependence of the drag force on the radius of a single particle (see, for example, Tabuteau et al. [58]). Additionally, the drag force depends on the shear rate [4]. Consequently, the effective viscosity of the medium varies with the shear rate and thus the mobility of particles and ions also varies. Since the nanofluids studied in this book are pseudoplastic or viscoplastic the electrical conductivity increases with increasing shear rate. Consequently, in real flows, where the shear rate varies, the mobility of particles and ions is different, and therefore the electrical conductivity is inhomogeneous along (or more precisely across) the flow. The electrical conductivity increases with an increase in the flow non-Newtonicity. Indeed, for example, for a power-law fluid [see Eq. (3.29)] at a given shear rate, the mobility of charge carriers is $l_i \sim 1/\mu = 1/\left(k_v\dot{\gamma}^{n-1}\right)$. Since for the nanofluids under consideration $n < 1$, the mobility of the charge carriers and, consequently, the electrical conductivity increase with an increase in the shear rate $\sigma \sim \dot{\gamma}^{1-n}/k_v$. In general, however, starting from certain values of shear rate, the viscosity ceases to change and hence the conductivity reaches a certain limit value.

The above considerations provide a qualitative explanation of the observed electrical conductivity. The first model of the electrical conductivity of a dispersed liquid

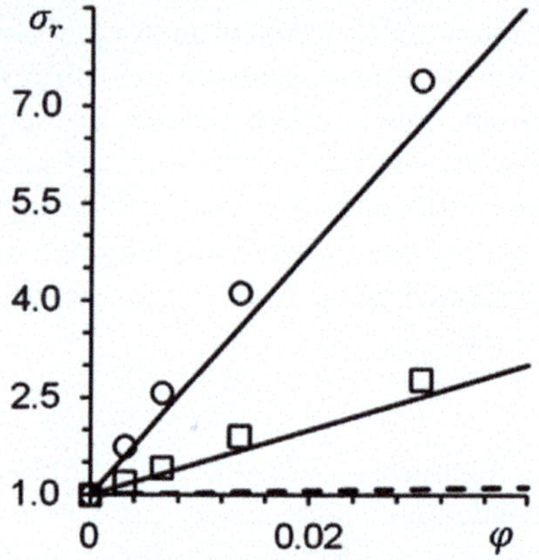

Fig. 5.5 Comparison of electrical conductivity measurement data of ethylene glycol-based nanofluids with CuI (circles) and CuII (squares) particles with calculation by formula (5.9) (solid lines)

was proposed by Maxwell [32]. The presence of nanoparticles introduces an additional mechanism associated with their movement. This movement is in fact electrophoresis. Consequently, the electrical conductivity of nanofluids can be described by the equation proposed by Dong et al. [12]

$$\sigma = \frac{1 + 2\omega + 2(1 - \omega)\varphi}{1 + 2\omega - (1 - \omega)\varphi}\sigma_{bf} + \frac{8\varepsilon_r^2\varepsilon_0^2\zeta^2}{\mu d^2}. \tag{5.9}$$

where ε_r, ε_0 are the relative dielectric constants of particles and vacuum, respectively.

A comparison of calculations by formula (5.9) with experimental data for ethylene glycol-based nanofluids with CuI (circles) and CuII (squares) particles is presented in Fig. 5.5. The data corresponding to formula (5.9) for these nanofluids are depicted by corresponding lines. The dotted line represents the calculation according to Maxwell's theory (the first term of Eq. (5.9)), which does not align with the experimental data even at low concentrations of particles. In contrast, formula (5.9) describes the experimental data within measurement accuracy.

It is of interest to compare the electrical conductivity of nanofluids based on water and ethylene glycol containing metal particles and metal oxide particles. This comparison is shown in Fig. 5.6. In the presented plot, the experimental data obtained in the previous section are compared with data for Al_2O_2 (20–30 and 80 nm) and Cu (80 nm) nanoparticles [52], as well as data for Al_2O_2 nanoparticles (15 nm) [23]. The electrical conductivity of ethylene glycol-based nanofluids with copper particles [52] is consistent with the data presented in the previous section (see Fig. 5.6b). It is noteworthy that in almost all cases, the electrical conductivity of nanofluids with oxide particles is significantly higher than that of nanofluids with metal particles. This is primarily due to the fact that the nanoparticles in our experiments exhibited a significantly larger size. However, this is not the sole factor.

Let us now formulate the main factors that determine the electrical conductivity of nanofluids with ordinary spherical particles. The primary mechanisms that determine the electrical conductivity of such nanofluids are the transport of ions of the base

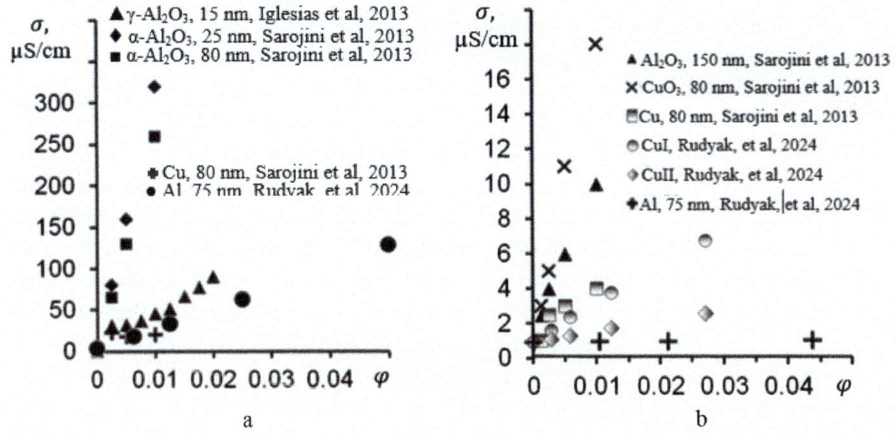

Fig. 5.6 The electrical conductivity of nanofluids based on water (**a**) and ethylene glycol (**b**), containing metal and metal oxide nanoparticles depending on concentration

fluid and nanoparticles due to the creation of the DEL. These mechanisms diverge significantly from those previously considered by J.C. Maxwell. The derivation of formulas (5.4) is based on the hydrodynamic account of disturbances of the temperature field in the base fluid, which is generated by a particle with other physical characteristics (density and thermal conductivity). In this case, the transfer of this disturbance is determined by the thermal conductivity of the base fluid. In nanofluids there is also a contribution from this Maxwell's mechanism (see formula (5.9)), but new factors appear. It is now possible to sum factors which differently determine and affect the electrical conductivity of nanofluids.

It is first necessary to note that the electrical conductivity of nanofluids is dependent on the size of the nanoparticles, with an inverse relationship observed between conductivity and nanoparticle size. This indicates that the mechanisms of electrical conductivity and thermal conductivity of nanofluids are distinct. Furthermore, the electrical conductivity of nanofluids is dependent on the concentration of nanoparticles. At relatively low concentrations, this dependence is linear. However, the degree of accrue in electrical conductivity decreases with a further increase in the concentration of particles. This decrease is due to several factors. On the one hand, an increase in concentration complicates the free formation of DEL, and on the other hand, increases the emerging Coulomb and relaxation deceleration. In accordance with the DLVO theory, at sufficiently high concentrations or for small particles, Coulomb repulsion of nanoparticles also appears. It is important to note that the corresponding critical concentration is not universal and is determined by many circumstances (including the particle material, particle size, and base fluid properties). At the beginning of this chapter, the Wiedemann–Franz–Lorenz law (5.1) for metallic materials was referenced. Its emergence is attributed to the fact that the electrical conductivity and thermal conductivity mechanisms in metals were similar.

However, this is not the case in nanofluids. It is evident that there is no correlation like (5.1) between these two types of transport processes.

The next crucial factor is the base fluid. First of all, its electrical conductivity, if we are talking about dielectrics, is determined by the dielectric constant. As this value increases, the number of ions also rises. The dielectric constant of water is approximately twice that of ethylene glycol. Another significant aspect is the viscosity of the base fluid. As this value grows, the mobility of both ions and nanoparticles declines. It is evident that the nanoparticle material plays also an important role. On the one hand, with an increase in the electrical conductivity of nanoparticles, their contribution to the electrical conductivity of the nanofluid will change (see the formulas (5.4), (5.5) and (5.9)). However, the electrical conductivity of nanoparticles is not the sole determining factor. In this regard, it is pertinent to concur with the authors [52], who correctly posit that the formation of DEL is contingent upon the polarization of nanoparticles, which is influenced by the dielectric constant and particle density. It is evident that the dielectric constant of oxide particles is considerably higher than that of copper particles, whereas the density, conversely, is lower.

Finally, it is necessary to discuss the effect of temperature on the electrical conductivity of nanofluids. Temperature is an important factor influencing the electrical conductivity of nanofluids. This is determined by two circumstances. With increasing temperature, the thermal velocities of the particles increase, while the viscosity, on the contrary, decreases. Thus, with increasing temperature, the electrical conductivity of nanofluids always increases. However, this increase is not universal; it depends on both the material of nanoparticles and the base fluid.

5.1.4 A Brief Review of Electrical Conductivity Data of Nanofluids with CNTs

The electrical conductivity of nanofluids with CNTs has been the subject of considerable research only in the last five to six years, with the first studies appearing approximately ten years ago. It is noteworthy that the majority of available publications on the electrical conductivity of nanofluids with CNTs are devoted to nanofluids with MWCNTs. One of the earliest publications on the electrical conductivity of nanofluids with MWCNTs is a paper by Kumar et al. [29]. This paper presents the results of electrical conductivity measurements of solar glycol-based nanofluids. Gum Arabic (GA) was employed as a surfactant. The inner and outer diameters of the MWCNTs were 5–10 and 30–50 nm, respectively, with a length of 10–20 μm. The weight concentration of MWCNTs ranged from 0.2 to 0.6%, while the weight concentration of GA ranged from 0.25 to 1.25%. It was observed that in all cases, the electrical conductivity exhibited a linear increase with increasing concentrations of MWCNTs and GA, as well as temperature. At the maximum concentration of MWCNTs and GA, the electrical conductivity of the nanofluid increased by 190.57% at a temperature of 50 °C. Later, Kumar and other co-authors [30] investigated the electrical conductivity of a nanofluid based on a mixture of solar glycol and water (50:50) with the same surfactant. It was demonstrated that the electrical conductivity increased by approximately 62% at a MWCNT volume concentration of 0.6% and a temperature of 50 °C.

The impact of surfactants on the electrical conductivity of nanofluids was investigated in Shanbedi et al. [53]. The base fluid consisted of water with the addition of GA, cetyltrimethylammonium bromide (CTAB) and SDS. The MWCNTs had a diameter of 10–20 nm and a length of approximately 30 μm. The weight concentration of the MWCNTs was 0.1%, and that of the surfactant ranged from 0.05 to 2%. It was demonstrated that in all cases, the electrical conductivity increased with increasing temperature. For any used surfactant, the electrical conductivity of nanofluids was observed to increase with increasing their concentration. The maximum values were obtained using SDS, at the same time its use allowed for the most efficient isolation of MWCNTs. The results of the study [57] yielded qualitatively similar outcomes, wherein the electrical conductivity of water-based nanofluids with MWCNTs and SDS was also investigated. It was demonstrated that the electrical conductivity of the W + SDS base fluid escalates with increasing of the surfactant concentration. The electrical conductivity of nanofluids also increased with an increase in the concentration of SDS. In all cases, the electrical conductivity of these nanofluids was significantly higher than that of nanofluids with oxide particles. At the same time, it is demonstrated that the electrical conductivity of this nanofluid exhibited a nonlinear growth with an increase in the MWCNT concentration. This non-linear growth can be attributed to the effect of percolation, with the critical weight concentration of percolation for MWCNTs being estimated to be 0.1%.

Finally, it is worth mentioning the article by Kumar et al. [31], which examined the electrical conductivity of a nanofluid based on a water-solar glycol mixture (70:30) containing also GA surfactant (weight concentration was 0.25%). The weight concentration of MWCNTs took three values: 0.15, 0.3 and 0.45%. The length of the CNTs was 10–30 μm, with inner and outer diameters of 5–10 and 30–50 nm, respectively. However, given CNTs were processed in a mill for an hour, therefore it is difficult to draw any definitive conclusions about their size in nanofluids. The maximum enhancement of electrical conductivity was approximately 93% and was observed at the highest concentration of MWCNTs and a temperature of 50 °C.

In the paper [40], an attempt was made to study the mechanism of electrical conductivity in nanofluids. The electrical conductivity of ethyl alcohol-based nanofluids with MWCNTs with an average hydrodynamic size of 344 nm was studied. The measurement results indicated that the electrical conductivity of nanofluids is determined by the number and quality of tunnel contacts between nanotubes rather than by electrical properties of CNTs. These contacts enable conductivity to be realized by a jumping mechanism. This mechanism is associated with the formation of the percolation structure of the MWCNTs in the nanofluid, which occurs at a critical concentration.

In practice, in addition to surfactants, various types of CNT functionalization are often employed when preparing nanofluids. On occasion, surfactants are also utilized in conjunction with functionalization. It is important to note that the physical properties of CNTs (for example, their thermal conductivity or electrical conductivity) undergo a transformation following functionalization. Consequently, such nanofluids are not considered in this book. Nevertheless, it is beneficial to provide some illustrative examples. Thus, Naddaf and Heris [37] investigated the electrical conductivity of nanofluids based on diesel oil with MWCNTs functionalized with hexylamine. In addition, two types of surfactants were employed: non-covalent oleic acid (OA) and covalent hexylamine (HA). The weight concentrations of MWCNTs varied from 0.05 to 0.5%, while the temperature ranged from 5 to 100 °C. The diameter of the MWCNTs varied in the range of 20–30 nm with lengths ranged between 5 and 10 μm. The electrical conductivity of diesel oil was found to be so low that it could not be measured. However, the electrical conductivity of all nanofluids was well measurable, and it increased with increasing MWCNT concentration and temperature. It is noted that the electrical conductivity of the functionalized MWCNTs was found to be lower than that of the original ones. The authors posit that the formation of DEL around CNT is the primary mechanism responsible for the electrical conductivity of nanofluids.

Interesting data is presented in the paper [54]. The electrical conductivity of a water-based nanofluid comprising MWCNTs with a diameter of 10–20 nm and a length of 5–15 μm was investigated. A rather complex technology was employed for the synthesis of MWCNTs, which involved the initial preparation of carboxylated MWCNT-COOH. Subsequently, the CNTs were functionalized with aspartic acids (Asp). In order to impart magnetic properties to the MWCNT surface, iron ions were covalently attached through the carboxylic Asp acid group. It was found that in all cases, the electrical conductivity of corresponding nanofluids increased with increasing temperature and concentration of MWCNTs. The maximum electrical conductivity of 56.8 μS/cm was obtained at a weight concentration of 0.2% and a temperature of 80 °C. The electrical conductivity of three water-based nanofluids was compared. In the first case, the base fluid was prepared using a GA surfactant, in the second and third cases, the base fluid was water, however, MWCNT-Asp and MWCNT-Fe were used. The latter nanofluids exhibited the greatest electrical conductivity.

As previously stated, only a few studies have been conducted to assess the electrical conductivity of nanofluids containing SWCNTs. One of the earliest publications on this topic is the paper by Harish et al. [20], in which nanofluids were prepared using water as the base fluid and sodium deoxycholate as a surfactant. The diameter of the SWCNTs was between 0.8 and 1.6 nm. The maximum weight concentration of SWCNTs was 0.48%. The electrical conductivity of the base fluid was approximately 10 μS/cm. With an increase in the SWCNT concentration, the electrical conductivity of the nanofluid initially increased sharply (almost linearly), and then the growth gradually slowed down. The maximum increase in electrical conductivity reached eight times. The authors attribute the observed slowdown to the formation of the percolation structure of the SWCNTs. They were able to demonstrate the formation of such a structure in nanofluids using atomic force microscopy [19]. At the same time, it is indicated that the critical weight concentration of SWCNTs was

0.025%. It is regrettable that no information is provided on the temperature at which the measurements of electrical conductivity were performed.

Somewhat earlier, the electrical conductivity of nanofluids was measured also in the paper [18]. The nanofluid was prepared using a W/EG mixture (50:50) with up to 0.5% weight concentration of sulfonated SWCNTs. The results have demonstrated a significant increase in electrical conductivity, with a 13-fold enhancement observed. This increase was found to be nearly linear. This prompted the authors to hypothesize that in this case, there was no percolation effect (or no percolation occurred at all). They proposed that the linear increase in conductivity depending on the weight concentration of SWCNTs indicated that the dominant mechanism was ionic conductivity, which was consistent with the chemical structure of the functionalized SWCNTs, where the anionic SO_3^- group is present.

The presented overview of the electrical conductivity of nanofluids indicates that the data obtained are very contradictory. Nevertheless, it is possible to draw certain conclusions about the magnitude of the electrical conductivity of nanofluids and formulate questions that require answers. First of all, it should be noted that in almost all nanofluids with CNTs, a sufficiently high electrical conductivity was recorded, although CNT concentrations usually did not exceed 0.5%. It was revealed that the electrical conductivity of nanofluids increased with an increase in temperature. This increase was found to be associated with an increase in the electrical conductivity of the base fluid when increasing its temperature. Furthermore, it was demonstrated that the electrical conductivity of nanofluids with CNTs increased with an increase in their concentration. The majority of studies have demonstrated that electrical conductivity increases linearly with increasing concentration. However, there are also studies where a nonlinear increase in electrical conductivity was observed. The authors of these publications have attributed this phenomenon to the presence of CNT percolation. It is assumed that percolation impedes the growth of electrical conductivity with an increase in the concentration of CNTs. Furthermore, it can be observed that surfactants influence the electrical conductivity of nanofluids. However, the underlying mechanisms of this phenomenon remain unclear. Additionally, there have been almost no attempts to compare data on the magnitude of the excess electrical conductivity of nanofluids. The latter is understandable, since virtually all of the nanofluids considered were quite different.

Thus, the following questions can be formulated, the answers to which must be obtained in order to make the possible application of nanofluids predictable. (1) What is the electrical conductivity enhancement of nanofluids containing MWCNTs and SWCNTs? (2) How does the electrical conductivity enhancement of nanofluids increase with increasing CNT concentration? (3) How does the use of surfactants affect the magnitude of the electrical conductivity enhancement of nanofluids? (4) What is the reason for the influence of surfactants? (5) What are the main mechanisms of the electrical conductivity of nanofluids with CNTs? The following section is devoted to answering these questions (see Rudyak et al. [49]).

5.1.5 *Electrical Conductivity of Nanofluids with SWCNTs and MWCNTs*

At the conclusion of the preceding section, a series of questions have been posed that demand a prompt response. In order to objectively compare the electrical conductivity of nanofluids with MWCNTs and SWCNTs, it is essential to use the same base fluids. To ascertain the influence of surfactants, it is necessary to compare the electrical conductivity of nanofluids prepared with and without surfactants. These requirements have influenced the selection of the nanofluids under study. Nanofluids with CNTs based on isopropyl alcohol (IPA), ethylene glycol (EG), EG + S, and W + S were considered. The weight concentrations of CNTs varied from 0.05 to 1%, and the surfactant concentration was either equal to that of CNTs or twice as high. Since the temperature dependence of the electrical conductivity of nanofluids has been studied quite well and does not raise any particular questions, the data presented below are performed at a temperature of 25 °C.

The manufacturing technologies of nanofluids with CNTs present significant challenges. In some cases, the use of surfactants is mandatory. This is particularly relevant in the context of water-based nanofluids. The presence of surfactants significantly alters the electrical conductivity of base fluids. Therefore, it is advisable to first study the electrical conductivity of nanofluids prepared on base fluids with no surfactants. In this study, IPA and EG were used as such base fluids.

CNTs exhibit very high electrical conductivity, and the conductivity of SWCNTs and MWCNTs can differ quite significantly. For this reason, it can be expected that the electrical conductivity of nanofluids with SWCNTs will also be higher than that of the base fluid. This hypothesis has been confirmed by experiments. The specific electrical conductivity of the IPA used in this series of measurements was 0.06 μS/cm, while that of the EG was 0.91 μS/cm. Table 5.1 presents the specific electrical conductivity of nanofluids with SWCNTs based on EG and IPA. It can be observed that the electrical conductivity of both nanofluids increases significantly with an increase in the weight concentration of SWCNTs. This increase is nearly linear (see Fig. 5.7a).

It is evident that the most crucial factor influencing the electrical conductivity of nanofluids with SWCNTs is the nature of the base fluid. The electrical conductivity of

Table 5.1 Electrical conductivity of EG- and IPA-based nanofluids with SWCNTs

NF, w (%)	σ (μS/cm)	σ_r
EG/SWCNT, 0.05	3000	3297
EG/SWCNT, 0.1	7500	8242
EG/SWCNT, 0.25	18,900	20,769
IPA/SWCNT, 0.05	550	9167
IPA/SWCNT, 0.1	1060	17,667
IPA/SWCNT, 0.25	2590	43,167

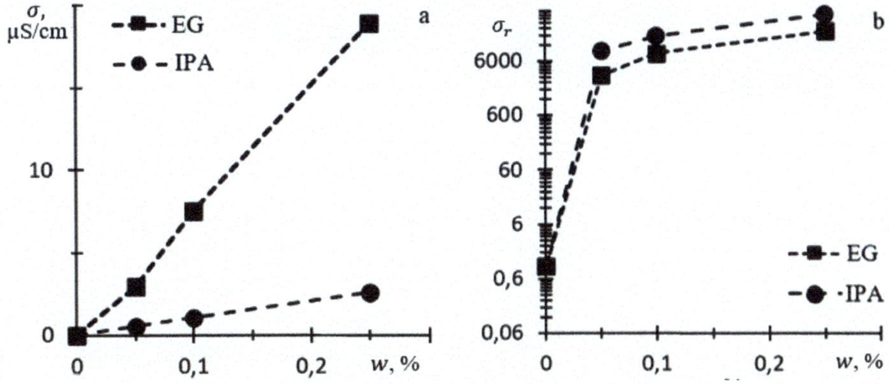

Fig. 5.7 Specific electrical conductivity of a nanofluid based on IPA and EG (**a**) and the relative electrical conductivity of these nanofluids (**b**) depending on the weight concentration of SWCNTs

these nanofluids varies by a factor of more than 15. Therefore, it is important to understand how the properties of the base fluid affect the enhancement of the electrical conductivity of these nanofluids. Figure 5.7b illustrates the dependence of the relative electrical conductivity of nanofluids with SWCNTs based on EG and IPA. The corresponding numerical values are also provided in Table 5.1. Although the absolute values of electrical conductivity of IPA-based nanofluids are several times lower than those of EG-based nanofluids, the corresponding relative values are significantly higher for IPA-based nanofluids than for EG-based nanofluids. The electrical conductivity enhancement of these nanofluids, even at such low concentrations, is considerable, reaching tens of thousands of times at maximum concentrations. In all cases, the relative coefficient of specific electrical conductivity of IPA-based nanofluid is significantly higher than that of EG-based, although this difference decreases slightly with increasing concentration of SWCNTs. In conclusion, it should be noted that the classical Maxwell's theory [see Eq. (5.4)] for the considered nanofluids predicts electrical conductivity values that are orders of magnitude lower than the experimental values given in Table 5.1.

Since the use of dispersants is typical in the preparation of nanofluids, it is imperative to investigate the impact of these additives on the electrical conductivity of the initial base fluids before studying the electrical conductivity of the corresponding nanofluids with CNTs. Table 5.2 presents the data on the electrical conductivity of EG and water, as a function of the weight concentration of three types of surfactants, namely, PVP, SDBS, and SDS. Here, Bf stands for the base fluid. Note that in this series of measurements, EG and water had slightly different values of electrical conductivity equal to 1.1 µS/cm for EG, and 3 µS/cm for water. The weight concentrations of surfactants were 0.1, 0.2 and 0.4%.

The conducted measurements allow drawing two conclusions. The first is that surfactants in all cases increase the electrical conductivity of initial base fluids. At the same time, the degree of electrical conductivity enhancement differs and increases

Table 5.2 Electrical conductivity of base fluids

Bf, w (%)	σ (μS/cm)	Bf, w (%)	σ (μS/cm)
EG + PVP 0.1	1.6	W + PVP 0.1	3.8
EG + PVP 0.2	1.9	W + PVP 0.2	5.6
EG + PVP 0.4	2.2	W + PVP 0.4	7.6
EG + SDBS 0.1	17	W + SDS 0.1	195
EG + SDBS 0.2	35	W + SDS 0.2	380
EG + SDBS 0.4	67	W + SDS 0.4	577
		W + SDBS 0.1	340
		W + SDBS 0.2	616
		W + SDBS 0.4	1157

with increasing concentration of the surfactant. At the maximum weight concentration of PVP the electrical conductivity of the EG + PVP base fluid increases more than twice, while that of the EG + SDBS – almost seventy times. On the other hand, the electrical conductivity of the W + PVP base fluid increases by more than twofold, while that of W + SDS rises by approximately two 100-fold, and that of W + SDBS by almost four 100-fold. Such a significant variation is due to the different ionization of the surfactant molecules and the fact that PVP is steric, while the latter two are ionic.

Let us now consider the effect of surfactants on the electrical conductivity of nanofluids containing SWCNTs. Table 5.3 presents data for nanofluids with SWCNTs based on EG + PVP and EG + SDBS. The concentration of the surfactant in all cases was twice the weight concentration of SWCNTs. The analysis of these data and their comparison with the data presented in Table 5.1 shows that the use of surfactants in general significantly reduces the electrical conductivity of nanofluids compared with nanofluids prepared without them. Concomitantly, the incorporation of a surfactant into the EG results in an augmented electrical conductivity of the base fluid in direct proportion to its weight concentration. Furthermore, SDBS exhibits a markedly enhanced electrical conductivity compared to PVP (see Table 5.2). Such different influence of surfactants is also associated with the different nature of their interaction with SWCNTs. PVP largely encapsulates the SWCNTs, thereby reducing the electrical conductivity of the corresponding CNTs. Finally, it can be observed that the electrical conductivity of nanofluids with surfactants increases non-linearly with an increase in the concentration of SWCNTs (see Fig. 5.8).

The study of electrical conductivity of W + S-based nanofluids with SWCNTs yielded qualitatively similar results. The corresponding data are presented in Table 5.4. As observed, the W + SDBS-based fluids exhibited an electrical conductivity that was approximately two orders of magnitude higher than that of W + PVP-based nanofluid. It is to be expected that the electrical conductivity of the corresponding nanofluids would vary significantly. The electrical conductivity of W + PVP-based nanofluids exhibits a significant increase of approximately 70-fold at the maximum concentration of SWCNTs, whereas that of W + SDBS-based nanofluids

Table 5.3 Electrical conductivity of nanofluids with SWCNTs based on EG + PVP and EG + SDBS

NF, w (%)	σ (μS/cm)	σ_r
EG + PVP/SWCNT, 0.05	3.56	3.2
EG + PVP/SWCNT, 0.1	4.64	4.2
EG + PVP/SWCNT, 0.2	5.5	5
EG + SDBS/SWCNT, 0.05	170	154.5
EG + SDBS/SWCNT, 0.1	920	836.4
EG + SDBS/SWCNT, 0.2	6018	5470.9

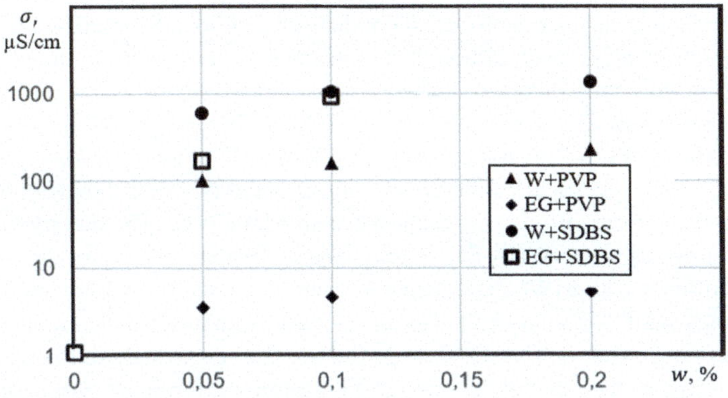

Fig. 5.8 Comparison of the relative specific electrical conductivity of EG- and W-based nanofluids with surfactants depending on the weight concentration of SWCNTs

shows a comparatively modest increase of only by about a factor of two. It is important to note that in these experiments, the weight concentration of the surfactant was approximately twice that of the corresponding values of the SWCNTs.

Nanofluids with SWCNTs and MWCNTs are significantly different dispersed fluids. They have different rheology and the thermal conductivity of a nanofluid with MWCNTs is significantly lower than that of the nanofluid with SWCNTs (see Chaps. 3 and 4). In this section, along with measurement data on the electrical conductivity of a nanofluid with SWCNTs, numerous measurement data on the electrical

Table 5.4 Electrical conductivity of nanofluids based on W + PVP and W + SDBS

NF, w (%)	σ (μS/cm)
W + PVP/SWCNT 0.05	100
W + PVP/SWCNT 0.1	160
W + PVP/SWCNT 0.25	230
W + SDBS/SWCNT 0.05	600
W + SDBS/SWCNT 0.1	1055
W + SDBS/SWCNT 0.2	1367

conductivity of a nanofluid with MWCNTs are discussed. The main data obtained are reduced in Table 5.5. The data presented in this section are qualitatively consistent with those described above for nanofluids with SWCNTs. The maximum excess electrical conductivity is achieved in an EG-based nanofluid. The electrical conductivity increases by more than thirty times at a weight concentration of 0.2%, and by almost 150 times at a concentration of 1%. Similarly as in the experiments observed previously, the electrical conductivity of an EG + SDBS-based nanofluid is significantly higher than that of an EG + PVP-based nanofluid. Thus, at the maximum concentration of the surfactant, the electrical conductivity of the former increases almost 350 times, whereas in a nanofluid with PVP it increases only about 3.5 times. On the other hand, the excess electrical conductivity of W + S-based nanofluid turns out to be significantly lower, although it is worth remembering that the presence of surfactants significantly increases the electrical conductivity of water (see Table 5.1). Finally, as observed in a nanofluid with SWCNTs, the electrical conductivity of a nanofluid with MWCNTs is not described by Maxwell's theory (5.4).

In all cases, the electrical conductivity enhancement of nanofluids with SWCNTs is found to be significantly higher than that with MWCNTs. The electrical conductivity of nanofluids with SWCNTs is found to be 2–3 orders of magnitude higher than that of nanofluids with MWCNTs prepared using the same base fluids (compare the data of Tables 5.2, 5.3, 5.4 and 5.5). This discrepancy can be attributed to the differing electrical conductivities of the CNTs themselves. The electrical conductivity of the SWCNTs is higher than that of the MWCNTs. Therefore, the electrization of the base fluids in their vicinity occurs differently. Furthermore, the densities and size of these CNTs differ, which also determines their different mobility. At the same time, it is crucial to comprehend the relative excess of electrical conductivity of the nanofluid in comparison to the base fluid. For two types of nanofluids based on EG and B + PVP this is illustrated in Fig. 5.9. The maximum excess is observed in nanofluids with minimal electrical conductivity of the base fluids. However, the differences between the electrical conductivity of nanofluids with SWCNTs and MWCNTs vary by several orders of magnitude.

Table 5.5 Electrical conductivity of nanofluids with MWCNTs

NF, w (%)	σ (μS/cm)	NF, w (%)	σ (μS/cm)
EG/MWCNT 0.1	28.4	W + PVP/MWCNT 0.05	4.8
EG/MWCNT 0.2	34.7	W + PVP/MWCNT 0.1	7.7
EG/MWCNT 1	157	W + PVP/MWCNT 0.2	10.1
EG + PVP/MWCNT 0.05	2.0	W + PVP/MWCNT 0.05	336
EG + PVP/MWCNT 0.1	2.7	W + PVP/MWCNT 0.1	659
EG + PVP/MWCNT 0.2	3.8	W + PVP/MWCNT 0.2	1216
EG + SDBS/MWCNT 0.05	130		
EG + SDBS/MWCNT 0.1	215		
EG + SDBS/MWCNT 0.2	366		

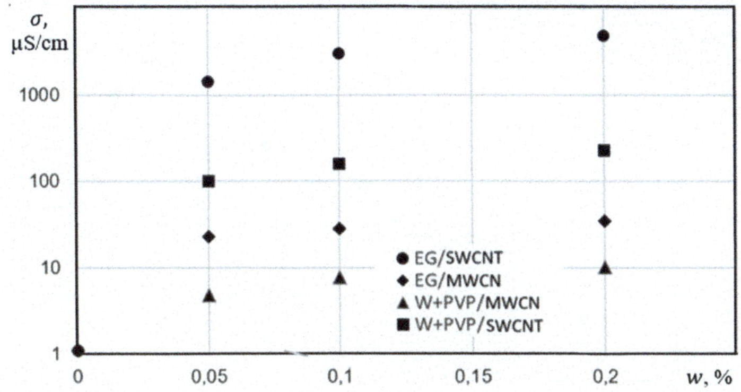

Fig. 5.9 Comparison of the relative electrical conductivity of nanofluids with SWCNTs and MWCNTs based on EG and W + PVP at different concentrations

CNTs have a fairly complex morphology, their diameter varies from 1 to 100 nm (or even more), and their length ranges from hundreds of nanometers to tens of micrometers. Additionally, they can be single-walled or multi-walled, with the latter exhibiting a variable inner diameter. For this reason, the dependence of electrical conductivity, as well as rheology, and thermal conductivity on CNT size is a crucial factor. The measurement of electrical conductivity in nanofluids with CNTs of varying aspect ratios does not permit drawing objective conclusions. The same aspect ratios can be achieved in nanofluids comprising either SWCNTs or MWCNTs. Adequate data can be obtained only by measuring the electrical conductivity of a nanofluid with CNTs of the same class and having the same diameter but different lengths. To achieve this, the length of the CNTs was deliberately altered through grinding in a Fritsch Pulverisette 6 spherical planetary mill. This enabled to identify the dependence of electrical conductivity on the length of the SWCNTs, since it can be assumed that the diameter of CNTs remained consistent during the grinding process. This means that grinding is actually equivalent to changing the aspect ratio.

The dependence of the electrical conductivity of EG-based nanofluids with SWCNTs on their effective hydrodynamic radius [see formula (1.1)] is illustrated in Fig. 5.10. The weight concentration of SWCNTs is 0.2%. The four points correspond to the different processing times of the nanofluid in the spherical planetary mill. The upper right point corresponds to the initial nanofluid that has not been subjected to grinding. The second, third, and fourth points on the left correspond to the grinding of CNTs for 1, 2, and 4 h, respectively. It can be observed that the electrical conductivity of the nanofluid decreases significantly with a decrease in the effective hydrodynamic size of the SWCNTs. The observed decrease in electrical conductivity is almost linear to a decrease in the size of the SWCNTs.

It is evident that this pattern persists at all concentrations of SWCNTs, as illustrated in Fig. 5.11. Here, the rightmost points correspond to the initial nanofluids, and the subsequent ones correspond to the grinded ones. For concentrations of 0.05 and 0.1%, the grinding time of SWCNTs was 1, 2, and 4 h. This is indicated by three

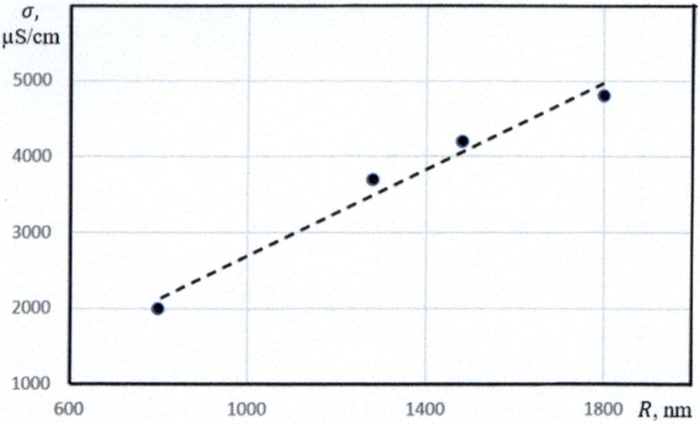

Fig. 5.10 The electrical conductivity (EC) of an EG-based nanofluid depending on the effective hydrodynamic radius of the SWCNTs

points in Fig. 5.11. At a concentration of 0.2%, a grinding process for over 0.5 h was also carried out. Accordingly, in Fig. 5.12, instead of four, five points correspond to this concentration. In all three cases, the electrical conductivity of the nanofluid decreases significantly with a decrease in the effective hydrodynamic radius of the SWCNTs, and hence the aspect ratio.

Finally, Fig. 5.12 compares the change in the coefficient of electrical conductivity of nanofluids with SWCNTs and MWCNTs after grinding both. Ethelene glycol was used as the base fluid. A decrease in electrical conductivity is observed with a decrease in the effective hydrodynamic radius of the CNTs. In the case of a nanofluid with MWCNTs, the initial hydrodynamic radius was almost twice as large as for a nanofluid with SWCNTs.

Summing up the presented measurement data, the first important conclusion can be drawn. The electrical conductivity of nanofluids with CNTs is markedly higher than

Fig. 5.11 The electrical conductivity of an EG-based nanofluid depending on the effective hydrodynamic radius of the SWCNTs

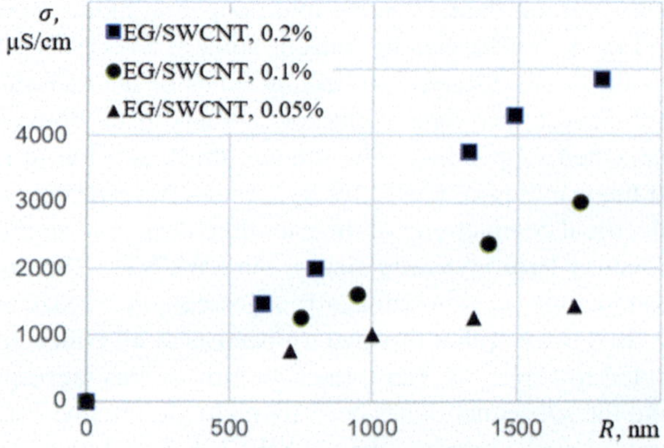

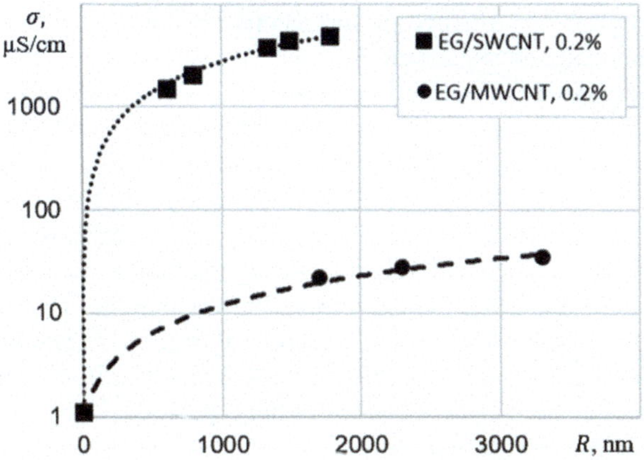

Fig. 5.12 Comparison of the electrical conductivity of an EG-based nanofluid with SWCNTs and MWCNTs depending on the effective hydrodynamic radius of CNTs

that of the corresponding base fluids. The degree of this excess, however, depends on the base fluid type, the type of CNT (whether SWCNTs or MWCNTs), the presence of surfactants, and the size of the CNTs (their aspect ratio). In all cases, the electrical conductivity of a nanofluid with CNTs is not described by Maxwell's theory (5.4). This implies that the electrical conductivity mechanisms of coarse dispersed fluids and nanofluids considered here are distinct. To explain the observed differences, it is necessary to understand these mechanisms. They are extremely complex and there are several of them. Let us consider them sequentially.

Section 5.4 examined the impact of the electrophoretic conductivity of the base fluid ions on the electrical conductivity of nanofluids in the presence of an electric field. Naturally, such ions are in nanofluids with CNTs. In addition, DEL is formed around CNTs in an electric field similarly as around conventional nanoparticles. Thus, the electrical conductivity of nanofluids with CNTs also depends on their mobility l_{CNT}. This mobility is also inversely proportional to the drag force acting on the CNTs in the fluid

$$l_{CNT} \sim 1/(6\pi R\mu), \tag{5.10}$$

where R is the effective hydrodynamic radius of CNTs (1.1).

Thus, the electrical conductivity of nanofluids is inversely proportional to the hydrodynamic radius of CNTs and the viscosity of the base fluid. The dimensions of the MWCNTs are significantly larger than those of the SWCNTs (see Fig. 5.12). Therefore, the mobility of the former, and hence the electrical conductivity of the corresponding nanofluids, is lower than that of nanofluids with SWCNTs. The viscosity of the base fluids also significantly affects the electrical conductivity of the nanofluids prepared on their basis.

It is now necessary to understand the effect of surfactants on the electrical conductivity of the nanofluids. As previously demonstrated (see Table 5.2), the electrical conductivity of base fluids containing surfactants is significantly higher than that of the pure initial base fluids (EG and W). It is also important to note that the degree of influence of surfactants on the electrical conductivity of the base fluid depends on surfactant type. The impact of steric (PVP) surfactants is significantly less pronounced than that of ionic (SDS and SDBS) surfactants. This is due to several circumstances. Firstly, the viscosity of the W + PVP solution is considerably higher than that of the W + SDBS solution [46, 50]. Secondly, the ionization of ionic solutions with SDS or SDBS occurs significantly more intensively than a solution with a steric dispersant PVP. Thirdly, it is notable that SDS and, in particular, SDBS form micelles and polymycelles in the base fluid, whereas PVP is just a fairly large molecule. This phenomenon is illustrated in Fig. 1.13, which shows the size distributions of SDBS (Fig. 1.13a) and PVP (Fig. 1.13b) molecules in water obtained using the DLS method. Micelle formation is a threshold phenomenon that begins when a certain their critical concentration is reached. Until this concentration is reached, the surfactant ions and sodium counterions contribute to the electrical conductivity of the base fluid in addition to the contribution made by the micelles. Micelles are nearly spherical formations [39] encircled by DEL formed in an electric field. The same occurs with polymicelles. Unlike the PVP solution, additional electricity carriers emerge in the SDBS (and SDS) solution. However, due to the substantial weight of micelles, their mobility is low, and thus the primary contribution to electrical conductivity is made by the Na^+ counterions.

It is now necessary to understand the reasons for the sharp decrease in the electrical conductivity of nanofluids prepared using surfactants compared to nanofluids with no surfactants (compare Tables 5.1 and 5.3). This is mainly due to a decrease in the effective charge of DEL, which occurs near CNTs due to the screening of their surface by surfactants. At the same time, the number of free micelles around which DES were formed in the nanofluid sharply decreases. In general, when utilizing surfactants, the electrical conductivity of CNTs, upon which these surfactants are adsorbed, also diminishes.

Finally, the role of CNTs in the formation of the total electrical conductivity of a nanofluid should be discussed. This is the most complex issue. Depending on the chirality, SWCNTs can be both conductors or semiconductors. It is usually extremely difficult (in fact, impossible) to determine the appropriate character of the ensemble (the source powder used) of the SWCNTs dissolved directly in the nanofluid. This significantly complicates the analysis of their contribution to the electrical conductivity of the nanofluid. Nevertheless, it can be argued that the electrical conductivity of a nanofluid with CNTs is directly related to the electrical conductivity of CNTs themselves. The electrical conductivity of the SWCNTs is primarily determined by the ballistic mechanism [13, 42, 56]. However, as the length of SWCNTs increases, the diffusion mechanism should also be considered. Therefore, in fact, we should talk about a quasi-ballistic mechanism. This mechanism is also typical for MWCNTs. Finally, it is necessary to take into account the electrical resistance that occurs at the CNT–fluid interface. This can be considered an analogue of the Kapitsa thermal resistance [28]. As the length of the CNT decreases, the total value of this resistance increases. In this regard, the conclusion drawn in the work [40] is pertinent: the electrical conductivity of a nanofluid is not determined by the electrical properties of

CNTs, but rather by the number and quality of tunnel contacts between nanotubes. These contacts allow for the realization of conductivity by a jumping mechanism.

Thus, the contribution of CNTs to the electrical conductivity of the corresponding nanofluids is very important. In fact, nanofluids can be considered as a composite material, whose electrical conductivity is determined by two factors: the electrical conductivity of the carrier fluid, including the mobility of CNTs, and the electrical conductivity of the resulting CNT skeletons of dispersed particles. From this point of view, it is necessary to understand the role of percolation, the critical concentration of which is quite small (especially for SWCNTs). Two processes occur during the formation of the percolation lattice. On the one hand, the electrical conductivity of such a lattice reaches a certain limit. On the other hand, the mobility of CNTs is significantly reduced. The viscosity of the base fluid increases sharply, which consequently results in a reduction in the mobility of the formed ions. Thus, an increase in CNT percolation in the nanofluid results in a notable slowing of its electrical conductivity growth rate. Finally, it should be noted that with increasing temperature, the mobility of all structural elements of the nanofluid increases and its viscosity decreases. This enhances its electrical conductivity.

References

1. Abdolbaqi MK, Sidik NAC, Aziz A et al (2016) An experimental determination of thermal conductivity and viscosity of BioGlycol/water based TiO_2 nanofluids. Int Com Heat Mass Transfer 77:22–32
2. Bagheli S, Fadafan HK, Orimi RL et al (2015) Synthesis and experimental investigation of the electrical conductivity of water-based magnetite nanofluids. Power Technol 274:426–430
3. Banisi S, Finch JA, Laplante AR (1993) Electrical conductivity of dispersions: A review. 6(4):369–385
4. Beaulne M, Mitsoulis E (1997) Creeping motion of a sphere in tubes filled with Herschel-Bulkley fluids. J Non-Newt Fluid Mech 72:55–71
5. Bhanvase B, Barai D (2021) Introduction to nanofluids. In: Nanofluids for heat and mass transfer. Elsevier, pp 3–42
6. Brunauer S, Emmett PH, Teller E (1938) Adsorption of gases in multimolecular layers. J Am Chem Soc 60(2):309–319
7. Chereches EI, Minea AA (2023) Experiments on the electrical conductivity of PEG 400 nanocolloids enhanced with two oxide nanoparticles. Nanomaterials 13:1555
8. Chereches M, Bejan D, Chereches EI et al (2021) An experimental study on electrical conductivity of several oxide nanoparticle enhanced PEG 400 fluid. Int J Thermophys 42:104
9. Chilambarasan L, Prakash R, Shanu JP et al (2019) Investigation on the electrical conductivity of aqueous glycol based ZnO nanofluids. J Appl Fluid Mech 12(3):865–870
10. Coelho MF, Rivas MA, Vilão G et al (2019) Permittivity and electrical conductivity of copper oxide nanofluid (12 nm) in water at different temperatures. J Chem Thermodyn 132:164–173
11. Cruz RCD, Reinshagen J, Oberacker R et al (2005) Electrical conductivity and stability of concentrated aqueous alumina suspensions. J Colloid Interface Sci 286(2):579–588
12. Dong M, Shen LP, Wang H et al (2013) Investigation on the electrical conductivity of transformer oil-based AlN nanofluid. J Nanomat 2013:842963
13. Ebbesen TW, Lezec HJ, Hiura H et al (1996) Electrical conductivity of individual carbon nanotubes. Nature 382:54–56

14. Fal J, Mahian O, Zyła G (2018) Nanofluids in the service of high voltage transformers: breakdown properties of transformer oils with nanoparticles, a review. Energies 11:2942
15. Fal J, Sobczak J, Stagraczyński R et al (2022) Electrical conductivity of titanium dioxide ethylene glycol-based nanofluids: impact of nanoparticles phase and concentration. Powder Technol 404:117423
16. Fal J, Wanic M, Malick M et al (2019) Experimental investigation of electrical conductivity of ethylene glycol containing indium oxide nanoparticles. Acta Phys Pol A 135:1237–1239
17. Fricke HA (1924) Mathematical treatment of the electric conductivity and capacity of dispersed systems, I. The electric conductivity of a suspension of homogeneous spheroids. Phys Rev 24:575–585
18. Glover B, Whites KW, Hong H et al (2008) Effective electrical conductivity of functional single-wall carbon nanotubes in aqueous fluids. Synth Met 158:506–508
19. Harish S (2014) Thermal conductivity enhancement of liquid and solid with single-walled carbon nanotubes. Doctoral thesis. The University of Tokyo, Japan
20. Harish S, Ishikawa K, Einarsson E et al (2012) Temperature dependent thermal conductivity increase of aqueous nanofluid with single walled carbon nanotube inclusion. Mater Express 2:213–223
21. Heyhat MM, Rajabpour A, Abbasi M et al (2018) Importance of nanolayer formation in nanofluid properties: equilibrium molecular dynamic simulations for Ag-water nanofluid. J Mol Liquids 264:699–705
22. Huang Z, Wang F, Wang Q et al (2019) Significantly enhanced electrical performances of eco-friendly dielectric liquids for harsh conditions with fullerene. Nanomaterials 9:989
23. Iglesias TP, Rivas MA, Iglesias R et al (2013) Electric permittivity and conductivity of nanofluids consisting of 15 nm particles of alumina in base Milli-Q and Milli-Ro water at different temperatures. J Chem Thermodyn 66:123–130
24. Iglesias R, Rivas MA, Reis JCR et al (2015) Permittivity and electric conductivity of aqueous alumina (40 nm) nanofluids at different temperatures. J Chem Thermodyn 89:189–196
25. Islam M, Shabani B, Rosengarten J et al (2015) The potential of using nanofluids in pem fuel cell cooling systems: a review. Renew Sustain Energy Rev 48:523–539
26. Jones W, March NH (1985) Theoretical solid state physics. Courier Dover Publications
27. Jóźwiak B, Boncel S (2020) Rheology of ion nanofluids—A review. J Mol Liquids 302:12568
28. Kapitza PL (1941) The study of heat transfer in helium II. J Phys USSR 4(1–6):181–210
29. Kumar PG, Kumaresan V, Velraj R (2017) Stability, viscosity, thermal conductivity, and electrical conductivity enhancement of multi-walled carbon nanotube nanofluid using gum Arabic. Fullerenes Nanotubes Carbon Nanostr 25(4):230–240
30. Kumar PG, Renuka M, Velraj R (2018) Thermal and electrical conductivity enhancement of solar glycol-water mixture containing MWCNTs. Fullerenes Nanotubes and Carbon Nanostr 26(12):871–879
31. Kumar PG, Sakthivadivel D, Meikandan M et al (2019) Experimental study on thermal properties and electrical conductivity of stabilized H_2O-solar glycol mixture based multi-walled carbon nanotube nanofluids: developing a new correlation. Heliyon 5:e02385
32. Maxwell JC (1881) A treatise on electricity and magnetism. Clarendon Press, Oxford
33. Minea AA (2013) Electrical and rheological behavior of stabilized Al2O3 nanofluids. Current Nanosci 9:81–88
34. Minea AA (2019) A review on electrical conductivity of nanoparticle-enhanced fluids. Nanomaterials 9:1592
35. Minea AA, Luciu RS (2012) Investigations on electrical conductivity of stabilized water based Al_2O_3 nanofluids. Microfluidics Nanofluidics 13(6):977–985
36. Minea AA, Murshed SMS (2018) A review on development of ionic liquid based nanofluids and their heat transfer behavior. Renew Sustain Energy Rev 91:584–599
37. Naddaf P, Heris SZ (2018) Experimental study on thermal conductivity and electrical conductivity of diesel oil-based nanofluids of graphene nanoplatelets and carbon nanotubes. Int Comm Heat Mass Transfer 95:116–122

38. Nurdin I (2017) Investigation on electrical conductivity enhancement of water based maghemite (γ-Fe_2O_3) nanofluids. Int J Mat Sci Appl 6:32–36
39. Palazzeci F, Calvareci M, Zerbetto F (2011) A molecular dynamics investigation of structure and dynamics of SDS and SDBS micelles. Soft Matter 7(19):9148–9156
40. Panin YuV, Prilepo YuP, Torba GN et al (2012) Investigations of electrical conductivity in alcohol suspensions of multilayer carbon nanotubes. Bulletin of the VorGTU 8(2):70–72
41. Poongavanam GK, Duraisamy S, Vigneswaran VS et al (2021) Review on the electrical conductivity of nanofluids: recent developments. Mat Today: Proc 39(4):1532–1537
42. Roche S, Triozon F, Rubio A et al (2001) Electronic conduction in multi-walled carbon nanotubes: role of intercell coupling and incommensurability. Phys Lett A 285:94–100
43. Pordanjani AH, Aghakhani S, Afrand M et al (2019) An updated review on application of nanofluids in heat exchangers for saving energy. Energy Convers Manag 198:111886
44. Rubbi F, Das L, Habib K et al (2021) State-of-the-art review on water-based nanofluids for low temperature solar thermal collector application. Solar Energy Mat Solar Cells 230:111220
45. Rudyak VYa, Belkin AA, Tomilina EA (2008) Force acting on nanoparticle in fluid. Tech Phys Lett 34:76–78
46. Rudyak VYa, Dashapilov GR, Minakov AV et al (2023) Comparative characteristics of viscosity and rheology of nanofluids with multi-walled and single-walled carbon nanotubes. Diamond Rel Mat 132:109616
47. Rudyak VYa, Minakov AV, Pryazhnikov MI (2019) The electrical conductivity of nanofluids with metal particles. Tech Phys Lett 45(5):457–460
48. Rudyak VYa, Pryazhnikov MI, Minakov AV (2024) Thermal conductivity, rheology and electrical conductivity of water- and ethylene glycol-based nanofluids with copper and aluminum particles. Phys Mesomechanics 27(2):205–216
49. Rudyak V, Pryazhnikov M, Minakov A et al (2024) Electrical conductivity of nanofluids with single- and multi-walled carbon nanotubes. Experimental study. Nano-Struct Nano-Obj 38:101143
50. Rudyak VYa, Tret′yakov DS (2021) Rheological properties of water- and ethylene glycol-based nanofluids with single wall carbon nanotubes. J Eng Phys Thermophys 94(5):1208–1216
51. Samira A, Hossein M, Branch A et al (2022) Investigation of thermophysical properties of Io nanofluids containing multi-walled carbon nanotubes and graphene. Iran J Chem Chem Eng 41(2):380–391
52. Sarojini K, Manoj SV, Singh PK et al (2013) Electrical conductivity of ceramic and metallic nanofluids. Colloids Surf A: Physicochem Eng Aspects 417:39–46
53. Shanbedi M, Heris SZ, Maskooki A (2015) Experimental investigation of stability and thermophysical properties of carbon nanotubes suspension in the presence of different surfactants. J Therm Anal Calorim 120(2):1193–1201
54. Shanbedi M, Heris SZ, Amiri A et al (2016) Synthesis of water-soluble Fe-decorated multi-walled carbon nanotubes: a study on thermo-physical properties of ferromagnetic nanofluid. J Taiwan Inst Chem Eng 60:547–554
55. Sharafeldin MA, Gróf G, Abu-Nada E et al (2019) Evacuated tube solar collector performance using copper nanofluid: energy and environmental analysis. Appl Therm Eng 162(13):114205
56. Shehzad K, Dang Z-M, Ahmad MN et al (2013) Effects of carbon nanotubes aspect ratio on the qualitative and quantitative aspects of frequency response of electrical conductivity and dielectric permittivity in the carbon nanotube/polymer composites. Carbon 54:105–112
57. Shoghl SN, Jamali J, Moraveji MK (2016) Electrical conductivity, viscosity, and density of different nanofluids: an experimental study. Experimental Thermal Fluid Sci 74:339–346
58. Tabuteau H, Oppong FK, de Bruyn JR et al (2017) Drag on a sphere moving through an aging system. EPL 78:68007
59. Traciak J, Sobczak J, Vallejo JP et al (2023) Experimental study on the density, surface tension and electrical properties of ZrO_2–EG nanofluids. Phys Chem Liquids 61:14–24
60. White SB, Shih AJ-M, Patrick K (2011) Pipe, investigation of the electrical conductivity of propylene glycol-based ZnO nanofluids. Nanoscale Res Lett 6:346

61. Zakaria I, Azmi W, Mohamed W et al (2015) Experimental investigation of thermal conductivity and electrical conductivity of Al_2O_3 nanofluid in water-ethylene glycol mixture for proton exchange membrane fuel cell application. Int Com Heat Mass Transfer 61:61–68
62. Zakaria I, Mohamed WANW, Azmi WH et al (2018) Thermo-electrical performance of PEM fuel cell using Al_2O_3 nanofluids. Int J Heat Mass Transfer 119:460–471
63. Zawrah MF, Khattab RM, Girgis LG et al (2016) Stability and electrical conductivity of water-based Al_2O_3 nanofluids for different applications. HBRC J 12:227–234
64. Zyła G, Fal J (2016) Experimental studies on viscosity, thermal and electrical conductivity of aluminum nitride ethylene glycol (AlN-EG) nanofluids. Thermochim Acta 637:11–16
65. Żyła G, Fal J (2017) Viscosity, thermal and electrical conductivity of silicon dioxide–ethylene glycol transparent nanofluids: an experimental study. Thermochim Acta 650:106–113
66. Żyła G, Fal J, Bikić S et al (2018) Ethylene glycol-based silicon nitride nanofluids: an experimental study on their thermophysical, electrical and optical properties. Physica E: Low-Dimens Syst Nanostruct 104:82–90

Chapter 6
Heat Transfer in Nanofluid Flows

6.1 Experimental Setup

The creation of modern heat exchange equipment and corresponding technologies requires intensification of heat transfer processes. In certain cases, this problem becomes a fundamental one, and only its solution will allow the desired goal to be reached. This concerns, for example, the development of next-generation processors and various microfluidic devices. Naturally, the solution to this problem is also relevance in all other industry branches, as it will reduce energy consumption and achieve maximum compactness of the corresponding devices at their minimum material intensity. One of the ways for enhancing heat transfer processes could be the use of nanofluids. Over the past three decades, hundreds of articles on this topic have been published. To date, the systematic experimental study of the heat transfer of nanofluids with ordinary spherical particles has been sufficiently comprehensive. The relevant literature can be found in recent reviews [5, 19, 22, 44, 56, 62, 63]. A notable enhancement of heat transfer is typically observed in these studies when nanofluids are employed. However, the resulting effect is significantly influenced by the flow regime (laminar or turbulent) [8, 13–15, 21, 38–41]. The situation with heat exchange of nanofluids with CNTs is much more complicated. This is due to the fact that nanofluids, for various reasons, tend to be incomparable. Nevertheless, numerous reviews on the use of nanofluids with CNTs have emerged in recent years [2, 6, 47, 53, 54].

Given the vast number of publications and the dissimilar nature of the results, it is not feasible to undertake a comprehensive and critical review of the available studies within the scope of this chapter. For this reason, the author presents a summary of the principal experimental works in which he was directly involved. The data obtained in these studies appear to be both relevant and representative. They provide a fairly complete and objective view of the possibilities of using nanofluids as heat carriers.

V. Rudyak, *Thermophysical Properties of Nanofluids*, Fluid Mechanics and Its Applications 140, https://doi.org/10.1007/978-3-031-97995-8_6

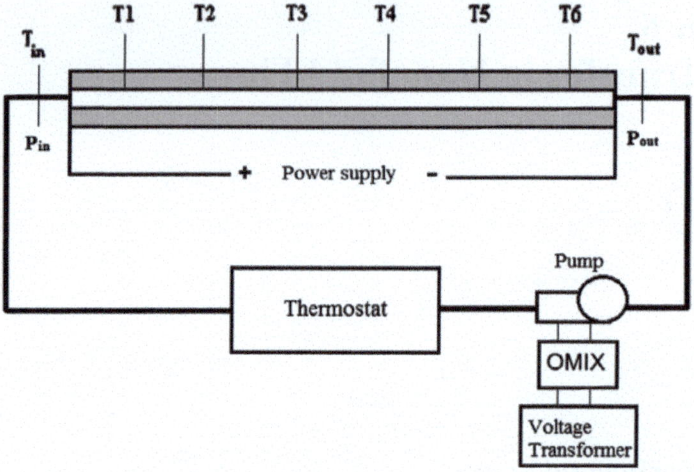

Fig. 6.1 Schematic diagram of the measuring setup for investigating convective heat transfer of nanofluids

Generally, the cooling of a heated surface is achieved by flowing of a heat carrier along the surface. Consequently, in order to study the heat transfer processes in nanofluids, it was necessary to create a setup that would allow the flow of the nanofluid under study to be implemented. A setup was developed by our team for the purpose of measuring the heat transfer coefficients of nanofluids. This setup has been used for the past ten years [39, 41]. The schematic diagram of the measuring setup for studying the heat transfer coefficient is presented in Fig. 6.1. The setup is a closed loop with a circulating coolant. A pump is used to supply the working fluid from the reservoir to the measuring heated section. The fluid then passes through the heat exchanger, where it transfers heat to the thermostat. The flow rate of the heat carrier in the loop is regulated by adjusting the pump power. The pump power is regulated by a laboratory current converter and monitored with the Omix™ measuring instrument.

The measuring heating section is a metallic tube with thermocouples embedded in the wall. The tubes were made of either copper or stainless steel. The diameter of the tubes ranged from 4 to 10 mm, while the wall thickness varied from 0.3 to 1 mm. The length of the measuring section was typically one meter. The tube was subjected to a controlled heating by electric current applied through thermally insulated current leads. This methodology ensured the constant heat flux density on the wall. To measure the local temperature of the channel, 6 chromel-copel thermocouples are fixed to its walls at equal distances from each other. The temperature measurements were conducted using TRM-200 m (T_k in Fig. 6.1). Furthermore, the temperature at the inlet and the outlet of the heated section was measured by thermocouples T_{in} and T_{out} (see Fig. 6.1). At the same time, in the experiments the pressure drop between the inlet (P_{in}) and outlet (P_{out}) of the investigated channel was also measured using Testo 510 differential manometer.

During the experiments, the local heat transfer coefficient was determined by formula $\alpha_c = GC_p(T_i - T_o)S^{-1}(T_w - \overline{T})^{-1}$, where $\overline{T} = (T_i + T_o)/2$ is the average liquid temperature, T_i, T_o are the temperatures at the inlet and outlet of the heated section, T_w is the local temperature of the channel wall, G is the mass flow rate, and S is the surface area of the heated section. The average heat transfer coefficient was also calculated along with the local one: $\overline{\alpha}_c = GC_p(T_i - T_o)S^{-1}(\overline{T}_w - \overline{T})^{-1}$, where $\overline{T}_w$ is the arithmetic average temperature of the channel wall obtained by averaging over the data of six thermocouples. The experimental setup described was tested using a variety of empirical data for water and ethylene glycol [38, 41]. Both laminar and turbulent flow were studied. The local and average heat transfer coefficients were compared. The hydraulic drag coefficients were compared with the Poiseuille and Blasius dependences for both the laminar and turbulent flow regimes. The test results have demonstrated that the accuracy of the local and average heat transfer coefficients and pressure drops in all experiments was not high than 5%.

6.2 Heat Transfer of Nanofluids in the Laminar Flow Regime

Since the thermophysical characteristics of nanofluids are dependent upon not only the nanoparticle concentration but also their size and material, it is essential to consistently analyze the effects of these parameters on heat transfer. In addition, some nanofluids have non-Newtonian rheology, therefore it is necessary to understand how this can affect heat transfer. Finally, convective heat transfer can occur in different flow regimes, namely laminar or turbulent. In contrast to the laminar flow regime, where heat transfer is proportional to the thermal conductivity of the heat carrier, the transfer processes in turbulent flow are determined not only by molecular mechanisms, but also by other factors. This is where efficient turbulent mechanisms become relevant. The results of this analysis will be presented in the following sections. All experiments were conducted at the setup described in Sect. 6.1. The nanofluids used in this study were prepared according to the methodology outlined in Chap. 1. The experimental results presented in this section are obtained in studies reported in the following papers [13, 15, 39–41].

6.2.1 Dependence of Heat Transfer Coefficient on Nanoparticle Concentration

The first example presented below concerns a water-based nanofluid with 150 nm aluminum oxide particles. The viscosity and rheology measurements conducted have demonstrated that these nanofluids are Newtonian at all concentrations. Table 6.1 presents a summary of the relative viscosity and thermal conductivity coefficients at different particle volume concentrations. The measurements were conducted at a temperature of 25 °C. The dependences of the relative viscosity and thermal conductivity coefficients of these nanofluids on temperature exhibit slight differences from the corresponding dependences for water.

Figure 6.2 a shows the dependence of the local heat transfer coefficient for a water-based nanofluid with 150 nm aluminum oxide particles on their volume concentrations (expressed as a percentage). Here the flow rate was fixed at 0.0011 kg/s, which corresponds to a Reynolds number for water of 250. The heat transfer coefficient increases monotonically with increasing nanoparticle concentration. The maximum excess is approximately 25% at a particle volume concentration of 6%. It is important to note the existence of a region of heat transfer stabilization within the flow. The enhancement of the heat transfer coefficient resulting from an elevated nanoparticle concentration in this region is nearly directly proportional to the rise in the heat conductivity coefficient of the nanofluid associated with an elevated particle concentration. This is consistent with the Gretz theory [20], which states that in a stabilized heat transfer section (at a constant heat flux at the channel wall), the Nusselt number $Nu = \alpha_c d_c/\lambda$ is constant and equal to 4.36 (where d_c is the channel diameter). This is indeed well confirmed in the present experiment. Figure 6.2 b illustrates the variation of the local Nusselt number along the test section. Indeed, at the test section portion, where heat transfer is stabilized, the Nusselt number within the experimental error corresponds to noted value.

The dependence of the average heat transfer coefficient of the same nanofluid on the Reynolds number $Re = (\rho U_G d_c)/\mu$ here $U_G = 4G/\left(\rho\pi d_c^2\right)$ is the velocity determined by the flow rate G is presented in Fig. 6.3. At a given Reynolds number, the heat transfer coefficient enhances with increasing nanoparticle concentration. The enhancement of the heat transfer coefficient of nanofluid at maximum particle concentration is greater than 28%. As the nanoparticle concentration is reduced, the heat transfer coefficients exhibit a monotonic tendency to approach the values corresponding to pure water. It is evident that as the flow rate is increased, the heat transfer coefficient of nanofluids also rises.

The most important characteristic of convective heat transfer is the pressure drop in the channel, which determines the hydraulic resistance of the flow. Figure 6.4 depicts the measured values of pressure drop in the channel for the same nanofluids, as a function of the Reynolds number (Fig. 6.4 a) and the flow rate (Fig. 6.4 b). It is

Table 6.1 Relative viscosity and thermal conductivity coefficients of water-based nanofluids with aluminum and titanium oxide particles

	Al_2O_3			TiO_2	
φ	μ_r	λ_r	φ	μ_r	λ_r
0.01	1.060	1.059	0.01	1.090	1.048
0.02	1.134	1.131	0.02	1.180	1.100
0.04	1.357	1.178	0.04	1.570	1.146
0.06	1.600	1.240	0.06	1.910	1.206

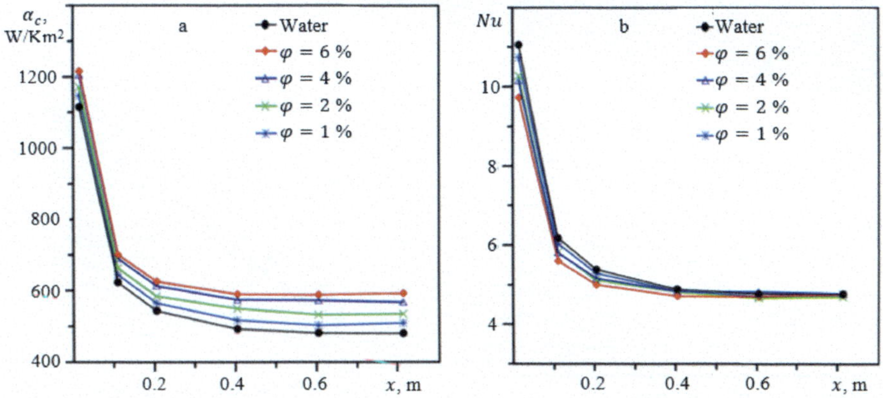

Fig. 6.2 Distribution of local heat transfer coefficient (**a**) and Nusselt number (**b**) along the test section for nanofluid with Al_2O_3 particles

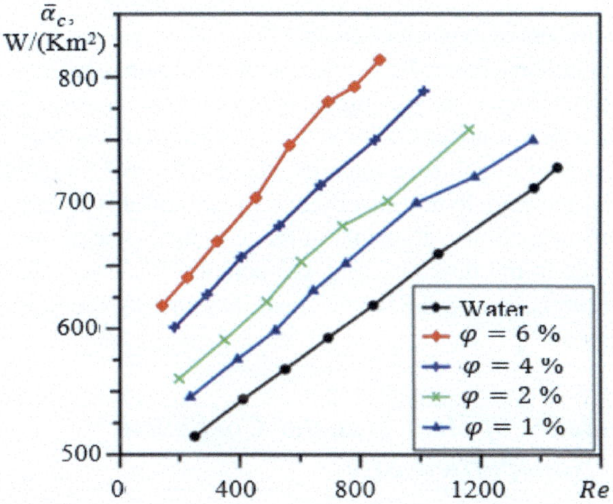

Fig. 6.3 Dependence of the average heat transfer coefficient of nanofluid with aluminum oxide particles on the Reynolds number

evident that as the concentration of nanoparticles in the fluid increases, the pressure drop in the channel also increases, as the viscosity of nanofluids also increases (see Table 6.1). A comparison of these two Figures reveals that the dependence of the pressure drop on the Reynolds number is significantly steeper than the dependence on the flow rate. This is to be expected, since at a given Reynolds number the following relation is valid $\Delta p \sim \mu^2/\rho$, while at a given flow rate $\Delta p \sim \mu/\rho$, It is also noteworthy that the dependence of the pressure drop on the flow rate is extremely important from a practical standpoint.

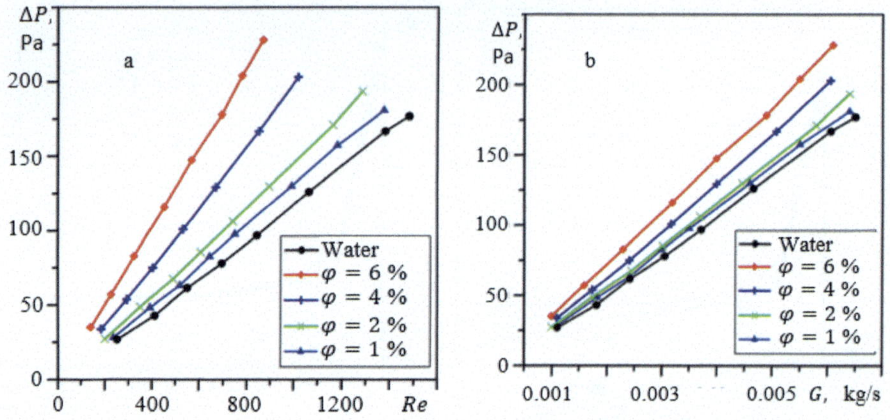

Fig. 6.4 Dependence of pressure drop in the channel depending on *Re* number (**a**) and flow rate (**b**)

Although the presented experimental data (which are typical) demonstrate significant heat transfer enhancement, from a practical standpoint, thermal–hydraulic efficiency is often more important parameter than just heat transfer. This is exemplified by the average heat transfer coefficient as a function of the pressure drop in the channel. The corresponding illustration is provided in Fig. 6.5 for the nanofluid under consideration. The thermal–hydraulic efficiency of nanofluids at all concentrations is found to be significantly higher than that of the base fluid (water). Since the pressure drop is proportional to the flow rate, Fig. 6.5 actually illustrates the dependence of the heat transfer coefficient on the power spent on pumping the liquid.

6.2.2 *Dependence of Heat Transfer Coefficient on the Nanoparticle Size*

From the point of view of organizing heat exchange, both the thermal conductivity and viscosity of the refrigerant are of paramount importance. In Chaps. 3 and 4, it was established that the viscosity and thermal conductivity of nanofluids depend on the size of the nanoparticles. It was revealed that as the size of the nanoparticles decreases, the viscosity increases, while the thermal conductivity decreases on the contrary. This enables the development of an effective strategy for the creation of nanofluids intended for use as refrigerant. However, in order to achieve this, it is necessary to gain an understanding of the effect of nanoparticle size on the nanofluid heat transfer coefficient. Below, the corresponding data will be analyzed using two water-based nanofluids with zirconium oxide particles of two sizes, 44 and 105 nm, as an example. The volume concentration of the particles in both nanofluids is 4%.

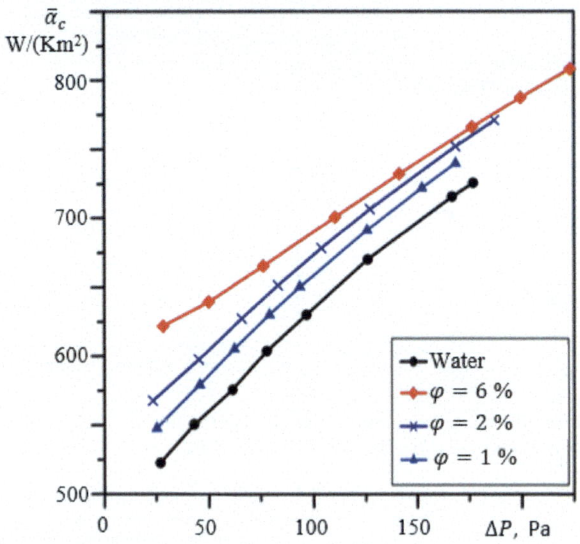

Fig. 6.5 Dependence of the average heat transfer coefficient on the pressure drop for water-based nanofluid with 150 nm aluminum oxide particles

The relative viscosity coefficient of these two nanofluids is 1.225 and 1.175, while the relative thermal conductivity coefficient is 1.143 and 1.169, respectively.

Figure 6.6 illustrates the dependence of the average heat transfer coefficient on the Reynolds number and flow rate of water-based nanofluids with zirconium oxide particles of different sizes. In both cases, the heat transfer coefficient enhanced with increasing particle size. At a given flow rate (flow velocity), the heat transfer coefficient for nanofluid with 44 nm particles was found to be enhanced by 8%, while that with 105 nm particles was enhanced by 14%. A similar picture is also recorded for the dependence on the Reynolds number. The heat transfer coefficient of the nanofluid with 44 nm particles enhanced by 14%, while that of the nanofluid with 105 nm particles enhanced by 19%.

Finally, it is beneficial to analyze the thermal–hydraulic efficiency of these two nanofluids. The corresponding objective information is provided in Fig. 6.7. The particle concentration in both nanofluids is identical and equal to 4%. The thermal–hydraulic efficiency is significantly influenced by the particle size. For nanofluids with particles of 105 nm in size, the thermal–hydraulic efficiency is consistently higher than that of water. This efficiency is even slightly enhanced with increasing flow velocity, whereas, for nanofluids with smaller particles, the thermal–hydraulic efficiency is also higher than that of water at low flow rates, however, then it becomes lower than that of water.

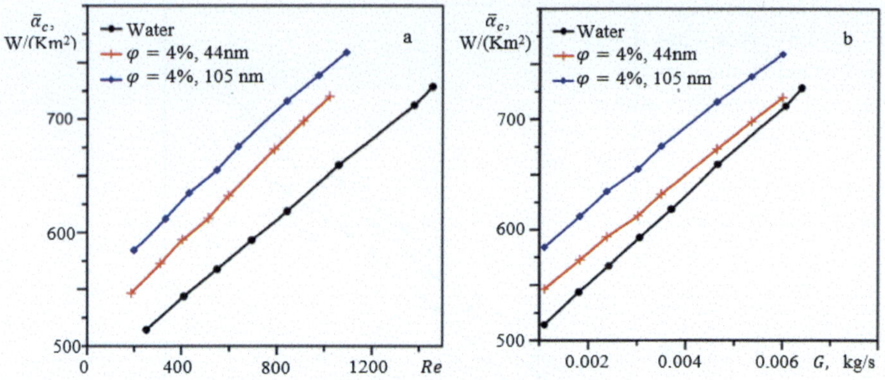

Fig. 6.6 Dependence of the average heat transfer coefficient of water-based nanofluid with zirconium oxide particles on the Reynolds number (**a**) and flow rate (**b**)

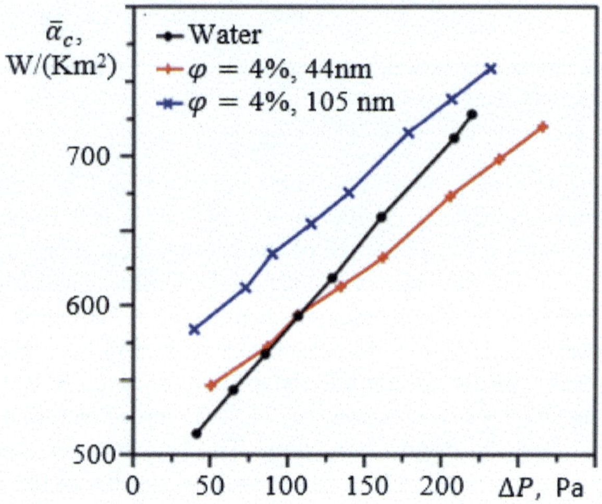

Fig. 6.7 Dependence of the average heat transfer coefficient on the pressure drop of water-based nanofluids with zirconium oxide particles of two sizes

6.2.3 Dependence of Heat Transfer Coefficient on Nanoparticle Material

The material of nanoparticles is a significant factor influencing the thermophysical characteristics of nanofluids. Naturally, nanofluids with particles of different materials will have different heat transfer coefficients. To investigate the nature of this dependence, it is necessary to consider two nanofluids with particles of the same size but different materials. In this section, water-based nanofluids with aluminum and

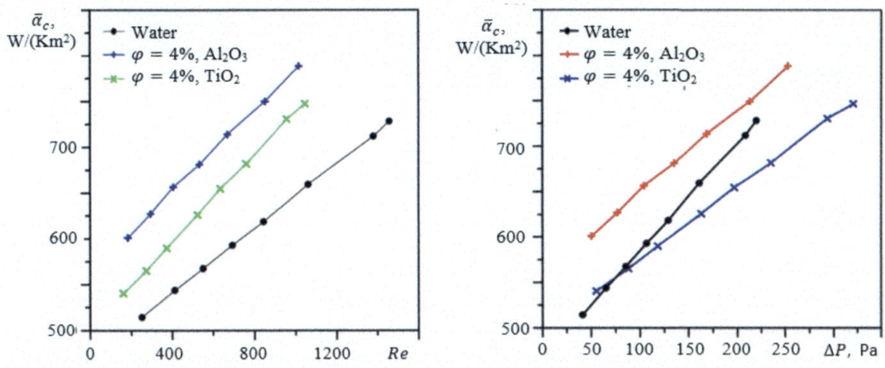

Fig. 6.8 Dependence of average heat transfer coefficient of two water-based nanofluids with aluminum and titanium oxide particles on Reynolds number and pressure drop

titanium oxide particles are considered for this purpose. Table 6.1 presents the relative viscosity and thermal conductivity coefficients for nanofluids under study. The data on the dependence of average heat transfer coefficient of these two water-based nanofluids on Reynolds number and pressure drop are presented in Fig. 6.8.

The average heat transfer coefficient of both nanofluids exceeds that of water at all Reynolds numbers considered (see Fig. 6.8). Nevertheless, the observed excess is notably smaller for the nanofluid comprising titanium oxide particles than for the nanofluid with aluminum oxide particles. This is due to lower values of the thermal conductivity coefficient (see Table 6.1). At this specific concentration, the relative thermal conductivity coefficient of the nanofluid with aluminum oxide particles is approximately 3% higher. On the other hand, the viscosity coefficient of the nanofluid with titanium oxide particles is considerably higher (about 16%) than that of the nanofluid with aluminum oxide particles. This indicates that the thermal–hydraulic efficiency of these two nanofluids will be significantly disparate. Indeed, the data presented in Fig. 6.8 this assertion. The average heat transfer coefficient of the nanofluid with aluminum oxide particles is higher than that of the water at any pressure drop considered. However, the corresponding data for nanofluid with titanium particles are lower than for water excepting the region of very low the pressure drop values. Thus the thermal–hydraulic efficiency of the nanofluid with titanium oxide particles is observed to be lower than that of water across the entire range of flow rates investigated.

6.2.4 Heat Transfer Coefficient of Non-Newtonian Nanofluids

In Chap. 3 it was demonstrated that many nanofluids become pseudo- or viscoplastic with increasing particle concentration or decreasing particle size. The rheology of

Table 6.2 Dependence of rheological parameters of nanofluid with copper oxide particles on their concentration

φ (%)	n	k_v
2	0.502	0.223
1	0.697	0.025
0.5	0.853	0.005
0.25	0.980	0.001

these fluids is well described by the power-law fluid model (3.29) or the Herschel–Bulkley model (3.30). These nanofluids make up about a quarter of the total number of nanofluids (see [42, 52]). It is therefore necessary to understand the peculiarities of heat transfer in such nanofluids. This section considers the heat transfer of water-based nanofluids with copper oxide nanoparticles [13, 39, 40]. The average size of the nanoparticles was 55 nm, and their volume concentration varied from 0.25 to 2%. Furthermore, a surfactant (xanthan gum biopolymer) with a weight concentration of no more than 0.3% was employed to stabilize the nanofluid. Consequently, the base fluid constituted a solution of surfactant in water. The nanofluid was stable, retaining its physical properties for at least two months.

The rheological properties of this nanoflu are well described by the power-law fluid model (3.29). The rheological parameters of nanofluids with different particle concentrations are presented in Table 6.2. As the concentration of nanoparticles in the nanofluid increases, the fluid index n decreases, while the consistency parameter k_v. increases. The presented measurements were carried out at a temperature of 25 °C.

The conducted measurements have demonstrated that the nanofluids significantly enhance both local and average heat transfer coefficients of the fluid. A comparison of the variation of the heat transfer coefficient of nanofluids and water along the channel at two different coolant flow rates is presented in Fig. 6.9. The concentration of particles in the nanofluid was 2%. In the first case, the flow rate was 104 g/min, while in the second case the flow rate was nearly twice as high, reaching 193 g/min. The marks in the plot correspond to measurements made by thermocouples placed sequentially on the channel walls (see Fig. 6.1). The distance between the thermocouples was 0.1 m. In both cases, the use of nanofluid allows for more than 10% intensification of local heat transfer along the entire length of the channel.

Similar dependences at these flow rates are obtained for other nanofluids, although the effect diminishes with decreasing nanoparticle concentration. At low flow rates (up to approximately 300 g/min), which correspond to the laminar flow regime, the degree of enhancement in the heat transfer coefficient grows almost proportionally to the volume concentration of nanoparticles

$$\Delta\alpha_c = (\alpha_c - \alpha_{cw})/\alpha_{cw} = \alpha_{cr} - 1 = b\varphi, \tag{6.1}$$

where α_{cw} is the heat transfer coefficient of water.

Due to the significant differences in the viscosities of the fluids under study, the flows of the carrier fluid and nanofluids at a given flow rate correspond to different Reynolds numbers, which in turn result in different flow regimes. For this reason, it

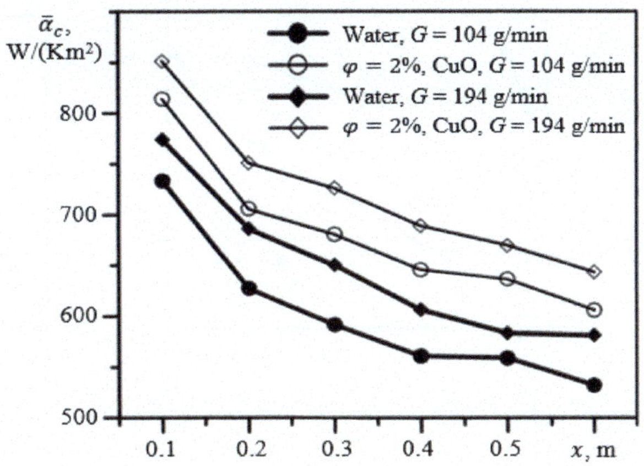

Fig. 6.9 Variation of local heat transfer coefficient along the channel at fluid flow rates of 104 and 193 g/min

is beneficial to examine the relationship between the heat transfer coefficient and the Reynolds number. Since the nanofluid under study is non-Newtonian, the Reynolds number is determined in accordance with the standard procedure for a power-law fluid [36]

$$\mathrm{Re} = \frac{\rho U_G^{2-n} d_c^n}{k_v 8^{n-1} \left(\frac{3n+1}{4n}\right)^n}. \tag{6.2}$$

The intensification of heat transfer when using nanofluids at a fixed Reynolds number turns out to be quite significant. Thus, the heat transfer coefficient for a 1% nanofluid is demonstrably more than 40% higher than that for water at almost all measured Reynolds numbers (see Fig. 6.10a). The pump used in these experiments was incapable of sustaining flow rates in excess of 500 g/min. Therefore, only Reynolds numbers on the order of 300 were achieved for the 2% nanofluid. Figure 6.10a clearly shows that the dependence of the heat transfer coefficient on the Reynolds number obtained for this nanofluid increases extremely steeply. Extrapolating the dependence of the heat transfer coefficient on the Reynolds number for a 2% nanofluid presented in this Figure, it is safe to say that heat transfer coefficient enhancement turns out to be at least twofold.

Figure 6.10 b illustrates the dependence of the average heat transfer coefficient on the flow rate for a given nanofluid at different nanoparticle concentrations. At a given flow rate, the average heat transfer coefficient of the 2% nanofluid is approximately 13% higher than that of water, which corresponds to an increase in the local heat transfer coefficient. As the nanoparticle concentration is reduced, the heat transfer coefficient declines monotonically. The non-monotonic behavior of the heat transfer coefficient for a nanofluid with a 0.25% CuO particle concentration at high flow rates, as well as the sharp increase in α_c for water are associated with the flow transition from laminar to the turbulent regime. Thus, the utilization of non-Newtonian nanofluids can result in a considerably enhanced heat transfer coefficient in comparison to

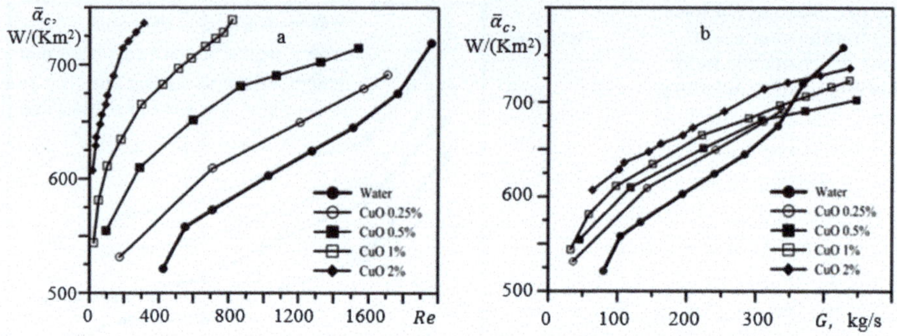

Fig. 6.10 Dependence of average heat transfer coefficient on Reynolds number and flow rate

Newtonian nanofluids. The reasons for this phenomenon will be analyzed in the subsequent chapter.

From the practical point of view, the pressure drop in the channel is an important parameter along with the heat transfer coefficient. The corresponding measurement data of the pressure drop depending on the Reynolds number are shown in Fig. 6.11a. It is evident that the pressure loss during the flow of nanofluid through a channel is considerably higher than that of water. This is due to the high viscosity of nanofluid. One of the criteria for the thermal–hydraulic efficiency of nanofluids is the dependence of the ratio of heat transfer coefficient to pressure drop on Reynolds number. This dependence is illustrated in Fig. 6.11b. A review of the presented data indicates that the thermal–hydraulic efficiency of tested nanofluid is relatively low. Nevertheless, the high heat transfer coefficient of this nanofluid indicates the necessity of investigating methods to reduce its viscosity, particularly, by employing an alternative surfactant.

Concluding the discussion of laminar convective heat transfer of nanofluids, it is worth highlighting that in this instance, the relative heat transfer coefficient exhibits

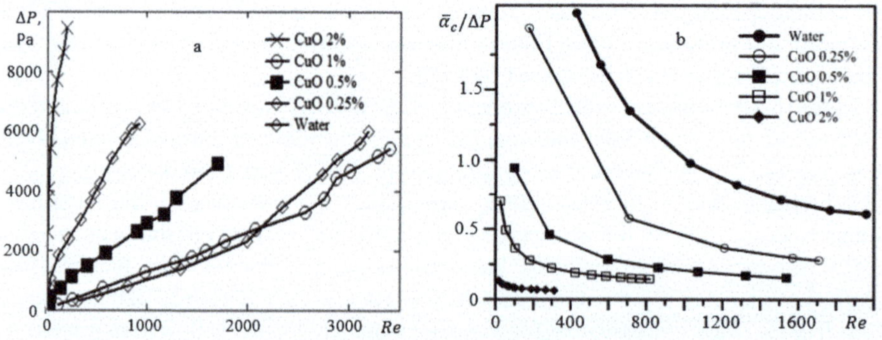

Fig. 6.11 The pressure drop in the channel (**a**) and the ratio of heat transfer coefficient to pressure drop (**b**) depending on the Reynolds number

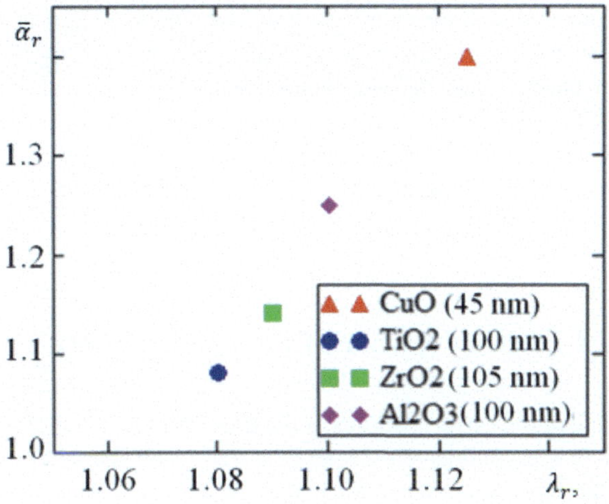

Fig. 6.12 Dependence of the relative heat transfer coefficient on the relative thermal conductivity coefficient of nanofluid

a nearly linear enhancement with increasing thermal conductivity of the nanofluid. This is demonstrated in Fig. 6.12, which presents the average relative heat transfer coefficient $\overline{\alpha}_r = \overline{\alpha}_{cr}/\overline{\alpha}_{cbf}$ (where $\overline{\alpha}_{cbf}$ is the heat transfer coefficient of the base fluid) as a function of the relative heat conductivity coefficient for the four nanofluids. Since the thermal conductivity coefficient of the nanofluid increases with increasing particle size, the coefficient b in (6.1) will also be the larger the larger the particle size.

6.3 Heat Transfer of Nanofluids in Microchannels

The rapid development of microsystem technology for various applications, where liquid flows often play a key role, requires a comprehensive investigation of flows in microchannels. In many practically important applications, there is a need to cool the working surfaces of these devices. Apparently, one of the earliest attempts to utilize microflows for the cooling of silicon integrated circuits was conducted by Tuckerman and Pease [59]. Today, microflows are widely used in a broad variety of thermophysical applications. The relevant literature can be found in specialized reviews. The use of nanofluids for this purpose also appears to be quite logical. Experimental studies of the heat exchange characteristics of nanofluid microflows have been actively pursued for over fifteen years. This resulted in the production of a considerable number of pertinent publications. Of course, it is impossible to make a sufficiently complete review of these publications within the framework of this section, but the readers can get acquainted with them using the reviews [1, 4, 7, 11, 26]. It should be note that there are not many actual experimental studies on investigating heat exchange and thermal–hydraulic efficiency of nanofluids in microchannels. This is due to the significant technical difficulties encountered in conducting the relevant experiments. However, a huge number of publications have appeared on numerical simulation of nanofluid flows in microchannels. These works will not be discussed here at all. One of the reasons for this is that the majority of these numerical simulations have significant shortcomings. For these reasons, discussing the physical outcomes of these numerical simulations doesn't make sense. The rationale behind this conclusion will be presented in detail in the subsequent chapter. With regard to the experimental data that are already known, the principal results obtained are presented below in a concise form.

6.3.1 Brief Review of Experimental Data

One of the earliest studies of the heat transfer of nanofluids in rectangular microchannels was conducted by Jung et al. [24]. The channel cross-section was $50 \times 50\ \mu m^2$. The base fluids employed were either water or a water-based solution of ethylene glycol (50:50). The nanoparticles were 170 nm in size and their volume concentration ranged from 0.6 to 1.8%. The highest excess heat transfer coefficient of 32% was observed at the highest volume concentration of nanoparticles. The enchantment of heat transfer coefficient increases with increasing the Reynolds number. Concomitantly, the pressure loss due to friction of the nanofluids exhibited just slight increase.

Later, the convective heat transfer of nanofluid flow in a wide rectangular microchannel heat sink with a height of 580 μm was investigated both experimentally and numerically [25]. A water-based nanofluid comprising 40 nm aluminum oxide particles was employed. The volume concentration of nanoparticles was 0.1 and 0.2%. It is regrettable that no direct measurement data on the local or average heat transfer coefficients are presented. The authors report only the dependence of the average Nusselt number on the Reynolds number. However, neither the viscosity nor the thermal conductivity of the nanofluids was measured. Instead, the authors employed Brinkman's formula and a correlation, whose applicability in this particular case is not evident.

Nanofluids with aluminum oxide particles 5 nm in size have also been used in experiments [61] to study heat transfer in a microchannel heat exchanger. In this study, however, the nanofluid was prepared based on water with the addition of PVP as a surfactant. The nanofluids were studied at three particle concentrations of 0.1, 0.5 and 1%. The maximum excess heat transfer was recorded at a volume concentration of nanoparticles of 1% and was about 30%, however, at a much smaller concentration of 0.1% the excess heat transfer was more than 10%. At the same time, the hydraulic resistance of nanofluid flows was increasing significantly, although this increase exhibited a slight decline as the Reynolds number increased.

The heat transfer of water-based nanofluids with aluminum (20 nm) and copper (40 nm) oxide particles was investigated in a microchannel heat exchanger comprising 17 microchannels of rectangular cross-section with dimensions of $400 \times 560\ \mu m^2$ [50]. All experiments were conducted in laminar flow regime at a constant heat flux. The excess heat transfer of nanofluids containing 0.2% CuO and 1% Al_2O_3

was found to be 27 and 49%, respectively. In this regard, the nanofluid with copper oxide particles exhibited superior efficacy, as it could be employed at lower concentrations. Moreover, it was demonstrated that the heat transfer coefficient increases with increasing nanoparticle concentration. However, the Nusselt number demonstrated a non-monotonic dependence on the Reynolds number. It should be noted that the viscosity and heat conductivity coefficients were not measured in this paper. Consequently, the conclusions drawn from this analysis should be regarded with a certain caution.

The heat transfer of two water-based nanofluids with aluminum and tin oxide particles in a microchannel heat exchanger was investigated by Putra et al. [51]. The volume concentration of the former particles varied from 1 to 5% and the concentration of latter particles was 1%. The hydraulic diameter of the heat exchanger was 852 μm. In this study, both the viscosity and thermal conductivity of the nanofluids under investigation were measured in advance. The excess heat transfer coefficient of the first nanofluid at 5% concentration was 13%, and that of the second nanofluid at 1% particle concentration was 14%. The pressure drop in both cases increased with increasing Reynolds number.

Heat transfer of water-based nanofluid with silver particles in a copper heat exchanger with seven microchannels of rectangular cross-section was investigated by Sarafraz et al. [55]. The nanofluid was prepared by adding a surfactant (nonylphenol ethoxylate) to water. The weight concentration of nanoparticles was 0.01, 0.05, and 0.1%. Unfortunately, the size of the nanoparticles is unknown. The pressure drop, drag coefficient, and heat transfer coefficient were measured. The measurements have demonstrated that the use of nanofluid did not result in a significant increase in pressure drop and drag coefficient in the system. Nevertheless, as the flow rate and nanoparticle weight concentration increased, the heat transfer coefficient demonstrated a notable enhancement. At a nanoparticle weight concentration of 0.1%, the heat transfer coefficient exhibited a 47% increase.

The thermal properties of a heat exchanger with 17 microchannels of rectangular cross-section ($400 \times 560\ \mu m^2$) were investigated in Chabi et al. [10]. A water-based nanofluid comprising 40 nm copper oxide particles was employed. The volume concentration of nanoparticles was 0.1 and 0.2%. The viscosity and thermal conductivity of the nanofluids were not directly measured, but rather estimated using the Einstein (3.29) and Hamilton–Grosser (4.9) formulas, respectively. The maximum enhancement of the average heat transfer coefficient was 40% at a maximum particle concentration and Reynolds number of 1150. Nevertheless, it is indicated that the heat transfer coefficient exhibited a decreasing trend with further increase in the Reynolds number. It is important to note that the aforementioned conclusions, as well as the dependences of pressure drop and Nusselt number on Reynolds number, should be treated with caution, given that the viscosity and thermal conductivity coefficients were not adequately evaluated.

The efficiency of convective heat transfer in a microchannel heat exchanger comprising 25 rectangular microchannels with rectangular cross-sections of $283 \times 800\ \mu m^2$ was investigated in Ho et al. [17]. A water-based nanofluid comprising 33 nm aluminum oxide particles was employed. The volume concentration was 1 and 2%. In addition to measuring the heat transfer coefficient, the viscosity of the nanofluids was also measured. It was shown that the use of nanofluids for cooling purposes is considerably more efficient than water cooling due to a notable reduction in thermal resistance, particularly at high pumping power. Despite the marked increase in dynamic viscosity, only a slight increase in the drag coefficient was observed during heat removal by nanofluid. The relative heat transfer coefficient exhibited a notable enhancement, reaching 64% for the 2% nanofluid (at Re of approximately 1000) and 70% for the 1% nanofluid (at Re exceeding 1500). It is important to note that, in this context, the relative heat transfer coefficient of the 2% nanofluid at a given Reynolds number is always higher (by 10–25%) than that of the 1% nanofluid.

In contrast to the majority of publications in this field, the paper by Manay and Mandev [33] investigated the heat transfer of a water-based nanofluid with aluminum oxide particles in a cylindrical microchannel with a diameter of 500 μm. The nanoparticles were 10–20 nm in size and their volume concentration was 0.2 and 0.4%. One advantage of this study is that the viscosity and thermal conductivity coefficients of nanofluids were also measured. As it was found, the Nusselt numbers increased almost linearly with increasing Reynolds numbers. The heat transfer coefficient increased in proportion to the nanoparticle concentration and the corresponding increase of the thermal conductivity. The Nusselt number was observed to increase by 14% when nanofluid with a particle concentration of 0.2% was used, and by approximately 32%, when nanofluid with a concentration of 0.4% was employed.

A very interesting and comprehensive study was done in Topuz et al. [58]. Heat transfer in stainless steel cylindrical channels with diameters of 400, 750, and 1000 μm was studied. Three distinct types of nanofluids were utilized. One was prepared by addition of aluminum oxide particles (13 nm) into water, while the other two nanofluids were water-based and contained titanium (10–25 nm) and zinc (18 nm) oxide particles, as well as surfactant (SDS). Moreover, the weight concentration of surfactant was quite high (15%). The volume concentrations of nanoparticles were 0.5, 0.7, and 1.0%. The most optimal thermal characteristics were observed for the nanofluid with aluminum oxide particles at a concentration of 1%. The excess heat transfer coefficient was 15.3%, and no discernible increase in pressure drop was observed in this case.

A study of the heat transfer of water-based nanofluids with aluminum oxide particles in a cylindrical microchannel with an internal diameter of 500 μm was conducted in Lodhi et al. [32]. The distinctive feature of this study is the size of the particles in the nanofluid, which were on average larger than 450 nm. The volume concentration of the nanofluid ranged from 1 to 3%. The maximum excess heat transfer coefficient of 24.5% was reached at the highest concentration of nanoparticles. The friction coefficient increased by 22.7% compared to that for water. As a result, the authors have shown that the thermal–hydraulic efficiency factor of nanofluid was greater than unity.

This brief overview allows drawing several conclusions. In fact, the heat transfer phenomenon has been extensively studied in heat exchangers consisting of a large number of microchannels. Therefore, the resulting characteristics of both the heat transfer coefficient and pressure drop represent integral properties of flows in these heat exchangers. The study of heat transfer in individual microchannels has been conducted in only a few papers, and the channel sizes were quite large. Furthermore, it is important to note that the nanoparticle concentrations in the fluids used did not exceed 1–2%. This has resulted in the observation that the pressure losses in nanofluid flows were, in most cases, slightly greater than those observed in base fluid flows. However, in some cases, non-monotonic character in the growth of the heat transfer coefficient with increasing nanoparticle concentration was observed. Rather significant discrepancies were also observed with respect to the heat transfer coefficients. These conclusions suggest the necessity for further experimental studies. One of such experiments will be described in the next two sections.

6.3.2 Experimental Setup

The experimental study of nanofluid microflows is not a trivial task, since appropriate instrumentation is required. A special experimental setup was designed for conducting experiments in microchannels. The heating of the test section (microchannel) was achieved by applying an electric current through the microchannel. To exclude the effect of external convective heat transfer on the experimental results, the test section was placed in a vacuum chamber (see Fig. 6.13). The chamber, with a volume of 1.3 dm^3, was made of stainless steel. The air in the chamber was pumped by a 2NVR-5DM vacuum pump to a pressure of 13.32 mm Hg. The side wall of the chamber was provided with holes for pumping air, measuring vacuum, supplying and discharging the working fluid, and was also equipped with six vacuum connectors of seven contacts each.

Fig. 6.13 Experimental setup. Vacuum chamber with test section

In order to obtain reliable data on the local heat transfer coefficient, it was necessary to determine a velocity profile corresponding to the developed flow at the inlet of the heated microchannel. To this end, the test section was constructed as a composite of a preparatory section and a heated section. The heated section was a cylindrical stainless steel channel with an outer diameter of 550 μm and an inner diameter of 358 μm. The length of the heated section was 25.7 mm. The preparatory section was comprised of two parts: a copper tube with an inner diameter of 0.9 mm and a length of 20 mm, and a stainless steel tube with a length of 27 mm and the same diameter as the working heated section. The tubes were joined using photopolymer in order to achieve a smooth mating profile of the inner diameters of the tubes. The pertinent technology is described in the paper [3].

To ascertain the temperature of the fluid at the inlet to the heated test section, a thermocouple was installed at the connection point and a hole was made in order to measure the pressure which was necessary to determine the hydraulic resistance coefficient of the channel. A chromel–alumel (L-type) thermocouple with a thickness of approximately 20 μm and a planar size of $80 \times 80\ \mu m^2$ was utilized. The temperature of the heated test section was measured by means of eleven thermocouples, which were positioned along its entire length. The thermocouples were affixed to the heated section with varnish, in addition were fixed with thermally conductive glue. This methodology ensured a reliable thermal contact between the thermocouples and the channel surface, while maintaining an electrical isolation. One more thermocouple with a thickness of 20 μm and a planar size of $60 \times 300\ \mu m^2$ was installed at a distance of approximately one caliber from the cut of the steel channel in its cross-section. This thermocouple was employed to ascertain the average temperature at the outlet of the heated test section.

The test section was subjected to heating by means of an electric current passing through the microchannel. A voltage was provided by a mains voltage and a system of step-down transformers with a rectifier bridge. The application of electric current ensured a constant heat flux from the channel wall into the liquid. The electrical resistance of the heated section was 0.5 ohms. The supplied power varied within a range from 0.2 to 0.5 W.

When the channel was heated, a thermal boundary layer was formed near its wall and its thickness increased downstream. The length of the section where the temperature profile in the channel cross-section was formed was determined by the empirical formula [28]: $L = 0.05\mathrm{Re}d_c Pr$, where Pr is the Prandtl number. In the experiments, this length varied from 20 to 60 mm, i.e., in almost all cases exceeded the length of the heated section. Consequently, the experiments were conducted under conditions of a steady flow velocity profile and a developing thermal layer.

The setup was verified by measuring the heat transfer coefficient in the flow of distilled deionized water. The fluid was pumped with a Gilson 305 piston pump. The fluid flow rate was varied between 3 and 10 ml/min, corresponding to Reynolds numbers ranging from 180 to 850. Consequently, in all instances, a laminar flow regime was observed. A series of experiments was conducted at varying Reynolds numbers within the specified range. The results obtained from these experiments corresponded to the data typically observed in experiments with macrochannels.

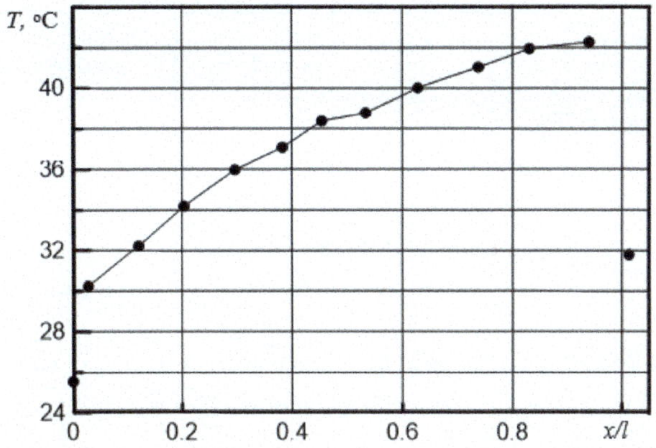

Fig. 6.14 Temperature distribution along the channel

Figure 6.14 illustrates the typical temperature distribution along the channel at a fluid flow rate of 7 ml/min. The terminal points indicate the liquid temperatures at the channel inlet and outlet.

When studying heat transfer, the local heat transfer coefficient was measured [9]

$$\alpha_c = \left\{ \frac{\pi dL\left[T_w(x) - T_f(x)\right]}{C_p G(T_o - T_i)} - h_w \right\}^{-1}, \tag{6.3}$$

where

$$h_w = \frac{d_e^2 d_i}{16\left(d_e^2 - d_i^2\right)\lambda_w}\left[2ln\left(\frac{d_e}{d_i}\right) + \left(\frac{d_e}{d_i}\right)^2 - 1\right]$$

here $h_{w\ w}$ is the heat flux across the wall due to its thermal conductivity, $T_w(x)$, $T_f(x)$ are the wall and liquid temperatures at a distance x from the inlet to the heated channel, T_o, T_i are the liquid temperatures at the channel inlet and outlet, respectively, d_e, d_i are the external and internal diameters of the heated channel, λ_w is the thermal conductivity coefficient of the channel wall.

Figure 6.15 illustrates the distribution of local heat transfer coefficient along the channel at different water flow rates. Here $G = 5.69$ ml/min (a) and $G = 7.78$ ml/min (b), which corresponds to Reynolds numbers of 460 and 620, respectively. The labels correspond to the experimental data obtained by the formula (6.3). The solid line, depicted for comparison, corresponds to the calculation by semi-empirical correlation describing the heat transfer coefficient in ordinary macroscopic flows [49]

$$\alpha(x) = Nu\frac{\lambda}{d_i},\ Nu(x) = 4.36 + 1.31\left(\frac{x}{d_i Pe}\right)^{-1/3}\left(\frac{\mu_w}{\mu_f}\right)^{-1/6} exp\left(-13\sqrt{\frac{x}{d_i Pe}}\right), \tag{6.4}$$

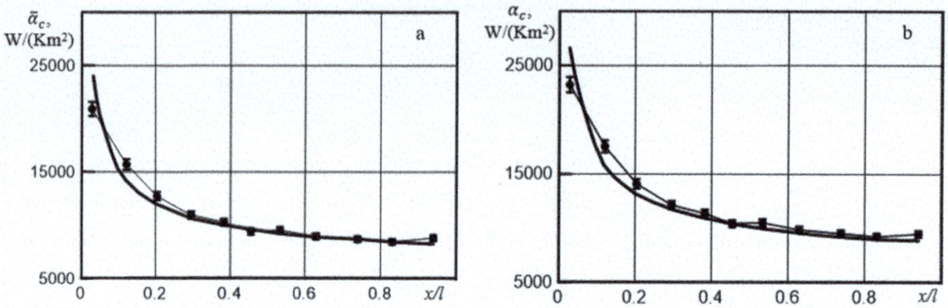

Fig. 6.15 Distribution of local heat transfer coefficient of water at different flow rates

Here *Pe* and *Nu* are the Peclet and Nusselt numbers, μ_w is the water viscosity coefficient at the wall temperature, μ_f is the water viscosity coefficient corresponding to the temperature at the channel axis. As it is seen, within the measurement error and accuracy of correlation (6.4) the experimental data well agree with each other.

6.3.3 Heat Transfer Coefficient of Nanofluids

This section presents a discussion on the heat transfer coefficient of a distilled water-based nanofluid with spherical silicon oxide particles with a diameter of 25 nm at a volume concentration ranging from 0.25 to 1%. A typical example of a comparison between the local heat transfer coefficient for water and the two nanofluids is shown in Fig. 6.16. The square symbols correspond to water, while the triangles represent the nanofluids. The triangles with the top up represent nanoparticle concentrations of 0.5%, while those with the top down represent a concentration of 1%. The presented data were obtained at a water flow rate of 9.95 ml/min. It is practically impossible to set the same flow rate for nanofluids. Therefore, for nanofluids with particle concentrations of 0.5 and 1%, the flow rates were 10.0 and 9.3 ml/min, respectively. In both cases, the heat transfer coefficient of the nanofluid exceeded the corresponding values for water (Fig. 6.16a). The excess heat transfer coefficient was higher the higher the volume concentration of nanoparticles.

An understanding of the actual excess heat transfer coefficient is provided by the dependence of the relative local heat transfer coefficient $\alpha_r = \alpha_c/\alpha_{wc}$, shown in Fig. 6.16b, where α_{cw} is the local heat transfer coefficient of water. The data presented here correspond to the same case illustrated in Fig. 6.16a. The notations employed are identical to those utilized in the aforementioned figure. The enhancement of heat transfer at a nanoparticle concentration of 0.5% is approximately 20%, while at a concentration of 1%, it is approximately 54%.

In conclusion, it should be noted that the data presented above indicate a notable enhancement in the heat transfer coefficient of nanofluids in comparison to the base fluid. Even a 1% nanofluid enhances the heat transfer coefficient by approximately

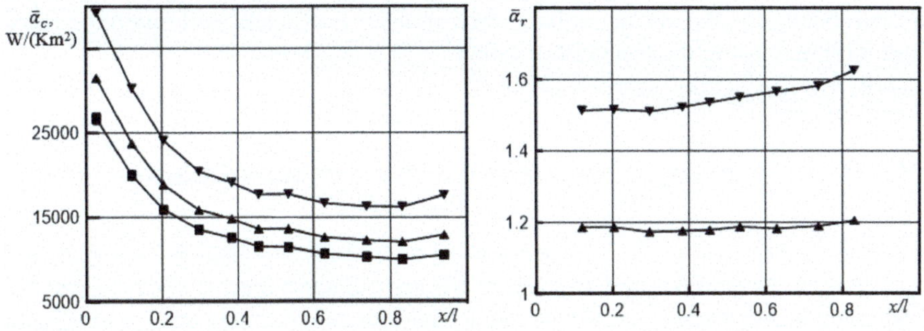

Fig. 6.16 Comparison of local (**a**) and relative (**b**) heat transfer coefficients for water and two water-based nanofluids with silicon oxide particles

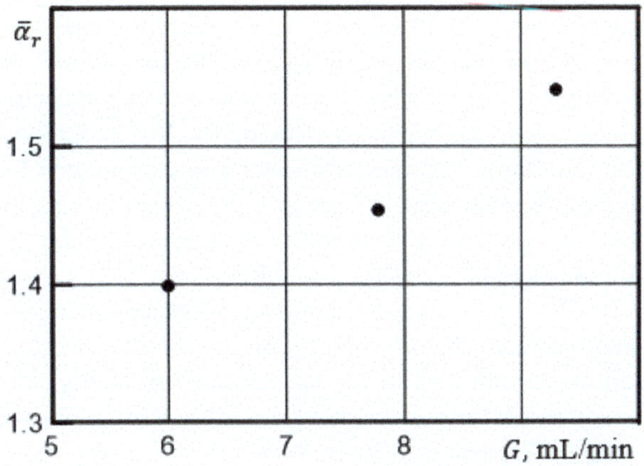

Fig. 6.17 Dependence of the average relative heat transfer coefficient on the flow rate of nanofluid at a concentration of 1%

60% in comparison to water. At the low nanoparticle concentrations considered, the excess heat transfer coefficient is proportional to the particle concentration. Therefore, the enhancement of heat transfer coefficient will increase in direct proportion to the nanoparticle concentration.

The heat transfer coefficient enhances when the flow rate increases. This is demonstrated in Fig. 6.17, which presents the data on the relative average heat transfer coefficient $\overline{\alpha}_r$ of 1% nanofluid as a function of the flow rate (the average heat transfer coefficient $\overline{\alpha}_r$ is obtained by conventional averaging of local values). In the range of flow rates under consideration, the growth of this coefficient is nearly proportional to the flow rate. It can be reasonably deduced that a twofold enhancement in the average relative heat transfer coefficient can be expected at a flow rate of approximately 10.3 ml/min, while a threefold enhancement can be expected at a flow rate of approximately 12.6 ml/min. It should be noted that in both cases, the flow remains laminar. Naturally, the average relative heat transfer coefficient will retain the same trend with increasing the Reynolds number.

6.4 Heat Transfer of Nanofluids in the Turbulent Flows

In the previous sections, the heat transfer of nanofluids in channels was analyzed with regard to the laminar flow regime. In all cases, enhancement of the heat transfer coefficient was observed and it was noted that this enhancement was proportional to the thermal conductivity coefficient of the nanofluid. The effect of nanoparticle concentration, size, and material on the enhancement of the heat transfer coefficient was also investigated. However, the situation may change significantly in the turbulent flow regime. This section addresses the heat transfer in turbulent flows.

The first investigation into the heat transfer of nanofluids in a turbulent regime was conducted by Pak and Cho [45]. In this study, the heat transfer of water-based nanofluids containing Al_2O_3 and TiO_2 particles in a horizontal circular tube at a constant wall heat flux was examined experimentally. It was demonstrated that the Nusselt number of nanofluids increased with increasing particle volume concentration and Reynolds number. However, at high particle concentrations, the heat transfer coefficient of nanofluids was lower than that of the carrier fluid.

Since then, a considerable number of publications have appeared (see, for example, [8, 12, 16, 21, 29, 30, 43, 57] and the references therein). The majority of studies have demonstrated that the incorporation of nanoparticles, even in relatively modest quantities (from 0.05 to 2% by volume), markedly enhanced the heat transfer coefficient with an improvement of 25–60%. Thus, when using a water-based nanofluid with TiO_2 particles, Duangthongsuk and Wongwises [12] observed a significant (over 30%) enhancement in heat transfer at a particle volume concentration of 1%, however, at a greater concentration of 2%, a decrease in heat transfer was observed comparing to water. In contrast, in the study by Williams et al. [60], which examined the heat transfer of two water-based nanofluids containing Al_2O_3 and ZrO_2 particles, it was found that the heat transfer did not exhibit a significant enhancement compared to that of the base fluid.

Moreover, in addition to the effect of nanoparticle concentration, attempts were made to ascertain the effect of nanoparticle size on heat transfer [16, 35, 43, 57]. However, the data obtained turned out to be quite contradictory. For example, in Nguyen et al. [43] it is indicated that the heat transfer diminishes with an increase in nanoparticle size. Conversely, in Timofeeva et al. [57], just the opposite conclusion was stated. The list of such rather contradictory conclusions can be continued.

Experimental data on the heat transfer of water-based nanofluids with silicon and aluminum oxide particles in turbulent flow regime are presented below. The average particle diameter ranged from 10 to 100 nm and the volume concentration varied from 0 to 2% [14, 38]. In the experiments, the average and local heat transfer coefficients were measured at the channel walls, as well as the pressure drop along the channel. Figure 6.18a presents a typical dependence of the average heat transfer coefficient

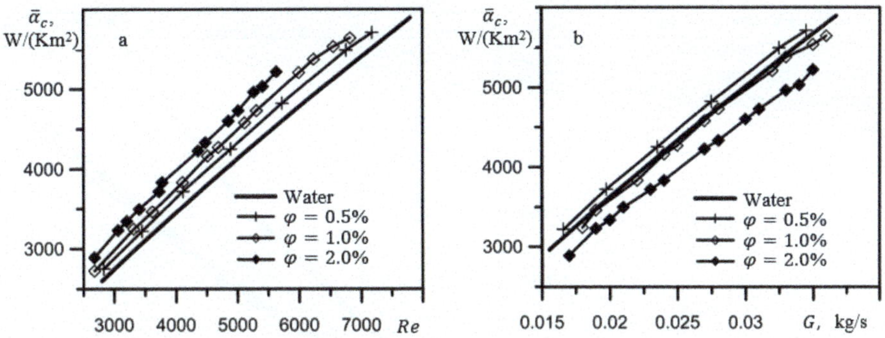

Fig. 6.18 Dependence of the average heat transfer coefficient of water-based nanofluid at different concentrations of 25-nm SiO_2 nanoparticles on the Reynolds number (**a**) and flow rate (**b**)

on the Reynolds number for a nanofluid comprising silicon oxide particles with an average particle size of 25 nm. The maximum particle volume concentration was 2%. In all cases, the heat transfer coefficient of nanofluid is found to be greater than that of water. The heat transfer coefficient increases with an increase in nanoparticle concentration. The degree of this enhancement increases, as in the laminar regime, almost proportionally to the volume concentration of nanoparticles. The heat transfer coefficient for a 2% nanofluid is more than 15% higher than that for water.

From a practical point of view, the power required to pump the fluid through a channel is an important factor. It was therefore interesting to analyze the dependence of the average heat transfer coefficient on the fluid flow rate. Such dependence is shown in Fig. 6.18 b. The average heat transfer coefficient at a fixed flow rate also depends on the nanoparticle concentration. However in this case, at a given flow rate, the heat transfer coefficient for a 2% nanofluid is about 10% lower than that for water. As the concentration of nanoparticles decreases, the heat transfer coefficient tends monotonically towards the value corresponding to pure water. At the lowest nanoparticle concentration of 0.5%, the heat transfer coefficient exceeds the corresponding value for water. The reason for this behavior is obvious. The viscosity coefficient of the nanofluid is significantly higher than that of the base fluid. Qualitatively, the effect is easy to understand using the formula $Nu = 0.021\mathrm{Re}^{0.8}Pr^{0.43}$ [37]. This formula is valid for nanofluids with accuracy of about 10%. For a given flow rate, the heat transfer coefficient is proportional to $\lambda^{0.57}/\mu^{0.37}$ ratio. It is evident that if the enhancement in the thermal conductivity of the heat carrier due to nanoparticles is considerably less than the increase in its viscosity, there may be a reduction in the heat transfer coefficient.

This is the fundamental distinction between turbulent and laminar heat transfer of nanofluids. In steady laminar flow, the heat transfer coefficient is proportional to the thermal conductivity of the fluid and does not depend on its viscosity. Since the presence of nanoparticles always increases the thermal conductivity of the fluid, in the laminar regime heat transfer always enhances regardless of how much the viscosity has increased. Since the viscosity coefficient of the nanofluid increases

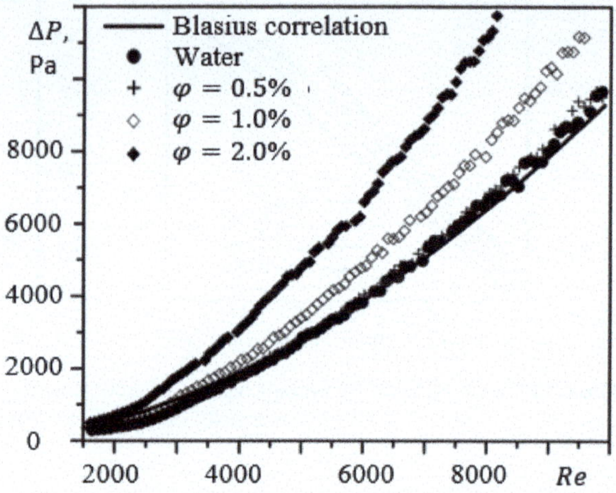

Fig. 6.19 Dependence of pressure drop on Reynolds number at different concentration of SiO_2 nanoparticles 25 nm in size

rapidly with increasing nanoparticle concentration, the pressure drop in the channel for a given Reynolds number should be higher than that for the base fluid. It is easy to understand what this dependence should be. The density of the nanofluid is a linear function of the volume concentration of the nanoparticles. On the other hand, the viscosity coefficient at the low concentrations considered is described by a quadratic dependence of the form (3.18). The Reynolds number is therefore a non-linear function of the concentration. Therefore, the ratio $\Delta P/\mathrm{Re}$ is also a non-linear function of the nanoparticle concentration and grows rapidly as it increases. This is illustrated in Fig. 6.19.

The next question, the answer to which had to be obtained experimentally, was whether the characteristics of turbulent heat transfer of a nanofluid depend on the size of the nanoparticles. Since the nanofluid transport coefficients are a function of particle size, such a dependence should be expected. Experiments have shown that such dependency really exists.

Figure 6.20a shows the dependence of the average heat transfer coefficient on the Reynolds number for water-based nanofluids with silicon oxide particles of four sizes, namely 10, 16, 25, and 100 nm. In all cases the volume concentration was 2%. The heat transfer coefficient for a given Reynolds number was maximum for the nanofluid with a particle size of 25 nm. In this case it is also about 15–20% higher than the corresponding values for water. For the nanofluid with 100 nm particles, the excess heat transfer coefficient is about 10%. On the other hand, the heat transfer coefficient with 10 and 16 nm particles is lower than that of water. The dependence of nanofluid properties on nanoparticle size is also evident in the pressure drop behavior. A comparison of the pressure drop data for water and two nanofluids with 25 and 100 nm silicon oxide particles at 1% volume concentration is shown in Fig. 6.20 b. As is seen, the pressure drop naturally decreases with increasing nanoparticle size.

As shown in Chaps. 3 and 4, nanofluid transport coefficients depend not only on the size of the nanoparticles, but also on their material. Therefore, the heat transfer

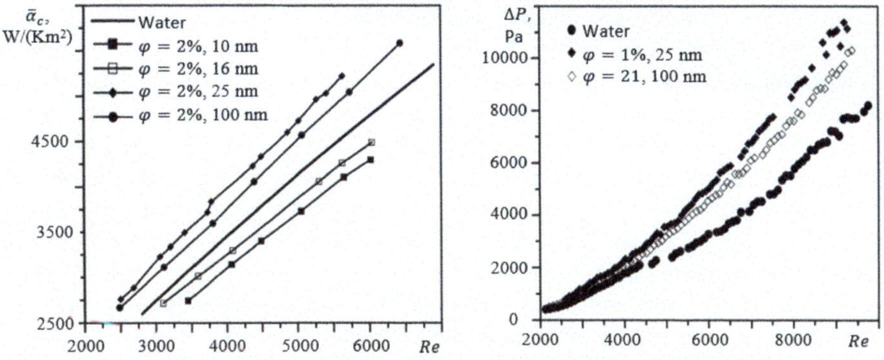

Fig. 6.20 Heat transfer coefficient (**a**) and pressure drop (**b**) of water and water-based nanofluids with SiO_2 particles as a function of Re number

coefficient will also depend on the material of the nanoparticles. To understand what this dependence is like, the heat transfer coefficients were measured for two nanofluids with the same concentration and size of silicon and aluminum oxide nanoparticles. For the data shown in Fig. 6.21, the concentration of nanoparticles was 2% and their size was 100 nm. The data presented demonstrate a clear impact of the nanoparticle material on the heat transfer coefficient. For all Reynolds numbers, the excess heat transfer coefficient of the nanofluid with aluminum oxide particles is almost twice that of the nanofluid with silicon oxide particles. On the contrary, the pressure drop along the channel in the nanofluid with silicon oxide particles is much greater than that of the nanofluid with aluminum oxide particles.

The heat transfer measurements presented show that, in general, the heat transfer coefficient of the nanofluid is significantly higher than that of the base fluid. At the same time, it is shown that the overall effect depends on the concentration of nanoparticles, their size, and the material. The heat transfer coefficient depends on

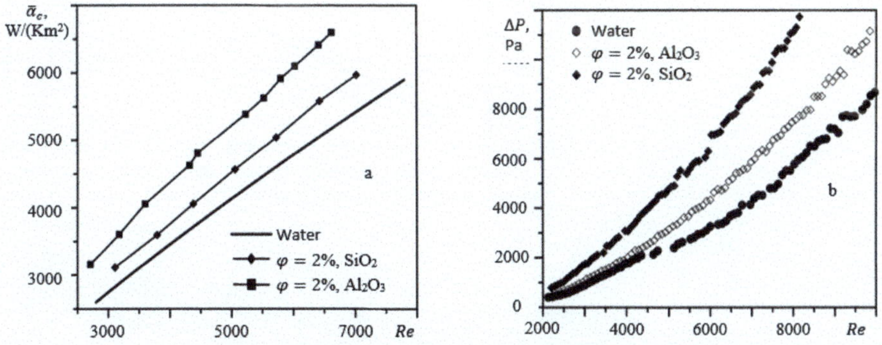

Fig. 6.21 Dependence of the average heat transfer coefficient (a) and pressure drop (b) of water-based nanofluids with copper and silicon oxide nanoparticles on Reynolds number

both the thermal conductivity and the viscosity of the fluid, just as it does for conventional fluids. Therefore, there may be situations where the heat transfer coefficient of the nanofluid is lower than that of the base fluid. Hence, the intensification of heat transfer by using nanofluids in a turbulent regime is a non-trivial task. The beneficial impact of heat transfer enhancement depends on the ratio between viscosity and thermal conductivity of the nanofluid. This explains the extensive dispersion and contradictions observed in the existing literature data, wherein the influence of the aforementioned factors (nanoparticle size and their material etc.) has not been actually considered or taken into account.

Since both viscosity and thermal conductivity are significantly dependent on temperature, it follows that the heat transfer coefficient should also be dependent on the temperature of the heat carrier. Nevertheless, the character of this dependency is not obvious. The measurements carried out show that such a dependence does indeed exist. Figure 6.22 shows the dependences of the heat transfer coefficient on the Reynolds number for a nanofluid comprising silicon oxide particles 100 nm in size with a volume concentration of 2% at two different temperatures at the inlet to the heated section. It can be observed that the heat transfer coefficient enhances with an increase in the temperature of the heat carrier. An increase in the temperature of the heat carrier at the inlet from 15 to 25 °C leads to an increase in the heat transfer coefficient by almost 30%. As the temperature rises, the heat conduction coefficient increases, while viscosity decreases. Consequently, the use of nanofluids is more efficient at higher heat carrier temperatures.

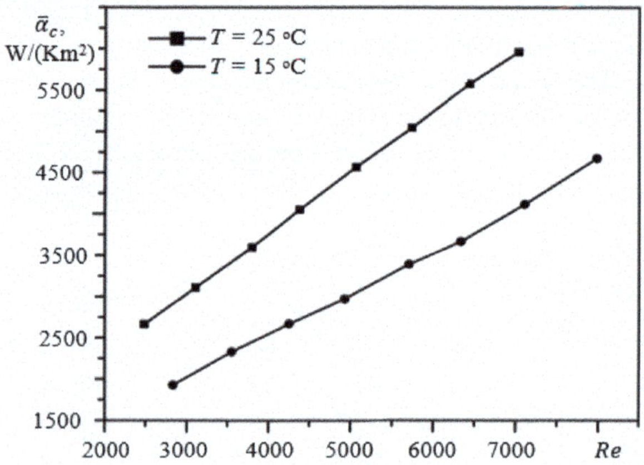

Fig. 6.22 The dependence of the heat transfer coefficient on the Reynolds number at various inlet temperatures of the heat carrier

6.5 Heat Transfer of Nanofluids with CNTs in Laminar Flows

Heat transfer in nanofluids with CNTs began to be actively studied about a decade later than with conventional nanofluids. Three-quarters of the publications on this topic are devoted to nanofluids with MWCNTs. Nanofluids with SWCNTs have been studied much less. To date, dozens of papers devoted to the experimental study of heat transfer of nanofluids with CNTs have been published. The data obtained are highly dissimilar. This is not surprising, since in almost all studies the experimental conditions vary greatly. First of all, the nanofluids used in these studies differed in both their base fluids and the type of CNTs used. In a number of works, different surfactants were usually used or CNTs are functionalized. MWCNTs had different internal and external diameters, as well as varying numbers of graphene layers. In addition, experiments were conducted on heat exchange devices made of different materials. Finally, the construction of these devices also turned out to be very different. In such conditions it is difficult to compare the obtained data. Nevertheless, several recent reviews provide the ability to navigate this vast array of dissimilar data with a high degree of reliability [2, 34, 46, 48].

It is worth noting the typical features of the aforementioned published papers. The weight concentrations of CNTs were typically fractions of a percent and did not exceed 1%. The majority of studies on heat transfer were conducted in the laminar flow regime. In almost all cases, a significant excess heat transfer coefficient of the nanofluids was observed with values ranging from 10 to 300% in comparison to the base fluid (see [2, 18, 23, 27, 31, 46, 48] and the references therein). The following examples also fall within specified range [54]. At the same time, they illustrate the problems that arise here. The data presented were obtained employing the setup described in Sect. 6.1.

The high thermal conductivity of nanofluids with SWCNTs demonstrated in Chap. 4 indicates that their use in various heat transfer devices is highly promising. As previously shown, this is quite evident in the laminar flow regime, where the heat transfer of nanofluids is primarily determined by the thermal conductivity of the working fluid. It is clear that the use of nanofluids with SWCNTs presents a number of specific features that require conducting appropriate experiments. This section presents the results of heat transfer measurements of an IPA-based nanofluid with SWCNTs flowing in a cylindrical channel. A comparison of the dependence of the average heat transfer coefficient on the mass flow rate of the IPA-based nanofluid at 0.1 and 0.25% SWCNT concentration is presented in Fig. 6.23a. The use of nanofluids significantly increases the average value of the heat transfer coefficient. At a weight concentration of 0.25% SWCNT, the heat transfer coefficient increased by almost one and a half times. Such enhancements of heat transfer coefficient using nanofluids with metal oxide particles are practically unattainable. Similarly, achieving this effect for nanofluids with metal nanoparticles is also challenging, as it would necessitate

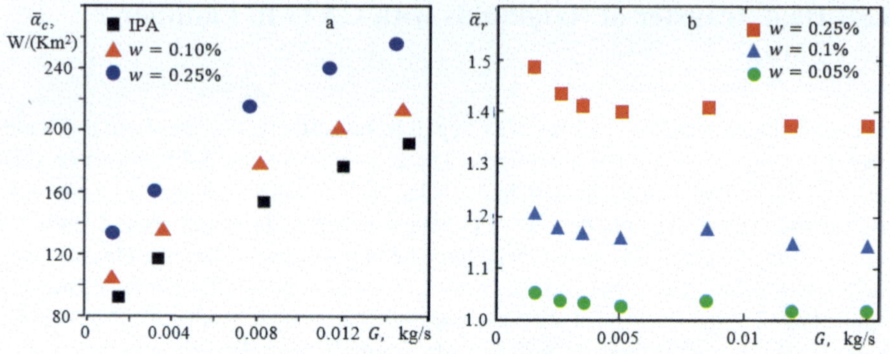

Fig. 6.23 Comparison of the dependence of the average and relative heat transfer coefficients on the mass flow rate for IPA and IPA-based nanofluids with SWCNTs

exceedingly high concentrations of particles (tens of percent by weight). Consequently, if the objective is to achieve the greatest possible heat transfer coefficient in convective heat transfer, the use of SWCNTs is unrivaled among other nanoparticles.

It is also of interest to examine the behavior of the relative heat transfer coefficient with increasing the flow rate. The corresponding dependence is illustrated in Fig. 6.23b. It is important to highlight that for a fixed weight concentration of SWCNTs, the relative heat transfer coefficient exhibits an increasing trend with decreasing flow rate. Thus, at a weight concentration of 0.25% and a mass flow rate of 0.0025 kg/s, the relative heat transfer coefficient is 1.47, while at a mass flow rate of 0.015 kg/s, the relative heat transfer coefficient is 1.38 (see Fig. 6.23b).

Information on the dependence of the heat transfer coefficient on the Reynolds number also turns out to be very useful for analysis. This dependence is presented in Fig. 6.24. It should be noted that the heat transfer coefficients for IPA and the corresponding nanofluid with SWCNTs are not directly comparable. The heat transfer coefficient of nanofluid at a fixed Reynolds number is significantly higher than that of IPA. Here it should be kept in mind that the IPA-based nanofluid is non-Newtonian and the Reynolds number values for nanofluids are given in Fig. 6.24 at fixed shear rates.

The rheological and viscosity properties of the IPA-based nanofluid with SWCNTs were investigated in Chap. 3. It was demonstrated that this nanofluid exhibits non-Newtonian behavior at all concentrations of SWCNTs. In all cases, this nanofluid is viscoplastic and its rheology is well described by the Herschel–Bulkley model. The pressure drop along the channel, as a function of mass flow rate, is presented in Fig. 6.25. As illustrated in figure, the pressure drop during the pumping of nanofluid with SWCNTs is considerably higher than that observed for pure IPA. From a practical standpoint, it is recommended that nanofluids with SWCNTs be used at a higher inlet temperature in order to enhance their thermal–hydraulic efficiency. In this instance, the viscosity of the nanofluid will be markedly reduced, while the thermal conductivity coefficient will be significantly enhanced.

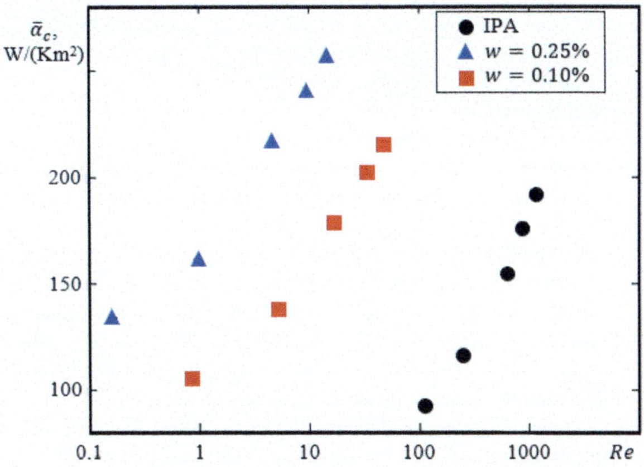

Fig. 6.24 Dependence of the average heat transfer coefficient on the Reynolds number

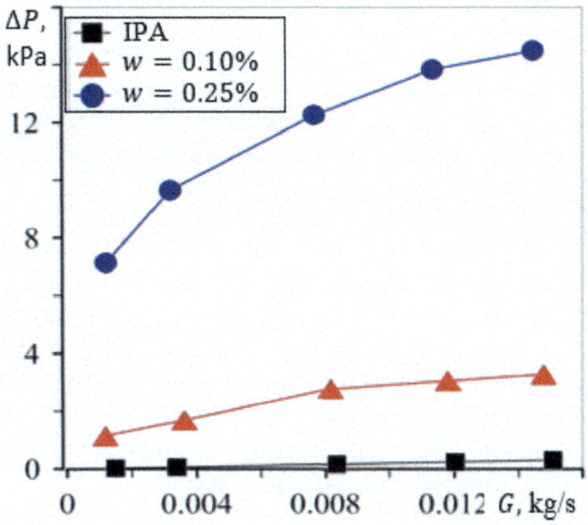

Fig. 6.25 Comparison of pressure drop dependences on mass flow rate for IPA and IPA-based nanofluids with SWCNTs

Thus, if the task is to maximize the value of the heat transfer coefficient during convective heat transfer, the use of SWCNTs has no competitors among other nanoparticles. The second circumstance pertains to the use of nanofluids with SWCNTs at higher inlet temperatures. In this case, the viscosity of the nanofluid will undergo a notable reduction, whereas the thermal conductivity coefficient will experience an opposing increase.

References

1. Alfaryjat A, Miron L, Pop H et al (2019) Experimental investigation of thermal and pressure performance in computer cooling systems using different types of nanofluids. Nanomaterials 9:1231
2. Ali N, Bahman AM, Aljuwayhel NF et al (2021) Carbon-based nanofluids and their advances towards heat transfer applications—a review. Nanomaterials 11:1628

3. Aniskin VM, Rudyak VYa (2016) Experimental study of heat transfer of nanofluid in a microchannel. Vestnik NSU Phys 11(2):5–11
4. Alkasmoul F, Asaker M, Almogbel A et al (2022) Combined effect of thermal and hydraulic performance of different nanofluids on their cooling efficiency in microchannel heat sink. Case Studies Thermal Eng 30(9):101776
5. Almurtaji S, Ali N, Teixeira JA et al (2020) On the role of nanofluids in thermal-hydraulic performance of heat exchangers—A review. Nanomaterials 10:734
6. Amin A (2018) A guideline towards easing the decision-making process in selecting an effective nanofluid as a heat transfer fluid. Energy Conversion Management 175:1–10
7. Apmann K, Fulmer R, Scherer B et al (2022) Nanofluid heat transfer: enhancement of the heat transfer coefficient inside microchannels. Nanomaterials 12:615
8. Aydın DY, Gürü M (2021) Nanofluids: preparation, stability, properties, and thermal performance in terms of thermo-hydraulic, thermodynamics and thermo-economic analysis. J Thermal Anal Calorim 47(60):1–34
9. Celata GP, Cumo M, McPhail SJ et al (2007) Single-phase laminar and turbulent heat transfer in smooth and rough microtubes. Microfuidics Nanofluidics 3:697–707
10. Chabi AR, Zarrinabadi S, Peyghambarzadeh SM et al (2016) Local convective heat transfer coefficient and friction factor of CuO/water nanofluid in a microchannel heat sink. Heat Mass Transfer 53(2):661–671
11. Chamkha AJ, Molana M, Rahnama A et al (2018) On the nanofluids applications in microchannels: a comprehensive review. Powder Technol 332:287–322
12. Duangthongsuk W, Wongwises S (2010) An experimental study on the heat transfer performance and pressure drop of TiO_2-water nanofluids flowing under a turbulent flow regime. Int J Heat Mass Transfer 53:334–344
13. Guzey DV, Minakov AV, Rudyak VYa et al (2014) Measuring the heat-transfer coefficient of nanofluid based on copper oxide in a cylindrical channel. Tech Phys Lett 40(3):203–206
14. Guzey DV, Minakov AV, Rudyak VYa (2016) Investigation of heat transfer of nanofluids in turbulent flow in a cylindrical channel. Fluid Dyn 51(2) 189–199
15. Guzey DV, Minakov AV, Rudyak VYa (2019) On efficiency of convective heat transfer of nanofluids in laminar flow regime. Int J Heat Mass Transfer 139:180–192
16. He Y, Jin Y, Chen H et al (2007) Heat transfer and flow behavior of aqueous suspensions of TiO_2 nanoparticles (nanofluids) flowing upward through a vertical pipe. Int J Heat Mass Transfer 50(11–12):2272–2281
17. Ho CJ, Wei LC, Li ZW (2010) An experimental investigation of forced convective cooling performance of a microchannel heat sink with Al_2O_3/water nanofluid. Appl Thermal Eng 30(2–3):96–103
18. Hosseini M, Sadri R, Kazi SN et al (2017) Experimental study on heat transfer and thermophysical properties of covalently functionalized carbon nanotubes nanofluids in an annular heat exchanger: a green and novel synthesis. Energy Fuels 31:5635–5644
19. Ibrahim I, Sharifpur M, Manca O et al (2023) Nanofluid's convective heat transfer for laminar, transitional, and turbulent flow. In: Sonawane SS, Sharifpur M (eds) Shriram S. Nanofluid applications for advanced thermal solutions, Elsevier Science, pp 151–192
20. Incropera FP (2007) Fundamentals of heat and mass transfer. Wiley
21. Jamali M, Toghraie D (2020) Investigation of heat transfer characteristics in the developing and the developed flow of nanofluid inside a tube with different entrances in the transition regime. J Thermal Anal Calorim 139:685–699
22. Jejurkar A, Singh P, Shaikh A et al (2018) Heat transfer enhancement using various nanofluids—A review. Int Res J Eng Technol (IRJET) 5(11):25–35
23. Jianli W, Jianjun Z, Xing Z et al (2013) Heat transfer and pressure drop of nanofluids containing carbon nanotubes in laminar flows. Exp Thermal Fluid Sci 44:716–721
24. Jung JY, Oh HS, Kwak HY (2009) Forced convective heat transfer of nanofluids in microchannels. Int J Heat Mass Transfer 52:466–472
25. Kalteh M, Abbassi A, Saffar-Avval M et al (2012) Experimental and numerical investigation of nanofluid forced convection inside a wide microchannel heat sink. Appl Thermal Eng 36:260–268

26. Kumar S, Kumar A, Kothiyal DA et al (2018) A review of flow and heat transfer behavior of nanofluids in micro channel heat sinks. Thermal Sci Eng Progress 8:477–493
27. Kumaresan V, Velraj R, Das SK (2012) Convective heat transfer characteristics of secondary refrigerant-based CNT nanofluids in a tubular heat exchanger. Int J Refrig 35:2287–2296
28. Li D (ed) (2008) Encyclopedia of microfluidics and nanofluidics. Springer Sci+Business Media, LLC
29. Fotukian S, Esfahany NM (2010) Experimental investigation of turbulent convective heat transfer of Al_2O_3/water nanofluid inside a circular tube. Int J Heat Fluid Fl 31:606–612
30. Li Q, Xuan Y (2002) Convective heat transfer and Flow characteristics of Cu-water nanofluid. Sci China E 45:408–421
31. Lin L, Peng H, Ding G (2015) Dispersion stability of multi-walled carbon nanotubes in refrigerant with addition of surfactant. Appl Thermal Eng 91:163–171
32. Lodhi MS, Sheorey T, Dutta G (2020) Single-phase fluid flow and heat transfer characteristics of nanofluid in a circular microchannel: Development of flow and heat transfer correlations. J Mech Eng Sci 234(18):095440622091653
33. Manay E, Mandev E (2019) Experimental investigation of mixed convection heat transfer of nanofluids in a circular microchannel with different inclination angles. J Thermal Anal Calorim 135:887–900
34. Mehta B, Subhedar D, Panchal H et al (2022) Synthesis, stability, thermophysical properties and heat transfer applications of nanofluid – a review. J Mol Liquids 364(8):120034
35. Meriläinen A, Seppälä A, Saari K et al (2013) Influence of particle size and shape on turbulent heat transfer characteristics and pressure losses in water-based nanofluids. Int J Heat Mass Transfer 61:439–448
36. Metzner AB, Reed JC (1995) Flow of non-Newtonian fluids—Correlation of the laminar, transition, and turbulent-flow regions. AIChE J B1 1(4):434–440
37. Mikheev MA, Mikheeva IM (1977) Basics of heat transfer. Energy, Moscow
38. Minakov AV, Guzei DV, Rudyak VYa et al (2016) Study of turbulent heat transfer of the nanofluids in a cylindrical channel. Int J Heat Mass Transfer 102:745–755
39. Minakov AV, Lobasov AS, Rudyak VYa et al (2015) The experimental and theoretical study of laminar forced convection of nanofluids in the round channel. Appl Thermal Eng 88:140–148
40. Minakov AV, Rudyak VYa, Guzei DV et al (2015a) Measurement of the heat transfer coefficient of a nanofluid based on water and copper oxide particles in a cylindrical channel. High Temp 53(2):246–253
41. Minakov AV, Rudyak VYa, Guzei DV et al (2015b) Measurements of the thermal conductivity coefficient of nanofluids by the hot-wire method. J Eng Phys Thermophys 88:149–162
42. Minakov AV, Rudyak VYa, Pryazhnikov MI (2018) Rheological behavior of water and ethylene glycol nanofluids with oxide nanoparticles. Coll Surf A: Physicochem Eng Aspects 554:279–285
43. Nguyen CT, Roy G, Gauthier C et al (2007) Heat transfer enhancement using Al2O3-water nanofluid for electronic liquid cooling system. Appl Thermal Eng 28:1501–1509
44. Okonkwo EC, Wole-Osho I, Almanassra IW et al (2021) An updated review of nanofluids in various heat transfer devices. J Thermal Anal Cal 145:2817–2872
45. Pak BC, Cho YI (1998) Hydrodynamic and heat transfer study of dispersed fluids with submicron metallic oxide particles. Exp Heat Transfer 11:151–170
46. Patil MS, Kim SC, Seo J-H et al (2016) Review of the thermo-physical properties and performance characteristics of a refrigeration system using refrigerant-based nanofluids. Energies 9:22
47. Pavía M, Emo M, Estellé P et al (2023) Controlled structural damaging of multiwalled carbon nanotubes and graphene nanoplatelets by oxidation for stable nanofluids with enhanced thermal conductivity. J Mol Liquids 390:123194
48. Pereira J, Moita A, Moreira A (2023) Fewer dimensions for higher thermal performance: a review on 2D nanofluids. Appl Sci 13:4070
49. Petukhov BS (1967) Heat transfer and resistance during laminar flow of liquid in pipes. Energia, Moscow

50. Peyghambarzadeh SM, Hashemabadi SH, Chabi AR et al (2014) Performance of water based CuO and Al_2O_3 nanofluids in a Cu–Be alloy heat sink with rectangular microchannels. Energy Convers Manag 86:28–38
51. Putra N, Septiadi WN, Julian G et al (2013) An experimental study on thermal performance of nanofluids in microchannel heat exchanger. Int J Technol 2:167–177
52. Rudyak VYa (2019) Thermophysical characteristics of nanofluids and transport process mechanisms. J Nanofluids 8:1–16
53. Rudyak VYa, Minakov AV, Pryazhnikov MI (2023) Rheology and thermal conductivity of nanofluids with carbon nanotubes. In: Wythers MS (ed) Advances in material science research, vol 66, Nova Science Publisher, New York, pp 1–92
54. Rudyak VYa, Minakov AV, Pryazhnikov MI et al (2022) Measurement of the thermal conductivity and heat transfer coefficient of nanofluids with single-walled nanotubes. High Temp 60(5):631–638
55. Sarafraz MM, Nikkhah V, Nakhjavani M et al (2018) Thermal performance of a heat sink microchannel working with biologically produced silver-water nanofluid: Experimental assessment. Exp Thermal Fluid Sci 91:509–519
56. Sajid MU, Ali HM (2019) Recent advances in application of nanofluids in heat transfer devices: a critical review. Renew Sustain Energy Rev 103:556–592
57. Timofeeva EV, Wenhua Y, France et al (2011) Base fluid and temperature effects on the heat transfer characteristics of SiC in ethylene glycol/H_2O and H_2O nanofluids. J Appl Phys 109:014914
58. Topuz A, Engin T, Özalp AA et al (2018) Experimental investigation of optimum thermal performance and pressure drop of water-based Al2O3, TiO2and ZnO nanofluids flowing inside a circular microchannel. J Therm Anal Calorim 131:2843–2863
59. Tuckerman DB, Pease RFW (1981) High-performance heat sinking for VLSI. IEEE Electron Device Lett 2(5):126–129
60. Williams W, Buogiorno J, Hu L-W (2008) Experimental investigation of turbulent convective heat transfer and pressure loss of alumina/water and zirconia/water nanoparticle colloids (nanofluids) in horizontal tubes. J Heat Transfer 130(4):042412
61. Xia GD, Liu R, Wang J et al (2016) The characteristics of convective heat transfer in microchannel heat sinks using Al_2O_3 and TiO_2 nanofluids. Int Comm Heat Mass Transfer 76:256–264
62. Younes HA, Mao M, Murshed SMS et al (2022) Nanofluids: key parameters to enhance thermal conductivity and its applications. Appl Thermal Eng 207:118202
63. Zhang X, Li J (2023) A review of uncertainties in the study of heat transfer properties of nanofluids. Heat Mass Transfer 59:621–653

Chapter 7
Simulation of Nanofluid Flows

7.1 Classification of Dispersed Fluids and the Concept of a Continuous Medium

The thermophysical properties of nanofluids are typically the subject of experimental investigation. However, practical applications require an understanding of the nature of nanofluid flow. The variety of possible applications makes their preliminary experimental study practically impossible. In this case, it is appropriate to use the simulation of nanofluid flows encountered in such applications. To date, thousands of relevant papers have been published. However, many of the published data must be treated with great caution, and some of them are even incorrect. For this reason, the present book will not include any literature review.

Certainly, it is necessary to immediately try to understand why such a situation arose when simulating nanofluid flows. One of the objectives of this chapter is to gain insight into this issue. Subsequently, several specific examples will be considered. An analysis of three classes of problems will be presented. Since heat transfer is an undoubted priority from the point of view of various thermophysical applications, the first task under consideration is the modeling of heat transfer in channels. The most significant issue that arises in this context is the necessity for nanofluid pumping through the channel. As previously discussed, the viscosity of nanofluids is markedly higher than that of the base fluid, which consequently results in an increase in hydraulic resistance. On the other hand, Chap. 6 demonstrates that the efficacy of nanofluid utilization in heat exchangers is largely determined by the flow regime, i.e. laminar or turbulent. It is therefore imperative to examine the laminar–turbulent transition conditions. Examining this problem is the second objective of this chapter.

The last few decades have been marked by the rapid development of microsystems technology (MST) for various applications (see, for example, [24, 44, 45, 56, 91, 106], and the literature cited therein). The rapid implementation of MST is associated with a number of factors, including their energy efficiency and low material intensity. However, it is also due to the fact that there are devices, systems, and various kinds of

V. Rudyak, *Thermophysical Properties of Nanofluids*, Fluid Mechanics and Its Applications 140, https://doi.org/10.1007/978-3-031-97995-8_7

objects where the use of such technology is the only possible option. One example of this is the creation of new-generation computers, and a number of biomedical applications. The majority of such microdevices operate realizing certain fluid flows. It is important to note that such flows are typical in a number of natural objects and systems, such as the nutrient delivery system in plants, as well as oil and gas reservoirs. It is understandable that an experimental study of microflows presents known difficulties. In most cases, it is feasible to quantify certain integral characteristics of flows, such as flow rate, pressure drop, etc. Concurrently, the use and optimization of disparate microfluidic devices requires more detailed information. Such information can be obtained through the use of microflow simulations.

One of the critical devices of this type are micromixers, the use of which is necessary in various chemical, biomedical, etc. technologies. The challenge lies in the fact that conventional mixing mechanisms, such as diffusion and turbulent mixing, are ineffective in this context. This is because diffusion coefficients are typically low, and flows are typically laminar. To organize mixing, special devices called micromixers are used [24, 45, 56, 64, 91, 106]. A variety of micromixers are employed, including those that are passive (e.g., operating due to change of channel geometry, wall surface structure, and various inserts) and those that are active (e.g., operating by change of flow rate and impact of external fields). In certain instances, microdevices, and particularly, micromixers can employ nanofluids. Providing a detailed description of the specific characteristics and applications of these devices in this context is the third objective of this chapter.

Furthermore, in order to comprehend the distinctive characteristics of nanofluid flows, a careful examination of the structure of nanofluids is essential, as it is this structure that ultimately determines the simulation methods. Nanofluids, as shown in previous chapters, are non-standard dispersed fluids. Therefore, hydrodynamic modeling of their flow requires special approaches and careful analysis. The initial sections of this chapter are dedicated to such an analysis. In general, dispersed fluids can be either single-phase or multiphase. A single-phase fluid is exemplified by an emulsion comprising two immiscible fluids, including a carrier fluid and droplets of a dispersed component. This book addresses the nanofluids, which are two-phase media comprising a base fluid and solid dispersed nanoparticles. If the particles of the dispersed phase have an identical size, the medium is called monodisperse; otherwise, it is called polydisperse. The classification of dispersed media according to their aggregate state is now widely accepted [65, 105]. However, such classification is ultimately insufficient. In certain instances, for example, mixtures of solid particles with fluids and gases exhibit analogous behavior, whereas in other instances, even the behavior of dispersed media belonging to the same class may diverge significantly. The diversity of heterogeneous media precludes the possibility of developing a universal model that would be equally applicable to the description of any heterogeneous medium. We can only talk about creating models to describe certain classes of such medium. The classification of heterogeneous media should identify such classes. Such classification is briefly described below (see also [76, 77]).

Heterogeneous media are defined as those with a developed internal structure. It can be concluded, therefore, that the most important classifying feature should be the ratio of internal structural elements sizes of the medium in addition to the aggregate state. In addition, the properties of the heterogeneous medium will change depending on the density of the dispersed phase. Accordingly, the following triad of classifying heterogeneous media can be proposed.

- Classification by aggregate state.
- Classification by the size ratio of internal structural elements of the medium.
- Classification by the density of the dispersed phase.

When developing a classification of heterogeneous media by aggregate state, it is necessary to consider the aggregate state of both the carrier and the dispersed phases. For the simplicity, let us consider monodisperse media. Gas suspensions and aerosols are commonly defined to as mixtures of gas with solid and liquid particles, respectively. Suspensions are mixtures of a carrier liquid with solid particles, whereas emulsions are mixtures of a carrier liquid with drops of another liquid. A gas–liquid medium can also be distinguished. Of course, this classification is quite conditional and should be taken with precision until the definition. In practice, as well as in scientific literature, the term 'aerosol' is often used to refer to a range of substances, such as gas suspensions, suspensions, and emulsions. In the meteorology, for example, aerosols are defined as substances dispersed in the atmosphere. For this reason, specific classification and terminology are used there. For example, in Dovgaluk and Ivlev [22] dust, water dust, smoke, smog are distinguished. A specific classification is also performed in accordance with the optical characteristics of the dispersed phase. In an excellent monograph [25] aerosols are defined as gas suspensions. Additionally, distinctions are made between various forms of aerosols, including smokes, soot, dusts, and fogs. Any systems comprising dispersed particles (of any nature) and carrier air or other gases are collectively referred to as aerosols also. Thus, according to the accepted classification nanofluids are suspensions.

In order to discuss the internal structure of a heterogeneous medium, it is first necessary to distinguish the characteristic scales of its elements. In the general case, the set of these structure scales is determined not only by the dimensions of the internal structural elements of the medium, the molecules of the carrier medium, and the dispersed particles, but also by other quantities. In a gas, these are the free path lengths of molecules or particles, whereas in plasma, the Debye radius is considered. The complete set of such structure scales differs in various heterogeneous media and is determined by the overall state of the carrier phase. Therefore, when developing a classification of heterogeneous media based on the ratio of the sizes of internal structural elements, it is necessary to separately consider the already identified classes of media.

This book addresses the suspensions. In this context, three distinct linear scales of length can be identified: the sizes of molecules and dispersed particles (σ_m, d), and the characteristic linear scale of the system under consideration (L_s). If the system is a flow in a pipe, this scale is pipe diameter; if it is a flow in a boundary layer, the characteristic scale is the thickness of the boundary layer, and so on. The construction of models to describe the suspension flow dynamics is dependent upon the ratio of these sizes. Following a brief interval of approximately the duration of a molecule's collision time, a carrier fluid can almost always be described at the hydrodynamic level.[1] Hydrodynamic mode is named a flow regime that is described using methods of continuum mechanics. In this case, the medium state is defined by a set of average macroscopic observables. Such quantities include density, velocity, and temperature (energy) of the fluid. The introduction of such macroscopic quantities implies that they are insensitive to changes in density, momentum, and energy due to individual molecular motions. This implies that the fluctuations of these quantities at the microscopic level, which are associated with the motion of molecules, can be disregarded. What is the scale of such fluctuations? The answer to this question is provided by the introduction of the continuity hypothesis, which offers a precise definition of the configuration space of the system. In particular, this means that a material point in space describing a given continuous medium (in particular, a liquid) is defined as a physically infinitely small volume, that is, a volume containing a sufficiently large number of molecules. Only in this case it is possible to neglect the mentioned fluctuations. Henceforth, this volume shall be referred to as the hydrodynamic physically infinitesimal volume. It is equal to $v_{hf} = (4/3)\pi r_{hf}^3$, where r_{hf} is the hydrodynamic physically infinitesimal length scale of the medium (fluid), defined by the relation [75]

$$r_{hf} \sim \sqrt{\sigma_m L_s}. \tag{7.1}$$

Thus, the corresponding hydrodynamic physically infinitesimal volume is $v_{hf} \sim (\sigma_m L_s)^{3/2}$. In equilibrium, the fluctuations of macroscopic quantities are inversely proportional to the square root of the number of particles in the system. Given that the number of molecules in this volume is $N_{hf} = n_f v_{hf} \sim n_f (\sigma_m L_s)^{3/2}$ (n_f is the numerical density of the liquid molecules), the following relation must be satisfied for the fluctuations of macroscopic quantities in this volume to be small: $\sqrt{N_{hf}} = \sqrt{n_f}(\sigma_m L_s)^{\frac{3}{4}} \gg 1$. This relation is almost always satisfied in a fluid. To illustrate, while assuming that the characteristic hydrodynamic scale of the flow is $L_s \sim 1$ cm, $\sigma_m \sim 10^{-8}$ cm, and $n_f \sim 10^{23}$, then under normal conditions we get $\sqrt{N_{hf}} \geq 10^5$. Nevertheless, it is evident that relation (7.1) may not be applicable in microflows and particularly in nanoflows.

Let us now consider a possible classification of suspensions according to the particle size of the dispersed phase. Two main situations can be distinguished. Finely dispersed suspensions, whose particles satisfy the ratio

$$\sigma_m/d < 1, \quad d \sim r_{hf}, \tag{7.2}$$

and coarse dispersed suspensions for which

$$L_s \gg d > r_{hf} \quad \text{or} \quad L_s \gg d \gg r_{hf} \tag{7.3}$$

[1] There is a possibility of strongly non-equilibrium flows of the shock wave type occurring; therefore, the term "almost always" is used in this context. Nevertheless, such flows remain exceptions, despite their significant importance.

The classification of suspensions according to the density of the dispersed phase is now to be discussed. In ordinary conditions, the dispersed particles can be regarded as a pseudogas. A necessary condition for a pseudogas of particles to be considered rarefied is the smallness of the corresponding virial parameter (volume concentration)

$$\varepsilon_p = n_p d^3 \sim \varphi \ll 1. \tag{7.4}$$

Here n_p is the numerical density of the dispersed particles.

Depending on the density of the dispersed phase, it is possible to identify a highly rarefied suspension, in which the interaction of particles with each other can be neglected. For this purpose, the average distance between particles l_{mp} should be much greater than their size

$$d/l_{mp} \sim dn_p^{1/3} \sim \varphi^{1/3} \ll 1. \tag{7.5}$$

It can be shown that in this case the weight concentration of the particle pseudogas should also be small.

Next, a rarefied suspension can be distinguished. It is for such a suspension that Batchelor [6] derived the formula for the viscosity coefficient (see Chap. 3). It is also pertinent to consider moderately dense and dense suspensions. In the latter case, the mean distance between particles is approximately equal to their size.

7.2 Methods for the Simulation of Dispersed Fluid Flows

A dispersed fluid is a combination of two distinct continua: a carrier fluid and a pseudogas of dispersed particles. Such a system can be in several different states. In the first case, the two continua indicated may have different macroscopic velocities and temperatures. On the other hand, a dispersed fluid can move with a single macroscopic velocity and have a single temperature. As a result, the dynamics of such flows is described in a variety of ways, which are briefly discussed in this section. The main attention is paid to the description of the dynamics of nanofluids.

It should be noted that more complex situations are also possible. A typical situation is when the carrier medium is described by hydrodynamically (by means of methods of continuum mechanics), however, the ensemble of dispersed particles is so rarefied that it must be described using some kinetic equation. This situation is typical when describing flows in microchannels and especially in nanochannels. Such situations will not be considered in this book. You can get to know them by the following work [74, 78].

7.2.1 Two-Fluid Description of Dispersed Fluid Flows

The description of dispersed fluid flows by means of continuum mechanics is only possible if a number of conditions are met. Indeed, let it be possible to introduce appropriate macroscopic variables for the carrier fluid. This implies that their fluctuations with respect to a physically infinitesimal volume with a linear scale (7.1) are negligibly small. In order to describe the particle pseudogas based on the hydrodynamic approach, it is first necessary to define the corresponding physically infinitesimal scale, $r_{hp} \sim \sqrt{dL_s}$ and volume $v_{hp} \sim (dL_s)^{3/2}$. If the fluctuations of the mean values of density, velocity, and temperature of the particle pseudogas with respect to this volume are small, its dynamics can be described by continuum mechanics methods. It can therefore be concluded that the dispersed medium in question is constituted of two distinct continuums, with interpenetrating motion of components. This approach was initially articulated with the greatest clarity by Rakhmatulin [71]. The fundamental principle is that the state of each component of a dispersed fluid is characterized by partial density ρ_α, velocity $\boldsymbol{u}_\alpha$, and energy E_α. . All components of the mixture fill the same volume, so that the densities, velocities and energies of all components of the mixture are determined at each point in space. Of course, an important question arises, but when is such a description possible? As demonstrated previously, the relationship (7.1) is typically satisfied for a carrier fluid, with the exception of flows in very confined condition. In dispersed fluids, similar relations must also be fulfilled for pseudogas of particles. In particular, the relation $r_{hp} \ll L_s$ must be fulfilled, while fluctuations of the corresponding macroscopic parameters with respect to the physically infinitesimal volume v_{hp} must be small. It is evident that the fulfillment of these relations in each particular case imposes rather rigid conditions on particle sizes and their volume concentrations.

In the event that a hydrodynamical description of each component are possible, it is necessary to formulate the corresponding transport equations. Such a description is referred to as a multifluid description. The equations of multifluid hydrodynamics, that is to say the transport equations for partial macroscopic variables, are obtained in the usual way as a consequence of the fulfillment of the mass conservation laws of mass, momentum, and energy of the whole system and their local conservation within a component. This can be done phenomenologically (see, for example, [65]) or by employing methods of nonequilibrium statistical mechanics [78, 81]. In the absence of phase transitions, chemical reactions and other processes resulting in a change in the composition of the particle system, the equations for the l-component system in the locally accompanying frame of reference moving with velocity $\boldsymbol{u}_\alpha$ have the following form

$$\frac{d_\alpha \rho_\alpha}{dt} = -\rho_\alpha \nabla \cdot \boldsymbol{u}_\alpha,\ \rho_\alpha \frac{d_\alpha \boldsymbol{u}_\alpha}{dt} = -\nabla \cdot \boldsymbol{J}_{2\alpha} + \boldsymbol{j}_{2\alpha},\ \frac{d_\alpha}{dt} = \frac{\partial}{\partial t} + \boldsymbol{u}_\alpha \cdot \nabla,$$
$$\rho_\alpha \frac{d_\alpha E_\alpha}{dt} = -E_\alpha \nabla \cdot \boldsymbol{u}_\alpha - \boldsymbol{J}_{2\alpha} : \nabla \boldsymbol{u}_\alpha - \nabla \cdot \boldsymbol{J}_{3\alpha} + j_{3\alpha}$$
$$- \sum_{\varphi \neq \alpha} \left(\boldsymbol{J}_{\alpha\varphi} : \nabla \boldsymbol{u}_{\alpha\varphi} + \frac{1}{2} \boldsymbol{j}_{2\alpha} \cdot \boldsymbol{u}_{\alpha\varphi} \right), \tag{7.6}$$

where $\boldsymbol{u}_{\alpha\varphi} = \boldsymbol{u}_\alpha - \boldsymbol{u}_\varphi$ are the diffusion rates, $\boldsymbol{J}_{2\alpha}$ and $\boldsymbol{J}_{3\alpha}$ are the stress tensor and heat flux vector, respectively, $\boldsymbol{j}_{2\alpha}$ is the interphase interaction force, $j_{3\alpha}$ is the energy production due to the inter-component interaction.

The challenge of utilizing Eq. (7.6) is contingent upon the necessity to establish the constitutive relations, that is, the relations linking the momentum and energy fluxes to the gradients of macroscopic variables. Furthermore, it is necessary to define interphase forces. The derivation of constitutive relations for heterogeneous systems is complicated exactly by the heterogeneity of their composition. When investigating the nature of deformation of a selected volume of the mixture, it is essential to consider not only the reactions associated with the displacement of the outer boundary of the volume, as is typically done when deriving the Navier–Stokes equations [7, 105], but also those associated with the displacements of interfacial boundaries. It is also not easy to determine the interphase interaction forces. However, the situation is significantly simplified if the phase interface remains unaltered in shape, as is the case in suspensions. In this case, the constitutive relations can be formulated using the thermodynamics of irreversible processes or nonequilibrium statistical mechanics. Here it should be borne in mind, however, that different flow regimes can be realized depending on the relative magnitude of diffusion velocities and temperatures relative to the gradients of hydrodynamic variables. The interphase forces and fluxes for these modes will have different structures. In the case of small diffusion velocities, temperatures, and their gradients in the linear approximation, in binary systems they will have the following form

$$\boldsymbol{J}_{2\alpha} = p_\alpha \boldsymbol{U} - \mu_\alpha (\nabla \boldsymbol{u}_\alpha)^s - \mu_{\alpha\beta} (\nabla \boldsymbol{u}_\beta)^s - \sum_{\gamma,\varphi}^{1,2} \left[\mu^s_{\alpha\gamma\varphi} (\boldsymbol{u}_{\gamma\varphi} \nabla T_\gamma)^s + \mu^V_{\alpha\gamma\varphi} \boldsymbol{u}_{\gamma\varphi} \cdot \nabla T_\gamma \boldsymbol{U} \right], \tag{7.7a}$$

$$\boldsymbol{J}_{3\alpha} = -\lambda_\alpha \nabla T_\alpha - \lambda_{\alpha\gamma} \nabla T_\gamma, \tag{7.7b}$$

$$\boldsymbol{j}_{2\alpha} = -\eta_{\alpha\beta} \boldsymbol{u}_{\alpha\beta},\quad j_{3\alpha} = -\lambda_{\alpha\beta} T_{\alpha\beta}, \tag{7.7c}$$

$$\boldsymbol{J}_{\alpha\beta} = -\eta_{\alpha\beta} (\nabla \boldsymbol{u}_\beta)^s - \sum_{\gamma,\varphi}^{1,2} \left[\eta^s_{2\alpha\gamma\varphi} (\boldsymbol{u}_{\gamma\varphi} \nabla T_\gamma)^s + \eta^V_{2\alpha\gamma\varphi} \boldsymbol{u}_{\gamma\varphi} \cdot \nabla T_\gamma \boldsymbol{U} \right], \tag{7.7d}$$

where T_α is the partial temperature of the component α, and $(\boldsymbol{AB})^s = (1/2)\left(A_iB_j + A_jB_i\right) - (1/3)A_kB_k\boldsymbol{U}$, $\boldsymbol{U}$ is a unit tensor of the second rank.

All kinetic coefficients included are time integrals of the corresponding correlation functions [78], and thus, in principle, can be calculated. A distinctive feature of the constitutive relations (7.7) is the emergence of a novel stress tensor $\boldsymbol{J}_{\alpha\beta}$ which is associated with the presence of diffusion velocities in the system. This tensor is not present in phenomenological theories. The second feature is the presence of cross kinetic coefficients in these ratios, which determine the contribution of the nanoparticles and molecules to the transport processes. Therefore, for example, the stress tensor of the carrier fluid depends on the velocity gradients of both components (see (7.7a)). The viscosity coefficient μ_α included here is the shear viscosity coefficient of the carrier fluid or particle pseudogas. Concurrently, the coefficient $\mu_{\alpha\beta}$ is associated with the transport of momentum within the system, which is a consequence of the interaction between dispersed particles and molecules. The difficulty is that this and similar transport coefficients cannot, in fact, be measured. As previously stated, these coefficients can be simulated in principle. For rarefied gas suspensions this is done in Rudyak [77]. Thus, the application of two-fluid hydrodynamic equations to simulate dispersed fluid flow dynamics is confronted with a multitude of fundamental difficulties, although such attempts are known in the literature.

7.2.2 Single-Fluid Description of Nanofluid Flows

In nanofluids with spherical particles, the typical volume concentration does not exceed 10%. In terms of size and density, it is typically a finely dispersed and rarefied or even highly rarefied suspension. A comparison of the nanoparticle size with the characteristic infinitesimal scale of the fluid (7.1) demonstrates that nanoparticles are a material points in the fluid metric, provided that the characteristic size of the flow is not too small. For this reason, nanofluid can be considered a fairly homogeneous fluid. Nevertheless, in order to describe the dynamics of nanofluid hydrodynamically, it is necessary to introduce a new physically infinitesimal scale for nanofluid. This physically infinitesimal scale can be determined by the following relation

$$r_{hnf} \sim \sqrt{(\sigma_m + d)L_s}. \tag{7.8}$$

If the nanoparticles are sufficiently small, then $r_{hnf} \sim r_{hf}$, otherwise $r_{hnf} \sim r_{hnp}$, , This implies that the physically infinitesimal scale for such a nanofluid is determined by the condition of applicability of the hydrodynamic description of the nanoparticle pseudogas. The fluctuations of the average macroscopic characteristics of the nanofluid with respect to a physically infinitesimal volume corresponding to the scale (7.8) must be sufficiently small. Here, generally speaking, one should keep in mind the volume concentration of nanoparticles. In the event that the concentration is exceedingly low, fluctuations in macroscopic variables relative to the physically infinitesimal volume may become discernible. This indicates that the dynamics of the nanoparticle pseudogas cannot be described hydrodynamically. In such case, the carrier fluid is described hydrodynamically, but the corresponding transport equations contain source terms associated with the interaction of this fluid with particles. The dynamics of nanoparticles can be described then by a set of Newton's equations, wherein the forces are determined by the interaction of the particles with the medium. This is the so-called Eulerian–Lagrangian method of describing the dynamics of a dispersed fluid. This approach will not be applied in this book and therefore will not be discussed in detail further. There is only one remark to be made. The efficacy of the method is related to the correct determination of the forces acting on particles in the liquid. Usually forces acting on coarse particles (Stokes, Basset forces, etc.) are considered. It should only be understood that these forces have been derived and experimentally measured for rather large particles, and their applicability to nanoparticles is not obvious. In Rudyak et al. [82, 83], Rudyak and Belkin[82] it is shown, for example, that the force acting on a sufficiently small nanoparticle differs significantly from the Stokes force, in particular, it is nonstationary (see also Chap. 2).

Nevertheless, in the majority of cases, the flow of nanofluids can be simulated within the framework of single-fluid hydrodynamics. The fluid density is then defined in accordance with the following relation

$$\rho = \varphi\rho_p + (1-\varphi)\rho_f, \tag{7.9}$$

The temperature, pressure p and velocity $\boldsymbol{u}$ of the nanofluid are assumed to be unified, so that the transport equations are of the form [14, 91, 92]

$$\frac{\partial\rho}{\partial t} + \nabla\cdot(\rho\boldsymbol{u}) = 0,\ \frac{\partial\rho\boldsymbol{u}}{\partial t} + \nabla\cdot(\rho\boldsymbol{u}\boldsymbol{u}) = -\nabla p + \nabla\cdot\boldsymbol{\tau}, \tag{7.10}$$

$$\frac{\partial\rho w}{\partial t} + \nabla\cdot(\rho\boldsymbol{u}w) = \nabla\cdot(\rho D w) + \nabla\cdot(\rho D_{Th} w), \tag{7.10a}$$

here, as before, w is the weight concentration of nanoparticles, D and D_{Th} are the diffusion and thermodiffusion coefficients, and $\boldsymbol{\tau}$ is the stress tensor. In the general case, the nanofluid can be non-Newtonian; then the stress tensor is a function of the shear rate.

In the event that the flow is non-isothermal, it is advisable to introduce enthalpy

$$h(T) = \int_{T_0}^{T} C_p dT, \tag{7.11}$$

the corresponding energy conservation equation then takes the form

$$\frac{\partial\rho h}{\partial t} + \nabla\cdot(\rho\boldsymbol{u}h) = \nabla\cdot(\lambda\nabla T). \tag{7.12}$$

where the specific heat capacity of nanofluid is introduced as $C_p = \rho^{-1}\left[\varphi\rho_p C_{pp} + (1-\varphi)\rho_f C_{pf}\right]$, C_{pp} and C_{pf} are the specific heat capacities of nanoparticles and base fluid, respectively [119]. In a non-isothermal problem, all fluid variables, including transport coefficients, are functions of temperature, and this should be taken into account when solving Eqs. (7.10)–(7.12).

7.2.3 Peculiarities of Solving the Hydrodynamic Problem for Nanofluid Flows

The solution of the Eqs. (7.10–7.12) represents a significant challenge. Nevertheless, at least for laminar flows, a solution can be found. For this purpose, the standard ANSYS Fluent fluid simulation software or the σ-flow package can be employed. The latter package is based on the algorithm for solving hydrodynamic equations by the finite volume method, whose fundamentals are formulated in Rudyak et al. [83, 93], Gavrilov et al. [27]. This algorithm has been used many times in different tasks and is quite well described, including its testing, considered in Rudyak and Minakov [91], Rudyak et al. [93].

Usually, when solving almost any problem, the transport equations are reduced to dimensionless form, which causes to the emergence of a number of similarity parameters. In most problems of modeling nanofluid flows, a certain similarity parameter is fixed and the corresponding solution to the problem is found. Then, by varying the similarity parameter, the dependence of this flow characteristic on the similarity parameter is obtained. Unfortunately, the solution obtained in this way has no physical meaning at all. We should figure out why (see [80]).

First of all, the simplest thing. In many instances, the kinematic viscosity coefficient $\upsilon = \mu/\rho$ is employed in lieu of the dynamic viscosity coefficient when solving the momentum conservation equation. This cannot be done, since in this case it is impossible to obtain the correct dependence of the calculated flow parameters on the concentration of nanoparticles. Indeed, the nanofluid density (7.9) is a linear function of nanoparticle concentration, whereas the dynamic viscosity coefficient is a quadratic function of concentration (3.18). Naturally, the kinematic viscosity coefficient is also a quadratic function of the particle concentration, but it is not clear which one. This dependency must first be obtained.

For the same reason, it is important to exercise caution when using the Reynolds number $\mathrm{Re} = (\rho U L_s)/\mu$ (where U represents the characteristic velocity of the fluid and L_s denotes the characteristic scale of the flow). It is important to understand that the same Reynolds number may correspond to different nanofluids. As previously stated, the density is a linear function of nanoparticle concentration, whereas the viscosity coefficient is a quadratic function. Consequently, the Reynolds number is also a quadratic function of the particle concentration. Moreover, the viscosity coefficient is a function of particle size and particle material. Therefore, the dependences of certain parameters on the Reynolds number, which are often given in works on modeling nanofluid flows, without specifying the characteristics of the nanofluid, are simply impossible to physically interpret rationally. As mentioned, the same Reynolds number may correspond to completely different nanofluids. The situation can be further complicated. To illustrate, let us consider the Poiseuille flow in a pipe of radius h and length L_{ch}. In this case, the characteristic velocity of the steady-state flow can be expressed in terms of the flow rate and, as a consequence, in terms of the pressure drop: $U = (\Delta p h^2)/(2\mu L_{ch})$. Consequently, the Reynolds number calculated on the basis of this velocity is inversely proportional to the square of the viscosity coefficient, and its relationship with the particle concentration becomes even more complex.

Since the nanofluid under consideration often has non-Newtonian properties, for a power-law fluid the Reynolds number is defined by the formula (6.2). In this case, the dependence on the concentration of nanoparticles, as is well understood, becomes a non-trivial problem, since both the fluid index and the consistency parameter are its functions (see Chaps. 3 and 6). Thus, in this case to solve the problem by introducing the Reynolds number is almost impossible.

In many instances, the viscosity coefficient is determined through a variety of methods, including the utilization of a corresponding experimental value, which is a highly accurate approach. Nevertheless, the thermal conductivity coefficient is calculated introducing the Prandtl number ($\mathrm{Pr} = (C_p \mu)/\lambda$). The Prandtl number for nanofluids, however, is a rather complex quadratic function of nanoparticle concentration, size, and material

$$\mathrm{Pr}_{nf} = \mathrm{Pr}_{bf}\left[1 + k_1(d, m)\varphi + k_2(d, m)\varphi^2\right]. \tag{7.13}$$

An illustrative example of dependence (7.13) is presented in Fig. 7.1a for a water-based nanofluid with ZrO_2 particles of two different sizes [95]. Here, the dependence of the relative Prandtl number $\mathrm{Pr}_r = \mathrm{Pr}_{nf}/Pr_{bf}$ on the nanoparticle concentration is presented. As the concentration of nanoparticles increases, the Prandtl number of the nanofluid rises significantly. Furthermore, the k_i coefficients in (7.13) are not universal, since they are dependent on the size and the material m of the nanoparticles. The dependence of the relative Prandtl number $\mathrm{Pr}_r = \mathrm{Pr}_{nf}/Pr_{bf}$ on the nanoparticle concentration is presented in Fig. 7.1a. The second one for the special case, is illustrated in Fig. 7.1b. This figure depicts the data for a water-based nanofluid with Al_2O_3 and TiO_2 particles, as a function of their concentration. In both cases, the particle sizes are identical.

It is important to note that, in general, the coefficients k_i in Eq. (7.13) can be either positive or negative. Consequently, at specific values of the nanofluid parameters, the Prandtl number of the nanofluid may be less than that of the base fluid. Figure 7.1b illustrates this instance for nanofluids with low aluminium oxide concentrations.

The observed increase in the Prandtl number with growth in the particle concentration is due to the fact that with its growth, the viscosity coefficient of nanofluid increases much faster than the thermal conductivity coefficient. Indeed, an examination of the provided measurement data reveals that the maximum excess thermal conductivity coefficient is recorded for nanofluid with ZrO_2 particles (above 28%),

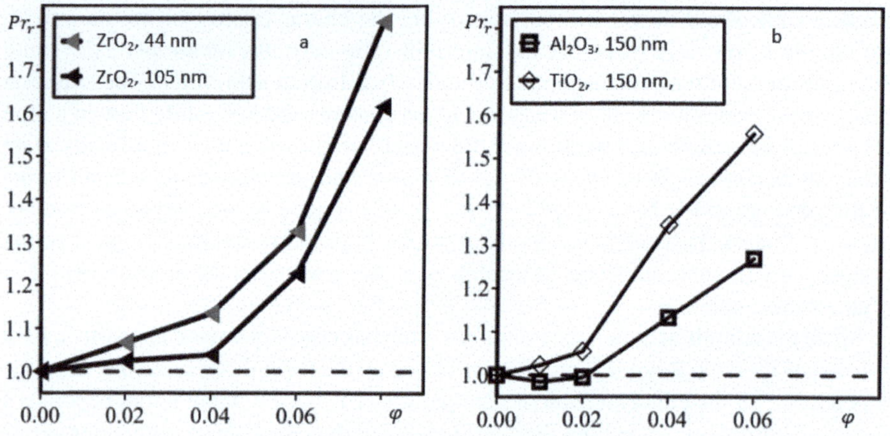

Fig. 7.1 Dependence of the relative Prandtl number of a water-based nanofluids on the concentration of ZrO_2 (**a**), Al_2O_3 and TiO_2 (**b**) particles

while the viscosity coefficient more than doubles with increasing particle concentration. Nevertheless, the opposite effect is also possible. A notable enhancement in the thermal conductivity coefficient of nanofluids is observed when metallic particles are employed even at their minimal concentrations (see the review by [113]). The viscosity coefficient and heat capacity of such a nanofluid change very little and a decrease in the Prandtl number observed in comparison to that of the base fluid.

The dependence of the viscosity and thermal conductivity coefficients of nanofluids on the nanoparticle sizes is opposite, i.e. viscosity increases with decreasing particle size, while thermal conductivity decreases. Therefore, the maximum values of the relative Prandtl number should be observed for nanofluids with small particles. As the particles increase in size, the Prandtl number will decrease. This is confirmed by Fig. 7.2, which was plotted based on the data presented in [61, 70].

An even more complex situation is implemented when using the Nusselt number $Nu = \alpha_c L_s/\lambda$. As demonstrated in Chap. 6, in the laminar flow regime, the heat transfer coefficient increases proportionally to the concentration of particles, but at the same time, it depends on their size and viscosity. It is exceedingly challenging to identify the dependence on these parameters, even in the presence of relevant experimental data. The complexity increases when trying to construct an adequate dependence of the Nusselt number on the Reynolds number. This is quite typical in studies on modeling heat transfer of nanofluid. In order to construct an accurate dependence, it is essential to fix the nanofluid, i.e. the size and material of the nanoparticles, and their concentration in each case. Then it is necessary to sequentially vary the particle concentration, determining in each case the Nusselt number and the corresponding Reynolds number.

In the event of heat exchange occurs during convection in the gravity field, the natural similarity parameter is the Grashof number

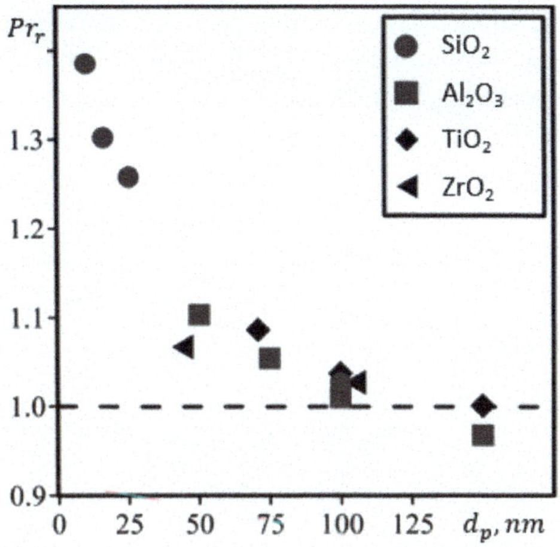

Fig. 7.2 Dependence of the relative Prandtl number of the nanofluid on the nanoparticle size at a concentration of 2%

$$Gr = \frac{gL_s^3\beta}{\upsilon^2\lambda^2}(T_s - T_0), \tag{7.14}$$

where β is the volume expansion coefficient, g is the acceleration of gravity, T_s is the temperature of the heat exchanger surface, and T_0 is the temperature of the coolant. Here again the volume expansion coefficient, kinematic viscosity, and thermal conductivity are functions of the nanoparticle concentration, while the last two are also functions of nanoparticle size and material. Thus, the similarity criterion (7.14) is an extremely complex function of nanoparticle concentration, size, and material. Therefore, by its specific numerical value, one cannot predict which nanofluid it will correspond to.

When solving the problems of magnetic fluid dynamics, the characteristic similarity parameter is the Hartmann number *Ha*, which is defined as $Ha = BL_s\sqrt{\sigma/\mu}$, where B is the magnetic field induction. In the context of nanofluid dynamics, the Hartmann number represents a complex parameter that is dependent on the concentration, size, and material of the nanoparticles. It is notable that these dependencies exhibit significant differences with regard to the viscosity and electric conductivity coefficients (refer to Chaps. 3 and 5). Therefore, when fixing the value of this parameter, it is impossible to determine which specific nanofluid it will correspond to. Moreover, there is ambiguity, different nanofluids can correspond to the same Hartmann number.

There is no theory that could predict the thermophysical characteristics of nanofluids. Thus, when simulating nanofluid flows, it is necessary to use correlations derived on data from experimental measurements. At a fixed temperature, the dependences of the viscosity and thermal conductivity coefficients are of the form (3.18) and (4.10), respectively. Numerous examples of experimental correlations can be found in Chaps. 3 and 4, as well as in the literature. The dependence on particle size has been demonstrated also in a number of studies. At the same time, numerous correlations describing the dependence of viscosity and thermal conductivity on both the concentration of nanoparticles and temperature are also available in the literature (see, for example, [32]). It should be noted that all such correlations have a certain area of applicability and cannot be used beyond this area. Moreover, correlations of the following type have been derived to describe the thermal conductivity coefficient of nanofluids [39]

$$\lambda = \lambda_{bf}(1-\varphi) + \lambda_p\varphi + 3C\left(\frac{\sigma_m}{d}\right)\varphi\lambda_{bf}Re^2Pr, \tag{7.15}$$

or [18]

$$\lambda = \lambda_{bf}\left[1 + 64.7\left(\frac{\sigma_m}{d}\right)^{0.369}\left(\frac{\lambda_{bf}}{\lambda_p}\right)^{0.7476} Pr^{0.9955}Re^{1.2321}\right]. \tag{7.15a}$$

Subsequently, correlations such as (7.15) and (7.15a) were frequently employed in the simulation of nanofluid flows. Nevertheless, it is evident that these correlations include similarity criteria that are themselves intricate functions of nanofluid parameters. Therefore, it is practically impossible to use these correlations when simulating and varying these parameters.

Finally, when simulating non-Newtonian fluid flows, there is one more problem to overcome which is extremely difficult. However, it is crucial to at least grasp the existence of such difficulty. The effective viscosity of a non-Newtonian fluid is dependent upon the shear rate. However, the shear rate is not uniform throughout the flow. For example, in a Poiseuille flow within a pipe, the maximum shear rate is observed in the vicinity of the wall, while the minimum is observed at the pipe axis. This indicates that the dissipation of momentum across the flow is heterogeneous, and at high flow rates this heterogeneity can be considerable. A further similar issue arises in non-isothermal flows. The rheology of a nanofluid can change with a change in temperature (and does change! See Sect. 3.4). In heat exchange devices, particularly large temperature gradients occur near the heated wall, but in this case the rheology of the nanofluid near the wall and far from it can be different. This discrepancy can be critical in the flow within the boundary layer.

7.3 Simulation of Heat Transfer in Nanofluids

The preceding chapter presented data demonstrating the efficacy of nanofluids for heat transfer applications. However, the sheer diversity of nanofluids makes it unfeasible to study all the necessary situations experimentally. In addition, the experiment provides some integral information, but detailed information on the flow dynamics is often needed. It can be obtained using numerical simulation. Experimental investigation does not allow revealing the effects of numerous significant factors present in nanofluid flows on heat transfer. Thus, for example, nanofluids are two-phase media. In non-isothermal flows, in addition to the conventional diffusion of nanoparticles, there is also thermodiffusion. How significantly does it affect heat transfer? The response to this question can be ascertained using the numerical simulation. This section is devoted to the description of the data obtained by the author and his colleagues over the past ten years [33, 34, 47–49, 59, 92, 94].

7.3.1 Heat Transfer of Newtonian Nanofluids

This section will examine the heat transfer in a water-based nanofluid with 150 nm aluminum oxide particles, as it flows in a circular minichannel. It is assumed that the flow is laminar and axisymmetric. The nanofluid is introduced into the tube at temperature T_0 and nanoparticle volume concentration φ_0. The axial velocity profile at the inlet is parabolic, given by the equation: $v(r) = 2v_0\left[1 - (r/R)^2\right]$, where v_0 is the cross-sectional average velocity of the nanofluid. In the interval $0 \le z \le L_{ch}$, the pipe is insulated thermally, while the section at $L_{ch} < z \le 2L_{ch}$ is heated at a constant heat flux $q > 0$. The following parameters are employed in the computations: $R = 0.003$ m, $L_{ch} = 0.905$ m, $q = 5865$ W/m^2, $T_0 = 23$ °C. This is consistent with the experimental conditions outlined in the paper [33] (see also Chap. 6).

The simulation is conducted using the aforementioned model (7.10)–(7.12). The temperature of the nanofluid at each point is calculated by Eq. (7.12) using the enthalpy calculated from Eq. (7.11). The thermophysical properties of the water-based nanofluid with aluminum oxide nanoparticles depend on the properties of the base fluid, nanoparticles, and their volume concentration. The dependence of density, viscosity and thermal conductivity coefficients, heat capacity, and thermal expansion coefficient of water on temperature in the range from 0 to 100 °C are described by interpolation of experimental data employing the following polynomials [66]

$$\begin{aligned}
\rho(T) &= a_0 + a_1T + a_2T^2 + a_3T^3 + a_4T^4 + a_5T^5 \\
\mu(T) &= b_0 + b_1T + b_2T^2 + b_3T^3 + b_4T^4 + b_5T^5, \\
\lambda(T) &= c_0 + c_1T + c_2T^2 + c_3T^3 + c_4T^4, \\
C_p(T) &= d_0 + d_1T + d_2T^2 + d_3T^3 + d_4T^4 + d_5T^5, \\
\beta_p(T) &= g_0 + g_1T + g_2T^2 + g_3T^3 + g_4T^4.
\end{aligned} \tag{7.16}$$

where T is the temperature, °C, and the corresponding coefficients are given in Table 7.1.

The temperature dependence of the density ρ_p, thermal conductivity coefficient λ_p, and heat capacity C_{pp} of aluminium oxide nanoparticles were given as follows

$$\rho_p(T) = 3921.71 - 8.5625 \cdot 10^{-2}T, \quad 0 \le T \le 100\ ^\circ\text{C}$$

Table 7.1 Coefficients in formulae (7.16)

a_0	998.86	b_0	$1.78 \cdot 10^{-3}$	c_0	0.561	d_0	4218.79	g_0	-6.35×10^{-5}
a_1	6.125×10^{-2}	b_1	-5.84×10^{-5}	c_1	1.95×10^{-3}	d_1	-3.167	g_1	1.68×10^{-5}
a_2	-8.31×10^{-3}	b_2	1.26×10^{-6}	c_2	-1.01×10^{-6}	d_2	9.5×10^{-2}	g_2	-1.98×10^{-7}
a_3	6.42×10^{-5}	b_3	-1.7×10^{-8}	c_3	-1.28×10^{-7}	d_3	-1.39×10^{-3}	g_3	1.69×10^{-9}
a_4	-3.95×10^{-7}	b_4	1.25×10^{-10}	c_4	6.21×10^{-10}	d_4	1.07×10^{-5}	g_4	-5.77×10^{-12}
a_5	1.08×10^{-9}	b_5	-3.75×10^{-13}			d_5	-3.2×10^{-8}		

$$\lambda_p(T) = 5.5 - 34.5e^{-0.0033T}, \quad 0 \leq T \leq 1300\ ^\circ\mathrm{C}$$

$$C_{p_p}(T) = 1044.6 + 0.1742(273.15 + T) - \frac{2.796 \times 10^7}{(273.15 + T)^2}, \quad T \leq 1500\ ^\circ\mathrm{C}$$

In order to solve the Eqs. (7.10)–(7.13), it is necessary to have data on the viscosity and thermal conductivity coefficients. In order to determine the viscosity coefficient of the nanofluid, an experimental correlation was constructed based on measurements taken at 25 °C [95]

$$\mu = \mu_w\left(1 + 6\varphi + 72\varphi^2\right) \tag{7.17}$$

The variations in the viscosity coefficient of the nanofluid with increasing temperature at the nanoparticle concentrations discussed below are consistent with the change in the viscosity coefficient of the base fluid, and therefore are determined by formula (7.16).

The thermal conductivity coefficient was also given by the correlation constructed based on experimental data and given in the same paper

$$\lambda = \lambda_w\left(1 + 6.42\phi - 39.5\phi^2\right). \tag{7.18}$$

The diffusion coefficient of nanoparticles in water was determined from Einstein's equation

$$D = kT/(3\pi\mu d). \tag{7.19}$$

As shown by the molecular dynamics method [82, 83] (see also Chap. 2), the Einstein–Stokes drag coefficient included in (7.19) is quite suitable for describing the diffusion of nanoparticles, unless their sizes are too small (larger than 10–15 nm).

The relationship proposed in Ryzhkov and Minakov [101] was used to calculate the thermodiffusion coefficient:

$$D_T = \alpha \beta_T \lambda_f \left[\mu_f \left(2\lambda_f + \lambda_p \right) \right]^{-1}, \tag{7.20}$$

Here, the proportionality factor α was selected to ensure that the D_T values were in accordance with the experimental data, therefore it was taken equal to $\alpha = 10^{-9}$.

One of the principal objectives of the computation was to ascertain the influence of thermodiffusion. Consequently, computations were conducted with and without incorporating thermodiffusion. Figure 7.3a illustrates the velocity profiles in the channel at varying nanoparticle concentrations and a constant flow rate of 0.0011 kg/s. As the nanoparticle concentration rises, the velocity maximum declines. This is due to the increase in fluid density with increasing nanoparticle concentration at a fixed flow rate. The corresponding viscosity profiles during the flow are illustrated in Fig. 7.3b, here the particle concentration increases bottom-up. As can be observed, the viscosity coefficient of the nanofluid increases in accordance with growing of nanoparticle concentration. This phenomenon occurs throughout the entire channel. On the other hand, a reduction in viscosity is observed in the radial direction. This decrease is attributed to an increase in nanofluid temperature at the channel wall.

The temperature profiles of nanofluids at varying particle volume concentrations, without consideration of thermodiffusion, are illustrated in Fig. 7.4a. As the concentration of particles increases, the temperature profile rises. This is due to the increasing thermal conductivity of nanofluids with rising particle concentration and decreasing heat capacity. Furthermore, it can be observed that the temperature gradient in the radial direction decreases with increasing nanoparticle concentration. The distribution of the thermal conductivity coefficient along the radius of the pipe is presented in Fig. 7.4b. As the nanoparticle concentration increases, the thermal conductivity coefficient of the nanofluid increases along the entire cross-section of the channel.

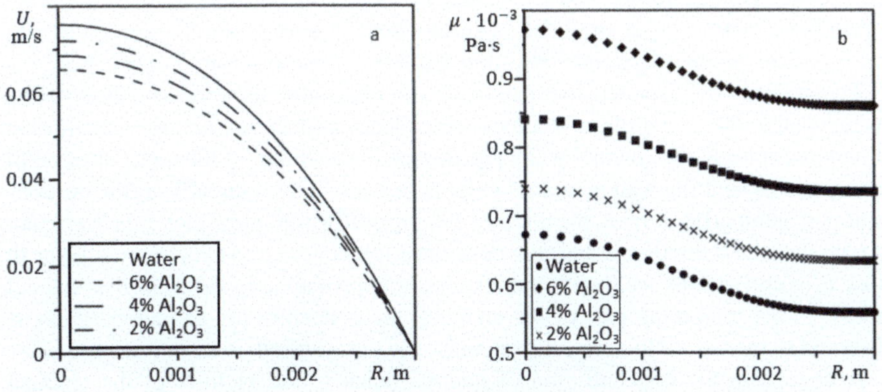

Fig. 7.3 Velocity (**a**) and dynamic viscosity (**b**) profiles in the channel cross-section for different particle concentrations without consideration of thermodiffusion

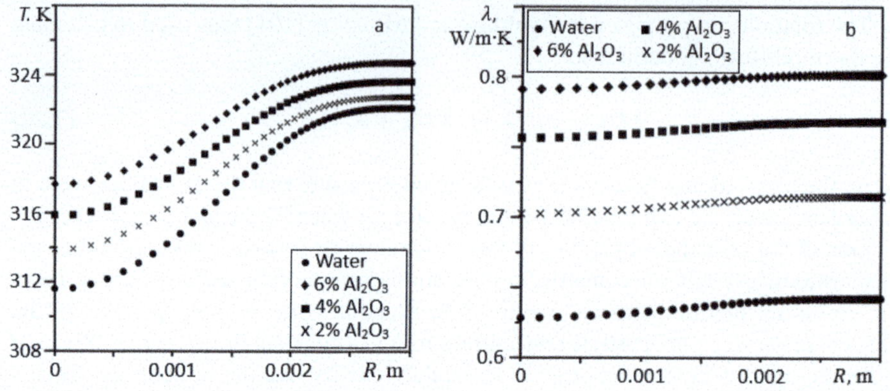

Fig. 7.4 Temperature (**a**) and thermal conductivity coefficient (**b**) profiles in the channel cross-section for different particle concentrations without consideration of thermodiffusion

In addition, a slight increase is observed at the channel wall. The latter is due to the higher wall temperature.

Nanofluids significantly change the local heat transfer coefficient. The corresponding data of the distribution of the local heat transfer coefficient along the channel are shown in Fig. 7.5. The lower curve corresponds to the base fluid and the subsequent ones represent nanofluids with particle concentrations of 2, 4, and 6% (bottom-up). As the nanoparticle concentration increases, the local heat transfer coefficient enhances. Analysis of the calculated data shows that the excess of the heat transfer coefficient with an increase in the particle concentration in the steady-state section is directly proportional to the increase in the thermal conductivity coefficient of the nanofluid.

Let us now analyze the effect of thermodiffusion. Its presence leads to a change in the concentration of nanoparticles in the channel, the particles distribution becomes heterogeneous. Figure 7.6 illustrates the nanoparticle concentration profiles for varying concentrations at the inlet, with the impact of thermodiffusion taken into account. The nanoparticle concentrations exhibit a variation from 2 to 6%.

The effect of thermodiffusion is to cause the migration of nanoparticles from the heated wall to the center of the channel, which leads to decrease in the concentration of nanoparticles at the channel wall. The degree of reduction of the concentration of nanoparticles near the wall increases with its increase. Consequently, as the concentration of nanoparticles rises, the impact of thermodiffusion intensifies. It is important to note that the reduction in particle concentration occurs in a very narrow layer in close proximity to the wall, with a thickness of approximately 2–3% of the channel radius. In the remaining part of the channel, the concentration of nanoparticles is observed to increase. However, this increase, as you can see, is insignificant.

A change in the concentration of nanoparticles in the vicinity of the channel wall has a marked effect on the thermophysical properties of the nanofluid in this region. As a consequence of the thermodiffusion of nanoparticles within the fluid,

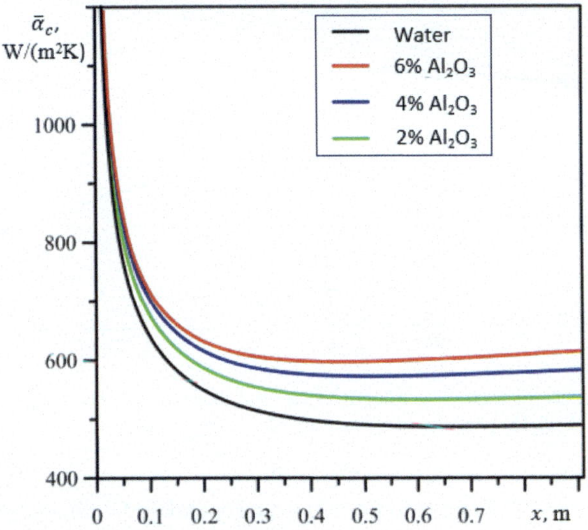

Fig. 7.5 The distributions of the local heat transfer coefficient for varying nanoparticle concentrations

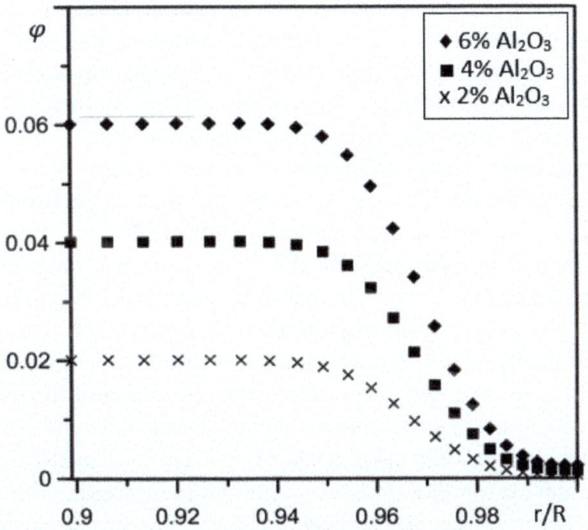

Fig. 7.6 Nanoparticle concentration profile near the channel wall

there is a notable reduction in the viscosity and thermal conductivity coefficients of the nanofluid in the vicinity of the channel wall. The corresponding dependencies are illustrated in Fig. 7.7. Here, in both cases, the liquid flow rate was the same and

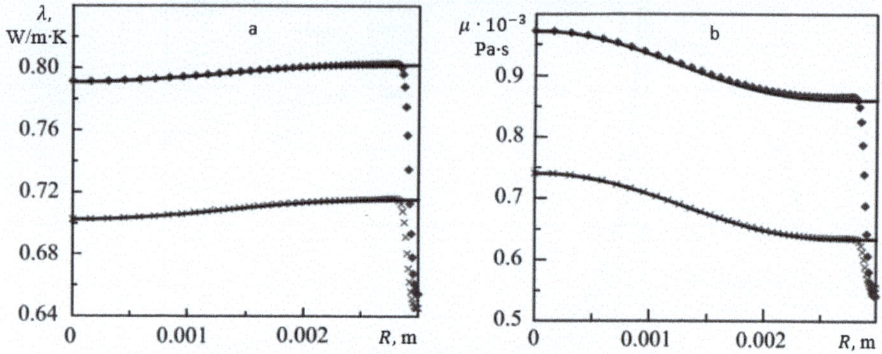

Fig. 7.7 Dependences of thermal conductivity (**a**) and viscosity (**b**) coefficients of nanofluid on the concentration of Al_2O_3 particles at a flow rate $G = 0.0011$ kg/s

equal to 0.0011 kg/s. The lines represent the data calculated without considering thermodiffusion, while the symbols indicate the computation points where nanoparticle thermodiffusion has been included. The lower curves correspond to a concentration of 2%, while the upper curves correspond to a concentration of 6%. The viscosity and thermal conductivity of the nanofluid at the channel wall decrease sharply.

Figure 7.8a illustrates the impact of thermodiffusion (TD) on the velocity profile at a nanoparticle concentration of 6%. Curve 1 represents the nanofluid with thermodiffusion taken into account, while curve 2 represents the nanofluid without it. The flow rate is equal to 0.0011 kg/s. Taking into account thermodiffusion results in a slight elevation in the flow velocity in the vicinity of the channel wall, accompanied by a decline in the center. This phenomenon is due to a decrease in the viscosity of the nanofluid caused by thermodiffusion of nanoparticles. Comparison of the velocity profiles at different coolant flow rates showed that the influence of thermal diffusion decreases with increasing flow rate. On the other hand, the temperature profile of the nanofluid differs significantly from that of the base fluid. A comparison of these profiles is presented in Fig. 7.8b, wherein the lower curve corresponds to the base fluid (water), and the upper symbols and lines correspond to nanofluids with particle volume concentrations of 1 and 6% (bottom-up). In this case, lines correspond to calculations in which thermodiffusion is not considered, and symbols correspond to calculations in which thermodiffusion is taken into account. The discrepancy between the velocity profiles of the base fluid and nanofluids increases with rising particle concentration. Thermodiffusion, however, has very little effect on the temperature profile in the channel. The phenomenon of thermodiffusion results in a slight increase in fluid temperature at the channel wall.

Thus, the thermodiffusion increases the flow velocity in the vicinity of the wall, which results in an additional enhancement of the heat transfer coefficient. It is important to note that the viscosity in the vicinity of the hot wall decreases even in the absence of thermodiffusion. A comparison of the calculated and measured data for the average heat transfer coefficient and pressure drop in the channel is presented

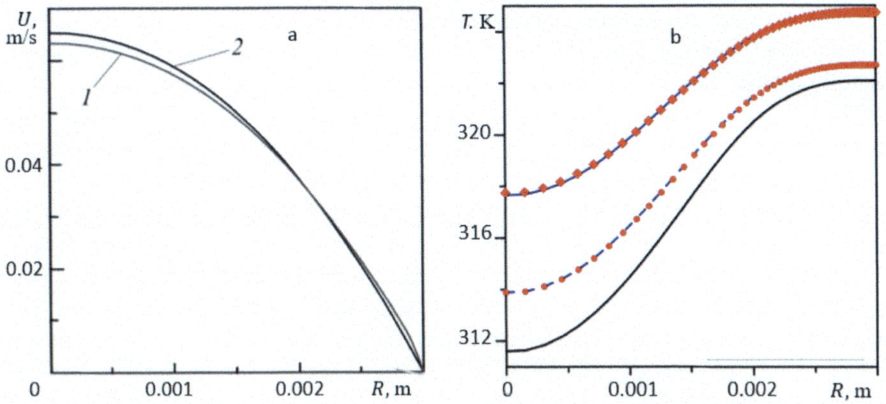

Fig. 7.8 Effect of thermodiffusion on velocity (**a**) and temperature (**b**) profiles of nanofluids

in Fig. 7.9. In this case, the dark symbols represent the experimental data, while the dashed and solid lines correspond to calculations without and with consideration of thermodiffusion, respectively. In all cases, the calculated data are in a good agreement with the experimental data. The maximum discrepancies do not exceed 3–5%. Comparison of the calculated data shows that taking into account thermal diffusion has a small effect only at high volume concentrations of nanoparticles, while at low concentrations of particles (1–2%) thermodiffusion practically does not affect the average heat transfer coefficient. Only at the highest particle concentration of 6%, the contribution of thermodiffusion becomes discernible, yet not exceeding 5%.

Thermodiffusion has an equally weak effect on the calculation of the pressure drop along the heated section (see Fig. 7.9b), and therefore on the hydraulic resistance. Taking it into account reduces the pressure drop by approximately 3% due to the decrease in the viscosity of the nanofluid near the channel walls. The anticipated

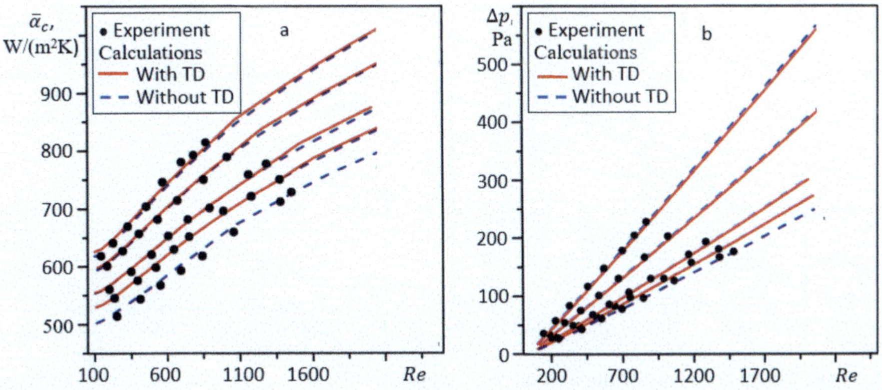

Fig. 7.9 Dependence of heat transfer coefficient (**a**) and pressure drop (**b**) on Reynolds number

pressure drop in nanofluid flows is considerably higher than that in water, since the viscosity coefficient of a nanofluid with a particle concentration of 6% is sixty percent higher than that of water.

The significant increase in viscosity of the nanofluid compared to the base fluid raises the question of the efficiency of its use both from the point of view of heat exchange intensification and energy costs for its pumping. In this context, a reasonable criterion is the energy input required for pumping, which is proportional to the pressure drop in the channel. If, at a fixed pressure drop, the nanofluid enhances the heat transfer coefficient in comparison to that of the base fluid, then the use of the nanofluid is effective. Figure 7.10a shows the average heat transfer coefficients as a function of pressure drop in the channel for water (lower curve) and nanofluids at particle concentrations of 1, 2, and 6%. For a fixed pressure drop, the heat transfer coefficient of the nanofluid enhances monotonically with increasing particle concentration. In practice, this means that for a given power required for pumping the heat carrier, the flow rate of the nanofluid will decrease compared to that of the base fluid. However, the heat transfer coefficient will still be higher than that of water. This means that the nanofluid in question is a much more efficient heat transfer medium than water.

Thus, thermodiffusion leads to a decrease in the concentration of nanoparticles near the channel walls, resulting in a decrease in local viscosity in this vicinity and an increase in flow velocity. This leads to a noticeable intensification of heat exchange, which enhances with increasing particle concentration.

Since the laminar flow regime has been considered, it is important to understand what primarily determines the heat transfer of nanofluids. Heat transfer coefficient enhances with nanoparticle concentration (see Figs. 7.5 and 7.9a). As the concentration of nanoparticles increases, so does the thermal conductivity of the nanofluid. From this point of view it is important to understand how the heat transfer coefficient and the thermal conductivity are related in this case. The answer to this question is

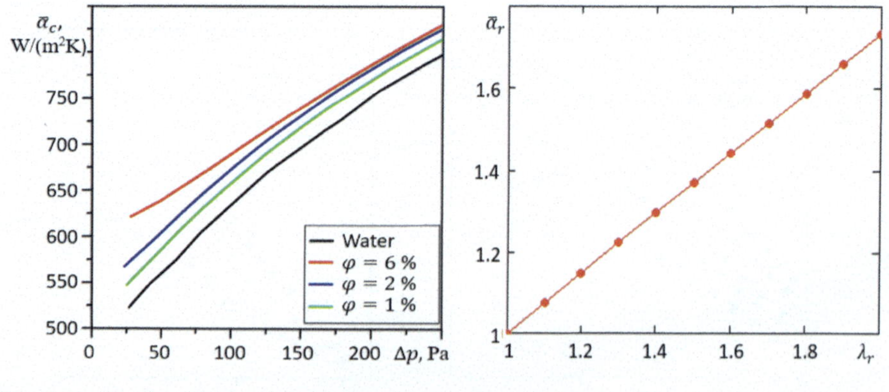

Fig. 7.10 The dependence of the average heat transfer coefficient of nanofluids on the pressure drop at different nanoparticle concentrations (**a**), and the dependence of the relative heat transfer coefficient on the relative thermal conductivity coefficient (**b**) at constant flow rate

given in Fig. 7.10b which demonstrates the dependence of the average relative heat transfer coefficient on the relative thermal conductivity coefficient at different particle concentrations and a fixed flow rate of 0.00302 kg/s. In this case, the thermal conductivity coefficient was determined using formula (7.18). The heat transfer coefficient increases in proportion to the nanoparticle concentration. It should be noted that this dependence is maintained up to concentrations of the order of 8–10%, since according to (7.18) at high nanoparticle concentrations the excess thermal conductivity starts to slow down.

7.3.2 Heat Transfer of Non-Newtonian Nanofluids

As demonstrated in Chap. 3, a considerable number of nanofluids with spherical particles are non-Newtonian. In this case, their flow and heat transfer can also be described by Eqs. (7.10)–(7.12), however the stress tensor, and hence the effective viscosity coefficient are functions of the shear rate.

As an example of describing the flows of such a fluid the forced convection of water-based nanofluids with 55 nm copper oxide particles is considered below. The volume concentration of the particles was varied from 0.25 to 2%. The corresponding experiment was described in detail in Sect. 6.2.4, and a simulation was performed in Minakov et al. [60]. The laminar flow regime is considered again. The rheology of the nanofluid is described by the power fluid model (3.29), whose parameters are provided in Minakov et al. [59].

The first effect, which manifests in both the present case and that of Newtonian nanofluid flows (see Fig. 7.8), is a transformation in the velocity profile. Figure 7.11 illustrates the velocity profiles of water and two nanofluids with particle concentrations of 1 and 2%. Here again, as in Newtonian fluids, the maximum velocity decreases with increasing concentration of nanoparticles, while on the other hand, the velocity on the wall increases slightly. This behavior is observed along the entire length of the channel. It should be noted, however, that in contrast to the Newtonian fluid, where the increase in velocity was due to the high temperature of the heated wall, another effect takes place in concerned case. The viscosity decreases also due to the presence of shear rate. For this reason, the flow velocity near wall in the non-Newtonian fluid increases much more than in case of the Newtonian fluid. For this reason, the heat exchange during the flow of a non-Newtonian fluid is significantly more efficient than during the flow of a Newtonian fluid under the same external conditions. Thus, the use of nanofluids with non-Newtonian rheology is promising for improving heat exchange.

A comparison of the numerical and experimental dependences of the average heat transfer coefficient on the Reynolds number is presented in Fig. 7.12. In this instance, the experimental data correspond to Fig. 6.10, while the colored symbols represent the calculated data. Since the fluid is non-Newtonian, the Reynolds number is determined in accordance with the relation (6.2). In general, computation results are in satisfactory agreement with the experimental data. The maximum discrepancy

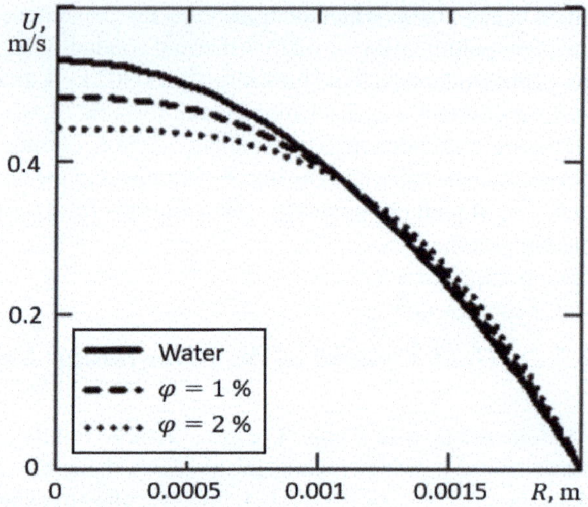

Fig. 7.11 Velocity profiles of the base fluid and nanofluids at the outlet of the heated section of the channel at different particle concentrations

between the calculated and experimental data is less than 5%. The enhancement of the heat transfer coefficient increases with increasing nanoparticle concentration and Reynolds number. At the maximum concentration of nanoparticles, the heat transfer coefficient of the nanofluid at a Reynolds number of 500 is approximately 1.7 times greater than that of water. It is important to note that the degree of excess heat transfer increases rapidly with increasing concentration of nanoparticles, and the curves in Fig. 7.12 become increasingly steep.

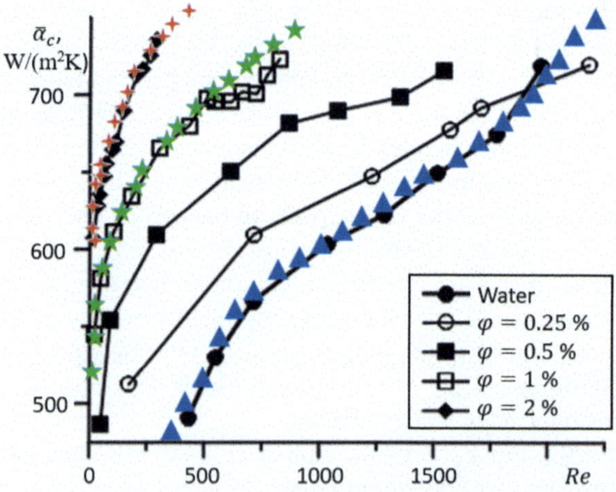

Fig. 7.12 Dependence of average heat transfer coefficient of water-based nanofluid with copper oxide particles on Reynolds number

7.3.3 Heat Transfer of Nanofluids with CNTs

There are relatively few attempts to simulate the heat transfer of nanofluids with CNTs. Here the approach is typical where the transport coefficients are described by conventional classical theories. Relevant examples can be found in Khan et al. [41], Arani et al. [4], Abedin et al. [3], Esmaeil et al. [38]. It has already been shown above that the key point of modeling is the use of adequate data on the transfer coefficients. Therefore, below heat transfer in a cylindrical channel of nanofluids based on isopropyl alcohol (IPA) and ethylene glycol (EG) with MWCNTs and SWCNTs is studied. To close the transport equations, measurements of the viscosity and thermal conductivity coefficients were used, which were carried out with the participation of the authors in the works [87, 94, 97–100]. Equations (7.10)–(7.12) were used for simulation. The nanofluids under consideration can be either Newtonian or non-Newtonian. In the latter case, the rheology of these fluids is described by the power law model (3.29), with the parameters determined empirically and presented in Table 7.2. Table 7.3 presents the values of the measured thermal conductivity coefficient and density of CNT nanofluids prepared based on above mentioned two fluids.

A numerical simulation was conducted to investigate the forced convection of a series of nanofluids with CNTs in a circular cylindrical channel. The weight concentrations of CNTs were varied between 0.05 and 0.25%. The outer and inner diameters

Table 7.2 Rheological characteristics of nanosuspensions with SWCNTs and MWCNTs

w%	IPA				EG			
	SWCNT		MWCNT		SWCNT		MWCNT	
	k_v, Pa s	n	k_v Pa s	n	k_v, Pa s	n	k_v, Pa s	n
0	0.00207	1.000	0.00207	1.000	0.0171	1.000	0.0171	1.000
0.05	0.162	0.543	0.00757	0.875	0,321	0.669	0.0344	0.956
0.1	0.679	0.409	0.0639	0.644	0.901	0.535	0.183	0.678
0.25	4.22	0.205	0.177	0.506	6.57	0.320	1.016	0.493

Table 7.3 Thermal conductivity coefficients and density of nanosuspensions with SWCNTs and MWCNTs

w %	Isopropanol (IPA)				Ethylene glycol (EG)			
	SWCNT		MWCNT		SWCNT		MWCNT	
	λ, W/(m K)	ρ, kg/m^3	λ, W/(m K)	ρ, kg/m^3	λ, W/(m K)	ρ, kg/m^3	λ, W/(m K)	ρ, kg/m^3
0	0.138	782.5	0.138	782.5	0.260	1112.0	0.260	1112.0
0.05	0.154	782.7	0.139	782.7	0.271	1112.2	0.262	1112.2
0.1	0.171	782.9	0.141	783.0	0.291	1112.4	0.263	1112.5
0.25	0.207	783.6	0.145	783.7	0.324	1113.1	0.267	1113.2

of the channel were 6 and 4 mm, respectively, resulting in a wall thickness of 1 mm. The length of the heated section was 0.9 m with a one-meter-long hydrodynamic flow stabilization section situated upstream of the measuring section. In the computations, the nanofluid flow rate was varied from 0.05 to 3 L/min, corresponding to Reynolds numbers ranging from 15 to 1500. The temperature of the nanofluid at the inlet to the computational domain was 23 °C.

The calculation domain consists of the flow part of the tube and the solid wall. The fluid velocity and temperature were defined at the inlet to the computational domain. A zero-pressure condition was imposed at the outlet. Furthermore, Neumann conditions were set for all considered quantities at the outlet, defined as: $\partial \Phi / \partial \boldsymbol{n} = 0$, where $\boldsymbol{n}$ is the vector of the external normal to the computational domain, and Φ is the set of all scalar quantities used in solving the problem A constant heat flux density was imposed at the outer wall of the tube. The problem was solved in axisymmetric formulation. The conditions of equality to zero of the derivative along the normal to the symmetry plane of all scalar quantities Φ, as well as the tangential $\boldsymbol{v}_{||}$ and normal $\boldsymbol{v}_{\perp}$ velocity components were set at the symmetry plane. The no-slip and non-leakage boundary conditions, $v_{\perp} = 0$, $v_{||} = 0$, were imposed at the solid walls. The conditions of no-diffusion fluxes for the volume fraction of the fluid and solid phases were set at the wall. In the computations, the Reynolds number (6.2) was varied by changing the fluid velocity at the inlet to the computational domain.

In Rudyak et al. [94, 98], the solution to the considered problem for forced convection of IPA-based nanofluid with CNTs was tested by comparing computational results with specially obtained experimental data (see also Chap. 6). During the experiments, the local heat transfer and average heat transfer coefficients were determined. Comparison of the experimental data (symbols) of the average heat transfer coefficient with the calculated data (solid lines) as a function of the mass flow rate of IPA-based nanofluid with SWCNTs is presented in Fig. 7.13a. The experimental heat transfer characteristics have already been discussed in the previous chapter. Note only that the simulation accuracy is quite high, with an error no more than 6%. The good agreement of experimental and calculated data is also evidenced by Fig. 7.13b, where the pressure drops in the channel are compared. Here, the deviation of the calculated pressure values from the corresponding experimental ones does not exceed 7%.

Since the primary objective of this study is to simulate the heat exchange properties of nanofluids, it is essential first to compare the velocity and temperature profiles in the channel of the nanofluid and base fluid. Below, we will consider an EG-based nanofluid with SWCNTs and MWCNTs. Figure 7.14 illustrates the variation in velocity profile within the channel during the flow of pure EG and EG-based nanofluids with SWCNTs and MWCNTs. The volume flow rate is 0.4 l/min. The data analysis for EG-based nanofluid with SWCNTs (Fig. 7.14a) demonstrates that with increasing CNT concentration, the maximum of the velocity profile decreases while the velocity near the wall increases. Thus, as the concentration of SWCNTs increases, the velocity profile flattens out. This effect is associated with alterations in the rheological properties of the nanofluid. As illustrated in Table 7.2, an increase in the concentration of SWCNTs is accompanied by an increase in the consistency

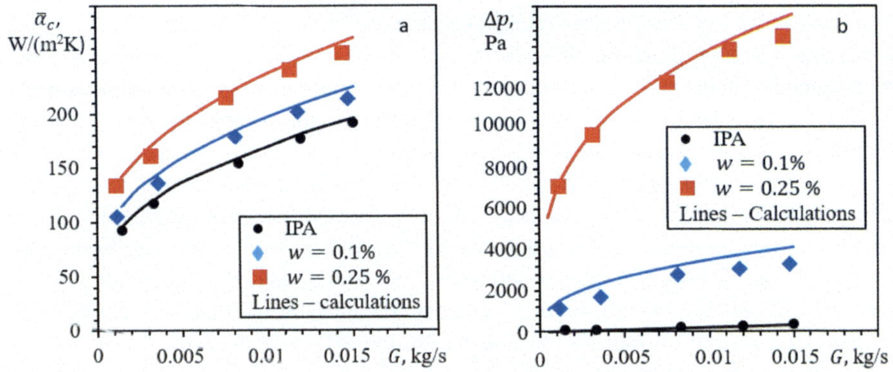

Fig. 7.13 Dependence of the average heat transfer coefficient (**a**) and pressure drop (**b**) on the nanofluid flow rate

index of the nanofluid. Conversely, the fluid index declines, resulting in the nanofluid becoming increasingly non-Newtonian.

The pseudoplastic nature of the rheology of nanofluids significantly changes the characteristics of flow and heat transfer in the channel. The distribution of nanofluid viscosity is inhomogeneous along the channel radius, exhibiting a notable decrease from the channel center to its walls. To illustrate, at a 0.25% SWCNT concentration, the viscosity in the center of the channel is approximately one thousand times higher than that near the wall. This is attributable to the fact that as the channel walls are approached, the shear rate increases significantly, resulting in a corresponding decrease in viscosity. The lower the fluid index, the more pronounced this effect is.

As a consequence of the flattening of the velocity profile, a region of quasi-solid flow is formed at the center of the channel. Since the velocity within the central part of the channel declines while the fluid flow rate remains constant, the velocity in the

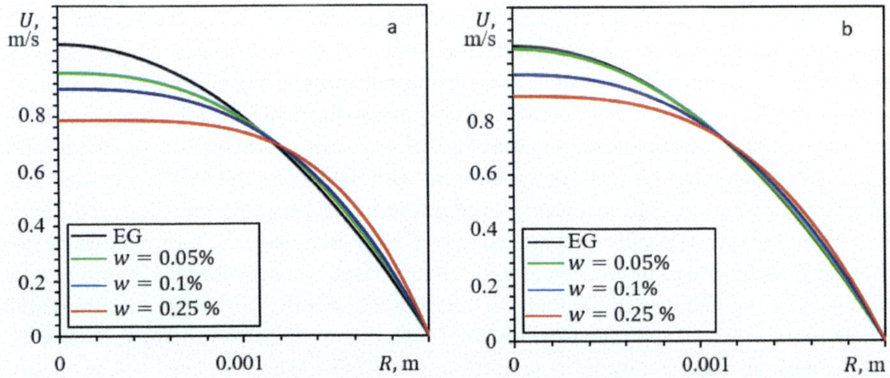

Fig. 7.14 Velocity profile at the outlet of the heated section for different concentrations of SWCNTs (**a**) and MWCNTs (**b**) at a flow rate of 0.4 l/min

near-wall region of the channel rises. The increase in velocity in the vicinity of the wall results in an enhancement of the heat transfer coefficient. This effect is more pronounced with increasing nanoparticle concentration. Therefore, the reinforcement of the non-Newtonian characteristics of nanofluids can notably contribute to the enhancement of heat transfer.

The flow of EG-based nanofluids with MWCNTs displays the same effects (Fig. 7.14b). Nevertheless, the degree of non-Newtonian behavior exhibited by these nanofluids is considerably less pronounced than that observed in nanofluids with SWCNTs (see Table 7.2). In consequence, the anticipated enhancement of heat transfer will also be less pronounced. Finally, it can be noted that with an increase in flow rate, due to an increase in shear rate, the effect of the nonlinearity of the nanofluid viscosity diminishes slightly. For this reason, the influence of the CNT concentration on the deformation of the velocity profile of the nanofluid will also be slightly reduced.

The deformation of the nanofluid velocity profile in comparison to the base fluid should also lead to a change in the temperature profile. This expectation is confirmed by the data presented in Fig. 7.15. The character of the temperature profile variation is significantly dependent on the nanofluid flow rate. At low flow rates (Fig. 7.15, top plots), an increase in CNT concentration results in a notable rise in fluid temperature at the center of the channel, accompanied by a decline near the wall. This effect is especially pronounced in the case of nanofluids with SWCNTs (see Fig. 7.15, top plots). Thus, in nanofluids with a 0.25% concentration of SWCNTs, the wall temperature decreases by approximately 4 degrees at a flow rate of 0.05 l/min and by 2 degrees at a flow rate of 0.4 l/min. In nanofluids with an equivalent concentration of MWCNTs, the wall cooling effect is approximately one degree.

At the same time, with increasing fluid flow rate (see Fig. 7.15, bottom plots), the temperature in the center of the channel practically does not differ from that in ethylene glycol, although a notable decline is observed at the channel wall, with a more pronounced reduction occurring at higher CNT concentrations. In this instance, the enhancement of heat transfer when using nanofluids with CNTs is accomplished by enhancing their thermal conductivity coefficient. Furthermore, an increase in the fluid flow rate results in a reduction in the thickness of the boundary layer. This is evident from a comparison of the lower and upper plots in Fig. 7.15.

A further key question, which is directly related to the heat transfer of nanofluids, is how the temperature changes along the channel wall. The corresponding distributions (T_w) are presented in Fig. 7.16 for a nanofluid with SWCNTs and MWCNTs at a flow rate of 0.05 l/min. In both instances, a reduction in temperature was observed along the channel length x. As the concentration of CNTs increases, the temperature of the channel wall decreases monotonically for all flow rates considered. This phenomenon is caused by the increase in the thermal conductivity coefficient of the nanofluid and the increase in its velocity near the channel wall due to the flattening of the velocity profile with an increase in the concentration of CNTs. The observed effect occurs for nanofluids comprising both types of CNTs, although it is significantly lower for nanofluids with MWCNTs than those with SWCNTs.

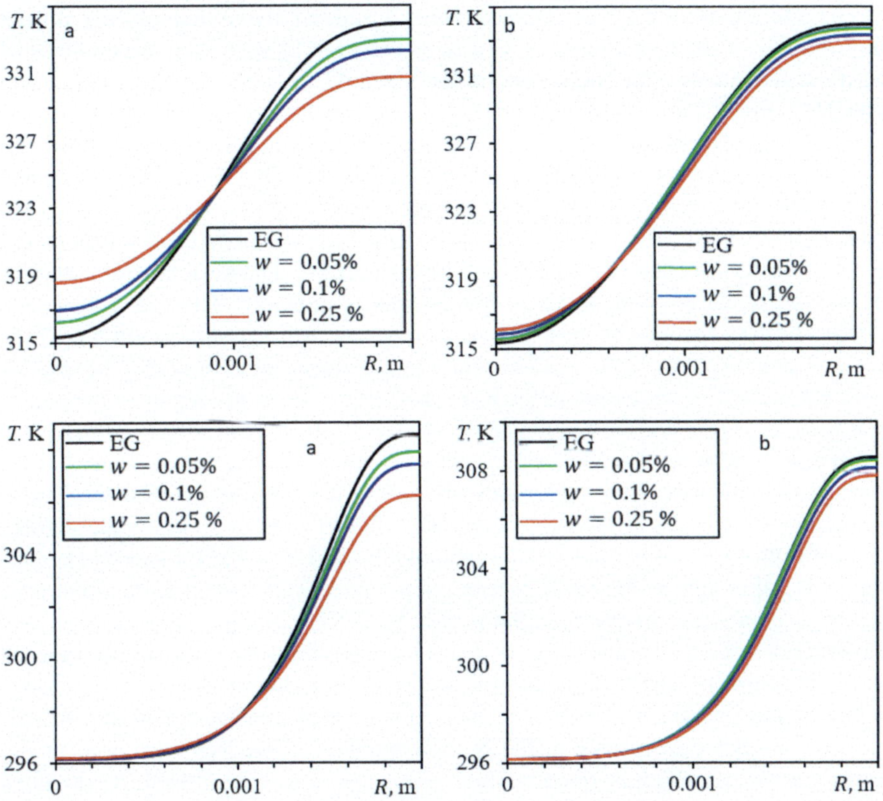

Fig. 7.15 Temperature profile at the outlet of the heated section for ethylene glycol-based nanofluid with SWCNTs (**a**) and MWCNTs (**b**) at flow rates of 0.05 l/min (top plots) and 0.4 l/min (bottom plots)

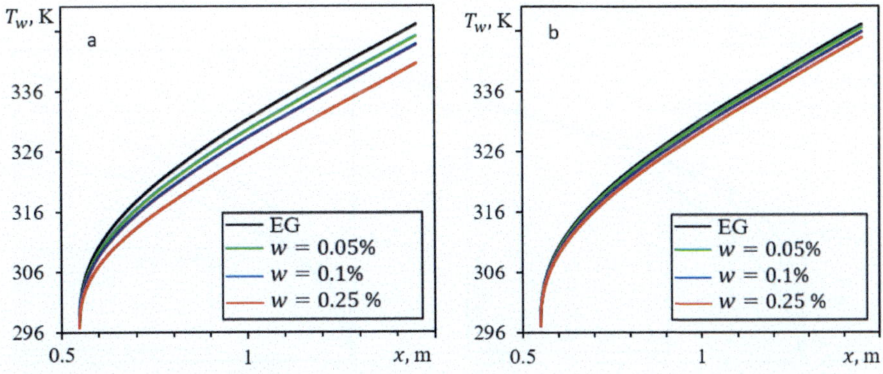

Fig. 7.16 Heated section wall temperature for different concentrations of SWCNTs (**a**) and MWCNTs (**b**) at a flow rate of 0.05 l/min *x*

The alterations in the flow characteristics described above undoubtedly impact the behavior of the heat transfer coefficient. Figure 7.17 shows the dependence of the average heat transfer coefficient on the Reynolds number for a nanofluid with SWCNTs (Fig. 7.17a) and MWCNTs (Fig. 7.17b). In all cases, the heat transfer coefficient increases monotonically with increasing CNT weight concentration. However, this enhancement depends significantly on the type of CNTs. Thus, within the range of Reynolds numbers from 10 to 1200, an excess heat transfer coefficient is observed for nanofluids with SWCNTs at all concentrations used. The greatest enhancement of heat transfer is observed at a concentration of 0.25%. The heat transfer enhancement is particularly pronounced at low Reynolds numbers. Thus, at a Reynolds number of 200, the excess heat transfer coefficient of a nanofluid with SWCNTs at a concentration of 0.05% is 23%, while at a concentration of 0.25%, it reaches approximately 95%. However, as the Reynolds number rises, the degree of the excess heat transfer coefficient of nanofluids decreases at all concentrations of SWCNTs, and at concentrations of 0.05 and 0.1 it even becomes negative.

The situation with nanofluids with MWCNTs (Fig. 7.17b) is somewhat more complex. In this case, at moderate Reynolds numbers (up to approximately 200), the heat transfer coefficient behaves qualitatively similar to that for nanofluids with SWCNTs, although its excess heat transfer is lower. Then the growth of the heat exchange coefficient of the nanofluid with MWCNTs sharply slows down. The maximum excess heat transfer is observed for nanofluids with a concentration of MWCNTs equal to 0.05%. The excess heat transfer is approximately 10% compared to that of pure ethylene glycol. Such behavior of heat transfer is associated with a rather complex dependence of the Nusselt number on the Reynolds number for power law fluids, which is given by the relation $Nu \sim Re^{1/(6-3n)}$ [10]. As the fluid index decreases, the Nu(Re) dependence becomes weaker, as illustrated in Fig. 7.17.

A similar picture of heat exchange is obtained when using isopropyl-based nanofluids. The corresponding data are presented in Fig. 7.18. There are notable differences in the physical properties of isopropyl alcohol and ethylene glycol. In

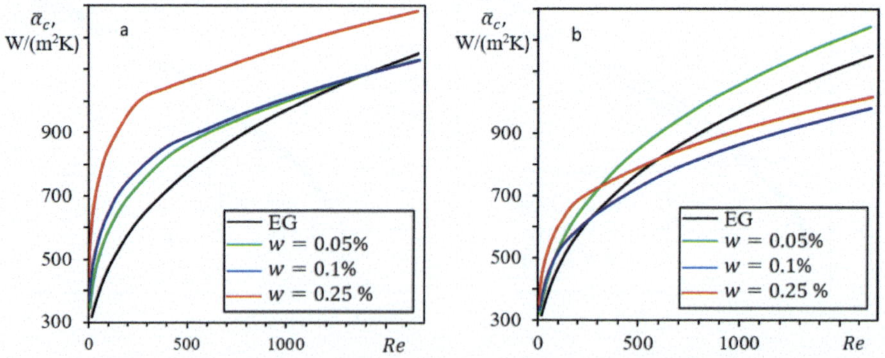

Fig. 7.17 Average heat transfer coefficient as a function of Reynolds number for EG-based nanofluids with SWCNTs (**a**) and MWCNTs (**b**)

particular, isopropyl alcohol has almost half the thermal conductivity of the ethylene glycol (see Table 7.3), and a density that is 40% lower. For this reason, it is a much worse heat carrier. As can be seen from a comparison of the graphs in Fig. 7.18, the heat transfer coefficient of isopropyl alcohol is almost three times less than that of ethylene glycol. On the other hand, the excess heat transfer coefficient of IPA-based nanofluids is observed to be greater than that of EG-based nanofluids. Thus, the IPA-based nanofluid with a 0.25% concentration of SWCNTs has been observed to enhance the heat transfer coefficient by a factor of 2.7, while the addition of MWCNTs has been found to increase the heat transfer coefficient by approximately 1.3 times. This behavior is due to the fact that the excess of the thermal conductivity coefficient of the nanofluid is higher, the lower the thermal conductivity of the base fluid. Indeed, at a concentration of 0.25% SWCNTs, the thermal conductivity coefficient of the EG-based nanofluid increases by 30%, while that of the IPA-based nanofluid increases by 53%.

The practical application of nanofluids, along with determining their heat transfer coefficient, also requires estimating the energy expended on pumping them. The latter is determined by the pressure drop along the length of the channel during the fluid flow. Figure 7.19 shows the dependence of the pressure drop on the Reynolds number for ethylene glycol and nanofluids based on it with CNTs. The dependences of pressure drop on Reynolds number for EG-based nanofluids with SWCNTs and MWCNTs differ. At moderate and medium Reynolds numbers the pressure drop in nanofluids with SWCNTs is higher than that in pure ethylene glycol (see Fig. 7.19a). But further, as the Reynolds number rises, the pressure drop in ethylene glycol surpasses that in the nanofluid. Concurrently, at a 0.25% SWCNT concentration, the heat transfer coefficient within this regime is 65% higher than that observed for ethylene glycol.

In nanofluids with MWCNTs, the situation is somewhat more complicated (see Fig. 7.19b). However, the behavior of the dependence of the pressure drop on the Reynolds number is consistent with the dependence of the heat transfer coefficient

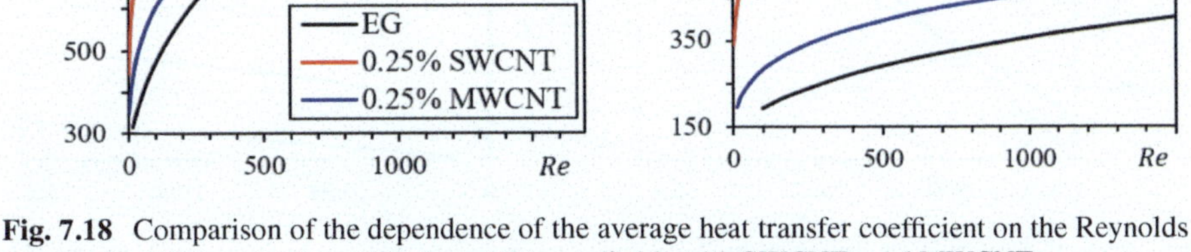

Fig. 7.18 Comparison of the dependence of the average heat transfer coefficient on the Reynolds number for EG-based (**a**) and IPA-based (**b**) nanofluids with SWCNTs and MWCNTs

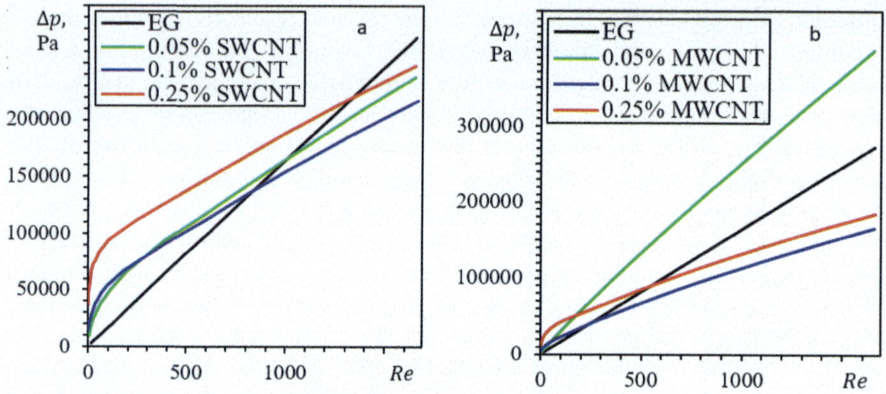

Fig. 7.19 Pressure drop as a function of Reynolds number for ethylene glycol and EG-based nanofluids with SWCNTs (**a**) and MWCNTs (**b**)

(see Fig. 7.18b). Thus, the nonlinear dependence of the heat transfer coefficient and pressure drop on the SWCNT concentration and Reynolds number allows identifying optimal flow regimes when applying nanofluids. In these regimes, nanofluids will not only enhance the heat transfer coefficient but also have lower pumping losses.

It is useful to compare the pressure drop when using ethylene glycol and isopropyl alcohol based nanofluids. The relevant data are presented in Fig. 7.20. It should also be noted that the viscosity of isopropyl alcohol is 8.5 times lower than that of ethylene glycol. Consequently, the pressure loss during the flow of IPA-based nanofluids is an order of magnitude lower than during the flow of EG-based nanofluids.

Since, as shown above, the use of nanofluids, on the one hand, increases the heat transfer coefficient, and on the other hand, leads to an increase in the pressure drop in the channel, a natural question arises about the effectiveness of their use as a heat carrier. Traditionally, when answering this question, the thermal–hydraulic efficiency

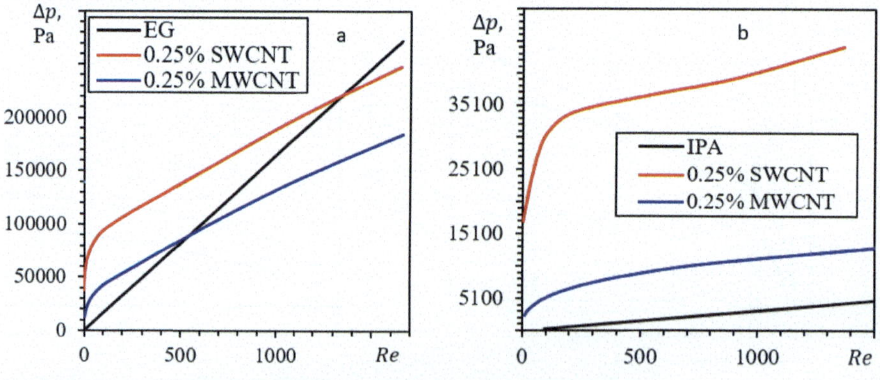

Fig. 7.20 Comparison of the pressure drop dependence on the Reynolds number for EG-based (**a**) and IPA-based (**b**) nanofluids comprising SWCNTs and MWCNTs

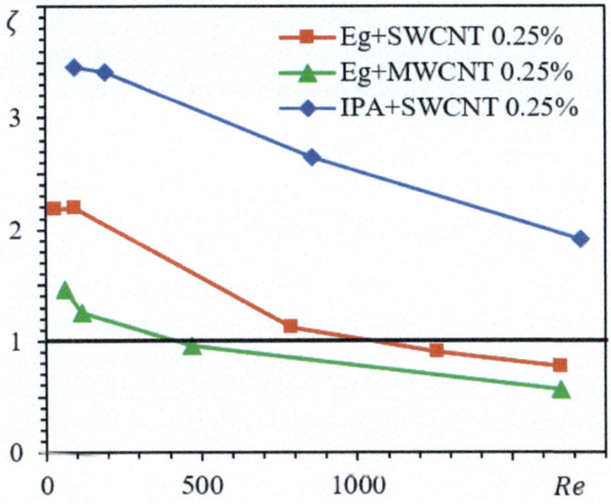

Fig. 7.21 Thermal–hydraulic efficiency of EG-based and IPA-based nanofluids with SWCNTs and MWCNTs

is studied, which is determined using the following coefficient

$$\zeta = \left(\frac{Nu_{nf}}{Nu_{bf}}\right)\left(\frac{\xi_{nf}}{\xi_{bf}}\right)^{-1}, \tag{7.21}$$

where the indices *nf* and *bf* refer to the nanofluid and the base fluid, respectively, and ξ is the hydraulic resistance coefficient.

Figure 7.21 shows the dependence of thermal–hydraulic efficiency (7.21) on the Reynolds number for nanofluids with CNTs at their weight concentration equal to 0.25%. The analysis of the behavior of the thermal–hydraulic efficiency of nanofluids demonstrates that the latter declines with an increase in the Reynolds number. However, with increasing CNT concentration, the range of Reynolds numbers at which the thermal–hydraulic efficiency exceeds unity expands. The most effective in this case is the use of nanofluids with SWCNT. At low Reynolds numbers, the hydraulic efficiency of such nanofluids can reach 3.5, which makes the application of such nanofluids for heat transfer enhancement extremely effective. A comparison of the thermal–hydraulic efficiency of IPA-based and EG-based nanofluids shows that, under equivalent conditions, the former exhibits a markedly higher efficiency.

7.4 Effect of Slip Flow Regimes on Heat Transfer

The thermal conductivity coefficient of nanofluids is markedly higher than that of the corresponding base fluids. It is therefore reasonable to conclude that nanofluids can be used to enhance the efficiency of heat transfer. However, nanofluids also have significantly higher viscosity. Consequently, from a practical perspective, it is necessary to maximize heat transfer on the one hand, and to minimize drag as much as possible on the other. To this end, hydrophobic or even superhydrophobic coatings are used (see [5, 13, 19, 35, 36, 115, 117] and the literature cited therein). Such coatings may prove particularly effective in microchannels. In the simulation, the presence of hydrophobic coatings is accounted for by setting slip conditions instead of the no-slip boundary conditions that are typical in fluid dynamics. At the slip conditions the velocity of the fluid on the channel wall differs from the wall velocity. The slip conditions are typical for rarefied gas flows. In this case, the slip velocity is proportional to the free path length of the molecules and depends on the accommodation coefficients that determine the efficacy of the interaction between the gas molecules and the wall [15, 26]. Nevertheless, slip at hydrophobic walls in microchannels occurs also when using liquids. Moreover, the slip length has been observed to reach tens of micrometers [8, 17, 43, 69, 111]. It is to be expected that the presence of slip at the wall will result in a significant reduction in hydraulic resistance. At the same time, it is important to understand the effect of slip on heat transfer in the channel. There is another important nuance here.

In general, flow slip on the channel wall occurs not only due to the presence of a flow velocity gradient near the surface, but also due to the presence of a temperature gradient in non-isothermal flows. So that the flow velocity at the boundary is determined in general by the Maxwell-Smoluchowski relation [53, 77, 103]

$$u_f = u_w + k_1 \left(\frac{\partial u}{\partial y} \right)_w + k_2 \left(\frac{\partial T}{\partial x} \right)_w, \tag{7.22}$$

where u_f and u are the fluid velocities, u_w is the velocity of the channel wall, T is the temperature, the index w denotes the corresponding quantity at the wall. x, y are the coordinates along and perpendicular to the channel wall, k_1 and k_2 are the slip and thermal slip coefficients, respectively. Later in this book they are also referred to as slip length and thermal slip length, which is also referred to as thermal creep in the literature.

The second significant factor is that similarly as in the case of rarefied gases [15, 77], a temperature jump generally occurs on the channel wall due to the presence of a temperature gradient across the flow

$$T_f = T_w + k_3 \left(\frac{\partial T}{\partial y} \right)_w. \tag{7.23}$$

The slip coefficient k_1 in a rarefied gas is dependent upon the momentum accommodation coefficient. Similarly, the coefficient k_2 is determined by the law of gas molecule interaction with the surface and is dependent upon the energy (temperature) accommodation coefficients. Furthermore, it is dependent upon the specific heat capacity of the gas and the Prandtl number. The coefficient k_3, which determines the temperature jump, depends on the same parameters.

As previously stated, in fluid mechanics, the no-slip boundary conditions on the wall are typically employed, which corresponds to the absence of slip. Nevertheless, in a number of cases, especially in microchannels, the presence of slip has been experimentally recorded. Its appearance is associated with various mechanisms including hydrophobic and ultrahydrophobic coatings, the presence of so-called gas lubrication, roughness, etc. [17, 35, 36, 43, 52, 67, 69, 112, 114, 116]. The dependence of the k_i coefficients on the flow parameters is poorly understood. It is only known that they can vary over a rather wide range. The objective of this section is to examine the impact of slip (7.22) and temperature jump (7.23) on the heat transfer characteristics. In this instance, the coefficients k_i are assumed to be constants, exhibiting variation over a sufficiently wide range. The complexity of simulating such non-isothermal flows also stems from the fact that the transport coefficients are dependent on temperature. Henceforth, these coefficients will be determined by relations of the type (7.16).

The study of the effects of slip and temperature jump on heat transfer is presented below with the example of fluid flow in a straight plane microchannel of rectangular cross-section [50]. The height and width of the channel was 10 μm and the length was 100 μm. An initial flow stabilization section of 50 μm length with adiabatic walls was installed upstream of the investigated channel. Given the significant influence of temperature on fluid properties, two series of computations were conducted, each with a distinct temperature head (defined as the maximum temperature difference between the wall and the fluid at the inlet). In the first case, the wall temperature was set to 300 K, while in the second case it was set to 350 K. The inlet fluid temperature was maintained at 290 K in both cases. Additionally, a parabolic velocity profile was set at the inlet. The boundary conditions were defined by relations (7.22) and (7.23), and the effect of k_i coefficients on heat transfer was studied sequentially.

7.4.1 Effect of Flow Slip on the Wall on Heat Transfer

In this section, we will examine the impact of the standard slip on the wall, characterized by the coefficient k_1, on the heat transfer process. In the computations, the coefficient was varied between 10^{-7} and 2.5×10^{-5} (in appropriate units). In this instance, the coefficients k_2 and k_3 were set to zero. The results obtained were then compared with the data from the corresponding computations performed with no-slip boundary conditions on the wall. In all cases, the velocity, temperature, and density profiles were analyzed.

Figure 7.22 illustrates the velocity distributions in the cross-section at the microchannel outlet. The upper line corresponds to the velocity profile that is obtained under no-slip conditions. The subsequent lines (from top downward) correspond to the slip coefficients equal to $k_1 = 5 \times 10^{-7}, 10^{-6}, 2.5 \times 10^{-6}, 10^{-5}, 2.5 \times 10^{-5}$. Taking slip into account leads to obvious consequences: the flow velocity on the wall becomes non-zero, and the velocity profile flattens out. This effect increases with increasing k_1. It is useful to understand how the velocity profile is affected by

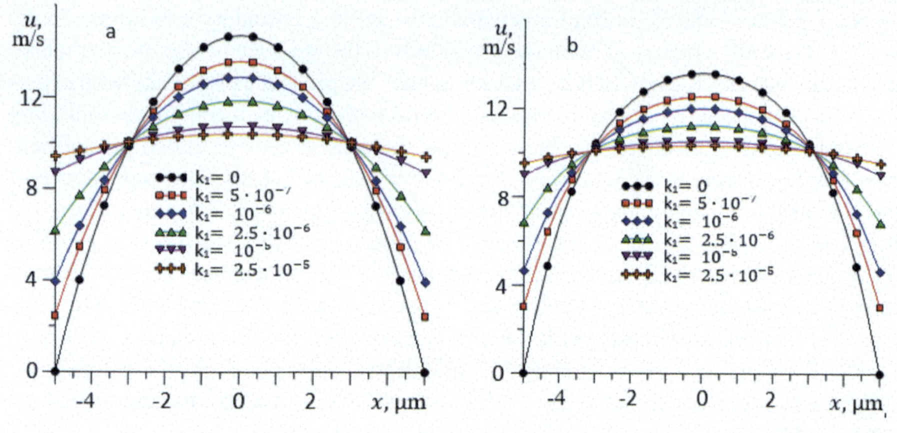

Fig. 7.22 Velocity profile in the microchannel cross-section at different values of k_1. Temperature head: 10 (**a**) and 60 K (**b**)

the temperature head. Figure 7.22a presents data for a temperature head of 10 K, while Fig. 7.22b shows those for a temperature head of 60 K. It can be observed that as the temperature difference increases, the velocity profile becomes wider and the maximum velocity decreases.

The presence of slip on the wall should result in a reduction in hydraulic resistance, which in turn should lead to a reduction in pressure drop along the channel. This behavior is confirmed by conducted computations. Figure 7.23 shows the distributions of pressure drop along the microchannel length for both temperature heads. The lines from top downward correspond to the same values of the slip coefficient as in Fig. 7.22. As the coefficient k_1 increases, the pressure drop naturally decreases. The presence of slip on the walls in the investigated microchannel affects the pressure drop in the whole microchannel. Comparison of Fig. 7.23a, b reveals a significant influence of the temperature difference between the wall and the liquid at the inlet on the pressure drop. At high values of temperature head, the pressure drop in the channel is significantly less than at low values. This is due to the observed decrease in both the viscosity coefficient and density with increasing temperature. Finally, it is interesting to note that when $k_1 \geq 10^{-5}$, the pressure drop becomes negative or close to zero (see Fig. 7.23a). A similar effect is observed at high temperature heads (Fig. 7.23b). Thus, the use of such ultrahydrophobic coatings can lead to the appearance of reversing flows in non-isothermal flows, or even the microchannel can begin to work as a pump.

Finally, it is necessary to study the variation of heat flux q in the microchannel. The corresponding data are presented in Fig. 7.24. The lines from bottom to top correspond to the same values of the slip coefficient as in Fig. 7.22. In general, varying the slip length has little effect on the heat flux profile (and temperature) in the channel. The pattern is almost identical for both small and large temperature heads. However, the heat flux density increases with an increase in slip length. This

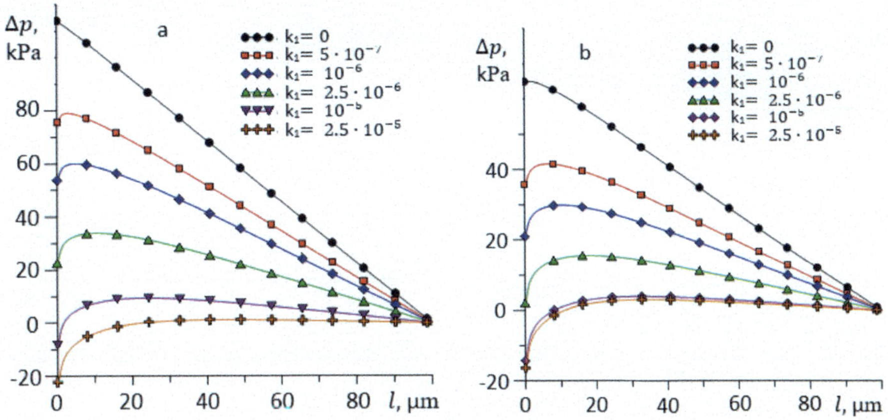

Fig. 7.23 Variation of pressure drop in a microchannel. Temperature head: 10 K (**a**) and 60 K (**b**)

increase is particularly significant at the initial part of the microchannel. At a distance of one caliber, the heat flux at maximum slip length exceeds the corresponding value at no-slip condition by approximately 40% (see Fig. 7.24). Nevertheless, change in the temperature head has a negligible impact on the relative increase in heat flux density (compare Figs. 7.24a, b).

It can be reasonably assumed that the notable impact of slip on the heat flux density will also be reflected in the heat transfer. The dependence of the average heat transfer coefficient on the slip coefficient k_1 is presented in Fig. 7.25. The excess of the average heat transfer coefficient at large values of k_1 compared to the corresponding value for flow without slip is almost 50%. This excess is practically independent of the temperature head. However, it is important to note that starting from certain values of k_1, the heat transfer coefficient practically ceases to change. In this case

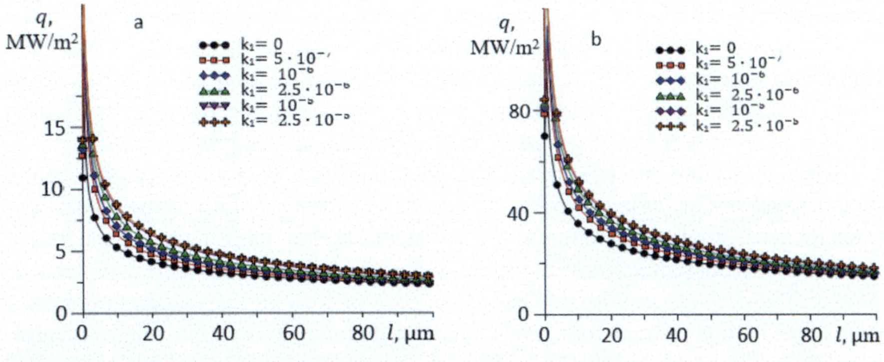

Fig. 7.24 Distribution of heat flux density along the microchannel at different values of k_1. Temperature head: 10 (**a**), 60 K (**b**)

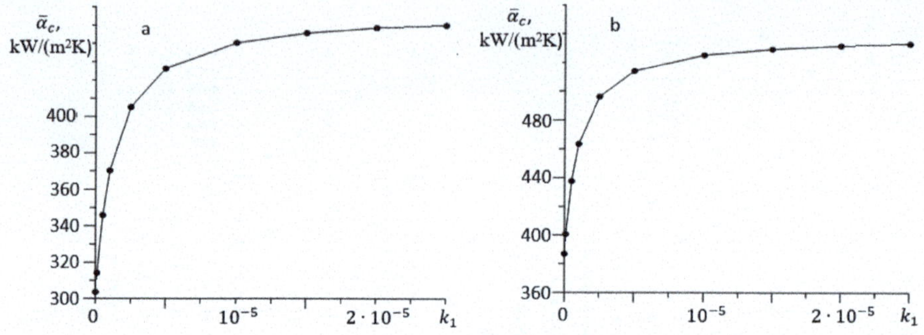

Fig. 7.25 Dependence of the average heat transfer coefficient on the slip coefficient k_1.Temperature head: 10 (**a**) and 60 K (**b**)

the temperature at the microchannel outlet also practically ceases to depend on the slip coefficient.

7.4.2 *Effect of Thermal Slip on the Wall on Heat Transfer*

Let us now consider the effect of thermal slip on the flow characteristics (the second term in Eq. (7.23)). The coefficient k_2 was varied from zero to 10^{-5}. The coefficient k_1 was fixed and equal to 10^{-6}, while $k_3 = 0$. The results obtained were compared with the data of the corresponding computations carried out with no-slip boundary conditions on the wall. It is shown that changing the k_2 coefficient has practically no effect on the velocity distribution in the microchannel outlet cross-section, and the differences do not exceed a few tenths of a percent. At the same time, in the case of a large temperature head, a smoother velocity distribution is observed, with a decrease in the maximum value in the center of the microchannel, as is the case for usual slip. Although the k_2 coefficient has almost no effect on the velocity distribution in the microchannel, its variation significantly affects the pressure drop (see Fig. 7.26a). The significant influence of thermal slip occurs in the initial section and continues until approximately the middle of the microchannel. As a result, the velocity distribution along the microchannel is also significantly transformed (see [50]).

When the thermal slip coefficient is varied, the fluid temperature profiles at the microchannel outlet undergo slight alterations. The temperature at the center and at the microchannel walls remains almost identical. For large temperature heads, the differences in temperature distribution become somewhat more pronounced; however, their deviation with respect to the case of a small thermal head does not exceed 5%. At low temperature heads, the heat flux density exhibits minimal variation with increasing thermal slip coefficients, as illustrated in Fig. 7.26b. The main change occurs only at the channel inlet, and this change is qualitative. Up to distances of the order of 15 μm, the heat flux density increases with increasing k_2 coefficient,

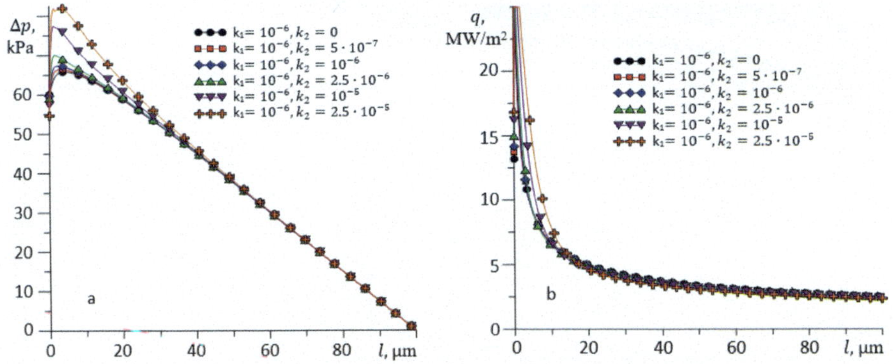

Fig. 7.26 Dependence of the pressure drop in the microchannel (**a**) and distribution of heat flux density along the microchannel (**b**) at different values of k_2, and temperature head of 10 K

and subsequently decreases. This qualitative behavior persists even at a large temperature head, although the change in the direction of heat flux evolution occurs at larger distances from the channel inlet. Certainly, the most important issue is the effect of thermal slip on heat transfer. It has been shown that the average heat transfer coefficient rises almost linearly with the thermal slip coefficient, however the maximum increase in the heat transfer coefficient is approximately 10%.

7.4.3 Effect of Temperature Jump at the Wall on Heat Transfer

Finally, let us consider the effect of a temperature jump on the wall on the heat transfer. It is assumed that slip also takes place with $k_1 = k_2 = 10^{-6}$ while the coefficient k_3 varies from zero to 2.5×10^{-7}. From general considerations it is evident that the existence of a temperature jump should have no discernible impact on the velocity profile. The calculations carried out confirmed this; the differences do not exceed a few tenths of a percent, both at low and high thermal pressure. At the same time, the temperature jump has an impact on the pressure drop in the microchannel, although this is only discernible at high temperature heads. In this case, the pressure drop increases in a linear fashion with rising k_3, reaching a maximum value of nearly 20%. Qualitatively, this is quite explainable and is associated with a certain decrease in temperature on the wall and, as a consequence, with a corresponding increase in the temperature of the fluid compared to the case of slip or no-slip flow. This conclusion is also confirmed by the computation of temperature isolines along the microchannel.

The possible influence of a temperature jump on the temperature profile distribution is quite predictable. The temperature decreases almost proportionally to the

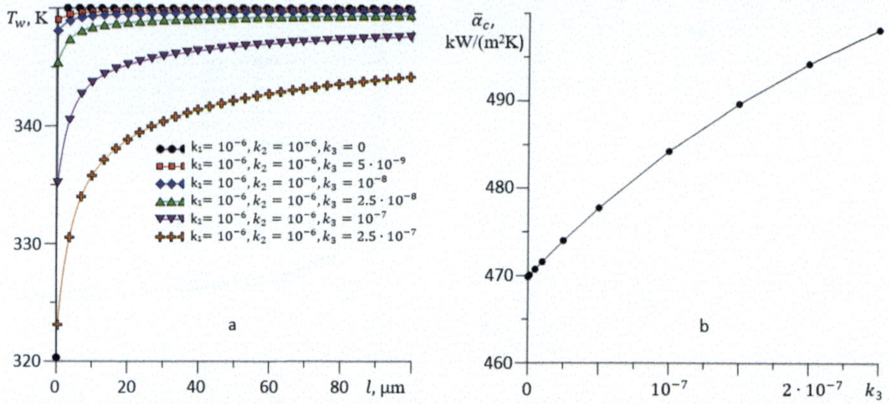

Fig. 7.27 The distribution of the wall temperature along the length of the minichannel (**a**), and the dependence of the average heat transfer coefficient (**b**) at different values of k_3. The temperature head is 60 K

increase in the coefficient k_3. Since this coefficient was quite small in the calculations, the temperature drop on the wall did not exceed 9 K. In addition, a slight broadening of the temperature profile was observed. It is natural that this broadening was greatest in the immediate vicinity of the wall, reaching approximately 8%. However, the change in the channel wall temperature with an increase in the coefficient k_3 nonlinearly decreases along the channel. The temperature distribution along the channel wall at high thermal head is presented in Fig. 7.27a. At low temperature head, the behavior is qualitatively similar, but the magnitude of the temperature decrease is appropriately reduced.

The heat flux and heat transfer coefficient are most significantly influenced by the temperature jump at the wall. A reduction in fluid temperature at the wall results in a corresponding decline in heat flux density with an increase in coefficient k_3. At its maximum values this decrease reaches 14%. At the same time, the difference in temperature heads has practically no effect. Nevertheless, the heat transfer coefficient increases with increasing k_3 (see Fig. 7.27b). The effect of different temperature heads is just quantitative, however, in nature they are qualitatively similar. At high temperature heads the maximum enhancement in the heat transfer coefficient reaches about 6%, while at low temperature heads it is about 3%.

7.4.4 Concluding Remarks

In conclusion, first of all, it can be noted that physically the factors determining different types of slip and temperature jump are significantly different. The occurrence of usual slip (the first term in Eq. (7.23)) is due to the velocity gradient perpendicular to the channel wall, which is almost always present in shear flows. Such slip is achieved when hydrophobic or even superhydrophobic coatings are employed, in the event of the presence of residual gas in wall roughness, and so forth.

The slip caused by the second term in Eq. (7.23) is a thermal creep, which is associated with inhomogeneous temperature distribution along the channel wall. In practical situations, such inhomogeneity during the cooling of a heated surface almost always occurs. Finally, the temperature jump at the wall (formula 7.24) is caused by the temperature gradient across the channel and also always occurs when the working fluid is employed for cooling the channel walls. Moreover, there are also various physical mechanisms. In a rarefied gas, this is the presence of a Knudsen layer near the surface [15, 104]. In liquids near the surface, liquid structuring is observed [110], Kapitsa resistance is manifested [40], and some other factors.

Thus, the influence of each of the aforementioned factors on heat exchange is extremely important, since conditions can be realized when they influence heat exchange independently. It is also worth noting that there are instances when these factors exert their influence on heat transfer simultaneously. As the slip length increases, the heat flux density rises and can reach a value that is approximately 40% higher than that corresponding to no-slip conditions. As a result, the excess average heat transfer coefficient in comparison to the corresponding value in no-slip flow can reach up to 50%. Naturally, in this case, the pressure drop is significantly reduced, thereby resulting in two beneficial effects occurring concurrently. Furthermore, at sufficiently large slip lengths reversing flows can be generated or even the microchannel can commence operation in a pump regime. The latter represents a noteworthy physical phenomenon that would be intriguing to investigate experimentally.

The presence of thermal creep also affects the pressure drop in the channel, resulting in its decrease. However, this influence is significantly smaller than that of the slip length. The mean heat transfer coefficient at thermal creep increases almost linearly with the rise in the k_2 coefficient, reaching a maximum of 10%. This effect increases with the degree of heating nonuniformity of the channel walls.

The existence of a temperature jump at the wall in contrast to slip mode can lead to some increase in the pressure drop. However, the heat transfer coefficient enhances. It is worth noting that when modeling heat exchange, it is necessary to take into account the dependence of all transport coefficients and other thermophysical properties of fluids on temperature in the calculation process. Otherwise, the resulting data will be significantly disparate in terms of both their value and quality. To illustrate, when the k_3 coefficient is increased at a constant fluid thermophysical characteristics, the microchannel pressure is observed to decrease, while in fact it increases.

Finally, it is important to note that the pressure drop in the channel is expected to decrease with an increase in temperature head. This decrease is found to be almost proportional to the increase in slip length. The impact of thermal creep on the pressure drop is only discernible at the maximum temperature head and exhibits a minimal dependence on the k_2 coefficient. On the other hand, the presence of a temperature jump at the wall surface results in a slight increase in pressure drop.

7.5 Simulation of Flow and Mixing Regimes in a Micromixer

As previously noted, microsystems technology is developing rapidly for a variety of purposes. One of the dominant research areas in modern microfluidics is the development of methods for controlling flow and heat transfer in microchannels. The surface-to-volume ratio in microfluidics devices is several orders of magnitude greater than that of their macroscopic counterparts. As a result, it is feasible to achieve markedly enhanced heat transfer coefficients, a sharp increase in transverse temperature and concentration gradients, a diminution in apparatus dimensions, material intensity, and etc. Mixing is one of the key processes in a multitude of such applications. In order to achieve this, specialized devices, known as micromixers, have been developed [2, 45, 64, 73]. A variety of micromixers are employed, including those that are passive (e.g., change of channel geometry, wall surface structure, and various inserts) and those that are active (e.g., change of flow rate and effect of external fields). The Y-mixer, or its special case, the T-mixer, represents the base-type micromixer (see, e.g., [16, 72, 73, 91, 109]). In the study conducted by Minakov et al. [57, 58, 62], the flow regimes in T-type microchannels at Reynolds numbers ranging from 1 to 1000 were systematically investigated. The study revealed five significantly different flow regimes, as follows: (i) stationary vortex-free flow at $\mathrm{Re} < 5$; (ii) stationary vortex symmetric flow at $5 < \mathrm{Re} < 145$; (iii) stationary vortex asymmetric flow at $145 < \mathrm{Re} < 240$; (iv) unsteady periodic flow a $t240 < \mathrm{Re} < 400$; and (v) practically stochastic flow regime at $\mathrm{Re} > 400$. It was shown that the mixing efficiency increased significantly when the flow regime was changed. Furthermore, an increase in Reynolds number resulted in a corresponding enhancement in mixing efficiency. These findings were subsequently validated through experimental investigation [62]. A comprehensive review of this subject can be found in the monograph [86]. In many cases, nanofluids can also be used in micromixers. An example of flow and mixing regimes implemented in this case is given below.

7.5.1 *Simulation of Fluid and Nanofluid Mixing*

The mixing of water and a water-based nanofluid with aluminum oxide particles in a T-type micromixer is considered in this section [48]. Nanofluids with particle sizes of 50 and 150 nm were used at volume concentrations of 2.5, 5, and 10%. The height of the micromixer channels was 200 μm, and the widths of the supply and mixing channels were 200 and 400 μm, respectively. The problem was solved using Eqs. (7.10) and (7.10a) with the viscosity coefficient determined by formula (3.18). The coefficients a and b were determined experimentally [48], and the diffusion coefficient of nanoparticles was described by relation (7.19). The isothermal problem at a temperature of 25 °C was considered.

Comparing to the base fluid the nanofluid has higher density and viscosity. The density increases in proportion to the volume concentration of nanoparticles and at the highest concentration exceeds the density of water by almost 30%. On the other hand, the viscosity of the fluid increases non-linearly with concentration and is much higher than that of water. It is evident that the simultaneous action of these two factors will markedly alter the flow characteristics of the mixed fluids within the micromixer. In addition, the nanofluid viscosity depends on the nanoparticle size, whereas the density remains unaffected. It would be prudent, therefore, to initially examine the impact of nanoparticle volume concentration and size on the flow and mixing behavior separately.

Thus, the mixing of water with a water-based nanofluid comprising aluminium oxide particles with an average particle size of 150 nm is considered. Water is introduced into the mixer at a flow rate G_{in} through one of the inlets, while the nanofluid is introduced through another inlet at the same flow rate and given volume concentration of nanoparticles. Therefore, the total flow rate within the mixing channel is $G = 2\,G_{in}$. The transition from one flow regime to another within the micromixer is defined by the Reynolds number

$$\mathrm{Re} = (\rho U d_h)/\mu, \tag{7.24}$$

where $U = G/(2\rho H^2)$ is the average velocity in the mixing channel determined from the flow rate, H is the channel height, and $d_h = 267$ μm is the hydraulic diameter of the channel.

When determining the density of the nanofluid by means of the formula (7.9), it was assumed that the density of the nanoparticles was 3920 kg/m^3. Consequently, at the maximum particle concentration, the density of the nanofluid is 1.29 times that of water, while the viscosity coefficient of the nanofluid is 2.35 times that of water. Similarly as with the mixing of two homogeneous fluids [47, 57, 62], several flow regimes can be identified when a conventional fluid is mixed with a nanofluid. Small Reynolds numbers ($\mathrm{Re} < 10$) were observed to result in a creeping vortex-free flow with rather poor mixing. In the region of Reynolds numbers between 10 and 140, a stationary vortex-symmetric flow was observed, exhibiting a lower degree of mixing efficiency than that observed in the preceding regime. The mixing of water and nanofluid has characteristic features caused by the diffusion of nanoparticles. At the initial stage, water is introduced into the micromixer via one inlet, while the nanofluid, with a specified particle volume concentration φ is introduced through the other inlet. Upon reaching a state of complete mixing, the mixing channel will produce a nanofluid with some average concentration. If the inlet flow rates are identical, the final concentration of the mixed nanofluid will be $\varphi/2$. Consequently, the mixing process can be proved through the volume concentration of nanoparticles in resulted fluid. As an example of such mixing at low Reynolds numbers, Fig. 7.28 shows the isolines of the concentration of water and nanofluid at different volume concentrations of particles at distances of 300, 800, 1300 and 2200 μm from the entrance to the mixing channel. Henceforth, the blue color denotes pure water,

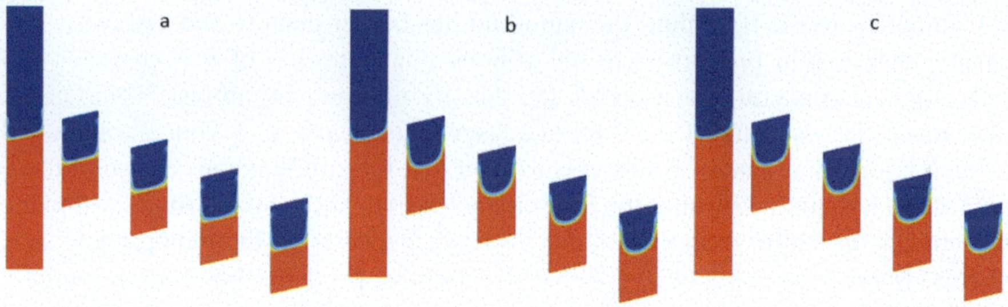

Fig. 7.28 Isolines of nanoparticle concentration in mixing channel cross-sections at $Re = 50$ and volume concentration of 2.5 (a), 5 (b), and 10% (c)

while the red color represents the maximum nanoparticle concentration. As can be observed, the mixing of the fluid and nanofluid in this regime is indeed minimal and is only evident at the contact line between the fluids. As the nanoparticle concentration increases, the mixing intensity increases slightly, although the diffusion coefficient decreases due to the rise in the viscosity coefficient.

Further, as the Reynolds number in the mixer increases, two symmetrical horseshoe-shaped vortices are formed at the left end section of the mixer, which propagate into the mixing channel. These vortices appear due to the development of secondary flows caused by the action of centripetal acceleration associated with the flow turn. These vortices are usually called Dean vortices. The structure of Dean vortices in this regime is illustrated in Fig. 7.29 by means of isosurfaces λ_2, which demonstrate the vortex structures that are formed. Here λ_2 is the second eigenvalue of the $(\mathbf{S}^2 + \mathbf{\Omega}^2)$ tensor, where $\mathbf{S}$ is the strain rate tensor, and $\mathbf{\Omega}$ is the vorticity tensor.

The velocity field for this complex vortex structure is illustrated in Fig. 7.30a. In this regime, four vortices are clearly visible in the channel cross-section. These are, in fact, different branches of two horseshoe-shaped vortices entering the mixing channel like two horseshoes. It is noteworthy that with increasing volume concentration of nanoparticles and, consequently, viscosity, the horseshoe vortex in nanofluid notably decreases in comparison with the horseshoe vortex in water flow (see Fig. 7.29). When mixing two homogeneous fluids with the same physical characteristics, these

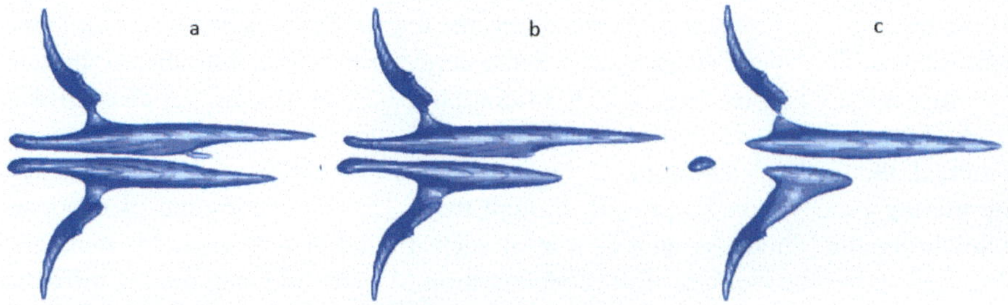

Fig. 7.29 Isosurfaces λ_2 at $Re = 140$ and nanoparticle concentration of 2.5 (**a**), 5 (**b**), and 10% (**c**)

vortices are symmetric [57]. Naturally, due to the difference in viscosities, here the mixing pattern is non-symmetrical.

Figure 7.31 illustrates the isolines of nanoparticle concentration at Re = 140 at distances of 300, 800, 1300, and 2200 μm from the inlet section of the mixing channel. It is clearly visible that the more viscous and dense nanofluid displaces water from the walls of the channel, as if enveloping it. As the concentration of nanoparticles and Reynolds number increase, this effect becomes more pronounced. This indicates that as the concentration of nanoparticles and the Reynolds number increase, the contact area between the mixing fluids expands, thereby enhancing the mixing efficiency (see Fig. 7.32a).

With a further increase in the Reynolds number, a restructuring of the flow regime is observed. Starting from a Reynolds number of approximately 145 (for pure fluids), due to the appearance of the Kelvin–Helmholtz instability caused by the shear between fluid layers, and the subsequent development of the Taylor–Görtler instability resulting from a 90° turn in both fluid flows, the pair of horseshoe vortices deviates from the center longitudinal plane of the mixer by approximately 30°. As a consequence of such reversal in flow, one branch of the horseshoe vortex undergoes a gradual decay, while the other branch exhibits an increase in intensity. Consequently, two intense vortices are formed within the mixing channel. The corresponding alteration in the flow structure is effectively demonstrated in Fig. 7.30. In the mixing channel, the formation of two vortices (Fig. 7.30b), rather than the four observed in

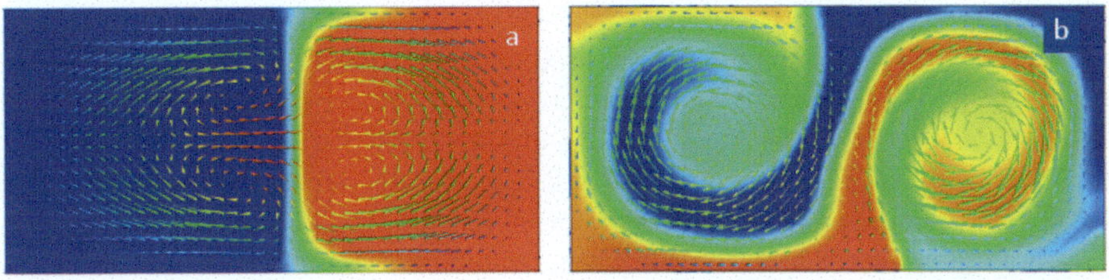

Fig. 7.30 Isolines of nanoparticle concentration and velocity field in the cross-section of the mixer at a distance of 300 μm from the inlet section, and a concentration of 2.5%. Re = 140 (**a**), 250 (**b**)

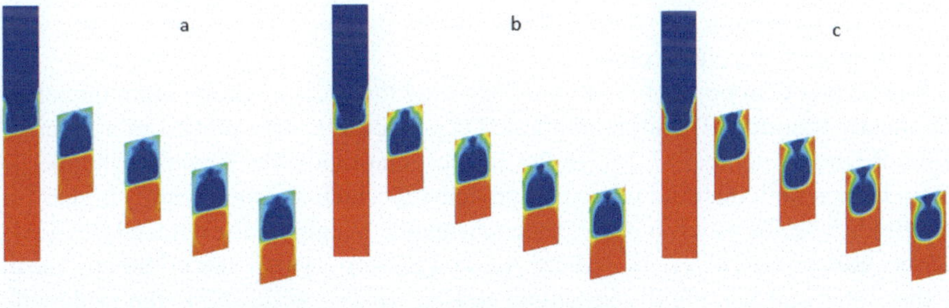

Fig. 7.31 Isolines of nanoparticle concentration in cross-sections of the channel at Re = 140 and particle concentration of 2.5 (**a**), 5 (**b**), and 10% (**c**)

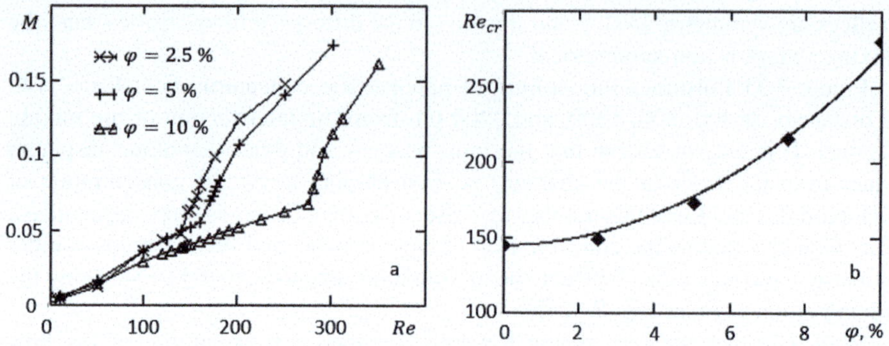

Fig. 7.32 Dependence of mixing efficiency on Reynolds number of nanofluid at particle concentration of 2.5, 5, and 10% (**a**), and critical Reynolds number on volume concentration of nanoparticles (**b**)

Fig. 7.30a is evident. Mixing fluids form layered S-shaped structures. The contact surface area of the mixing fluids in such a layered structure is very developed, which leads to a sharp increase in the mixing efficiency.

A significant increase in mixing efficiency is observed during the flow transition from a regime with two horseshoe vortices to a regime with two single vortices. This is well illustrated by Fig. 7.32a, which represents the dependence of the mixing efficiency on the Reynolds number. The mixing efficiency was estimated by the parameter $M = 1 - \sqrt{\iota/\iota_0}$, where $\iota = V^{-1}\int_V (w - \langle w \rangle)^2 dV$ is the root-mean-square deviation of the weight fraction of the mixture component from its mean value w, while $\iota_0 = \langle w \rangle (1 - \langle w \rangle)$ is the maximum root-mean-square deviation. Figure 7.32a illustrates the dependence of the mixing efficiency on the Reynolds number for nanofluids with particle concentrations of 2.5, 5, and 10% (from top to bottom). The observed shift in the mixing efficiency is explained by an increase in the viscosity of the nanofluid resulting from the rise in the volume concentration of nanoparticles. Thus, the critical Reynolds numbers Re_{cr} responsible for the changing flow regime depend on the nanoparticle concentration. This dependence is illustrated in Fig. 7.32b. The critical Reynolds number is defined as the value at which the transition from a flow with two horseshoe-shaped vortices in the mixing channel to a flow with two single vortices occurs. The critical Reynolds number increases monotonically with rising nanoparticle concentration.

From a practical point of view, data regarding the pressure drop within the mixer is typically required. The aforementioned dependencies were plotted as a function of the Reynolds number (7.24). In the case of nanofluids, the Reynolds number is dependent on both the fluid viscosity and density, which are significantly higher for nanofluids than for water. It is therefore reasonable to employ the reduced Reynolds number $\mathrm{Re}_m = (\rho_m U d_h)/\mu_m$, where $\mu_m = (\mu_w + \mu_{nf})/2$ is the arithmetic mean of the water and nanofluid viscosities, and $\rho_m = (\rho_w + \rho_{nf})/2$ is the arithmetic mean of their densities. The dependence of pressure drop in the mixing channel of the considered micromixer on the Reynolds number defined in this way is shown in

Fig. 7.33. The pressure drop increases with increasing nanoparticle concentration, which is associated with an increase in the viscosity and density of one of the fluids. However, the curves presented in Fig. 7.33 are qualitatively different. The initial two demonstrate a monotonic increase with rising Reynolds number, while the latter ($\varphi = 10\%$) exhibits an inflection point. Its appearance is associated with the laminar–turbulent transition, which in nanofluids is observed at lower Reynolds numbers than in the base fluid, which is discussed in the following section.

In addition to particle concentration, the viscosity of nanofluids depends significantly on the particle size, with a decrease observed with increasing particle size. The effect of nanoparticle size on the flow and mixing regimes is qualitatively quite predictable. A reduction in particle size has the same effect as an increase in particle concentration. Nevertheless, another crucial factor to consider is that when the nanoparticle size is diminished, the diffusion coefficient experiences a notable increase. When the particle size is reduced from 150 to 50 nm the diffusion coefficient increases by a factor of 2.7. Because of this, the mixing layer thickness of nanofluid with smaller nanoparticle size is significantly greater than that of nanofluid with larger particles. This is clearly demonstrated in Fig. 7.34, which compares the nanoparticle concentration isolines for nanofluids with particle sizes of 50 nm (Fig. 7.34a) and 150 nm (Fig. 7.34b). In both cases, the particle volume concentration is the same and equal to 10%. In this instance, the significantly higher viscosity of the nanofluid with smaller particles results in higher critical Reynolds numbers. Naturally, this causes a shift in the efficiency of the mixing process towards higher Reynolds numbers.

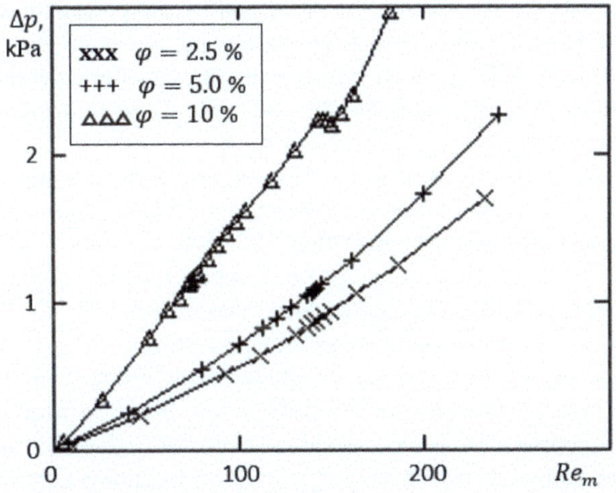

Fig. 7.33 Dependence of pressure drop in the mixing channel on the reduced Reynolds number at concentrations

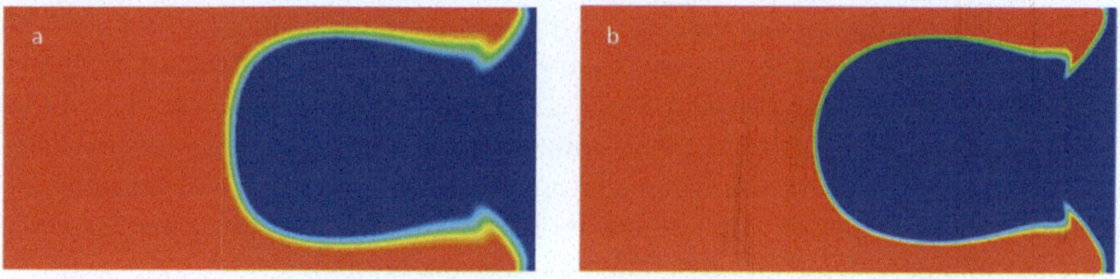

Fig. 7.34 Concentration isolines of nanoparticles 50 (**a**) and 150 nm (**b**) in size, Re = 100

7.5.2 Simulation of the Mixing of Non-Newtonian Fluids

As previously stated in Chap. 3, nanofluids frequently have non-Newtonian rheological properties.

It is therefore useful to examine the mixing behavior of non-Newtonian fluids in a micromixer. This task, of course, has independent significance, since mixing non-Newtonian fluids in micromixers is relevant in various biomedical and chemical technologies (blood, as is known, is a non-Newtonian fluid). The study of mixing non-Newtonian fluids in a T-type micromixer was previously performed in Afzal and Kim [1]. The Reynolds number was varied between 10^{-2} and 10^{2}. The viscosity of blood was determined using the Carreau–Yasuda and Casson models. It was shown that the mixing efficiency of a T-mixer declined with increasing Reynolds number due to a corresponding increase in the Peclet number. In paper [49], the mixing of typical nanofluids whose rheology is described by the power fluid model (3.29) was investigated. However, the consistency parameter was fixed at 0.001, while the fluid index was varied. A T-type micromixer with the same dimensions as previously described was used. The Reynolds number was calculated in accordance with the relation (6.2).

At very low Reynolds numbers (Re < 10), the flow structure and mixing regime diverge slightly from those observed in Newtonian fluids. The mixing efficiency of non-Newtonian fluids is lower than that of Newtonian fluids, and it decreases with decreasing fluid index (see Fig. 7.35a). Despite this, the qualitative structure of the flow remains virtually unchanged [49]. The transition to the regime, which is characterized by the formation of two symmetric vortices in the mixing channel, occurs approximately in the same region of Reynolds numbers as for Newtonian fluids. However, with a decrease in the fluid index, this transition shifts towards higher Reynolds numbers. This can be observed from the analysis of Fig. 7.35a which demonstrates the dependence of mixing efficiency on Reynolds number. As the fluid index decreases, the structure of the forming Dean vortices changes, their central core becomes more and more elongated. With a further increase in the Reynolds number a reversal of the pair of horseshoe vortices occurs similar to that observed in Newtonian fluid flow. As a result, two intense vortices are formed within the mixing channel. The transition from a symmetric to an asymmetric flow regime is accompanied by a notable increase in mixing efficiency (see Fig. 7.35a). Furthermore, this transition

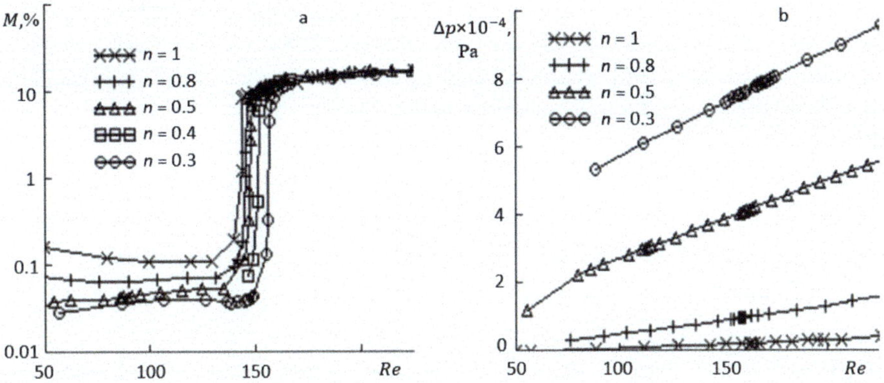

Fig. 7.35 Mixing efficiency at different fluid indices (**a**), and pressure drop between inlet and outlet of the mixer (**b**), as a function of Reynolds number

shifts towards higher Reynolds numbers when decreasing fluid index. The vortex structures of flows at different values of the fluid index are similar, except that the length of the vortices along the mixing channel decreases as the fluid index increases. This behavior is due to the fact that with an increase in the fluid index, the effective viscosity of the medium increases, resulting in a faster decay of the vortices.

Variation of pressure drop between the inlet and outlet of the mixing channel with increasing Reynolds number is shown in Fig. 7.35b. A reduction in the fluid index is accompanied by an increase in the pressure drop. This behavior is attributable to the essentially unsteady character of the flow, which is caused by the formation of Dean vortices following the merging of two flows.

7.6 Laminar–Turbulent Transition

As demonstrated in Chap. 6, the nature of heat transfer and the excess heat transfer coefficient of nanofluids are largely dependent on the flow regime. In the laminar flow the heat transfer coefficient of nanofluids is always fairly large, whereas in many instances this is not the case in the turbulent regime. The positive effect of using nanofluids depends on many factors in turbulent flow. In this regard, an investigation of the laminar–turbulent transition is quite important. Despite this fact, the laminar–turbulent transition in nanofluid flows has not been the subject of practical study for a considerable period of time. The conclusions presented in the few known papers are quite contradictory. For example, the study [46] has demonstrated that the laminar–turbulent transition in nanofluid flow occurs at higher Reynolds numbers, indicating enhanced flow stability compared to the flow of the basic fluid. On the other hand, in a paper [118] the exact opposite opinion is stated.

The author's research team has conducted theoretical studies on the hydrodynamical stability of various model flows of nanofluids [12, 84, 85, 91]. Furthermore, a dedicated experiment [96] was conducted to validate the theoretical findings. The objective of the present section is a review of the data obtained in the above mentioned papers.

7.6.1 Laminar–Turbulent Transition in a Plane Poiseuille Flow

The primary objective of the theoretical investigations presented in this section is to ascertain the flow parameters at which the laminar regime transits to the turbulent mode. For this reason, the solution of the linear problem of hydrodynamic stability, that is to say, the stability of the flow with respect to small disturbances, is further considered. In this section the stability of flows of ethylene glycol-based nanofluids with silicon dioxide particles is investigated. The particle size was varied from 10 to 210 nm, and the volume concentration of the particles was changed from 0 to 10%.

Let us consider the problem for a plane Poiseuille flow. It is assumed that the nanofluid is incompressible and homogeneous. The dynamics of the nanofluid is described by the Navier–Stokes equations

$$\nabla \cdot \boldsymbol{u} = 0,\ \rho\left(\frac{\partial \boldsymbol{u}}{\partial t} + \boldsymbol{u} \cdot \nabla \boldsymbol{u}\right) = -\nabla p + \nabla \cdot (\mu \boldsymbol{D}), \tag{7.25}$$

where $\boldsymbol{D}$ is the strain rate tensor

$$\boldsymbol{D} \equiv D_{ij} = \frac{1}{2}\left(\frac{\partial \boldsymbol{u}_i}{\partial x_j} + \frac{\partial \boldsymbol{u}_j}{\partial x_i}\right).$$

In this case, the viscosity coefficient included in the above formula is the effective viscosity coefficient. As is the case with dispersed fluids, it depends on the particle volume concentration and particle size. For ethylene glycol-based nanofluids with silicon dioxide nanoparticles, a following correlation was derived based on experimental data processing [79, 88]

$$\mu_r = 1 + 2.5\varphi + 6.2\varphi^2 + 13.427e^{-0.013d/\sigma_m}\left(\varphi + 38.33\varphi^2\right), \tag{7.26}$$

which approximates the experimental data almost with the accuracy of the experiment. In this formula σ_m is the size of the carrier fluid molecules. The size of the ethylene glycol molecule was assumed to be 0.3 nm.

For any flow, one can introduce the characteristic linear scale L_{ch}, velocity U, pressure p, and time T. Then, passing in Eqs. (7.25) to the dimensionless variables $x' = x/L_{ch}$, $u' = u/U$, $t' = t/T = tU/L_{ch}$, $p' = p/P$, $\mu' = \mu/\mu_{bf}$, and $\rho' = \rho/\rho_{bf}$, and omitting the strokes, we obtain

$$\nabla \cdot \boldsymbol{u} = 0, \frac{\partial \boldsymbol{u}}{\partial t} + \boldsymbol{u} \cdot \nabla \boldsymbol{u} = -Eu\nabla p + \frac{1}{\mathrm{Re}}\nabla \cdot \boldsymbol{D}, \tag{7.27}$$

where $Eu = p/(\rho_{bf}\rho U^2)$, and $Re = [(L_{ch}U\rho_{bf}\rho)/(\mu_{bf}\mu)]$ are the Euler and Reynolds numbers, respectively.

Equation (7.27) includes the Reynolds number as a similarity parameter, discussed in Sect. 7.2.3. This Reynolds number can be rewritten as follows

$$\mathrm{Re} = \mathrm{Re}_{bf}\frac{\rho(\varphi)}{\mu((\varphi, d_p)}. \tag{7.28}$$

From the analysis (7.28), a simple conclusion follows, the flows with different Reynolds numbers in the general case may simply correspond to the flows of different nanofluids.

The solution to the linear stability problem is constructed in a standard way [9, 23]. The linearization of Eqs. (7.27) is carried out according to the amplitude of small disturbances. The solution of the resulting linearized equations is presented in the form of a normal mode expansion, with the amplitude of pressure disturbances excluded by the introduction of the perturbation stream function. In the case of a plane flow between two infinite plates resulting from a pressure drop in the x-axis direction with velocity profile $U_0(y) = 1 - y^2$, the stream function Ψ is defined as follows: $u_x = (\partial\Psi/\partial y)$, $u_y = -(\partial\Psi/\partial x)$ Substituting the solution represented as normal modes $\Psi = \psi\exp[i\alpha(x - Ct)]$ into the linearized equations leads to an equation with respect to the disturbance amplitude. In the case of a plane flow with velocity profile U_0, the equation for the disturbance amplitudes has the form

$$\psi^{\mathrm{IV}} - 2\alpha^2\psi'' + \alpha^4\psi = i\alpha\mathrm{Re}\Big[(U_0 - C)(\psi'' - \alpha^2\psi) - U_0''\psi\Big]. \tag{7.29}$$

where, the dimensionless coordinate y is normalized by the channel halfwidth $L_{ch}\prime$.

The distinction between Eq. (4.29) and the classical Orr-Sommerfeld equation [9, 23] is that in the problem of the stability of the nanofluid flow, the dependence of the Reynolds number, the wave number of disturbances α, the phase velocity Re(C) and the growth coefficient of disturbances $\omega_i = \mathrm{Im}(\alpha C)$ on the concentrations and sizes of nanoparticles is taken into account. As previously stated, it is also important to exercise caution regarding the similarity in Reynolds number. Equation (4.29) turns into the equation of stability of pure liquid flow if the concentration of nanoparticles is equal to zero.

Equation (4.29) combined with the conditions for the stream function on the non-permeable motionless boundaries $\Psi(\pm 1) = \Psi'(\pm 1) = 0$ form an eigenvalue problem for the relative spectral parameter C. The solution of the stability problem consists in determining of the dependence of the wave number α on the Reynolds number at which the disturbance amplitude remains constant $\mathrm{Im}(C) = 0$. The set of parameter values (α, Re) that satisfy this condition form a neutral stability curve. As in the majority of studies on hydrodynamic stability, the solution to Eq. (4.29) was obtained numerically in [84]. The problem was solved by using the Galerkin method in the version developed by the authors. The method is based on constructing a non-divergent set of basis functions, which are combinations of Chebyshev polynomials. This method has previously been used to solve a broad range of hydrodynamic stability problems [11, 42, 89, 90].

The primary objective of the hydrodynamic stability theory of nanofluid flows is to ascertain whether nanofluid exhibits greater or lesser stability than the basic fluid. The response to this question is provided by neutral stability curves. Figure 7.36a illustrates the neutral stability curves of the plane Poiseuille flow of an ethylene glycol-based nanofluid with 17.6 nm SiO_2 nanoparticles. The curves 1, 2, and 3 are obtained at nanoparticle volume concentrations of $\varphi = 0.05, 0.01$, and 0, respectively. Therefore, curve 3 corresponds to the base fluid. In the classical paper of S. A. Orszag, it was shown that for a normal fluid, the critical Reynolds number is 5772 [68]. The nanofluid is observed to exhibit greater instability than the base fluid. The degree of destabilization increases with increasing nanoparticle concentration, leading to a decrease in the critical Reynolds number at a concentration $\varphi = 0.05$ by more than twice.

The stability properties of the Poiseuille plane flow are determined by the viscosity of the fluid. Given that the viscosity of nanofluids depends not only on the concentration of particles, but also on their size, it follows that the critical Reynolds numbers of the laminar–turbulent transition should also dependent upon particle size. The influence of nanoparticle size on the stability characteristics of this flow is illustrated in Fig. 7.36b. Here, the presented curves were obtained for nanofluid flows containing nanoparticles of varying sizes, namely, 17.6, 45.0, and 90.0 nm. Curve 4 represents the base fluid flow. In all cases, the volume concentration of particles is identical and equal to 0.05. It is evident that in flows with the same concentration of nanoparticles, the degree of flow destabilization depends on their size, and increases with its decreasing. The critical Reynolds numbers for nanofluids with particles of different

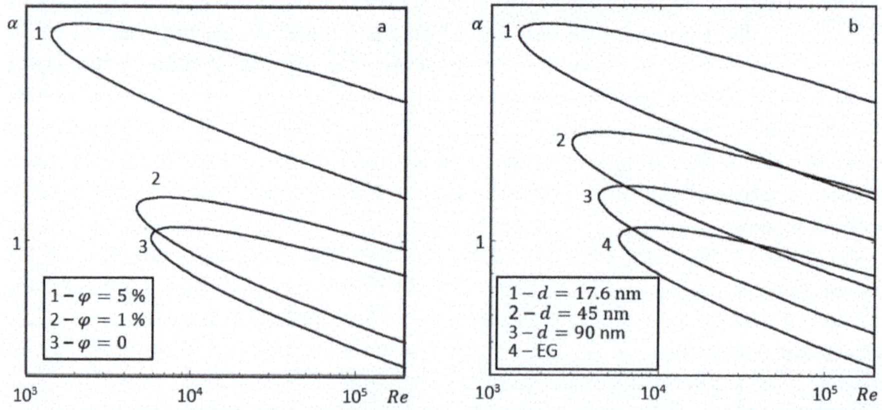

Fig. 7.36 Neutral stability curves for the flat Poiseuille flow of nanofluids at different particle concentration and $d = 17.6$ nm (**a**); different particle size and volume concentration $\varphi = 0.05$ (**b**)

sizes vary several times. The obtained data also indicate a monotonic dependence of the stability characteristics on the particle concentration. The precise values of critical Reynolds numbers and corresponding wave numbers can be found in [84].

7.6.2 Laminar–Turbulent Transition in a Cylindrical Poiseuille Flow

Now, let us examine the problem of flow stability in an annular channel between two concentric cylinders with radii $R_i \leq R \leq R_0$, resulting from a pressure gradient along the axis (Poiseuille flow). Here R_i and R_0 represent the radii of the inner and outer cylinders, respectively. The dimensionless radius, defined as $r = 2R/(R_o - R_i)$ varies within the range $[2\theta/(1-\theta),\ 2/(1-\theta)]$, where $\theta = R_i/R_o$. The considered flows are characterized by a common length scale, which is equal to the channel halfwidth $h_* = (R_o - R_i)/2$. The stationary enforced flow in an annual channel between two coaxial cylinders along the z-axis of the channel is defined by the velocity profile [51]

$$W_0(R) = -\frac{1}{4\mu}\frac{\partial p}{\partial z}\left[R_i^2 - R^2 + \frac{R_o^2 - R_i^2}{\ln(R_o/R_i)}\ln(R/R_i)\right], \tag{7.30}$$

and average velocity over the channel cross-section is equal to

$$\overline{W} = \frac{1}{8\mu}\left|\frac{\partial p}{\partial z}\right|\left[R_i^2 + R_o^2 + \frac{R_o^2 - R_i^2}{\ln(R_o/R_i)}\right],$$

while the Reynolds number $\mathrm{Re} = (\rho\overline{W}h_*)/\mu$ is determined by the velocity $\overline{W}$, channel halfwidth h_*, nanofluid density $\rho = \rho_{bf}\rho'$ and viscosity $\mu = \mu_{bf}\mu'$.

Below, the stability of flow (7.30) is studied in relation to axisymmetric perturbations. In this case, there are no azimuthal modes of velocity perturbations, that is $u_\theta = 0$, while the radial u_r and axial u_z velocities are expressed through the stream function as follows: $u_r = -(\partial\Psi/\partial z)$, $u_z = \partial\Psi/\partial\ r + \Psi/r$. In this case, the z-axis coincides with the channel axis, and r is the radial coordinate perpendicular to it. By substituting the solution represented in the form of normal modes $\Psi = \psi\exp[i\alpha(z - Ct)]$ into the linearized equations, an equation with respect to the perturbation amplitude is obtained

$$\left(L - \alpha^2\right)\left[L - \alpha^2 + i\alpha\mathrm{Re}(C - W_0)\right]\psi + i\alpha\mathrm{Re}\left(W_0'' - W_0'/r\right)\psi = 0, \tag{7.31}$$

where the operator $L \equiv d^2/dr^2 + (1/r)d/dr - 1/r^2$ is introduced. As in the case of plane flows, Eq. (7.31) transforms to the stability equation of pure fluid flow when the nanoparticle concentration is zero.

Equation (7.31), combined with the boundary conditions

$$\psi\left(\frac{2\theta}{1-\theta}\right)=\psi\left(\frac{2}{1-\theta}\right)=\psi'\left(\frac{2\theta}{1-\theta}\right)=\psi'\left(\frac{2}{1-\theta}\right)=0$$

forms an eigenvalue problem with respect to the spectral parameter $C = \omega/\alpha$.

The flow regimes and laminar–turbulent transition conditions depend on the ratio of radii θ. At $\theta \to 1$ (in the limit of a narrow channel), the flow velocity profile between the cylinders tends towards the velocity profile of plane Poiseuille flow. Conversely, at $\theta \to 0$, the flow becomes increasingly similar to flow in a tube, although with a singularity at the axis. It is clear that the stability of the cylindrical flow also varies with change of θ. As is known, the flow in pipe is generally stable with regard to small disturbances.

When considering the fluid flow in a cylindrical channel, both disturbances propagating along the flow axis and those with an azimuthal component (spiral perturbations) can be taken into account in the general case. It is known that the stability characteristics of spiral disturbances diverge considerably from those of axial disturbances [108]. The study of spiral disturbances is relevant in cylindrical flows, where one of the cylinders (internal or external) is rotated. In the absence of such rotation, the emergence of these disturbances is either random or attributable to the influence of boundary asymmetry (e.g., the presence of roughness). Thus, in a conventional cylindrical Poiseuille flow, axial disturbances play the primary role in the laminar–turbulent transition. This section considers exactly such disturbances. To illustrate, the stability characteristics of two cylindrical Poiseuille flows with a radius ratio of 0.5 are considered. The neutral stability curves for nanofluid flows are illustrated in Fig. 7.37. In Fig. 7.37a, the average particle size was 27.9 nm, while in Fig. 7.37b, it was 45 nm. The bottom-up curves correspond to particle volume concentrations of 0 (base fluid), 1, 2, 3, 4, and 5%. In all instances, the flow of nanofluids is observed to be more unstable than that of the base fluid. Similarly to the plane Poiseuille flow, the critical Reynolds number increases with increasing nanoparticle concentration and decreasing nanoparticle size. This stability character is also maintained at other values of θ [84]. As the parameter θ decreases, the flow becomes increasingly stable. However, even in this case, the critical Reynolds numbers for nanofluids remain significantly lower.

The most significant characteristic of flow stability is the disturbance growth rate ω_i. The dependence of the growth rate on the wave number α is illustrated in Fig. 7.38 for flows with particles of different sizes. Here the nanoparticle concentration is $\varphi = 0.05$, while the curves from right to left are obtained for nanofluid flows containing particles 17.6, 27.9, 45, 60, and 90 nm in size, respectively. Curve 6 represents the base fluid. Similarly, as in the behavior of the neutral stability curves, a similarity between $\omega_i(\alpha)$ dependences in flows with different nanoparticles was also noted. The impact of nanoparticles can be reduced to a mere stretching of the $\omega_i(\alpha)$ dependence along the wave number axis. This effect is most pronounced when the particle size is minimal.

The change in the values of the wave numbers corresponding to the maximum growth rates testifies that the wave length of unstable disturbances depends on the volume fraction of particles. Obviously, this dependence also changes with increasing

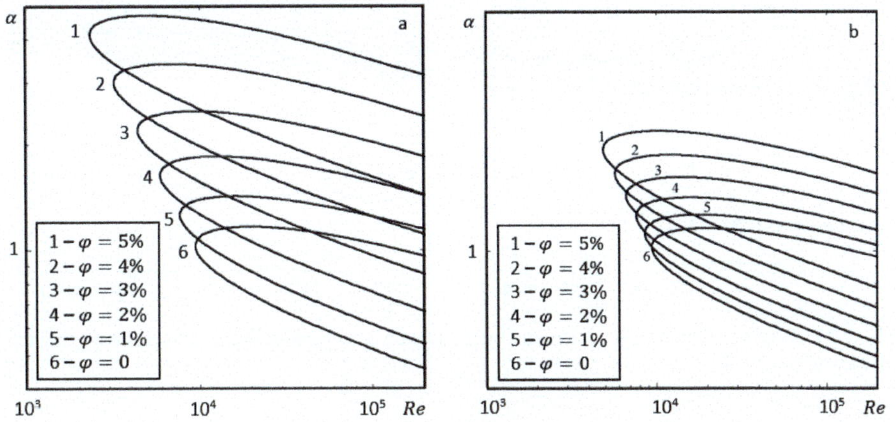

Fig. 7.37 Neutral stability curves for the nanofluid flow in a cylindrical channel at $\theta = 0.5$ and $d = 27.9$ nm (**a**); d = 45.0 nm (**b**)

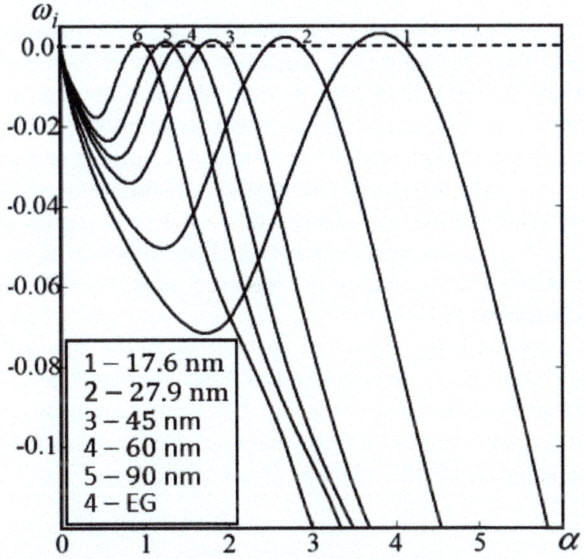

Fig. 7.38 Dependence of the unstable perturbation amplification factor on the wave number for the flow in a cylindrical channel, $\theta = 0.5$

particle size in nanofluids. This fact is illustrated in Fig. 7.39a, which shows the dependence of the dimensionless wavelength Λ of unstable disturbances in a cylindrical channel with a radius ratio of 0.5. The wavelength is normalized to the corresponding wavelength of the base fluid. The curves represent the results for fluid flows with nanoparticle sizes of 90 nm (curve 1), 60 nm (2), 45 nm (3), 27.9 nm (4), and 17.6 nm (curve 5). As the concentration of nanoparticles increases, the wavelength

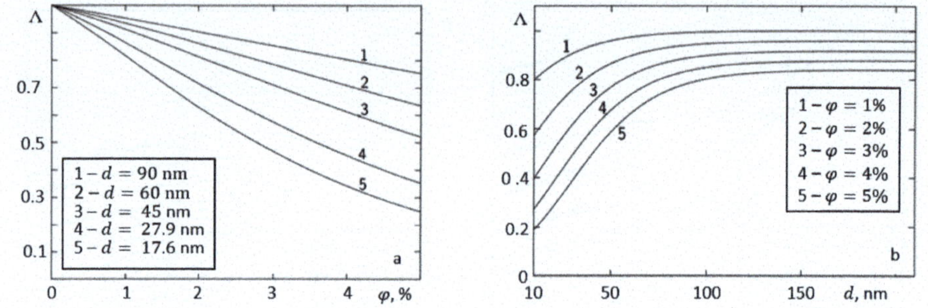

Fig. 7.39 Dependence of the wavelength of the unstable perturbation on the concentration (**a**), and nanoparticle size (**b**)

of the most unstable disturbance decreases monotonically, almost proportionally to the particle size.

The wave length of the most unstable disturbance is plotted in Fig. 7.39b as a function of the particle size. The wave length is normalized to the wave length of the most unstable disturbance for the nanofluid with the particle size 210 nm. It should be noted that, by definition, these are no longer nanoparticles. Indeed, this is the minimum size of a Brownian particle (this was the size of the minimum particles in J. Perrin's famous experiments on Brownian motion). In this figure, the curves from top to bottom correspond to nanoparticle volume concentrations of 1, 2, 3, 4, and 5%. The wavelength of the unstable disturbances is correlated with both the concentration and the size of the nanoparticles. In particular, it decreases with increasing concentration of nanoparticles. The influence of nanoparticles on the change in stability characteristics is expressed more strongly, the smaller the size of the nanoparticles. However, there is a threshold value of nanoparticle size $d \approx 180.0$ nm, upon reaching which further increase in the nanoparticle size does not lead to a significant change in stability. For flows with nanoparticles of different sizes, the degree of flow destabilization increases monotonically with increasing concentration. It is also important to note that the wavelength of the most unstable perturbation is smaller the smaller the size of nanoparticles.

7.6.3 Laminar–Turbulent Transition in a Cylindrical Couette Flow

Let us now consider the flow in the channel between the same concentric cylinders, when one of them rotates for certainty, the inner one. Such a flow is known as a Couette flow. In the cylindrical Couette flow caused by the rotation of the inner cylinder with angular velocity Ω, the stationary velocity profile also depends only on the radial coordinate and has the following form

$$V_0(\eta) = \frac{\theta}{1-\theta^2}\left(\frac{1}{\theta\eta} - \theta\eta\right) \tag{7.32}$$

where $\theta\eta = (1-\theta)r/2$, and the corresponding Reynolds number is defined as follows

$$Re = \frac{\rho\theta\Omega h^2}{\mu(1-\theta)}.$$

Since the publication of the seminal paper [21], the Couette flow between coaxial rotating cylinders has become one of the most popular in the theory of hydrodynamic stability of flows. In the problem of the Couette flow stability, it is necessary to consider all velocity components and pressure disturbances. As in previous case, axisymmetric disturbances are taken into consideration, with the solutions to the resulting linear stability equations presented in the form of a normal mode expansion $[u_z, u_r, u_\theta] = [u(r), v(r), w(r)]\exp(i\alpha z - i\omega t)$. In this case, the linear stability problem is reduced to the following system of equations

$$\begin{gathered} u_r + u/r + i\alpha w = 0, \quad (L - \alpha^2 + i\omega)u + 2\mathrm{Re}V_0 v/r = p', \\ (L - \alpha^2 + i\omega)v - \mathrm{Re}(V_0' + V_0/r)u = 0, \quad (L - \alpha^2 + i\omega)w = i\alpha p. \end{gathered} \tag{7.33}$$

Equations (7.33) are supplemented by the conditions of equality to zero disturbances at the channel boundaries and are subsequently solved numerically in the same way as that employed in previous problems [85]. In this work, the stability characteristics of such flow of pure fluid are systematically studied. In particular, it is demonstrated that the flow becomes markedly unstable with an increase in channel width. However, when the parameter θ reaches a threshold value equal to 0.5, a further increase in width does not result in a discernible change in stability characteristics. At $\theta \to 0$ the flow velocity profile is more dependent on the summand having singularity at the axis. In nanofluids, both the concentration of nanoparticles and their sizes influence the stability characteristics. The relevant data are provided in Tables 7.4 and 7.5 at $\theta = 0.5$. In all cases, the critical Reynolds numbers decrease with increasing particle concentration and decreasing particle size.

Table 7.4 Critical Reynolds numbers and wave numbers for nanofluid flow in a channel $\theta = 0.5$; nanoparticle size $d_p = 17.6$ nm

φ	0	0.01	0.02	0.03	0.04	0.05
Re_c	34.0937	30.9901	27.3552	23.7243	20.4034	17.5137553
α_c	1.567	1.7239	1.9530	2.2519	2.6184	3.05045

Table 7.5 Critical Reynolds numbers and wave numbers for nanofluid flow in a channel $\theta = 0.5$; nanoparticle concentration $\varphi = 0.05$

d_p	90	60	45	27.9	17.6
Re_c	30.4786	27.9711	25.34675	20.852087	17.5137553
α_c	1.75286	1.90999	2.10776	2.56209	3.05045

7.6.4 Stability of the Spiral Flow

The spiral flow represents a superposition of the Poiseuille and Couette flows in an annular channel. It is evident therefore that this flow is rather complex. The first comprehensive experimental investigation into the stability of spiral flow in a cylindrical channel with a cylinder radii ratio of $\theta = 0.5$ was conducted in [107]. In this study, the critical values of the azimuthal Reynolds number Re_ϕ and critical wave numbers were determined for axial Reynolds numbers Re_z within the range $\mathrm{Re}_z = 1 \div 150$. It was found that variety of azimuthal modes, including disturbances with azimuthal wave number equal to zero, can be unstable. The critical azimuthal Reynolds numbers did not exceed 200. Practically simultaneously, the stability of spiral flows at $\theta = 0.77$ and 0.95 was studied numerically [63]. Neutral stability curves were constructed for a series of Taylor numbers, which serve as an analogue for the azimuthal Reynolds number.

The instability mechanisms of the spiral flow and similar flows occurring when the inner cylinder slides in the axial direction were investigated in Meseguer and Marques [54, 55]. Neutral stability surfaces were constructed using variables $(\mathrm{Re}_\phi, \mathrm{Re}_z)$. The found boundaries of the instability region qualitatively corresponded to the experimental data presented in Takeuchi and Jankowski [107]. Nevertheless, the critical Reynolds numbers obtained differed significantly from those found experimentally. In a further study [20], neutral stability curves were also obtained numerically at a given axial Reynolds number $\mathrm{Re}_z = 100$. The investigation revealed the existence of closed instability regions and the formation of separate branches on the neutral stability curves. The most comprehensive study was conducted by Bord and Rudyak [12], who investigated also the stability of nanofluid flows.

The spiral flow is described by a time-independent axisymmetric solution of Eqs. (7.25) where $\boldsymbol{u} = (0, V_0(r), W_0(r))$. In the cylindrical coordinate system (r, ϕ, z) the circumferential $V_0(r)$ and axial $W_0(r)$ velocity components have the following form [51]

$$V_0(r) = \frac{\theta V_{\max}}{1-\theta^2}\left[\frac{2}{(1-\theta)r} - \frac{(1-\theta)r}{2}\right],$$

$$W_0(r) = W_{\max}\left[1 - \left(\frac{1-\theta}{2\theta}r\right)^2 - \frac{1-\theta^2}{\theta^2 \ln\theta}\ln\left(\frac{1-\theta}{2\theta}r\right)\right].$$

where $V_{\max} = V_0(R_i) = \Omega R_i$, $W_{\max} = (\Delta p R_i^2)/(4\mu\eta_{\max})$, Δp is the pressure drop along the cylinder axis, and the parameter $\eta_{\max}$ is defined by the expression

$$\eta_{\max} = 1 + \frac{1-\theta^2}{2\theta^2 \ln\theta}\left[1 - \ln\left(\frac{\theta^2-1}{2\theta^2 \ln\theta}\right)\right].$$

Two distinct Reynolds numbers correspond to independent fluid motions, described by the circumferential and axial components of the main flow velocity (in twentieth century works, the pressure flow was commonly characterized by the axial Reynolds number, while the flow caused by cylinder rotation was characterized by the Taylor number). The pressure flow is characterized by the Reynolds number $\mathrm{Re}_z = (W_{\max}\rho h)/\mu$, while the flow caused by the rotation of the inner cylinder is characterized by the azimuthal Reynolds number $\mathrm{Re}_\phi = (V_{\max}\rho h)/\mu$.

Linearization of Eqs. (7.25) and representation of the perturbed solution in the form of normal waves leads again to linear stability problem

$$[u(r,\phi,z,t),\ v(r,\phi,z,t),\ w(r,\phi,z,t)] = [u(r),\ v(r),\ w(r)]\exp(i\alpha z + in\phi - i\omega t),$$

where α and n are the axial and azimuthal wave numbers, respectively, ω is the complex perturbation frequency. As a result, the problem is reduced to solving the following equations for the disturbance amplitudes

$$\begin{aligned}
&u_r + u/r + inv/r + i\alpha\omega = 0,\\
&(i\tilde{W} - \tilde{L})u + 2\mathrm{Re}_\phi V\frac{v}{r} + \frac{u + 2inv}{r^2} = -\partial_r p,\\
&(i\tilde{W} - \tilde{L})v + \mathrm{Re}_\phi\left(V' + \frac{V}{r}\right)u + \frac{v + 2inu}{r^2} = -\frac{inp}{r},\\
&(i\tilde{W} - \tilde{L})w + \mathrm{Re}_z W'u = -i\alpha p,
\end{aligned} \tag{7.34}$$

where $\tilde{L} = \partial_r^2 + r^{-1}\partial_r - n^2/r^2 - \alpha^2$, $W = -\omega + \alpha\mathrm{Re}_z\tilde{W} + n\mathrm{Re}_\phi(V/r)$.

The uniform boundary problem for Eqs. (7.34) leads to the problem on eigenvalues ω at a given values of axial and azimuthal Reynolds numbers, as well as axial and azimuthal wave numbers. The stability of spiral flow was investigated for water-based nanofluids with zirconium oxide nanoparticles with an average size of 44 and 105 nm. The viscosity coefficients measured at 25 °C are presented in [95]. Figure 7.40a shows, as an example, neutral stability curves with an azimuthal wave number $n = 0$ for a nanofluid with 44 nm particles at different particle concentrations. Lines 1, 2, 3, 4, and 5 correspond to volume concentrations of 0, 2, 4, 6, and 8%. The azimuthal Reynolds number Re_ϕ is equal to 75. The degree of flow destabilization increases monotonically with increasing nanoparticle concentration. The critical Reynolds number at maximum particle concentration is more than three and a half times smaller than that for water. Thus, in this case too, the laminar–turbulent transition of nanofluid flow is observed at significantly lower critical Reynolds numbers.

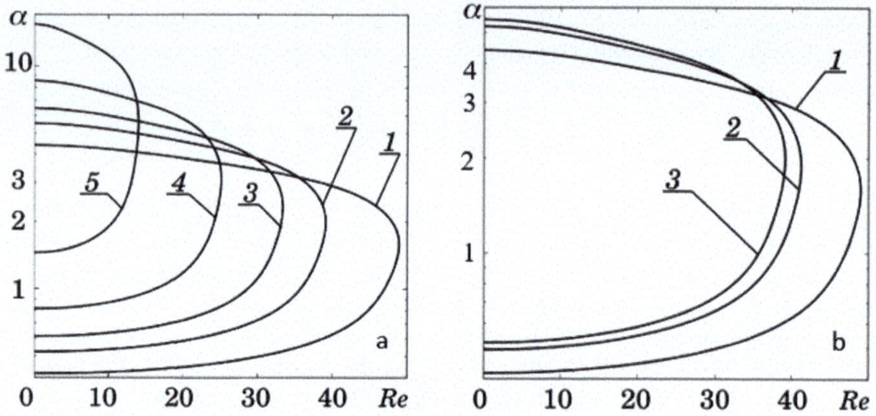

Fig. 7.40 Neutral stability curves of water-based nanofluids with zirconium oxide particles as a function of particle concentration (**a**), and particle size (**b**)

Since the viscosity of nanofluids depends not only on the concentration of nanoparticles but also on their size, it is evident that the stability characteristics will also change significantly. Figure 7.40b presents the neutral stability curves of nanofluids with particles of two sizes equal to 44 (curve 3) and 105 nm (curve 2). Curve 1 corresponds to the base fluid. In both cases, the particle volume concentration is equal to 2%. As anticipated, the nanofluid with smaller particles exhibits reduced stability, with a critical Reynolds number approximately 10% lower.

7.6.5 Laminar–Turbulent Transition of Nanofluid Flow in a Pipe

The simulation of the laminar–turbulent transition indicates that in nanofluids this transition occurs at significantly lower Reynolds numbers than in the base fluid. While four typical model flows have been discussed above, the corresponding experimental data are still lacking. On the other hand, the laminar–turbulent transition in a circular cylindrical channel has been studied experimentally in Rudyak et al. [96]. This is particularly noteworthy because the flow in the pipe is absolutely stable with respect to small disturbances. The measurement section was a brass tube with an inner diameter of 5 mm and a length of 1.045 m. Pressure drop measurements along the tube were conducted using an OVEN PD200 differential manometer. A 2.5-m-long inlet flow stabilization section was installed upstream of the measurement section. During the course of the experiments, the flow rate was observed to vary within the range from 0.1 to 2.4 l/min, which corresponds to Reynolds numbers for water ranging between 800 and 12,000. In addition to the integral pressure drop, the pressure pulsations were quantified utilizing 014MT piezoelectric pressure sensors, which were positioned at the outlet of the measurement section. The distilled water-based nanofluid flows with silicon oxide nanoparticles with average particle sizes of 25 and 100 nm were studied. The volume concentration of particles varied from 0.5 to 2%. The viscosity coefficient of the studied nanofluids with the accuracy of the measurements [96] are described by the formulas

$$\mu = \mu_{bf}\left(1 + 4.4\varphi + 552\varphi^2\right), \quad \mu = \mu_{bf}\left(1 + 3.9\varphi + 80\varphi^2\right).$$

One of the main markers that registers both different flow regimes (laminar and turbulent) and laminar–turbulent transition are data on the dependence of the hydrodynamic drag coefficient $\xi = (2d_c\Delta p)/\left(\rho U^2 L_{ch}\right)$ on the Reynolds number. Here d_c is the tube diameter, U is the average velocity calculated by the flow rate, ρ is the fluid density, Δp is the measured pressure drop, and L_{ch} is the length of the measuring section. This dependence at different nanoparticle concentrations is shown in Fig. 7.41a. For comparison, the Poiseuille (dashed line) and Blasius (solid line) drag coefficients for a circular channel are also plotted. The given experimental data demonstrate that the laminar–turbulent transition in nanofluid begins earlier, i.e. at lower Reynolds numbers than that of the base fluid (water). The degree of shift towards lower Reynolds numbers increases with increasing nanoparticle concentration. Accordingly, the deviation from the laminar dependence for water begins at Re ~ 2300, for a 1% nanofluid—at Re ~ 2100, and for a 2% nanofluid—at Re $\sim$ 2000.

Since the viscosity coefficient of a nanofluid depends not only on the concentration of nanoparticles, but also on their size, then one should expect a corresponding dependence for the drag coefficient. Experiments confirm this. Figure 7.41b shows the dependence of the drag coefficient on the Reynolds number for nanofluids with particles of average size 25 and 100 nm at a volume concentration of particles equal to 2%. Indeed, the smaller the nanoparticle diameter, the greater the deviation of the drag coefficient from the laminar dependence.

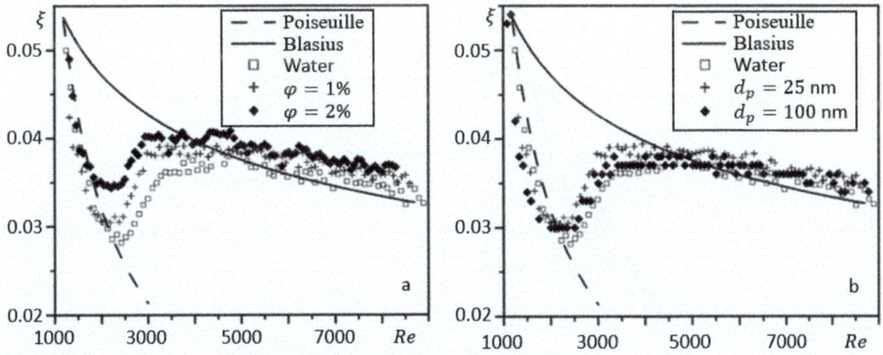

Fig. 7.41 Dependence of drag coefficient on Reynolds number for water-based nanofluids with silicon oxide particles on their concentration (**a**) and diameter (**b**)

In addition to the average pressure drop in the channel, the pressure pulsations at the channel outlet were also measured. The dependence of the maximum amplitude of the pulsations δp on the Reynolds number is shown in Fig. 7.42. The size of the nanoparticles was equal to 25 nm. The pressure pulsations in the nanofluid flow behave qualitatively different in the laminar and turbulent flows. The level of pressure pulsations in nanofluids is about 30–40% higher than in the base fluid at all subcritical Reynolds numbers, i.e. the particles destabilize the flow. On the other hand, the opposite is observed in the turbulent flow regime. Analysis of the spectra shows that nanoparticles significantly reduce pressure pulsations in the mid- and high-frequency parts of the spectrum, indicating suppression of small-scale turbulent pulsations. At the same time, pressure pulsations in the low frequency part of the spectrum increase with increasing particle concentration. This effect essentially depends on particle size. The destabilization of the flow by small particles has already been observed in gas suspension flows [37, 89, 90, 102]. Nanoparticles are mesoscopic objects. In the metric of the carrier fluid they are material points. Therefore, it is reasonably expected that nanoparticles of any size should destabilize the flow.

The influence of macroscopic particles on turbulence has also been studied for a long time. In [31] it was shown that the intensity of turbulent pulsations increases when the ratio of particle size d to the Kolmogorov scale δ_c meets the condition $d/\delta_c > 0.1$, and, conversely, decreases at $d/\delta_c < 0.1$. Following the logic of this work, it can be argued that the smaller the particles are, the greater the effect of turbulent pulsation reduction should be observed, which is illustrated by the experimental data shown in Fig. 7.42. It can be shown that for particles of 25 nm in size in the range of Reynolds numbers from 2200 to 10,000 the ratio of d/δ_c varies within the interval of $3 \times 10^{-5} \div 3.5 \times 10^{-4}$. Thus, according to the criterion [31], the turbulence intensity in the nanofluid should decrease in comparison with the base fluid over the entire range of Reynolds numbers studied.

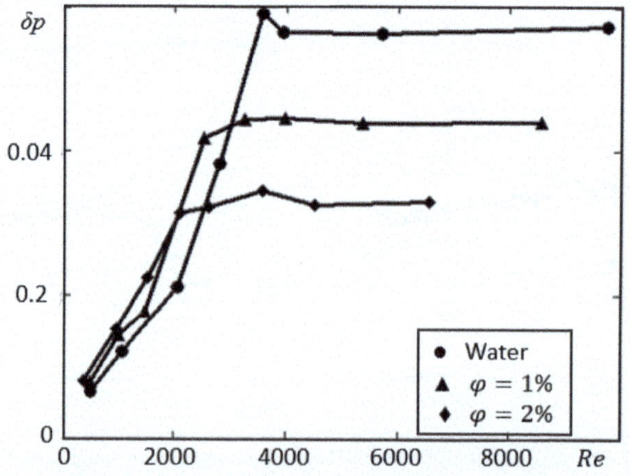

Fig. 7.42 Dependence of the maximum amplitude of pressure pulsations (in kPa) in nanofluids on the Reynolds number at different particle volume concentrations

7.6.6 Concluding Remarks

The principal conclusion that can be drawn from the above presented series of data is that nanofluid flows are less stable than that of their base fluids. This implies that the presence of nanoparticles serves to further destabilize the flow. The stability of ordinary dispersed liquids has been previously investigated using the simplest Saffman model [102] of a two-fluid medium. It was found that small particles actually destabilize the flow, while large particles, on the contrary, suppress the emerging disturbances. However, a noticeable effect occurred only for gas suspensions, when the mass concentrations of particles were large. In nanofluids, the mass concentrations of particles are usually not so high, but the effect is nevertheless similar. Nanoparticles are mesoscopic objects. They are material points in the metric of the carrier fluid. Therefore, nanoparticles of all sizes will destabilize the flow. It is important to note, however, that in flows with a noticeable inhomogeneity of particle distribution (e.g. in centrifuges), the situation is significantly more complex and requires special consideration.

The theoretical results obtained permit the initial conclusions to be drawn about the nature of nanoparticles' influence on the initial stage of the laminar–turbulent transition, which is described by the linear stability theory. The interaction between the carrier medium and nanoparticles results in the destabilization of the flow. The destabilization mechanisms appear to be common across different flows. A relationship is observed between the disturbance wavelength, which has a hydrodynamic scale, and the size of the nanoparticles. Destabilization of flows from the point of view of hydrodynamic theory is natural. In the considered flows, so-called viscous instability occurs when the nature of disturbances growth is determined by the viscosity of the medium. In nanofluids, viscosity is significantly higher than in the base fluid, and it increases with rising particle concentration and decreasing particle size. These effects were observed in all of the above considered examples.

In conclusion, it is important to note one further circumstance. This chapter has not considered the simulation of turbulent flows in any way. This is due to two underlying reasons. The first is that simulation of turbulent flows of Newtonian nanofluids can be performed in the usual way and examples of such calculations exist. Significant challenges emerge when attempting to simulate turbulent flows of non-Newtonian fluids. In this case, only non-Newtonian fluids described by the power fluid model can be reliably simulated (see [28–30]).

References

1. Afzal A, Kim K-Y (2014) Flow and mixing analysis of non-Newtonian fluids in straight and serpentine microchannels. Chem Eng Sci 116:263
2. Ansari M, Kim K-Y, Kim S (2018) Numerical and experimental study on mixing performances of simple and vortex micro T-Mixers. Micromachines 9:204

3. Abedin A, Arash K, Davood T (2017) The study of heat transfer and laminar flow of kerosene/multi-walled carbon nanotubes (MWCNTs) nanofluid in the microchannel heat sink with slip boundary condition. J Therm Anal Calorim 131:1553–1566
4. Arani AAA, Akbari OA, Safaei MR et al (2017) Heat transfer improvement of water/single-wall carbon nanotubes (SWCNT) nanofluid in a novel design of a truncated double-layered microchannel heat sink. Int J Heat Mass Transfer 113:780–795
5. Austin W, Lee H, Gates BD (2016) Rapid covalent modification of silicon oxide surfaces through microwave-assisted reactions with alcohols. Langmuir 32(29):72847293
6. Batchelor GK (1977) The effect of Brownian motion on the bulk stress in a suspension of spherical particles. J Fluid Mech 83:97–117
7. Batchelor GK (2012) An introduction to fluid dynamics. Cambridge University Press, Cambridge
8. Besson E, Gue A-M, Sudor J et al (2006) A novel and simplified procedure for patterning hydrophobic and hydrophilic SAMs for microfluidic devices by using UV photolithography. Langmuir 22(20):8346–8352
9. Betchov R, Criminale WQ Jr (1967) Stability of parallel flows. Academic, New York
10. Bird RB, Armstrong RC, Hassager O (1987) Dynamics of polymeric liquids. Vol 1. Fluid mechanics. Wiley, New York
11. Bord EG, Isakov EB, Rudyak VYa (1997) The stability of laminar flows of dilute dispersed media. Fluid Dynam 32(4):495–499
12. Bord EG, Rudyak VYa (2018) Stability regimes of flow in a channel between coaxial cylinders. Fluid Dynam 53(6):729–737
13. Bounds CO, Upadhyay J, Totaro N et al (2013) Fabrication and characterization of stable hydrophilic microfluidic devices prepared via the in-situ tertiary-amine catalyzed Michael addition of multifunctional thiols to multifunctional acrylates. ACS Appl Mater Inter 5(5):1643–1655
14. Buongiorno J (2006) Convective transport in nanofluids. ASME J Heat Transfer 128:240–250
15. Cercignani C (1975) Theory and application of the Boltzmann equation. Scottish Academic Press, Edinburgh
16. Cetkin E, Miguel AF (2021) Asymmetric Y-shaped micromixers with spherical mixing chamber for enhanced mixing efficiency and reduced flow impedance. J Appl Fluid Mech 14(5):1389–1397
17. Choi C-H, Johan K, Westin A et al (2003) Apparent slip flows in hydrophilic and hydrophobic microchannels. Phys Fluids 15:2897–2902
18. Chon CH, Kihm KD, Lee SP et al (2005) Empirical correlation finding the role of temperature and particle size for nanofluid (Al_2O_3) thermal conductivity enhancement. Appl Phys Lett 87(15):153107
19. Chuang M-J, Chu A-K (2011) Formation of a high hydrophilic/hydrophobic contrast surface on PET substrates by ECR generated sulfur hexafluoride plasma. Appl Surface Sci 257(9):3943–3947
20. Cotrell DL, Pearlstein AJ (2006) Linear stability of spiral and annular Poiseuille flow for small radius ratio. J Fluid Mech 547:1–20
21. DiPrima RC, Eagles PM, Ng BS (1984) The effect of radius ratio on the stability of Couette flow and Taylor vortex flow. Phys Fluids 27(10):2403–2411
22. Dovgaluk YuA, Ivlev LS (1996) Physics of water and other atmospheric aerosols. Sankt-Petersburg University Pres, Sankt-Petersburg
23. Drazin PG, Reid WH (1981) Hydrodynamic stability. Cambridge University Press, Cambridge
24. Dyson P, Ransing RS, Williams PM et al (2008) Fluid properties at nano/meso scale: a numerical treatment. Wiley, New York
25. Friedlander SK (2000) Fundamentals of aerosol dynamics. Smoke, dust, haze. Oxford University Press, New York
26. Gad-el-Hak M (1999) The fluid mechanics of microdevices: the Freemans scholar lecture. ASME- J Fluids Eng 121:5–33

27. Gavrilov AA, Minakov AV, Dekterev AA et al (2011) A numerical algorithm for modeling laminar flows in an annular channel with eccentricity. J App Industrial Math 5(4):1–12
28. Gavrilov AA, Rudyak VYa (2016) Reynolds averaged modeling of turbulent flows of power-law fluids. J Non-Newtonian Fluid Mech 277:45–55
29. Gavrilov AA, Rudyak VYa (2016a) Direct numerical simulation of the turbulent flows of power-law fluids in a circular pipe. Thermophys Aeromech 23(4):473–487
30. Gavrilov AA, Rudyak VYa (2017) Direct numerical simulation of the balance of turbulent energy and shear stresses in pipe flows of the power law fluids. Fluid Dyn 52(3):423–430
31. Gore RA, Crowe CT (1989) Effect of particle size on modulating turbulent intensity. Int J Multiphase Flow 15:279–285
32. Guo Z (2020) A review on heat transfer enhancement with nanofluids. J Enhanc Heat Transfer 27:1–70
33. Guzei D, Minakov A, Rudyak V (2019) On efficiency of convective heat transfer of nanofluids in laminar flow regime. Int J Heat Mass Transfer 139:180–192
34. Guzey DV, Minakov AV, Rudyak VYa (2022) Numerical modeling of forced convection of a nanofluid during its laminar flow taking into account thermal diffusion. J Eng Phys Thermophys 95:516–526
35. Hong L, Pan T (2011) Surface microfluidics fabricated by photo patternable superhydrophobic nanocomposite. Microfluidics Nanofluidics 10(5):991–997
36. Huang Q-Z, Fang Y-Y, Liu P-Y et al (2018) A novel strategy for durable superhydrophobic coating on glass substrate via using silica chains to fix silica particles. Chem Phys Lett 692:33–37
37. Isakov EB, Rudyak VYa (1995) The stability of rarefied dusty gas and suspension flows in a plane channel. Fluid Dyn 30(5):708–712
38. Esmaeil J, Omid AA, Sarafraz MM et al (2019) Heat Transfer of oil/MWCNT nanofluid jet injection inside a rectangular microchannel. Symmetry 11(6):757
39. Jang SP, Choi SUS (2004) Role of Brownian motion in the enhanced thermal conductivity of nanofluids. Appl Phys Lett 84(21):4316–4318
40. Kapitza PL (1941) The study of heat transfer in helium II. J Phys USSR 4(1–6):181–210
41. Khan WA, Khan ZH, Rahi M (2014) Fluid flow and heat transfer of carbon nanotubes along a flat plate with Navier slip boundary. Appl Nanosci 4(5):633–641
42. Kozlov V, Zverkov I, Zanin BYu et al (2007) Development of laminar-separating-flow perturbations on a wavy wing. Thermophys Aeromech 14(3):329–326
43. Lauga E, Brenner MP, Stone HA (2005) Microfluidics: the no-slip boundary condition. In: Foss J, Tropea C, Yarin A (eds) Springer handbook of experimental fluid dynamics. Springer, Berlin
44. Li D (ed) (2008) Encyclopedia of microfluidics and nanofluidics. Springer Science+Business Media, LLC
45. Li PCH (2005) Microfluidic lab-on-a-chip for chemical and biological analysis and discovery. CRC Press
46. Liu D, Yu L (2011) Single-phase thermal transport of nanofluids in a minichannel. J Heat Transfer 133:031009
47. Lobasov AS, Minakov AV, Rudyak VYa (2016) Viscosity effect on the flow patterns in T-type micromixers. Fluid Dyn 51(3):381–388
48. Lobasov AS, Minakov AV, Rudyak VYa (2018) Study of the mixing regimes of a fluid and a nanofluid in a T-shaped micromixer. J Eng Phys Thermophys 91:124–135
49. Lobasov AS, Minakov AV, Rudyak VYa (2018a) Flow modes of non-Newtonian fluids with power-law rheology in a T-shaped micromixer. Theoret Found Chem Eng 52(3):393–403
50. Lobasov AS, Minakov AV, Rudyak VYa (2022) The investigation of the velocity slip and the temperature jump effect on the heat transfer characteristics in a microchannel. Case Studies Thermal Eng 31:101791
51. Loitsyanskii LG (1966) Mechanics of liquids and gases. Pergamon Press, Oxford
52. Martini A, Roxin A, Snurr RQ et al (2008) Molecular mechanisms of liquid slip. J Fluid Mech 600:257–269

53. Maxwell CJ (1897) On stresses in rarefied gasses arising from inadequacies of temperature. Phil Trans Roy Soc I 170:231–256
54. Meseguer A, Marques F (2000) On the competition between centrifugal and shear instability in spiral Couette flow. J Fluid Mech 402:33–56
55. Meseguer A, Marques F (2002) On the competition between centrifugal and shear instability in spiral Poiseuille flow. J Fluid Mech 455:129–148
56. Michaelides EE (2014) Nanofluidics. Thermodynamic and transport properties. Springer
57. Minakov AV, Rudyak VYa, Gavrilov AA et al (2012) Mixing in a T-shaped micromixer at moderate Reynolds numbers. Thermophys Aeromech 3:385–395
58. Minakov A, Rudyak V, Dekterev A et al (2013) Investigation of slip boundary conditions in the T-shaped microchannel. Int J Heat Fluid Flow 43:161–169
59. Minakov AV, Lobasov AS, Guzei DV et al (2015) The experimental and theoretical study of laminar forced convection of nanofluids in the round channel. Appl Thermal Eng 88:140–148
60. Minakov AV, Rudyak VYa, Guzei DV et al (2015) Measurement of the heat transfer coefficient of a nanofluid based on water and copper oxide particles in a cylindrical channel. High Temp 53(2):246–253
61. Minakov AV, Rudyak VYa, Pryazhnikov MI (2020) Systematic experimental study of the viscosity of nanofluids. Heat Transfer Eng 41(5):457–460
62. Minakov A, Yagodnitsina A, Lobasov A et al (2013) Study of fluid flow in micromixer with symmetrical and asymmetrical inlet conditions. La Houille Blanche 5:12–21
63. Ng BS, Turner ER (1982) On the linear stability of spiral flow between rotating cylinders. Proc Roy Soc Lond A 382:83–102
64. Nguyen N-T, Wu Z (2005) Micromixers—a review. J Micromech Microeng 15(2):R1–R16
65. Nigmatulin RI (1990) Dynamics of multiphase media V I. CRC Press
66. Linstrom PJ, Mallard G (2011) NIST Chemistry Web Book. NIST standard reference database number 69. National Institute of Standards and Technology, Gaithersburg
67. Nojoomizadeh M, D'Orazio A, Karimipour A (2018) Investigation of permeability effect on slip velocity and temperature jump boundary conditions for FMWNT/Water nanofluid flow and heat transfer inside a microchannel filled by a porous media. Physica E: Low-Dimens Syst Nanostr 97:226–238
68. Orszag SA (1971) Accurate solution of the Orr-Sommerfeld stability equation. J Fluid Mech 50(4):689–703
69. Ou J, Perot B, Rothstein P (2004) Laminar drag reduction in microchannels using ultrahydrophobic surfaces. Phys Fluids 16:4635–4643
70. Pryazhnikov MI, Minakov AV, Vya R et al (2017) Thermal conductivity measurements of nanofluids. Int J Heat Mass Transfer 104:1275–1282
71. Rakhmatulin KhA (1956) Fundamentals of gas dynamics of interpenetrating movements of compressible media. Prikl Mat Mekh 20(2):184–195
72. Rao LT, Goel S, Dubey SK et al (2019) Performance investigation of T-shaped micromixer with different obstacles. J Phys: Conf Ser 1276:012003
73. Raza W, Hossain S, Kim K-Y (2020) A review of passive micromixers with a comparative analysis. Micromachines 11(5):455
74. Rudyak VY (1992) Kinetics of finely dispersed gas suspension. Sov Tech-Phys Lett 18(10):681–682
75. Rudyak VY (1995) Nonlocal solution of Boltzmann equation. Eng Phys 65(11):29–40
76. Rudyak VY (1996) Classification principles of dispersed media. J Aerosol Sci 27(1):S271–S272
77. Rudyak VY (2004) Statistical aerohydromechanics of homogeneous and heterogeneous media. Vol. 1. Kinetic Theory. NSUACE, Novosibirsk
78. Rudyak VY (2005) Statistical aerohydromechanics of homogeneous and heterogeneous media. Vol. 2. Hydromechanics. NSUACE, Novosibirsk
79. Rudyak VY (2013) Viscosity of nanofluids. Why it is not described by the classical theories. Adv Nanoparticles 2:266–279

80. Rudyak VY (2024) Specific features of modeling nanofluid flows (a review). Fluid Dyn 59(5):1072–1108
81. Rudyak VY, Belkin AA (1996) Equations of multifluid hydrodynamics. Mat Model 8(6):33–37
82. Rudyak VY, Belkin AA (2013) Transport processes of nanoparticles in gases and liquids. Adv Struct Mater 4:135–168
83. Rudyak VY, Belkin AA, Tomilina EA (2008) Force acting on nanoparticle in fluid. Tech Phys Lett 34:76–78
84. Rudyak VY, Bord EG (2017) On stability of plane and cylindrical Poiseuille flows of nanofluids. J Appl Mech Tech Phys 58(6):1013–1020
85. Rudyak VY, Bord EG (2018) About hydrodynamical stability of Poiseulle and Couette flows in the channel between concentric cylinders. J Eng Phys Thermophys 91:1282–1290
86. Rudyak VYa et al (2018) Modeling of micromixers. Fluid Mech Appl 118:127–184
87. Rudyak VY, Dashapilov GR, Minakov AV et al (2023) Comparative characteristics of viscosity and rheology of nanofluids with multi-walled and single-walled carbon nanotubes. Diamond Rel Mat 132:109616
88. Rudyak VY, Dimov SV, Kuznetsov VV et al (2013) Measurement of the viscosity coefficient of an ethylene glycol–based nanofluid with silicon dioxide particles. Doklady Phys 58(5):173–176
89. Rudyak VY, Isakov E, Bord E (1997) Hydrodynamic stability of the Poiseuille flow of the dispersed fluid. J Aerosol Sci 28:53–66
90. Rudyak VY, Isakov EB, Bord EG (1998) Stability of two-phase flows. Thermophys Aeromech 5:51–57
91. Rudyak VY, Minakov AV (2016) Modern problems of micro and nanofluidics. Science, Novosibirsk
92. Rudyak VY, Minakov A (2018) The features of the modeling the nanofluid flows. AIP Conf Proc 1959:060013
93. Rudyak VY, Minakov AV, Gavrilov AA et al (2008) Application of new numerical algorithm of solving the Navier-Stokes equations for modeling the work of a viscometer of the physical pendulum type. Thermophys Aeromech 15(2):333–345
94. Rudyak VY, Minakov A, Guzey D (2021) Modeling the convective thermal heat transfer of nanofluids with carbon nanotubes in cylindrical minichannel. J Phys: Conf Series. 2131(2):022068
95. Rudyak VY, Minakov AV, Pryazhnikov MI (2017) Thermophysical properties of nanofluids and similarity criteria. Tech Phys Lett 43:23–26
96. Rudyak VY, Minakov AV, Guzei1 DV et al (2016) On laminar-turbulent transition in nanofluid flows. Thermophys Aeromech 23(5):773–776
97. Rudyak VY, Minakov AV, Pryazhnikov MI et al (2022) Measurement of the thermal conductivity and heat transfer coefficient of nanofluids with single-walled nanotubes. High Temp 60(5):631–638
98. Rudyak VY, Minakov A, Pryazhnikov M (2021) Preparation, characterization, and viscosity studding the single-walled carbon nanotube nanofluid. J Mol Liquids 329:115517
99. Rudyak VY, Pryazhnikov MI, Minakov AV et al (2023) Comparison of thermal conductivity of nanofluids with single-walled and multi-walled carbon nanotubes. Diamond Rel Mat 139:110376
100. Rudyak VY, Tret′yakov DS (2021) Rheological properties of water- and ethylene glycol-based nanofluids with single wall carbon nanotubes. J Eng Phys Thermophys 94(5):1208–1216
101. Ryzhkov II, Minakov AV (2014) The effect of nanoparticle diffusion and thermophoresis on convective heat transfer of nanofluid in a circular tube. Int J Heat Mass Transfer 77:956–969
102. Saffman PG (1962) On the stability of laminar flow of a dusty gas. J Fluid Mech 13:120–128
103. von Smoluchowski M (1898) Ueber warmeleitung in verdunten gasen. Ann Physik Chemi 64:10130
104. Sone Y (2006) Molecular gas dynamics: theory, techniques, and applications. Birkhäuser, Boston

105. Soo SL (1967) Fluid mechanics of multiphase systems. Blaisdell Pub. Co, Blaisdell, Waltham
106. Tabeling P (2005) Introduction to microfluidics. Oxford University Press, Oxford
107. Takeuchi DI, Jankowski DF (1981) A numerical and experimental investigation of the stability of spiral Poiseuille flow. J Fluid Mech 102:101–126
108. Taylor GI (1923) Stability of a viscous liquid contained between two rotating cylinders. Phil Trans Roy Soc Lond A 223:289–343
109. Telha M, Mahammedi A, Naas TT et al (2022) Numerical investigation into mixing efficiency of T-micromixers with elliptic barriers. Int J Adv Stud Comp Sci Eng 11(9):12–19
110. Tovbin YuK (2012) Molecular theory of adsorption in porous bodies. Fizmatgiz, Moscow
111. Tretheway DC, Meinhart CD (2002) Apparent fluid slip at hydrophobic microchannel walls. Phys Fluid 14(3):L9–L12
112. Vinogradova OI (1998) Implications of hydrophobic slippage for the dynamic measurements of hydrophobic forces. Langmuir 14:2827–2837
113. Wang X-Q, Mujumdar AS (2007) Heat transfer characteristics of nanofluids: a review. Int J Thermal Sci 46:1–19
114. Wang Z, Sun S, Cui Y et al (2021) Molecular dynamics investigation of the slip flow liquid–solid interfacial thermal conductance. AIP Adv 11:055204
115. Watanabe M (2008) Microchannels constructed on rough hydrophobic surfaces. Chem Eng Technol 31(8):1196–1200
116. Wong HF, Ahmad N, Siri Z et al (2021) Viscous heating and cooling process in a mixed convection cavity with free-slip effect. Case Stud Thermal Eng 28:101349
117. Yamasaki Y, Kariyasaki A, Morooka S (2010) Hydrophilic and hydrophobic modifications of microchannel inner walls for liquid-liquid laminar layered flows. Int J Chem Reactor Eng 8:A140
118. Zhang J, Diao Y, Zhao Y et al (2015) Thermal-hydraulic performance of SiC-water and Al_2O_3-water nanofluids in the minichannel. J Heat Transfer 138(2):021705
119. Zhou S-Q, Ni R (2008) Measurement of the specific heat capacity of water-based Al_2O_3 nanofluid. Appl Phys Lett 92:93123

Chapter 8
Application of Nanofluids and Problems Not Previously Discussed in Detail

Nanofluids, whose thermophysical properties have been extensively discussed in previous chapters, are highly unusual fluids. Their properties are not described by classical theories, and I hope that the reasons for this are now largely clear. Nanoparticles are mesoscopic objects, and for this reason nanofluids are mesosuspensions. In the metric of the base fluid, nanoparticles are material points; this determines the very special mechanisms of the transport properties in nanofluids. The main point is that the presence of nanoparticles does not result in discernible hydrodynamic disturbances in the base fluid. However, its structure is notably changing. This phenomenon can be most clearly observed when the radial distribution functions of the base fluid and nanofluid are compared. The alteration in structure provides a satisfactory explanation for the observed increase in viscosity of the nanofluid. The mechanism of thermal conduction in nanofluids is more complex and less evident; however, the structure alterations of this dispersed fluid also exert a pivotal role. Thus, the experimental study of the transport properties of nanofluids and their modeling by the molecular dynamics method is extremely important from a fundamental point of view, since it opens a new page in the physics and mechanics of what is commonly called soft media.

At the same time, we must not forget that the enduring interest in the study of nanofluids was primarily initiated by their possible applications. So far, this issue has hardly been mentioned in this book. Nevertheless, this is an extremely important aspect and should be discussed in detail, which will be done in this chapter.

Any book is a statement of what has already been established. In above chapters, when discussing the diffusion of nanoparticles and CNTs, the viscosity, thermal and electrical conductivities of nanofluids, we have presented reliable experimental data largely obtained by the author and his research teams. In parallel, in order to determine the main qualitative characteristics of these transport processes and to identify the corresponding mechanisms, molecular dynamics modeling data were also provided. Naturally, the author offers a very definite point of view. However, nanofluids subject field is so extensive that a number of questions yet remain unanswered. Therefore,

V. Rudyak, *Thermophysical Properties of Nanofluids*, Fluid Mechanics and Its Applications 140, https://doi.org/10.1007/978-3-031-97995-8_8

the concluding section of the book will address some issues that have not yet been considered in detail and need to be addressed as a matter of priority. This chapter is beginning from the detailed discussion of the existing or perspective applications of nanofluids. The following three sections discuss the problems of phase transitions, boiling crisis and optical properties of nanofluids. Finally, the last two sections briefly discuss the features of the thermophysical characteristics of nanofluids in microchannels and hybrid nanofluids.

8.1 Applications of Nanofluids

The study of nanofluids has been a prominent research area for approximately three decades, largely due to their potential and existing applications. As illustrated in Fig. 8.1, the field of potential applications of nanofluids is quite extensive. While additional petals may be added to the presented "flower," it is already clear that the field of possible applications of nanofluids is quite extensive.

8.1.1 Thermophysical Applications of Nanofluids

The idea of using dispersed fluids as heat carriers and refrigerants is quite natural and was emerged long ago. However, conventional coarse dispersed fluids have proven unsuitable for this purpose due to the sediment of particles and their abrasive nature. On the other hand, nanofluids do not have these disadvantages. As previously stated, the initial experiments on measuring the thermal conductivity of nanofluids

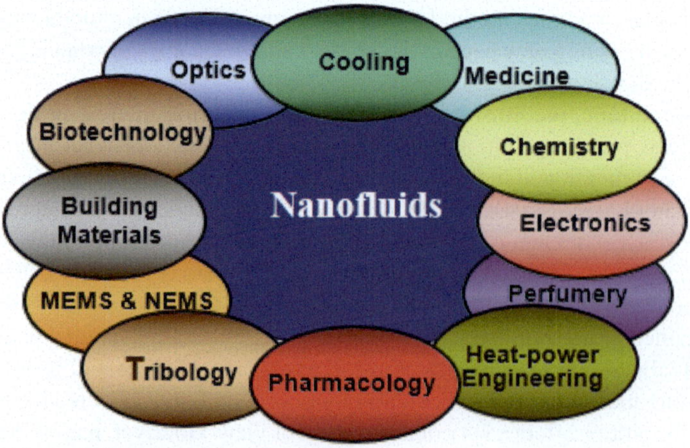

Fig. 8.1 Existing and potential applications of nanofluids

yielded highly promising results. The addition of even fractions of a percent of metallic nanoparticle concentrations increased the thermal conductivity of the base fluid by percent or even tens of percent. Today, it has become evident that such optimistic expectations are not attainable in practice. Nevertheless, the enhancement of thermal conductivity and heat transfer of nanofluids is indeed quite significant. This is especially true for nanofluids with SWCNTs. For this reason, the use of nanofluids in various thermophysical applications appears to be not merely a viable option, but in certain instances, an indispensable necessity. The question arises as to where this is primarily necessary and appropriate. Numerous articles and reviews provide an answer to this question (see, for example, one of the most recent reviews by [208]).

One of the markers of the beginning of this century is the rapid development of compact electronic devices for various purposes (including mobile phones, smartphones, tablets, game consoles, laptops, etc.) that require a sustainable, long-term and well-monitored energy supply. The growing power of existing computers necessitates a transition towards increasingly smaller chip sizes. This trend gives rise to an increase in heat generation, which in turn results in the necessity to remove generated heat within the range of 100–300 W/cm^2 or higher. The currently existing air cooling techniques are used at their limit of capacity. Therefore, fluid cooling methods are currently being developed. A variety of radiators, heat pipes, and other devices are employed for heat removal through natural convection [184, 212, 217]. Naturally, conventional fluids, such as water, various oils, and ethylene glycol, are employed primarily as cooling agents. The use of nanofluids for this objective is exceedingly promising. Back in 2006, Jang and Choi [75] have developed a novel cooler design based on the use of a microchannel heat exchanger with nanofluid. Significantly higher characteristics was achieved when nanofluids were employed as compared to using ordinary water as a coolant. The nanofluids diminished both the thermal resistance and the temperature difference between the heated walls of the microchannel and the cooling fluid. Research on the use of a wide variety of nanofluids for this purpose has continued in recent years, and impressive results have been obtained in all cases [44, 69, 99, 149, 155, 224]. As a result of all the studies conducted, it was established that nanofluids are significantly more effective in reducing the temperature of electronic components compared to base fluids. At the same time, possible deposition of nanoparticles on the walls of devices is noted. For this reason, it is recommended to use predominantly oxide particles at relatively low concentrations.

Nanofluids can be effective in a range of other microchannel devices (micro system technics (MSTs), microelectromechanical systems (MEMS), lab-on-chip analysis systems, and etc.). MEMS generate a lot of heat during operation, so conventional coolants don't work well with high power MEMS. In this case, the potential of nanofluids as a cooling fluid is becoming quite promising. Here, however, in contrast to macroscopic systems, where nanofluid can almost always be considered (and simulated) as an ordinary fluid, a certain caution should be exercised, since the presence of particles can significantly affect both the flow regimes and the processes of heat and mass transfer. Such flows may have other important features. As demonstrated in study [237], even minimal concentrations (approximately 3×10^6) of bismuth telluride can significantly alter the surface wettability characteristics. This effect increases with decreasing nanoparticle size, the smaller the nanoparticle size, the more the contact angle increases.

The use of nanofluids as a coolant in automobiles has great prospects, since it can lead not only to more efficient operation of cooling and braking systems, but also by increasing its efficiency, simplifying thermal control systems, reducing dimensions, greatly saving materials, and even changing the overall design of the car [1]. For example, this may result in a reduction in the vehicle's radiator area, and potentially even a change in its arrangement, which could lead to a decrease in aerodynamic drag. The application of nanofluids is particularly useful for improving the cooling rate of high-power engines. In recent years, there has been a great deal of research focused on the potential applications of nanofluids in the vehicles. The application of various nanofluids based on water, ethylene glycol, as well as water and ethylene glycol mixture with copper, silver, as well as copper, silicon, titanium, and aluminium oxide particles and MWCNTs has been studied in many works [31, 39, 66, 108, 198, 199, 208]. In all cases, the use of nanofluids for cooling was found to be markedly more efficient than all base fluids under investigation. The dependence of heat transfer characteristics of nanofluids on their concentration, flow conditions and temperature was established.

In general, the potential applications of nanofluids as a cooling fluid are extensive and diverse. This is applicable to the majority of commonly used equipment, including automotive radiators, refrigerators, air conditioning units, building heating systems, and various military and technical devices and systems [140, 208, 210]. In addition to conventional military equipment, a variety of miniaturized military and space vehicles are currently under active development in the world, the use of which is impossible without an effective and compact cooling system. The use of nanofluids here seems not only promising, but also possible and necessary.

J. Routbort and colleagues back in 2010 [185] developed a project on the use of nanofluids as refrigerants in industrial installations. The implementation of this technology promised significant profits and savings in the use of traditional materials and energy resources (see [238]). However, the resulting effect may also depend on the specific type of heat exchanger employed [154, 172]. The effectiveness of using nanofluids in different types of heat exchangers has been systematically studied in a review [199].

A novel approach to heating buildings in low-temperature regions utilizing nanofluids was proposed in Kulkarni et al. [107]. The base fluid was a mixture of ethylene glycol and water (60:40 by weight). The use of nanofluids in heat exchangers facilitates a reduction in the volume and weight flow rate of the coolant, thereby achieving overall energy savings in pumping, as well as results in a more compact and cost-effective system that allows reducing the initial cost of the equipment. Moreover, this also has the beneficial effect of reducing environmental pollution.

A systematic study was conducted at the Massachusetts Institute of Technology (MIT) on the potential use of nanofluids in all water-cooled nuclear installations, including pressurized water reactors, safety systems, plasma diverters, etc. [238]. It was established that the use of certain nanofluids (e.g., those with aluminum particles) can result in a notable enhancement in the critical heat flux. Preliminary experiments conducted at MIT research nuclear reactor have demonstrated that the use of nanofluids markedly enhances the reactor's operational efficiency. As a cooling agent, nanofluids could be successfully used in security systems as well [24, 101], where there is a need to rapidly reduce the heat flux. However, there are certain problems with the possible change in the concentration of nanoparticles due to their entrainment by steam from the working area. In particular, for this reason, although not only, the use of nanofluids in boiling water reactors is likely to be limited.

In recent years, it has been demonstrated by numerical simulations [205] that the use of nanofluids enhances the thermal efficiency of nuclear reactors while concurrently mitigating the incidence of associated hydraulic issues. Here, the effect of a water-based nanofluid with silicon dioxide particles on the thermal performance of a light water nuclear reactor was studied at different particle concentrations. As anticipated, the heat transfer enhanced in proportion to the rising concentration of particles, while the temperature of the cladding wall decreased. A substantial body of research has been dedicated to the simulation of the performance of different nuclear reactors using nanofluids for various purposes [46, 77, 96, 191]. In all cases, it was determined that the reactor operating time was extended, and the neutron multiplication factor was diminished, while the safety of the reactor operation was increased.

The potential applications of nanofluids in the energy sector are diverse, extending beyond cooling systems. In particular, nanofluids can be employed in the creation of thermal energy storage and accumulation systems. It is estimated that currently there are 13,000 GJ of geothermal energy reserves, while just less than 2% of this energy is recoverable. However, by improving existing technologies, this value could be increased by an order of magnitude. This requires the ability to extract this energy from the earth's crust from depths of 5–10 km at temperatures from 500 to 1000 °C. Nanofluids can be used for the cooling of pipes subjected to high temperatures, as well as in drilling processes, where they can also be employed for cooling purposes. An illustrative example of the use of nanofluids in a geothermal borehole heat exchanger can be found in the paper [34]. It has been shown that heat release increases with increasing volume concentration of aluminum oxide nanoparticles. And finally, nanofluids can serve as heat carriers for the extraction and transport of substantial quantities of energy. At the same time, nanofluids can be employed to cool employed tools including measuring instruments. Thus, it has been demonstrated that in certain cases nanofluids can serve as a control tool. Some nanofluids can act as a thermal relief valve regulating the heat flow, as it is shown in Donzelli et al. [42].

Heat pipes are an effective and compact systems of transferring heat. The success of their application is dependent upon many individual factors, including their geometry, the characteristics of the working fluid, the composition of the tube material, the shape of the tube, etc. [127]. The studies indicate that nanofluids can be effectively employed in heat pipes. As demonstrated in Tsaia et al. [236], the use of nanofluids results in a significant reduction in thermal resistance. This effect can be attributed to a notable enhancement of the critical heat flux in the nanofluid [131, 137, 174]. The thermal characteristics of the heat pipe are also dependent on the nanoparticle size and material. Research into the possibility of using nanofluids in heat pipes continues unabated. In study [93] water-based nanofluids with aluminium oxide particles used, while Gandomkar et al. [54] tested magnetic nanofluid. Kerosene-based nanofluid with iron oxide particles ranging in size from 10 to 30 nm used in heat pipes was studied in Goshayeshi et al. [59]. In recent years, hybrid nanofluids have been extensively used [159, 226]. This list can easily be continued, however, it should be emphasized that the use of nanofluids has had a positive effect in all cases, although their application in specific heat pipes certainly requires further systematic experiments and subsequent optimization.

The development of solar energy has also actively stimulated research into the possibility of using nanofluids here. Solar collectors capture and convert incident solar radiation into heat through the use of thermal devices or into electricity through the employment of photovoltaic panels [19]. The use of nanofluids is dependent upon a number of their distinctive properties, including high thermal and electrical conductivity, and non-standard optical properties (see reviews [187, 207]). Accordingly, the absorption of solar energy is maximized as the nanoparticle size, and concentration increase. This is the primary reason why nanofluids can be used in thermal solar collectors. The high thermal conductivity of nanofluids ensures a high heat transfer coefficient in the heat exchanger. Laboratory studies on the development of appropriate solar collectors started more than a decade ago. Thus, in the work of Otanicar et al. [156], water-based nanofluids with CNTs, as well as graphite and silver particles, were used for this purpose. The employment of nanofluids was found to be efficacious within a specific range of nanoparticle concentrations, which depends on the particle material or type of CNTs. It has also been observed that the presence of nanoparticles increases the absorption of incident radiation by times compared to the absorption by pure water [234]. A number of other nanofluids have been employed in similar studies [29, 37, 48, 192, 233]. In all cases, the efficiency of using nanofluids was sufficiently high. The advantages of employing nanofluids in various thermal solar collectors are elucidated in a review [196].

In addition to use of nanofluids in solar collectors, they are also employed in photovoltaic systems. Nanofluid in photovoltaic heating system absorbs waste heat, enabling efficient energy conversion [195, 255]. For example, in the study [6], a nanofluid with Si–C particles was employed to absorb heat generated by battery. It was shown that the overall efficiency of the photovoltaic thermal system has increased by 85%. Finally, the use of thermoelectric devices should be mentioned. Several different technologies have been developed here and it has been shown that using nanofluids for cooling not only improves the thermal performance but also increases overall power generation [219, 238]. In conclusion, it can be said that the use of nanofluids is certainly promising, but further direct technological development is currently required. The optimal performance of various solar collectors can be achieved using different nanofluids. To obtain definitive recommendations here, a specific engineering order is required, which could facilitate the corresponding experiments, including the development of the suitable nanofluids.

Another important challenge facing the thermal power industry is the accumulation and conservation of thermal energy (ATE). There are three principal types of energy storage: chemical energy storage, sensible heat storage, and latent heat storage. The distinction between the latter two types of thermal energy storage is contingent upon the phase transition of the heat carrier used for the storage process. In latent ATEs, a phase transition occurs prior to the energy release or storage, whereas sensible ATEs operate based on change in the temperature of the material [154]. The potential of nanofluids in the creation of ATEs is also a subject of active investigation [23, 40, 124, 150].

In recent decade, a number of studies have been conducted to examine the possibility of using nanofluids to increase the breakdown voltage of transformer oils [52]. Finally, it is point out a curious, and perhaps even interesting, application of nanofluids in the creation of fuel cells with a proton exchange membrane (PEMFC) [72, 261, 262]. A further thermophysical application of nanofluids, albeit still quite exotic, is the use of nanofluid-based microbial fuel cells (MFCs). To generate electricity, the energy of carbohydrates, proteins and other energy-rich natural products is used here. The efficacy of such MTCs is contingent upon the quality of the electrodes and the intermediary medium connecting them. As the latter, nanofluids are a highly promising medium [206].

To conclude this section, it is useful to integrate the described applications of nanofluids into a simple schematic diagram, shown in Fig. 8.2.

The application of nanofluids as a working fluid for the cooling of various devices and systems is due to their high thermal conductivity. The thermal conductivity enhancement of the nanofluid with oxide particles is 25–30% at a volume concentration of nanoparticles of 5–8%. Nanofluids with metal particles will have even higher thermal conductivity, although their weight concentration will be several times or even an order of magnitude higher than their volume concentration. On the other hand, the excess thermal conductivity of nanofluids with CNTs, particularly SWCNTs, reaches tens of percent at a weight concentration of merely 0.2–0.3%, with the volume concentration remaining approximately twice as low. What prevents them from being actively used?

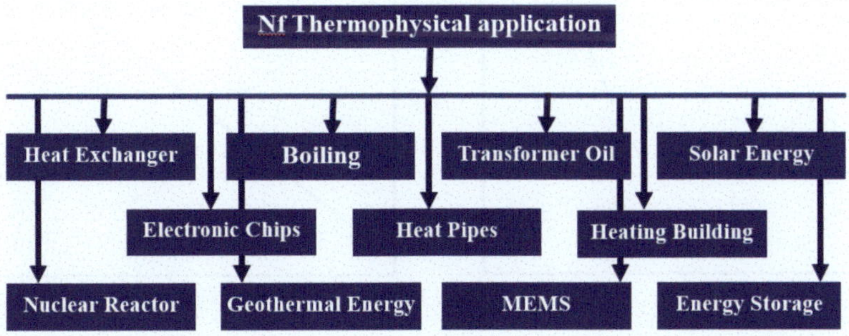

Fig. 8.2 Schematic diagram of the main thermophysical applications of nanofluids

To answer this question, it is necessary to note two main factors. The first is that for practical applications, nanofluids must be sufficiently stable and long-lived. The industrial production of nanofluids was initiated over a decade ago by a research team at Stanford University. Currently, nanofluids based on diverse base fluids with various nanoparticles and CNTs are being manufactured on a commercial scale in the USA, China, India, and the UK. One disadvantage of such nanofluids is the lack of sufficient information on their thermophysical properties. Besides, there is a dearth of data regarding the employed dispersants or functionalization of particles and CNTs. Nevertheless, this is a resolvable issue. It is possible to negotiate with manufacturers to divulge pertinent information or, at the very least, to furnish sufficiently comprehensive data on thermophysical properties in the certificate of the produced nanofluid.

The second factor is predominantly psychological in nature. It is a well-established fact that nanofluids, which possess high thermal conductivity, also exhibit high viscosity and frequently display non-Newtonian characteristics. Accordingly, there is a well-established opinion that the energy expended on pumping nanofluids will significantly exceed the benefit of its high thermal conductivity. This is, of course, a hasty conclusion. Firstly, there are situations when it is important to cool the surface of the device as much as possible, regardless of the amount of energy spent on pumping the coolant. However, this is not the most significant factor. As previously demonstrated, the viscosity and thermal conductivity of nanofluids depend on their size. As the size of the nanoparticles increases, the thermal conductivity rises while the viscosity declines on the contrary. The qualitative representation of this dependence is provided in Fig. 8.3a. The analysis of the presented dependences demonstrates that in practice it is feasible to identify a specific range of nanoparticle sizes at which the thermohydraulic effect will be optimal.

There is another circumstance. The viscosity of the nanofluid decreases with increasing temperature, while in contrast its thermal conductivity increases

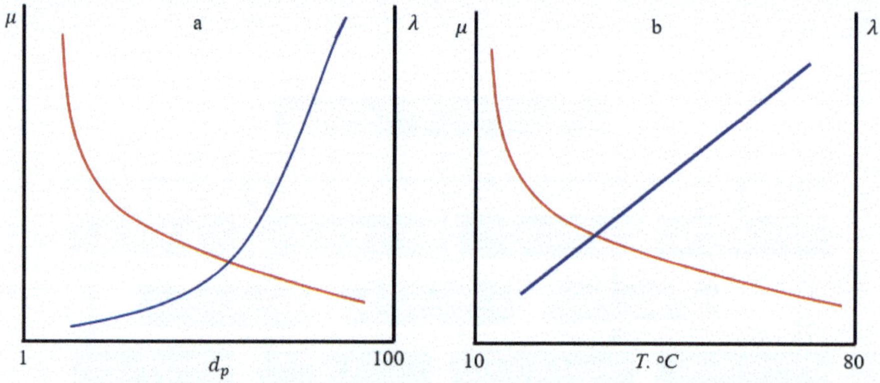

Fig. 8.3 Qualitative dependence of viscosity (red) and thermal conductivity (blue) coefficients on nanoparticle size (**a**) and on temperature (**b**)

(Fig. 8.3b). At the relatively small concentrations of nanoparticles under consideration, this growth is almost linear. The employment of nanofluid as a cooling fluid is expedient upon the existence of sufficiently high temperatures within the operational zone. For this reason, the energy spent on pumping will be relatively small. On the other hand, thermal conductivity, and therefore heat transfer, will be high.

8.1.2 Tribological Applications

The problem of friction in technology, and in everyday life, is a key issue. The role of friction can be both beneficial and detrimental. On the one hand, the presence of friction enables humans to walk and vehicles to drive. On the other hand, the rubbing of tires (and shoes!), machine parts, for instance, results in wear and tear. This is due to the action of frictional forces arising between parts (things, objects, etc.) that are in contact and rubbing against each other. It is evident that the presence of friction leads to significant energy losses. In order to regulate the friction and wear of contacting parts, a variety of lubricants are employed. In a number of devices, it is necessary to lubricate the rubbing parts in very small gaps (in particular in different MEMS). A magnetic recorder is a typical example of such a device, wherein the gap between the recording head and the recording medium is on the order of a few tens of nanometers. In order to achieve higher recording densities, it is necessary to reduce the gap.

Various oils, liquids, emulsions and sometimes even gases are commonly used as lubricants. The unique properties of nanofluids also permit their use in tribological applications. The tribological applications of nanofluids have different target functions. These include (i) nanofluid-based lubricants, (ii) friction-reducing nanofluids, (iii) wear-reducing nanofluids, (iv) brake nanofluids. In the process of vehicle braking, the fundamental properties of brake fluids are their viscosity, high thermal conductivity, and high boiling point. All these characteristics are significantly higher in nanofluids than in carrier fluids. For this reason, they are already being used successfully. Experimental studies on the use of such nanolubricants can be found in reviews [20, 164, 213–215]. Nanoparticles of various metals, oxides, and sulfates, as well as carbon particles including carbon nanotubes and graphene, are commonly used as nano-additives to lubricants. The earliest brake nanofluids, with copper and aluminum oxides, were developed as early as in 2008 [81]. Both these nanofluids have a higher boiling point, significantly higher viscosity and electrical conductivity than traditional brake fluid. As a consequence, the use of nanobrake oil reduces the probability of vapor lock formation and thus enhances safety while driving.

Nano-additives improve tribological properties by rolling of nano-spheres, the formation of a tribofilm, the filling nano- and micro-cracks on the surfaces of rubbing parts, as well as polishing contacting surfaces [104, 214, 256]. Paraffin oil-based lubricants with diamond particles, as well as standard oils with CuO, MoS_2, TiO_2, and Cu nanoparticles have shown excellent outcomes. Magnetic nanoparticles $Mn_{0.78}$ $Zn_{0.22}$ Fe_2O_4 are also effective lubricating additives. It has been established that nanofluids play an important role, especially in the formation of a cutting layer near the surface. The use of nanofluids leads to a considerable diminution of the tear-off force and a reduction in surface roughness. Nanofluids are a more favorable alternative to the use of solid lubricants, because they markedly reduce the temperature of the part.

The following key factors can be identified when using nanolubricants [11, 61, 116, 173]: (i) nanoparticles essentially change the viscosity of the base lubricant; (ii) the thermal conductivity of the nanolubricants enhances with increasing temperature and particle concentration; (iii) the specific heat capacity decreases with increasing particle concentration and increases with increasing temperature. Typically, the weight concentration of investigated nanoparticles varied from 0.05 to 1.5%. Hundreds of fulfilled experiments have shown the high efficiency of nanolubricants. Nevertheless, the obtained data turned out to be quite contradictory, which is largely due to the dissimilarity of the nanolubricants studied. They were prepared on the basis of various oils, using different nanoparticles (various sizes, materials and morphology). Furthermore, in numerous instances, surfactants were employed during the preparation process. It has been established that, in general, adding nanoparticles to the lubricant reduces the coefficient of friction and wear of rubbing parts [33]. This effect depends on the nanoparticle concentration and enhances with increase in concentration. However, this increase is non-monotonic, the beneficial effect may decrease starting from certain concentrations [94]. There is no definitive clarity on the effect of nanoparticle size, although it is noted that nanoparticles smaller than 30 nm are preferred [193, 213]. It has been shown in Muzakkir and Hirani [146] that the addition of zinc oxide nanoparticles to low viscosity lubricant reduces wear by about 94%, while the friction coefficient is reduced by 59%. On the other hand, their adding into high-viscosity lubricant diminishes wear by approximately 92%, but increases the friction coefficient by up to 71%.

The inability to provide unambiguous responses regarding the efficacy of specific nanolubricants is not solely attributable to their considerable diversity. The complexity lies in the multiscale nature of the processes occurring during friction. Conventional experiments are conducted at the macroscopic level, and thus the resulting measurements yield certain integral characteristics (e.g., the friction coefficient). At the same time, these integral characteristics are a consequence of processes occurring at the meso and microscales. These processes are of great significance, since they dictate the underlying mechanisms of macroscopically observable friction and the associated phenomena [141, 240]. It is practically impossible to study these mechanisms experimentally. An alternative approach is the use of molecular dynamic simulations. The molecular dynamics method has been widely employed for this purpose. The method has been used to examine the characteristics of diverse lubricants in both unconfined and confined conditions. Quite detailed relevant data on the results obtained can be found in reviews [50, 51].

Recently, a new direction has emerged in tribology, namely, biotribology, which studies all aspects of tribology related to biological systems. The human body provides numerous examples of biotribology. These include the wear and tear of natural and artificial joints, heart valves, friction at blood vessel walls, and friction within bronchi. In biotribology, nanofluids are almost always present.

For an extended period, tribology was a purely phenomenological macroscopic science based on the Amontons–Coulomb friction law. It was only in recent decades, following the emergence of atomic force microscopes, that it became feasible to investigate the fundamental nature of friction forces at the atomic level. In fact, this determined the birth of nanotribology (see review by Dedkov [36]), the subject of which is not only the study of the adhesion, friction, wear and destruction processes at the atomic and molecular level, but also the phenomena that arise during contact at the micro- and nanoscale. This includes, for instance, the analysis of nanometer-thick lubricant films. As the field of nanotribology progresses, the underlying mechanisms governing the tribological applications of nanofluids will become increasingly clear. At the same time, it is already clear today that for the successful use of nanofluids for tribological purposes, their further systematic study is required. It is necessary to study nanofluids with a wide range of nanoparticle materials, sizes, concentrations and rheological properties. Particular attention should be paid to the mechanical properties of nanofluids and their influence on the wear of parts, the degree of increase or decrease in friction.

8.1.3 Biomedical Applications

The biomedical applications of nanofluids, along with thermophysical applications, represent a significant and rapidly evolving field of research and today are, probably, the most important. With the advent of nanotechnology in medicine, it becomes an exact science, since it is nanotechnology that will allow, firstly, to conduct local diagnostics and accurately determine the location of damage to the body, and, secondly, to perform pinpoint delivery of drugs to the damaged organ. Finally, it becomes possible to track the development of diseases and their treatment at the cellular level. At this time, a significant number of bio-MEMS have been developed and implemented (or are being implemented) for a variety of purposes. Such devices include microchips for drug delivery, MEMS systems for DNA stacking, and specialized microdevices for drug delivery through various channels [100, 109, 157, 209]. A bio-MEMS is a fairly complex system that incorporates a range of micro components, including microvalves, micromixers, micropumps, and microsensors, which are used to obtain data and perform the requisite measurements.

The most significant and rapidly evolving field of nanoparticle and nanofluid application in the past decade is their use in cancer therapy. The application of nanofluids here is developing in two main directions: the detection of cancer cells (nanofluids as a contrast agent in magnetic resonance scanning, radiopaque nanofluids based on yttrium tantalate allow obtaining images of internal organs) and the targeted delivery of drugs or anti-tumor DNA to affected areas (see, for example, [57, 145]. It is curious that the nanoparticles themselves can act as such drugs. For this purpose, magnetic nanoparticles are widely used, whose movement through the bloodstream is regulated by an external magnetic field, thereby facilitating rapid and precise delivery of the drug to the affected cells. Furthermore, the therapeutic effect is enhanced by the fact that other healthy organs are not affected during the delivery process. In addition, magnetic nanoparticles bind significantly better to cancer cells than to healthy cells. Their nanoscale size and extensive surface area enable them to absorb transmitted energy much more efficiency than micron-sized particles. This property enables their successful application in the treatment of cancer by hyperthermia [74].

It is important to note that it has long been established that certain naturally occurring substances, such as zinc and silver, possess intrinsic antibacterial properties themselves. However, nanoparticles, due to their diminutive size, elevated surface-to-volume ratio, distinctive structure, and other characteristics, are capable of forming novel, distinctive antibacterial mechanisms. Therefore, nanoparticles of materials that are typically neutral to bacteria may possess antibacterial properties (see, for example, the review by Seil and Webster [200]). It is important to note that the antibacterial properties of nanoparticles may be significantly influenced by their size [218]. However, there is currently a lack of comprehensive data on this effect.

Various organic antibacterial materials used in medical practice are often unstable, especially at high temperatures or pressures. On the other hand, it is well known that a number of metals and their oxides also have antibacterial properties. For example, nanofluids with ZnO particles possess antibacterial activity [71, 263]. These nanoparticles have a wide spectrum of antibacterial action on a number of bacteria and microorganisms. Their efficacy increases with increasing concentration. It has been established that ZnO nanoparticles alter the cell membrane components, including lipids and proteins. They can deform the bacterial cell membrane, resulting in the loss of intracellular components and ultimately to the death of these cells.

Silver and gold nanoparticles are used (or are planned to be used) very intensively in various biomedical applications. Silver nanoparticles are used as antimicrobials and disinfectants. Silver nanoparticles are used as antimicrobial and disinfectant agents and are components of food packaging. The dissolution of the nanoparticles results in the formation of silver cations that exhibit a biocidal effect against a multitude of bacterial strains. Additives of silver nanoparticles impart antibacterial properties to clothing, paints, enamels and other materials. As demonstrated in Skomorokhova et al. [218], silver nanoparticles of a specific size exhibit high antibacterial efficacy. At present, methodologies are being devised to synthesize such nanoparticles that exhibit minimal toxicity to mammals and possess high economic value as antibacterial agents.

Nanofluids with Au-198 gold isotope particles (with a half-life of 65 h) are successfully used for therapeutic purposes in oncology practice, as well as for diagnostic purposes. The principle of oncodiagnostics is based on the binding of specific antibodies conjugated and gold nanoparticles to the surface of cancer cells. This allows for the precise identification of the tumor with a resolution of a few cells. In addition to gold and silver nanoparticles, numerous other nanoparticles have demonstrated antibacterial properties. This includes ZnO, CuO, TiO_2, and fullerenes.

As previously stated, nanofluids can be used for drug delivery purposes [248]. Gold nanoparticles serve as non-toxic carriers for drug and gene delivery. In these systems, the gold core provides stability to the assembly, while the monolayer allows for the regulation of surface properties, including charge and hydrophobicity [216].

Another attractive feature of gold nanoparticles is their interaction with thiols, which offers an efficacious and selective approach to controlled intracellular release [56].

Magnetic nanofluids are successfully used for drug delivery. CNTs have become a new alternative and effective tool for transporting and moving therapeutic molecules. These nanomaterials can be readily functionalized with bioactive peptides, proteins, nucleic acids, and drugs, thereby enabling their use as means for the targeted delivery of biological agents to cells and organs. Since functionalized CNTs exhibit low toxicity and are not immunogenic, these systems show considerable promise in the nanobiotechnology and nanomedicine.

The range of possible applications of nanofluids in medicine is very large. In this regard, one can also mention some rather exotic applications, for example, in cryosurgery [238]. The prospects for medical applications of nanoparticles and nanofluids are almost limitless. It is essential to have just a clear understanding of the specific particles and their properties that are required for a particular purpose, as well as the methodologies for their fabrication. The latter is a completely solvable problem, given the extensive array of methods currently available for their manufacturing. However, along with that, there is a need for nanofluids with controllable properties. It is only necessary to note that the use of nanofluids and nanoparticles, in particular, in medicine requires some caution. It is known, for example, that nanoparticles of sufficiently small sizes can penetrate into a cell, destroying its membrane [3, 63, 86, 95, 126].

8.1.4 Application of Magnetic Nanofluids

In contrast to conventional nanofluids, magnetic nanofluids have been the subject of investigation for approximately fifty years. Magnetic fluids with iron oxide particles have appeared on the market as early as in the seventies of the last century. In different years and by different authors, they were called differently: ferrofluid, ferromagnetic liquid, magnetizing liquid and, finally, magnetic liquid. With the advent of the term "nanofluid," it is appropriate to refer to these fluids as "magnetic nanofluids." Indeed, this is the first known nanofluid that has been extensively investigated and successfully used over an extended period. The creation methods of these nanofluids are essentially identical to those of conventional nanofluids. Moreover, it can be said that the methods for creating conventional nanofluids were actually transferred from ferrohydrodynamics.

Magnetic fluids were studied very intensively until the beginning of this century (for example, see the monographs of Rosensweig [186], Bloom et al. [21], Taketomi and Chikazumi [229]. Subsequently, interest to magnetic fluids waned considerably, but has been growing again in the last two decades or so. This is due to several circumstances: firstly, intensive attempts to apply them to various biomedical technologies. The second factor is of a more fundamental nature. The studies of magnetic fluids, which were intensively developed until approximately the end of the last century, were based on classical theories, and the obtained results were interpreted within the framework of these theories. It is now evident that a number of these theories are not applicable to nanofluids. For example, the viscosity of nanofluids is not described by Einstein's theory, even at low concentrations of nanoparticles. Viscosity, in particular, depends on particle size. Accounting for this factor alone can result in a notable change in the comprehension of magnetic nanofluid behavior. Finally, the third factor results from the advent of numerous modern technologies for the creation of magnetic nanofluids, encompassing not only ferromagnetic particles but also composite ones.

The presence of an external magnetic field provides an effective means of controlling the magnetic fluid. When the volume concentration of nanoparticles is sufficiently high, the resulting ponderomotive forces become significant, particularly in comparison to the gravitational force, which they can exceed by several orders of magnitude. On the one hand, this can significantly change the nature of the fluid flow (for example, convection), and on the other hand, it creates new possibilities for their use.

One of the earliest applications of magnetic nanofluids was the development of reliable sealants between moving components of space suits, which were created and implemented during the American Apollo project (human mission to the Moon). The operational principle is extremely simple [229], but its creation became possible only with the advent of magnetic nanofluid, which is the key element here. The next stage of highly interesting technical developments is the creation of various kinds of separators based on the fact that, in a magnetic field, the effective density of a nanofluid can significantly exceed its physical density.

A special type of heat engines has also been developed using magnetic nanofluids. These engines were based on the fact that the magnetization of fluids decreases with increasing temperature, and at a certain temperature becomes equal to zero. Subsequently, due to the high viscosity of magnetic nanofluids and their dependence on the external magnetic field, a number of devices were developed, including dampers and disk playing equipment. Magnetic fluids are used in printing technology, they, like conventional nanofluids, are used as lubricants and for polishing, in the creation of hydraulic valves, various sensitive sensors (rotation, tilt sensors, accelerometers, etc.)

Magnetic nanofluids have special optical properties, more precisely magneto-optical ones. To illustrate, a magnetic fluid film may become transparent when an external magnetic field is applied, whereas in its absence it is opaque. In magnetic fluids, birefringence is observed. These phenomena have been employed in the development of magnetic field sensors, optical gates, modulators, stabilizers, and amplifiers of optical signals [229]. Optical filters based on the use of magnetic nanofluid have appeared relatively recently. These filters are used to select specific wavelengths of light. The employment of specific nanofluids in conjunction with an external magnetic field enables the tuning to a given wavelength [165]. Finally, it can be noted that the creation of water-based ferromagnetic nanofluids has made it possible to obtain a new type of colored magnetic ink for printers.

8.1.5 Application of Nanofluids in the Oil and Gas Industry

Traditional methods of oil production using thermal, gas and chemical means are effective, but are associated with many undesirable consequences (significant amounts of oil remaining in the rock, heat leakage into the reservoir, high cost of the process, etc.). The oil and gas industry can be conditionally subdivided into three main sectors. The first is the production sector, which includes the exploration and production of crude oil and natural gas from reservoirs. The second sector pertains to the processing of extracted oil or gas. The third sector involves storage, marketing, and transportation of gas, oil, and petroleum products. Nanotechnology, including nanofluids, is now being successfully implemented or at least the possibility of their application is being explored in the first two sectors of this industry.

Drilling occupies a central place in the first sector. The creation of drilling fluids is one of the extremely important elements in the development of production technology in the oil and gas industry. The complexity of these fluids is determined by the numerous functions they serve in the drilling process. Among the most crucial functions are reservoir pressure compensation, transportation of cuttings formed during drilling, and formation of filter cakes at the well walls. The diverse functions of drilling fluids necessitate a rather complex composition, because the fluids are required to possess a multitude of characteristics. The oil and gas industry typically employs drilling fluids that are either water-based or oil-based. The multiplicity of functions performed by drilling fluids necessitates a corresponding diversity of their properties. These include an increased density, the presence of various additives with a high dispersive capacity, a high viscosity, the preservation of basic characteristics at temperature and pressure changes within a wide range, specific rheological properties, etc. The creation of such properties is achieved through the inclusion of a multitude of additives in the base fluid. These additives include various weighting agents (e.g., chalk, barite, hematite, and so forth), flocculants, a variety of reagents, and etc. The leading global service companies utilize hundreds of different drilling fluids. However, the diversity of challenges encountered in the practice of field development and oil and gas production necessitates the creation of new drilling fluids with specific properties.

Filtration losses of drilling fluid while drilling, when it partially or completely goes into the formation, is one of the central problems, the solution of which requires the creation of drilling fluids with well-defined properties. To avoid such losses, various methods are used, including, for example, lowering an additional casing column. An alternative approach is creating low-permeability crust with a high enough ultimate strength on the surface of the well. For an extended period, a variety of microparticle additives with characteristic sizes ranging from 1 to 100 μm were employed for this purpose. This method is not effective in low-permeability formations, where pore sizes do not exceed 0.1 μm. In this regard, over the last decade, nanoparticles have been actively used. In fact, one or another nanofluid is used. One of the first attempts of this kind is reported in the paper [10].

Drilling fluids are extremely complex dispersed fluids that contain nanoparticles (if they are used) as well as much larger particles. Therefore, even if a drilling fluid contains nanoparticles, it cannot be called a nanofluid. All physical properties, especially thermophysical properties, can be completely different. For this reason, a systematic study of such properties is extremely important.

The rheology of drilling fluids is one of their most important properties. It is the fluid rheology that determines the energy required to pump the fluid, the drilling rate, and, to a large extent, its filtration, as well as the efficiency of cuttings removal. In fact, the creation of drilling fluids with nanoparticles is a composition of the initial drilling fluid and the nanofluid, i.e. a fluid composed of the base fluid and nanoparticles. In this case, the presence of nanoparticles significantly changes the properties of the initial drilling fluid, which allows such a fluid to be called a drilling nanosolution. It is now reliably established that the rheology of most drilling nanosolutions is well described by the Herschel–Bulkley model [132, 133]. However, the nature of the rheology depends significantly on the type of the base drilling fluid. Nevertheless, as a rule, the fluid index decreases with increasing nanoparticle concentration while the plastic viscosity, consistency parameter, and yield stress increase.

An increase in nanoparticle concentration results in a notable increase in the effective viscosity coefficient of drilling nano solutions. This increase is nearly proportional to the nanoparticle concentration, though it is influenced by the material and size of the nanoparticles. As the nanoparticle size decreases, the viscosity excess increases. Conversely, with increasing nanoparticle size, the viscosity coefficient increasingly approaches to that of the base fluid. At the same time, it should be emphasized that with a change in the concentration and size of nanoparticles, the rheological parameters of the model (3.30) will change, however this change in the general case may be non-monotonic. It is evident that the rheological characteristics of drilling nanosolutions are largely dependent on the properties of the initial drilling fluid, particularly, the weighting agent used. Thus, for example, use of microbarite as a weighting agent results in a markedly elevated viscosity and shear stress in comparison to the use of microcalcite.

From a practical point of view, the most important characteristic is the behavior of drilling nanosolutions with increasing temperature. The viscosity of base drilling fluids increases significantly with increasing temperature, which is a negative property from a technological perspective. The viscosity of drilling nanosolutions also increases with rising temperature, though this increase is not as pronounced as that observed in the base drilling fluid. This indicates that the rheological properties of drilling nanosolutions remain stable when the temperature changes, which is very important for their practical application.

Summarizing the above, it is necessary to emphasize that the use of drilling nanosolutions is extremely promising from the practical point of view. The use of such solutions leads to a significant reduction of their filtration into the rock. However, it is worth understanding that the reduction of filtration depends on the nanoparticle concentration, however, in general, this dependence is non-monotonic in nature [136]. It has been demonstrated that the nanoparticle size exerts a considerable influence on the filtration loss of drilling fluid. As the size of the nanoparticles decreases, the filtration loss exhibits a consistent decline. However, the dependence on size is ultimately determined by the porosity of the sample. In any case, the drilling nanosolution has increased stability and ensures the stability of the wellbore, which ultimately leads to increased efficiency of oil production [80].

Another important characteristic of drilling nanosolutions is their lubricating capacity. The use of drilling nanosolutions can significantly reduce the friction coefficient [18, 201, 254]. The dependence of the friction coefficient on the nanoparticle concentration also has a non-monotonic character [134]. At low nanoparticle concentrations, the friction coefficient decreases, however, a further increase in concentration can lead to an increase in the friction coefficient. As the size of the nanoparticles decreases, the lubricating capacity of the solutions generally increases. The reduction of friction in filter cakes is associated with the viscosity of nanosolutions. In this case, the same mechanism is observed as with nanolubricants.

One of the primary functions of drilling fluid is the removal of cuttings from the well. This process is largely dependent on the rheological properties, density, rate of penetration, and other characteristics of the drilling fluid. The use of drilling nanosolutions enhances the efficacy of cuttings removal [25, 135, 138, 202]. Since their rheology depends significantly on the size and concentration of nanoparticles, this provides real ways to control the process of cutting removal. It is necessary to understand, for example, that an increase in the viscosity of drilling nanosolutions leads to an increase in pressure losses during its pumping. At the same time, an increase in viscosity and yield stress contributes to better cleaning of the well from cutting and enhances well stability.

Certainly, the most significant application of nanofluids in the oil and gas industry is related to oil production. Primarily, this concerns the enhanced recovery and production of oil in reservoirs with residual oil. Dozens of articles appear in print every year, demonstrating the effectiveness of using nanofluids (see, for example, the reviews [28, 129, 160]). The enhanced oil recovery that is reached when nanofluids are injected is explained by a change in wettability, a decrease in interfacial tension and a change in viscosity. Several types of nano-catalysts with metal and metal oxide particles are used for the enhanced oil recovery in the production of heavy and extra-heavy oils. Finally, one can mention the use of paramagnetic nanoparticles for the assessment and determination of oil saturation of large volumes of oil reservoirs using magnetic resonance imaging.

8.1.6 Desalination and Water Purification

The rather dramatic climate change that has occurred in recent decades has already resulted in the emergence of new arid regions worldwide and a reduction in water levels in rivers and freshwater lakes. The contamination of underground water sources is a further consequence of environmental pollution, which is itself exacerbated by a rapidly growing global population, particularly in areas with hot climates. Today, water is becoming, and in some regions has long been, one of the main resources that determines not only human life, but also economic development and the state of the environment. Despite the promising advancement of environmentally friendly technologies, the issue of water scarcity and the persistent contamination of existing water sources has reached a critical point, necessitating immediate and effective solutions. This gives rise to the issue of fresh water scarcity. One potential solution to this problem is the desalination of sea water. In a number of countries, such as the countries of the Persian Gulf, this problem has been successfully solved through the implementation of various technologies, including reverse osmosis, solar stills, etc. [112, 162]. One of the most evident applications of nanofluids in desalination is their use as heat transfer agents to enhance the rate of heat transfer to the desalination unit. Other potential applications include the use of plane plate solar collectors with nanofluids as working fluids to generate heat for desalination [128]. Kabeel and El-Said [85] have developed an integrated system comprising a solar collector with nanofluid and a flash desalination unit. The generation of freshwater can be increased not only by enhancing the thermal functionality of solar cells with nanofluids, but also by virtue of the high optical characteristics of nanofluids, which facilitate greater direct absorption of solar radiation.

To date, the use of many different nanofluids for desalination purposes has been studied (see review [196]), and nanofluids that maximize the effect have been identified. Once again, it should be noted that the main efforts have focused on desalination technologies that typically utilize solar energy.

Although water constitutes a significant portion of the planet's surface area, the availability of freshwater is limited, with 70% of the total volume present in the form of perpetual ice. A mere 1% of the world's freshwater is suitable for human consumption [9]. A considerable amount of water is consumed by the public and industrial sectors of the economy, while simultaneously generating substantial volumes of wastewater [232]. From this point of view, the development of existing and new water treatment technologies is increasingly becoming in demand.

The nature of emerging water pollutions is diverse, encompassing a range of pollutants, including chemical, organic, and inorganic substances, as well as bacteriological, viral, and other forms of contamination. The heaviest metals are highly toxic [9]. Therefore, the methods of water treatment are diverse and vary according to the type of water pollution. The most important techniques are various types of filtrations, gravity separation, flotation, sedimentation, coagulation, oxidation, solvent extraction, evaporation, distillation, reverse osmosis, ion exchange, electrodialysis, electrolysis, adsorption, settling, centrifugal and membrane separation, etc. In general, water purification requires the coordinated application of multiple techniques. However, even in this case, in a number of situations it is not possible to achieve success.

As estimated by the Mordor Intelligence analytical agency, in 2023 the global market for water purification technologies has amounted to $62 billion. With the development of nanotechnology, it has become clear that they can also be used to purify water from a number of pollutants. Nanotechnology-based solutions to water scarcity include, in addition to desalination, water purification and wastewater reuse. The application of nanotechnology to these problems has been a subject of active investigation over the past two decades. The range of methods employed for water treatment is quite extensive, including physical, chemical, and physicochemical techniques. When using nanoparticles and CNTs, their adsorption properties are primarily applicated. Adsorption technologies are relatively inexpensive and quite universal. Adsorption is a process, whereby pollutants are adsorbed on a solid surface due to the action of intermolecular liquid–solid attractive forces. Consequently, the molecules of the dissolved substance are sediment on the solid surface of the adsorbent. The adsorption efficiency is therefore contingent upon the adsorbing surface area and the absence of selectivity. From this point of view, the use of both nanoparticles and CNTs is undoubtedly a promising and unique approach [55, 84]. It is evident that the degree of purification depends on two key factors: the concentration of pollutants and the concentration of nanoadsorbents. The adsorption capacity of CNTs is influenced by a number of factors, including the molecule size [249], aromaticity [257], configuration [158], and hydrophobicity [60].

CNTs are employed as a substitute for activated carbon. As a consequence of their hydrophobic nature, they are capable of adsorbing a multitude of organic substances that are dissolved in water [22]. This is associated not only with hydrophobicity, but also with the presence of π-π interactions, hydrogen and covalent bonds, electrostatic interactions. Oxidized CNTs possess a high adsorption capacity for metal ions. By now, it has been reliably shown that CNTs adsorb ions of many heavy metals well, including Cu^{2+}, Pb^{2+}, Cd^{2+}, Zn^{2+}, and others [30]. Moreover, CNTs demonstrate superior efficacy compared to activated carbon in the adsorption of diverse organic compounds [179]. The authors of the paper [120] have concluded that SWCNTs have a higher adsorption capacity for the removal of benzene and toluene than MWCNTs. The adsorption properties of CNTs can be markedly enhanced through the preliminary functionalization. The simplest functionalization method involves the pre-oxidation of CNTs by some means or other. In this process, functional groups such as hydroxyl and carboxyl can be formed. Therefore, oxidation can markedly enhance the absorption capacity of CNTs. Numerous studies have shown that CNTs are better adsorbents of heavy metals (e.g., Cu^{2+}, Pb^{2+}, Cd^{2+}, and Zn^{2+}) compared to activated carbon [30, 118, 120]. In some cases CNTs are subjected to doping with specific molecules, molecular groups, or even nanoparticles to activate the absorption of specific pollutants. This process allows for the effective removal of a range of heavy metals (As, Pb, Cr, Cd, Ni, Cu, Fe, and Zn) and nitrates from industrial wastewater [17, 62, 253].

Research on the use of graphene for various environmental applications has emerged just in the last decade. In contrast to CNTs, single-layer graphene particles possess two planes for the adsorption of pollutants [203]. In practice, graphene oxide is often employed. Fairly comprehensive data on the use of graphene particles for water treatment can be found in reviews [92, 111]. In all cases, the efficiency of their use is quite high. In a study [12], magnetic graphene oxide particles were effectively used for the removal of Sr(II) and Cs(I) ions from aqueous solutions. In general, graphene and various graphene-based composites are employed extensively for water purification purposes [177].

Nanoparticles can be also successfully used in the treatment of water and wastewater. Their extensive surface area enables the removal of detrimental metal ions, pathogenic microorganisms, and a range of organic and inorganic pollutants from water [228]. The high mobility of nanoparticles in solutions allows for their extensive use in the treatment of water and wastewater against toxic metals such as Hg(II), Pb(II), Cr(III), Cr(VI), Ni(II), Co(II), Cu(II), Cd(II), Ag(I), As(V), and As(III) [163]. Zero-valent metal nanoparticles (Fe, Zn, Al, Ni and other elements) are extensively used in wastewater treatment. Magnetic nanoparticles can be employed in water treatment to facilitate the removal of heavy metals, chemical effluents, charged particles, bacteria, and other pathogens. The use of nanoparticles in wastewater treatment offers a multitude of benefits, including reduced operating costs, lower energy consumption, sludge reduction, reduced pesticide release, and a reduction in the concentration of various pollutants such as heavy metals, nitrates, and sulfates, tannins, and turbidity. Thus, the use of nanoparticles can improve the quality of treated water and give it natural color.

Along with metal nanoparticles, various oxide nanoparticles are widely used [147, 235, 251]. Silver is a commonly used material due to its low toxicity and antibacterial properties [231]. The most effective antibacterial action is provided by small-sized silver nanoparticles. Additionally, they are used in filtration membranes for water treatment. Titanium dioxide is among the most chemically stable nanoparticles. This makes it attractive for use as an adsorbent in the removal of pollutants from wastewater. In addition, iron, magnesium, zinc, manganese, aluminium, and copper oxide nanoparticles can be successfully used for wastewater treatment [103, 247]. Numerous experimental studies performed so far have revealed the high efficiency of nanoparticles in treating water against biological contaminants [252].

8.1.7 Nanofluids in Cosmetology and Cosmeceutics

The use of nanoparticles in the production of cosmetics is growing every year, with an annual turnover of several hundred billion dollars. The cosmetics industry is a significant player in the nanotechnology market. Indeed, the majority of patents related to nanoparticles are accounted for by this industry, which encompasses a diverse range of products, including toothpastes, sunscreens, shampoos, eye shadows, hair dyes and conditioners, lipsticks, aftershave gels, moisturizers, and deodorants. In terms of the number of patents nanocosmetics has a greater number than other fields, including nanoelectronics, medicine, pharmaceuticals, textiles, and advanced weaponry. There is one more important circumstance. For quite a long time, the new directions of cosmeceutics and nanocosmeceutics has been formed at the junction of cosmetology and pharmacology, which is a bridge between cosmetics, pharmacology and medicine [16, 45, 63, 144, 223].

The latest generation of nanocosmetic products is based on nanotechnology and includes a variety of innovative ingredients, such as nanoemulsions, nanocapsules, liposomes, niosomes, solid lipid nanoparticles, carbon nanotubes, and nanomicelles.

For example, liposomes are spontaneously formed bubbles in mixtures of phospholipids with water. Their wall consists of one or more phospholipid bilayers and is two molecules thick, although the walls may be thicker. Additionally, other substances, such as proteins, can be incorporated into the wall. Liposomes contain water or a solution inside. The diameter of liposomes varies, starting from about 20 nm. In contemporary cosmetology, liposomes are carriers of bioactive compounds and fluids that are beneficial for the skin. Typically, these bioactive substances have a high molecular weight and are not able to penetrate into the deep layers of the skin. Moreover, they are unable to penetrate the cell. Liposomes are capable of readily penetrating the deepest layers of the skin, thereby facilitating the delivery of biologically active substances into cells. For this reason, liposomes are used primarily in the production of various creams. It is worth mentioning that along with conventional ones, lipid nanoparticles are widely used in cosmetics (see, for example, [204]). Usually, used lipid nanoparticles are solid at body temperature.

The employment of nanotechnology in cosmetology and cosmeceuticals involves the application of various nanoparticles. The most commonly used nanoparticles include titanium, silicon dioxide, zinc oxide, aluminum, silver, gold, copper, calcium fluoride, fullerenes, and a few others. In addition to nanofluids, nanoemulsions, which often possess distinctive rheological characteristics, are also used in the cosmetics industry. Nanoparticles are employed in a multitude of cosmetic products, including shampoos and conditioners, lipstick, eye shadows, moisturizing creams, anti-wrinkle creams, sunscreens, after-shave deodorants, and perfumes, to enhance their efficacy. It is noteworthy that all of the aforementioned products are, in a broad sense, nanofluids, including creams, although the latter have a rather complicated rheology. The advantages of nanocosmetics and nanocosmecuticals are related to the fact that the corresponding preparations are better retained on the skin and penetrate it, the corresponding creams and nanofluids are better dispersed and spread, have a significantly longer action time, etc. [148].

The use of nanoparticles enables the activation of the penetration of ingredients that are usually unable to penetrate the skin. It is evident that the huge surface area of nanoparticles, even at relatively low concentrations, is of critical importance. This ensures their active interaction with the medium into which they are introduced. Japan is already producing toothpaste with nanoparticles. In addition to its cleansing properties, this toothpaste can repair imperfections in tooth enamel. Titanium dioxide and zinc oxide nanoparticles are effective filters against ultraviolet radiation. Consequently, they are successfully employed in the production of nano-sponged sunscreens. These nanoparticles function as sunscreen blockers by absorbing light in the ultraviolet range, although the efficacy of this process is dependent on the nanoparticle size. It is of particular significance that these particles are capable of absorbing radiation across a range of wavelengths, with zinc oxide demonstrating efficacy at higher wavelengths. Therefore, a cream comprising a combination of these particles provides comprehensive protection against broadband ultraviolet radiation.

The well known antibacterial and antifungal properties of gold and silver nanoparticles, previously discussed, are employed extensively in the cosmetic and cosmeceuticals industry, particularly, in the production of deodorants, anti-aging creams, and various drugs for the treatment of skin inflammations, etc. [246]. In this regard, the advantages of gold nanoparticles are related not only to their chemical inertness, but also to their known biocompatibility.

Along with conventional nanoparticles, the possibility of using fullerenes is widely studied. Their excellent antioxidant properties make them suitable for incorporation into a variety of dermatological applications and anti-aging creams [119].

While summarizing brief overview, let us list the areas where nanocosmetology and nanocosmecuticals are most used. These products are primarily used as body cares, including the skin, hair, lips, oral cavity, and makeup. The skin serves as the primary natural barrier between human body and the external environment. It is usually impermeable to drugs. The use of nanotechnology ensures the delivery of therapeutic drugs without the application of chemical enhancers, organic solvents, and so forth [82, 258]. Great attention is paid to the development of hair care products, their treatment and restoration. Thematically, nanocosmetology can also include the creation of detergents based on nanofluids. The popularity of nanofluid-based detergents is determined by the fact that environmentally friendly detergents can be created [2, 245]. Putri et al. [176] synthesized a nanofluid detergent using titanium dioxide nanoparticles in palm kernel oil. A similar nanofluid-based detergent was also described in the paper [250].

8.1.8 Application in Chemical Technologies

The discovery of significant heat transfer enhancement in nanofluids can be successfully employed to intensify chemical reactor processes (see, for example, [53]). The use of nanofluids is particularly advantageous in compact multifunctional microreactors. The small size of nanoparticles and their huge reaction surface area make them highly promising catalysts. The potential applications of nanoparticles or nanofluids in catalysis are so diverse that a special term, 'nanocatalysis', has been coined to describe this emerging field. Nanocatalysis can be considered as a kind of bridge between homogeneous and heterogeneous catalysis providing a significant enhancement in the activity per unit weight of the catalyst employed [169]. In addition to the evident expansion in the contact area between the active components, the use of nanoparticles in certain instances gives rise to novel catalysis mechanisms. In particular, the catalytic properties can be affected not only by the size of nanoparticles but also by their shape. This is a consequence of size effects in the electronic and structural properties of nanoparticles [27]. However, systematic data are still insufficient to formulate some generalizing conclusions. It should also be borne in mind that the reactive and catalytic capacity of nanoparticles in general can either increase or decrease. This is probably not surprising. The processes of momentum and heat transport in nanofluids also depend on changes in the size of nanoparticles but are opposite in nature.

Nanocatalysts have unique catalytic and absorption properties due to their unusually active surface areas. In situ catalytic enrichment of heavy oil through the use of metallic nanocatalysts represents a promising cost-effective and environmentally friendly technology for the production of high quality oils [123]. Both noble and transition metal nanoparticles are employed primarily as catalysts in numerous petroleum refining processes, including catalytic reforming, in situ aqua-thermolysis, hydrocracking, and hydrodesulfurization. The use of nanoparticles is also associated with their emerging distinctive properties, due to their meso size. A notable illustration of this phenomenon is the finding that gold nanoparticles sediment on certain metal oxides possess effective catalytic properties, whereas macroscopic gold is known to be entirely inactive [161]. Gold nanoparticles sediment on a carrier were found to catalyze many reactions at much lower temperatures or with much higher selectivity than other metal catalysts.

It should be noted that we are talking not only about the use of nanoparticles as catalysts, but also nanofluids themselves. The high specific surface area of nanoparticles makes this type of catalysis quite similar to homogeneous catalysis. This is due to the fact that the surface of nanoparticles contains a significantly higher proportion of metal atoms than that of a conventional heterogeneous catalyst. It is precisely with this circumstance that the appearance of the term nanocatalysis is associated [65, 143]. The efficiency of nanocatalysis is determined by the special properties of nanoparticles (melting point, heat capacity, electrical conductivity, as well as optical, magnetic, and electronic properties). For this reason, the weight concentration of a nanocatalysts is much, often orders of magnitude, less than that of a conventional catalyst. Typical examples of nanocatalysis applications can be found in Nebykov et al. [151], Popov et al. [170, 171].

The small size of nanoparticles makes them highly chemically active. It is a well-established fact that coal dust is prone to spontaneous combustion. It is less known that dust explosions are also a risk in other industries, including flour milling, plastics, pharmaceuticals, fuel crushing plants, and textile production. The explosion of coal dust is a consequence of the ignition of volatile combustible gases released during the heating of the dust, which then react with the oxygen present in the surrounding gas-air environment. The heating of coal dust particles results from their collisions with air molecules. From this point of view, it is clear that the explosiveness of coal dust at a given concentration will be greater, the smaller the size of its particles. In the absence of sufficient oxygen, an explosion is impossible. As the particle size decreases, the surface area to volume ratio increases. The reduction in particle size acts as a catalyst. Obviously, this property can be exploited also for beneficial purposes. For example, aluminum is easily oxidized in the presence of even relatively small concentrations of oxygen. The oxidation capacity of nanoparticles is significantly higher than that of conventional macroscopic samples due to the high contact area. Thus, during the fuel combustion, the presence of aluminum nanoparticles in contact with water results in the release of a considerable amount of hydrogen. It has been established that the combustion of diesel fuel in the presence of a water-based nanofluid with aluminum particles results in an increase in the total combustion heat and a reduction in the concentration of smoke and nitrogen oxide [89, 90].

Numerous studies have shown that nanoparticles dispersed in fuels (diesel, biofuels) can potentially contribute to combustion processes. The resulting effect depends on the size of the particles, their shape, material, and concentration. For example, if nanoparticles with high energy activity, such as aluminium, are employed, these nanoparticles can also react during combustion. Even nanoparticles not involved in chemical reactions can support combustion if they result in a reduction in surface tension or evaporation of the fuel. If nanofluid fuel is designed correctly, it can increase the likelihood of droplet ignition, as well as fuel evaporation and mixing. Therefore, if particulate emissions can be controlled, fuel containing nanoparticles could improve engine efficiency [38, 241–243].

8.1.9 Ecology and Defense Against Biological Weapons

As previously stated, nanoparticles possess high adsorption properties. This enables their use in a variety of environmental applications. By now, a substantial number of referenced has been accumulated devoted to the use of nanoparticles and nanofluids in various applications. Such information can be found in the following reviews [8, 152, 168]. Here we will note a few of the most characteristic points of these applications. In particular, nanoparticles can be used to purify air. For this purpose, it is promising to use nanoporous composites. The introduction of active agents into waste at the storage or generation site is emerging as a principal method of environmental purification. It could also be photocatalytic purification. Nanoparticles can be used to purify gases released as a result of waste incineration. They can effectively capture and retain dangerous pollutants, preventing them from entering the environment [113, 167]. For example, titanium dioxide nanoparticles, following ultraviolet irradiation, can be used to clean the air against a range of pollutants, including hazardous organic compounds, cells, viruses, and poisonous chemicals.

It is proposed that iron nanoparticles (as known, there's a lot more iron on the planet than any other metal), be used to remediate soil contamination with a range of toxic substances. For this purpose, the nanoparticles are introduced into the soil via specially drilled wells. It is important to emphasize here that nanoparticles are a universal reagent, they are able to neutralize a large number of harmful substances, but usually each toxin uses its own neutralizer. The neutralization process occurs as a result of the oxidation of iron in groundwater (there is always a nanofluid!), which involves the formation of insoluble complexes with a range of toxic substances, including lead, nickel, mercury, uranium, and etc. Iron nanoparticles are easily transported along with groundwater and simultaneously cleanse the surrounding area. In the presence of hazardous pollutants such as trichloroethylene, carbon tetrachloride, dioxins, etc., iron oxidation facilitates the breakdown of complex molecules into simpler, less toxic carbon components.

The only competitors to iron nanoparticles as a cleaning reagent are special bacteria. However, iron nanoparticles are a more cost-effective option and are less sensitive to temperature fluctuations. In fact, in the same way it is possible to disinfect soils contaminated with harmful chemicals, nuclear waste, and so forth. The usual procedure of soil excavating and subsequently decontaminating is both extremely expensive and not very effective.

Investigations have shown that nanodisperse metal oxide powders can be used for the decontamination of warfare agents and certain types of bacteriological weapons. Spraying magnesium oxide nanoparticles, for example, destroys the spores of some pathogens of particularly dangerous infections.

8.1.10 Robotics and the Creation of Novel Measurement Techniques and Equipment

The most important development tendency of all nanotechnologies is the creation of nanorobots, which would not only monitor the systems being serviced, but would also actively participate in their rearrangement and optimization. The widespread use of nanosensors will significantly change the conditions of our existence. Currently, inorganic types of vectors are being actively developed, which involve the delivery of gene or other information within cells. Some of them have been already created. Another research area is the development of vectors for the delivery of medical drugs. The integration of microsensors with medical drug delivery systems will take medicine to the next level, offering precise dosing of essential medications while minimizing their associated toxicity.

CNTs exhibit markedly superior electrical properties compared to conventional electrodes, thereby enabling their use in the fabrication of various devices and measurement systems. In particular, CNTs have been employed in the development of displays, thermo-anemometers, sensors, and actuators. Using nanoparticles, new highly sensitive gas sensors have been created and nanosensors for complete chemical analysis are being developed. The development of nanoscale sensors for the detection of biological and chemical warfare agents is also in the same line.

The continuous modernization of measurement technologies and corresponding tools is necessary for the advancement of oil and gas forecasting and exploration. Reservoir mapping and characterization technology using magnetic nanoparticles is widely studied and partly already implemented [178]. The movement of immiscible fluids is tracked by magnetic nanoparticles exposed to external magnetic and sonic fields due to the adsorption of nanoparticles at the oil–water or air–water interface. This technique is called magnetoacoustic tomography [105]. Magnetic iron oxide nanoparticles serve as nanosensors in nuclear magnetic resonance (NMR) diagnostics, both within the borehole and directly in the formation.

Nanosensors are employed for the monitoring of environmental pollution [91]. They exhibit markedly enhanced sensitivity and selectivity in comparison to conventional sensors. Such sensors have been instrumental in the detection of a range of typical pollutants, including carbon monoxide, nitrogen dioxide, and volatile organic compounds. Sensors based on CNTs have been developed for the detection of nitrogen dioxide. Furthermore, nanosensors are suitable for monitoring heavy metals and organic pollutants in water.

8.2 Problems not Previously Discussed in Detail

It is evident that the review of potential applications of nanofluids presented in the preceding sections is not exhaustive. Studies have been also conducted on the potential use of nanofluids in other areas. However, one direction needs to be mentioned. This concerns the use of nanofluids in the development of novel materials. With the development of nanotechnology, materials science is experiencing a rebirth. Materials unknown to nature appear. This is accomplished by modifying the structure of the initial material through the introduction of nanoparticles. Such materials are designated as nanostructured. It is important to distinguish between two main approaches in this context. The first is nanostructuring of the material surface, which is achieved through applying nanoparticles in one way or another. The second involves changing the structure of the entire material. Given that these materials are predominantly solid, the incorporation of nanoparticles into their volume is achieved through the use of nanofluids. A typical example of this type is the creation of nanostructured concrete. In this process, a water-based nanofluid with nanoparticles (or CNTs) is added to a primary cement mortar and homogenized in one way or another. The creation of concrete then occurs as a result of a phase transition.

A similar process takes place during the synthesis of certain nanostructured metals and their alloys. Here, the situation is even more interesting, since the base material is a liquid metal into which a nanofluid is subsequently introduced. Following homogenization and subsequent cooling accompanied by phase transition, a new material is formed. There are great possibilities for the application of nanofluids (partly already implemented) in the creation of other materials such as varnishes and paints, building materials, drilling fluids, etc. [244]. The key factor in all cases is the concentration of the nanoparticles employed and their size.

All above mentioned applications of nanoparticles and nanofluids are of significant importance and necessitate a dedicated discussion, which is beyond the scope of this book. Nevertheless, there are a number of other significant areas of nanofluid research that have not been described in the above sections. This is mainly due to the fact that there is a lack of clarity in the study of these undescribed properties of nanofluids and further systematic research is required. Such properties include, for example, the optical characteristics of nanofluids and their phase transformations. Moreover, there are hybrid nanofluids, a novel type of nanofluid, which have been the subject of investigation for over a decade. In this section, the relevant issues are briefly announced, and problems requiring further study are formulated.

8.2.1 Fluid–Solid Phase Transition

The unusual properties of nanofluids have been repeatedly mentioned above. The thermophysical properties of nanofluids diverge significantly from those of their base fluids. These differences are determined by the mesoscopic size of the nanoparticles, which results in a transformation of the nanofluid structure. In all cases, however, the properties of nanofluids away from the phase transition region were considered. In the general case, we can talk about nanofluid–solid and nanofluid–gas phase transitions. The first type of phase transition was the subject of experimental studies that commenced approximately fifteen years ago. These studies investigated the phase transition of water-based nanofluids with different nanoparticles [64, 115, 264]. It was found that the phase transition in nanofluids occurs at a higher temperature than in the base fluid. In the paper [79], an attempt was made to study the effect of the concentration of titanium dioxide nanoparticles and the cooling rate on phase transition. The authors considered a water-based nanofluid. It was found that at a given cooling rate, the phase transition temperature, its time, and enthalpy of the nanofluid first increased and then decreased with increasing the nanoparticle weight concentration. In fact, this is the end of the experimental data.

There are also practically no data on molecular dynamic simulation. The only such work was published a quarter of a century ago [188]. The simplest model of a nanofluid was considered, whereby the molecules of the base fluid and the nanoparticle were modeled as hard spheres. In the initial configuration, the molecules were situated at the nodes of a dense hexagonal lattice. The existence of a fluid–solid phase transition in a homogeneous system of solid spheres was previously established by the molecular dynamics method [7]. The phenomenon of a fluid–solid phase transition in a system of solid spheres is well illustrated in Fig. 8.4. The solid line represents a typical isotherm of a single-phase fluid. This isotherm is represented by the dependence of the compressibility factor κ, defined as $\kappa = (p - p_0)/p_0$, on the packing density of molecules in the system

$$\alpha = \frac{V}{V_0} = \frac{V\sqrt{2}}{8r^3 N_m^0}, \tag{8.1}$$

where p is the pressure in the system, $p_0 = nkT$ is the pressure of an ideal gas, $N_m^0 = N_m + N_m^*$ is the number of molecules in the system, r is the radius of a molecule.

In a single-phase fluid, with an increase in the density of the system, the compressibility factor at a certain density value $\alpha_f^* = 1.48$ (see Fig. 8.4) suffers a rupture, and at high densities of the medium passes to a new branch corresponding to the solid state. This phenomenon is interpreted as a first-order phase transition. The physical

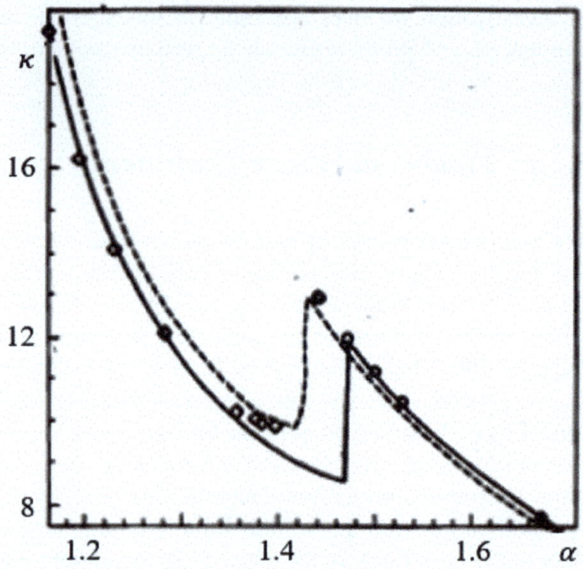

Fig. 8.4 Dependence of the compressibility factor of the single-phase fluid (solid curve) and nanofluid (dashed curve) on packing density

explanation of the phase transition in the solid sphere model is as follows. In the initial state, the molecules are arranged in the nodes of a dense hexagonal lattice. This configuration represents the maximum average distance between neighboring molecules at a given density of the medium. The addition to pressure caused by the interaction of molecules is correspondingly minimal. If the density of the system is large enough, the molecule remains near the lattice node all the time. Such medium may be considered an analogue of a solid body. As the concentration of molecules declines, the ordered structure is disrupted, resulting in a transition to a liquid state and a sudden increase in pressure. If, in the above mention molecular system, one of the molecules is replaced by its heavy isotope (for example, if the mass of the replaced molecule is equal to 100 m (m is the mass of the molecule), but the radii of the heavy and light molecules are identical), the character of the observed phase transition will remain essentially unchanged. The compressibility factor will be described by the solid curve in Fig. 8.4.

The nature of the phase transition, however, will change significantly if a nanoparticle is added to a homogeneous fluid. The nanofluid density is determined by the parameter α_{nf} (compare with Eq. (8.1))

$$\alpha_{nf} = \frac{V\sqrt{2}}{8r^3N_m + 8R^3}, \; N_m + N_m^* = N_m^0. \tag{8.2}$$

In the calculation of the dependence of the nanofluid compressibility factor presented in Fig. 8.4, the nanoparticle radius was set to $R = 4r$, while the mass ratio of particles and molecules was selected to be 100. The phase transition in nanofluid (the dashed curve in Fig. 8.4) occurs at higher densities and pressures. The density, at which the phase transition occurs, increases corresponding to a decrease in the parameter α_{nf} to a value of 1.43. At the same time, the compressibility factor increases from 11.9 to 12.95. Furthermore, in the liquid state, the nanofluid pressure is less than that of a single-phase fluid of the corresponding density. Indeed, in the liquid state, the entire region in proximity to the particle is accessible to molecules. The introduction of a nanoparticle into a homogeneous molecular fluid results in a reduction in the available cell volume by the particle volume, which subsequently leads to a decline in the number of molecules. Accordingly, the effective density parameter of the nanofluid in the liquid state can be expressed as follows

$$\alpha_{nf}^{l} = \frac{(V - V_p)\sqrt{2}}{8r^3 N_m}. \tag{8.3}$$

Conversely, if the considered system is remained in a solid state, screened regions are formed around the particle, into which the penetration of molecules is prohibited. These regions are screened by the surrounding molecules. In the solid state of the considered nano medium, the effective volume of the particle increases. Quantitatively, it is equal to the volume that would be occupied by N_{m}^{*} removed molecules. The corresponding density parameter is equal to

$$\alpha_{nf}^{s} = \frac{V\sqrt{2}}{8r^3(N_m + N_m^*)}. \tag{8.4}$$

Since $\alpha_{nf}^{s} < \alpha_{nf}^{l}$, it can be concluded that the dispersed system in the solid state is, in fact, denser than the homogeneous system. Thus, in the solid state, the cell volume available to molecules is reduced by greater extent than the particle volume. Therefore, the nano dispersed system in the solid state is effectively denser than the homogeneous one. This is precisely what is evidenced by the calculations (the left branch of the isotherm at $\alpha_{nf} < 1.43$ in Fig. 8.4). Note that if the isotherm branches in the liquid and solid states of a heterogeneous liquid are plotted as functions of the α_{nf}^{l} and α_{nf}^{s} parameters, respectively, then the calculation data will be on the isotherm branches of the homogeneous fluid. In Fig. 8.4, the mentioned branches are indicated by round and square symbols, respectively. Thus, it can be said that the α_{nf}^{l} and α_{nf}^{s} parameters are scaling factors that allow, knowing the isotherms of a homogeneous liquid, to construct the isotherms of a heterogeneous nanosystem.

The analysis of the data presented in Fig. 8.4 indicates that the presence of a nanoparticle in the molecular system significantly changes the nature of the phase transition in the system. On the other hand, from physical standpoint, it is clear that with a decrease in volume concentration of nanoparticles their influence on the phase transition in the system should diminish. This conclusion was confirmed by calculations [188]. However, it is important to note that as the radius of nanoparticles increases, their influence becomes increasingly significant. The size of the nanoparticle is a critical factor in the fluid–solid phase transition. This is due to the fact that the transition itself is largely geometric in nature.

The presented data provide important qualitative information about the nanofluid–solid phase transition. At the same time, the considered system is undoubtedly a model system. Today, it is possible to simulate fairly real nanofluids, although appropriate experiments are needed. This is significant not only from the perspective of understanding the physics of processes occurring in nanofluids but also from a practical point of view, since, for example, nanofluids can be used in so-called cold thermal energy storage systems.

8.2.2 *Boiling Crisis*

The nanofluid–gas phase transition, in contrast to the one described above, is being studied very actively. This is due to the fact that boiling represents one of the most effective mechanisms for heat removal from the heated surface. On the heated surface, depending on the magnitude of the supplied heat flow, either bubble boiling (individual steam bubbles are formed) or film boiling (a continuous steam film is formed) can occur. The most intense heat transfer occurs in the first case. The transition from bubble boiling to film boiling leads to a decrease in the heat transfer coefficient. This phenomenon is called the boiling crisis, and the corresponding heat flux is called the critical heat flux (CHF). There are various passive and active methods of increasing the CHF (see [211, 267] and the literature cited therein). Attempts of using nanofluids to enhance CHF fall within the same research line. The experimental investigation of the boiling crisis of nanofluids and the determination of the corresponding CHF was the main focus when studying phase transition in these fluids. The relevant references can be found in recent articles [97, 98, 142, 166].

Among the first works on the study of the boiling crisis of nanofluids, one can mention the papers [35, 260]. The first study examined the bubble boiling of a water-based nanofluid with Al_2O_3 particles at the surface of a cylindrical heating element with a diameter of 0.02 cm. It was noted that the presence of nanoparticles reduces the heat transfer coefficient, and this reduction increases with increasing their concentration. The authors attribute the decrease in the heat transfer coefficient to the change in the heater surface due to deposition of nanoparticles at the bubble nucleation site, which results in a smoother surface. The effect increases with increasing particle concentration. On the other hand, the authors of the second paper draw an entirely opposing conclusion. The boiling of water-based nanofluids with silicon and aluminum oxide particles was investigated revealing an increase in CHF by over 200%. Later, the increase in CHF was observed in numerous subsequent papers, with the increase attributed to the presence of nanoparticle deposits on the heater surface (see, e.g., [88, 102, 114, 121, 230]). It was noted that a porous layer of nanoparticles was formed at the heating surface, improving its wettability. The experimental data obtained with the participation of the author are briefly described below [131, 137, 174]. On the one hand, these data allow clarifying the principal mechanisms through which nanofluids exert an influence on the boiling crisis while on the other hand they allow formulating questions that demand further investigation.

The experimental setup is described in detail in the above mentioned papers. A nichrome wire was employed as a heating element, to which a voltage was applied, and the heat flux density was subsequently measured. The heat flux growth rate and the duration of boiling prior to the boiling crisis onset could be controlled. The boiling heat flux density at the heater was determined by the current in the electric circuit and voltage loss across the wire. The measurement error did not exceed 5%. The heater temperature was also measured with the same accuracy. The volume concentration of nanoparticles varied from 0.05 to 1%. In addition to CHF measurements, the surface morphology of the heaters was studied after boiling by electron microscopy.

The first series of experiments was conducted using water-based nanofluids with 25 nm silicon dioxide particles. The diameter of the heater was 0.2 mm. The resulting boiling curves are illustrated in Fig. 8.5a; the rhombs represent the boiling curve of water, while the other symbols correspond to boiling curves of nanofluids with varying particle concentrations (0.05, 0.1, and 0.05%). Upon reaching the CHF, bubble boiling changes to film boiling. The CHF of nanofluids increases with rising particle concentration. At the maximum nanoparticle concentration the CHF is approximately three times that of water. At the same time, it was not possible to obtain a stable transition from bubble boiling of the nanofluid to film boiling because the heater was burning out due to an exceedingly high heat flux.

Figure 8.5b shows the comparison of the relative CHF (normalized to the CHF of water) for different nanofluids as a function of the volume concentration of particles. All nanofluids were prepared based on water, with the addition of silicon and iron oxide particles (95 nm), as well as nanodiamond (5 nm). In all cases, the CHF of nanofluids exhibited a monotonic increase with rising nanoparticle concentration. The relationship between the CHF and the volume concentration of nanoparticles is described by a power law, which can be expressed as $q/q_w = 1 + a\varphi^b$. The coefficients included here for silicon oxide are $a = 2.169$, $b = 0.190$; for nanodiamond, they are

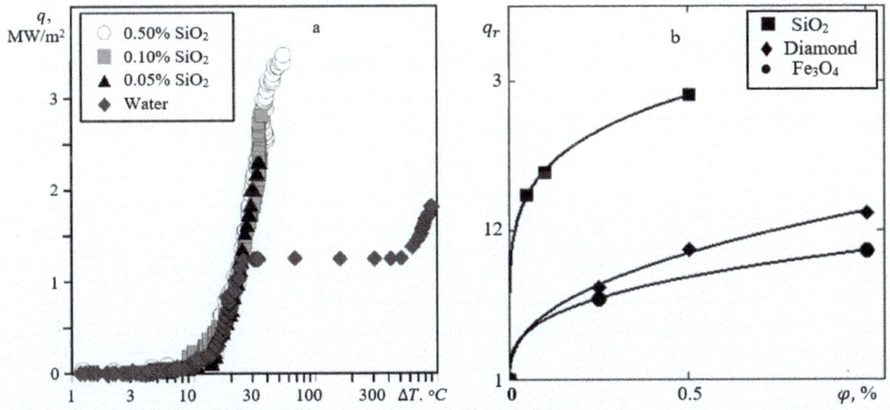

Fig. 8.5 **a** Boiling curves of nanofluids with different concentrations of SiO_2 nanoparticles; **b** dependence of relative CHF at boiling of nanofluids with different particles on their volume concentration

$a = 1.125$, $b = 0.412$; finally, for iron oxide, the coefficients are $a = 0.859$, $b = 0.329$. It can thus be concluded that the degree of growth of CHF differs for all three nanofluids considered. This suggests that the CHF may be dependent on both the size of the nanoparticles and their material.

In order to investigate the dependence of CHF on nanoparticle size, a series of measurements were performed with water-based nanofluids comprising silicon and aluminum oxide particles with sizes varying from 11 to 100 nm. The volume concentration was equal to 0.05% in all cases. As previously, a 0.2 mm diameter wire heater was used. The data obtained are presented in Fig. 8.6. In both cases, the CHF increased by approximately one and a half times when compared to that of water, with increasing nanoparticle size.

As previously stated, the observed increase of CHF in nanofluids can be explained by the influence of nanoparticle deposition on the heater surface. The larger the size of the nanoparticles, the more rapidly the requisite thickness of the capillary-porous coating is formed on the heater surface. This conclusion is confirmed by the electron microscopy data presented in Fig. 8.7. The degree of deposition coverage is higher for larger particle sizes.

In [137] it is shown that with an increase in the particle concentration in the nanofluid, the thickness of sediment layer and the degree of its coverage increases. Thus, the CHF of the nanofluid is directly correlated with the accumulation of nanoparticles on the heater surface. It is also important to note that the sediment layer formed on the heater is porous. The permeability of the sediment layer is dependent upon the size of the nanoparticles. As the size of the nanoparticles decreases, the

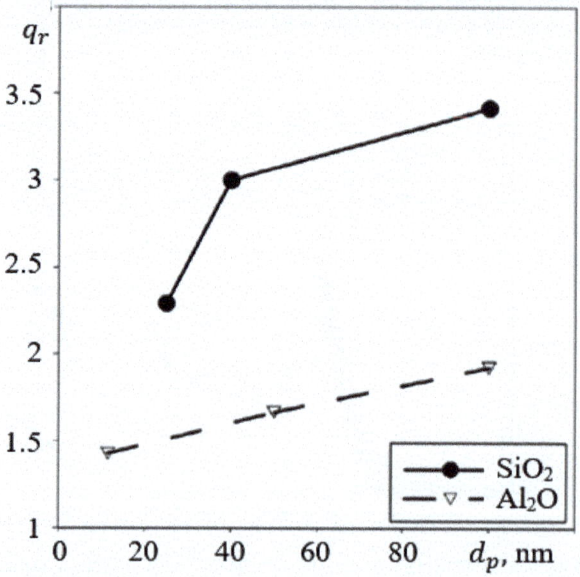

Fig. 8.6 Dependence of the relative CHF of water-based nanofluids on nanoparticle size

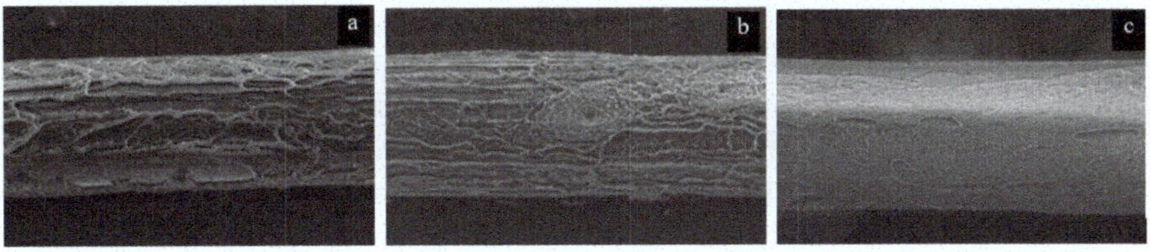

Fig. 8.7 Microphotographs of nichrome heater surface after boiling in nanofluid with SiO_2 nanoparticles of different sizes: **a** 25, **b** 40, **c** 100 nm

permeability of the sediment layer also decreases, which impedes the removal of vapor from the heater surface. Consequently, the CHF is reduced.

Since the formation of a sediment layer of nanoparticles on the surface of the heater depends on the size of the nanoparticles, the effectiveness of its formation will depend on the ratio of the diameter (surface) of the heater to the size of the nanoparticles. The resulting relationships of the relative CHF on the concentration of silicon oxide nanoparticles (25 nm) for various heaters are presented in Fig. 8.8a. With a decrease in the size of the heater (d_h), the relative CHF (all other things being equal) increases significantly (see Fig. 8.8b). Thus, for a given heater size, a nanofluid with a specific particle size can be selected to achieve an optimal CHF.

The formation of optimal nanoparticle sediment to achieve maximum CHF should depend on the duration of the boiling process. Indeed, such dependence was established in Minakov et al. [137]. At the initial stages of boiling, the CHF increases almost linearly with duration of boiling process. However, this growth subsequently decelerates.

Another significant factor is the dependence of the CHF on the nanoparticle material. This influence has been noted by many researchers. As illustrated in Fig. 8.6, the

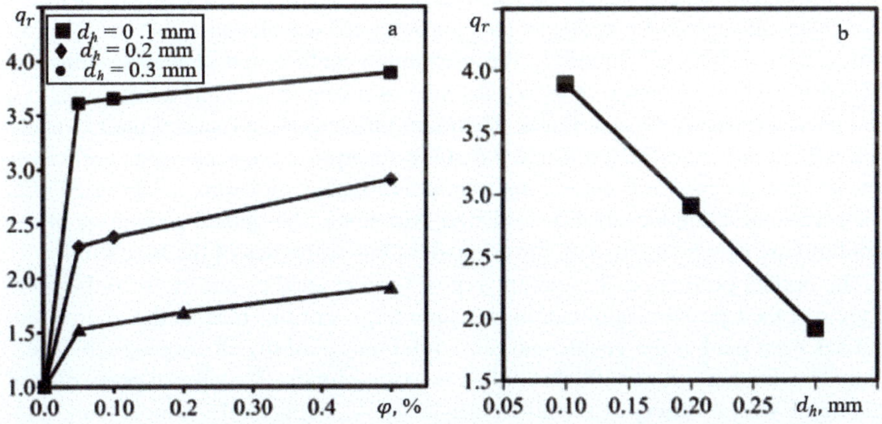

Fig. 8.8 Dependence of relative CHF on nanoparticle concentration for different heater sizes (**a**) and on heater size at particle concentration of 0.5% (**b**)

CHF at boiling of the nanofluid with aluminum oxide particles is approximately twice as low as with silicon oxide particles, under the same concentration and particle size. As demonstrated in Minakov et al. [137], the microstructure of the heater surface exhibits notable differences following boiling of these two different nanofluids. The formed sediment layer and its structure depend on the adhesion of nanoparticles to each other and to the heater material, as well as on the wettability of the particle material. Systematic research in this direction, however, still needs to be continued. The influence of the thermophysical properties of the nanofluid on the CHF is still unclear.

8.2.3 Optical Properties of Nanofluids

As previously stated, the optical properties of nanofluids are extremely important from the perspective of their use in solar energy applications. A comprehensive investigation of these properties and their systematic description requires a large number of experimental measurements. In the process of exposure to electromagnetic radiation on the nanofluid, it is weakened. This attenuation is due to the absorption and scattering of radiation by the nanofluid. In this case both the nanoparticles and the base fluid are absorbing electromagnetic radiation. On the other hand, scattering of radiation on nanoparticles and molecules of the base fluid leads to redistribution of radiation intensity in the bulk of nanofluid. This redistribution is dependent on both the range of the radiation and the properties of the nanofluid. Scattering by sufficiently large nanoparticles may be even greater than absorption.

The experimental data can be found in reviews [4, 58, 175].

The light scattering and absorption properties of nanoparticles are dependent on their shape, size, concentration, and refractive index (for further details, see [67, 130]). In addition, the optical properties can be affected by the clustering of nanoparticles. With an increase in the size of nanoparticles, the wavelength range of the dominant scattering shifts [5]. Naturally, the attenuation coefficient is changed in this case [43]. However, as observed in [197], the size of nanoparticles has a little effect on the optical properties of nanofluids. The main influence is associated their concentration. In contrast, in [220] it is noted that the absorption and scattering coefficients of water-based nanofluids containing titanium dioxide nanoparticles increased with rising nanoparticle concentration and size. Moreover, the optical properties of the nanofluid are dependent also on the corresponding properties of the base fluid.

The optical properties of nanofluids have been studied primarily from the point of view of their possible application in solar energy. For this reason, the main attention has been paid to the visible and ultraviolet ranges of the electromagnetic spectrum. The diversity of nanofluids studied is extraordinary. The optical properties of nanofluids with a variety of metallic particles (gold, copper, silver, nickel, and titanium), different oxides, and CNTs have been investigated. Additionally, various base fluids have been employed. For this reason, it is challenging to draw any general conclusions. Perhaps the easiest conclusion to draw is about the influence of the concentration of nanoparticles on the optical properties, since the amount of absorbed light depends on concentration. In [194] it is stated that the absorption coefficient of water-based nanofluid with titanium dioxide particles increases with their concentration in the wavelength range of 200–400 nm, whereas with aluminium oxide particles it is practically independent of their concentration. An increase in the extinction coefficient was observed with increasing nanoparticle concentration also in water-based nanofluids comprising copper oxide particles [239]. It is important to note that the range of electromagnetic radiation scattered or absorbed by nanofluids is largely dependent on the nanoparticle size [175].

In a theoretical study of the optical properties of water-based nanofluids with gold and silver particles, Rativa D, Gymez-Malagyn [183] investigated the influence of nanoparticle shape on optical properties. The nanoparticles were shaped as ellipsoids, with aspect ratios varying from 1 to 4. It was demonstrated that the absorption coefficients of ellipsoidal gold and silver nanoparticles were markedly higher (by 54 and 86%, respectively) than those of spherical nanoparticles. In addition, the dependence on the material of the nanoparticles is clearly visible here. In this regard, it should be mentioned that a considerable number of experiments have been conducted with nanofluids with CNTs. The results have shown a proportional increase in absorption with an increase in concentration. For example, for water-based nanofluids with CNTs, the absorption linearly depends on concentration [87]. However, at a sufficiently low concentration, the nanofluids turned out to be opaque to ultraviolet radiation. Here, only a few circumstances should be taken into account. CNTs are hydrophobic, therefore to create corresponding nanofluid it is necessary to use surfactants or perform CNT functionalization. On the other hand, nanofluids with CNTs typically have non-Newtonian rheology. Consequently, appropriate adjustments must be made when studying their optical properties. Additionally, MWCNTs and SWCNTs display markedly disparate optical properties. Finally, starting from certain concentrations (quite small) nanofluids with CNTs become opaque.

In conventional nanofluids, the transmittance coefficient is significantly lower than that of the base fluid, and it decreases as the size of the nanoparticles and their concentration increase. The degree of light scattering increases with both the volume concentration and the size of the metallic nanoparticles. However, rather small metal nanoparticles weakly scatter light [4].

It can also be noted that nanofluids can be used as optical filters to select different light wavelengths. The optical filter based on ferrofluid exhibits tunable properties. In particular, an optical filter based on magnetic nanofluid was developed in Philip et al. [165] for the selective detection of wavelengths in the ultraviolet, visible, and infrared ranges. The selection of the appropriate band is accomplished through the application of an external magnetic field. Mishra et al. [139] have developed water + PVP-based nanofluids with gold nanoparticles that are capable of detecting specific color bands of light.

In conclusion, it is important to note that the existing experimental data on the optical properties of nanofluids is fragmented and largely haphazard. It is imperative to conduct systematic measurements of the dependence of these properties on the size, material, and concentration of nanoparticles or CNTs. In order to draw conclusions about these properties with predictive ability, it is necessary to perform experiments with nanofluids prepared using the basic fluids of the same class under controlled external conditions.

8.2.4 Thermophysical Properties of Nanofluids in Nano- and Microchannels

The use of nanofluids as lubricants was discussed in Sect. 8.1.2. By now, quite a lot of experimental studies have been carried out. However, it remains challenging to establish definitive criteria for selecting a specific nanofluid for use as a lubricant. This is due to several factors. The film thickness of a lubricant typically ranges from micrometer to approximately ten nanometers. The characteristic film size naturally imposes a limit on the size of nanoparticles or CNTs. For instance, MWCNTs with sufficiently large diameter cannot be employed.

The second circumstance is of a fundamental nature. The point is that the thickness of the lubricant film is insufficient for direct measurement of its thermophysical characteristics. At the same time, the findings reported in Rudyak and Belkin [189, 190] show that the viscosity and thermal conductivity of fluids in confined conditions can exhibit notable discrepancies from their respective bulk values. On the other hand, the magnitude of these characteristics is of fundamental importance for the selection of an appropriate lubricant. The situation is further complicated in the case of nanolubricants. The typical size of the nanoparticles employed can be comparable to the thickness of the nanolubricant film. It is no longer possible to simulate the dynamics of such a film using continuous medium methods. Therefore, a real progress in understanding the tribological processes occurring in such nanolubricants can only be achieved through the use of the molecular dynamics method.

As previously stated in Sect. 8.1.2, such method has already been employed for this purpose. The first research cycle was dedicated to the investigation of the rheological properties of lubricants comprising large organic molecules. In this study, data were obtained on possible alterations in the rheological properties of these lubricants. The critical shear rate at which the transition from Newtonian to pseudoplastic behavior occurs is inversely proportional to the rotational relaxation time of the molecules [15]. The dependence of viscosity on pressure has been studied in Ramasamy et al. [180]. The paper [125] presents interesting data regarding the viscosity of 9,10-dimethyloctadecane, a primary component of PAO-2 oil-based lubricants. The viscosity was simulated over a broad range of pressures (up to 1 GPa) and temperatures (40–150 °C). The rheology of this lubricant was investigated as well. When studying the rheology of lubricants under confined conditions, it has been noted that the observed behavior is rather adequately described by the Carreau or Eyring versions of the power fluid models [32, 117, 153]. However, in [73] it is indicated that the selection of a specific model is dependent on the conditions, particularly, the system pressure and temperature. Viscosity rises sharply with increasing pressure, while an increase in shear rate leads to pseudoplastic behavior. At high temperature and low pressure, the Carreau version of the power fluid model demonstrates satisfactory results. However, as pressure increases or temperature decreases, it is recommended to change to the Eyring model.

The molecular dynamics method has provided valuable information on the behavior of diverse lubricant additives [141, 181, 182, 221]. This method is also used to investigate the properties of nanolubricants [222]. Typically, fairly large nanoparticles are used in experiments. The interaction of these nanoparticles with complex lubricants remains poorly understood. For this reason, studying nanoparticle additives was conducted typically by solving certain model problems. In particular, the dynamics of nanoparticles between solid surfaces have been investigated in the absence of lubricant but in the presence of shear rate [26, 47, 68, 110, 141]. Thus, the mechanical interaction between the nanoparticle and the substrate is frequently the subject of investigation. An illustrative example of such work is presented in Doan et al. [41], wherein the tribological characteristics of the interaction between a 6 nm nanodiamond and a CuAlNi alloy substrate were investigated. The abrasive nanoparticle slid along the substrate with sliding velocity ranging from 25 to 200 m/s, while the depth of penetration of the particle into the substrate was 2 nm.

One of the earliest attempts to model a nanolubricant was conducted in Lv et al. [122]. Argon-based nanofluid with copper particles (4 nm) was investigated when positioned between two copper plates. The upper plate was moving with translational velocity of 100 m/s. Furthermore, the upper plate was lowered to achieve a minimum gap of 7.5 nm. Unfortunately, the concentration of nanoparticles is unknown. However, based on the available screenshots, it can be reasonably inferred that it was quite high. The analysis of the screenshots yielded primarily qualitative information. It has been demonstrated that the nanoparticles underwent both translational and rotational motion. However, under the pressure of the top plate, the rotational movement of the nanoparticles was slowed, resulting in the formation of agglomerates. The agglomerating particles formed a solid layer that significantly reduced friction.

The role of nanoparticles in the sliding friction process was investigated in Ji et al. [78], where a water-based nanofluid with 2 nm silicon oxide particles in a plane channel with copper walls was considered. Notches, 3 nm in diameter were created in the upper wall to forma a rough surface. A coarse-grained model was used to simulate the behavior of water molecules, while the embedded atom method was employed to describe the interaction of copper atoms. The Tersoff potential was used to simulate the interaction of silicon oxide atoms. It was shown that rolling friction also occurred along with sliding friction.

In the work of Zheng et al. [265], an attempt was made to simulate the dynamics of a nanofluid based on C4-alkane with very small iron nanoparticles of 0.8 nm in size in a plane channel with iron walls that moved at a relative velocity of 40 m/s. In addition, the upper surface was subjected to normal pressure ranged from 0.25 to 1.0 GPa. The impact of the lubricant on surface contact and friction force was investigated. It was found that the nanoparticles function as ball bearings between the contacting surfaces, resulting in a transformation of the sliding friction mode to rolling friction mode. It should be noted, however, that the base medium was not a fluid or lubricant in the conventional sense, since the number of C4-alkane molecules present in the gap varied and their density was relatively low. This was a pseudogas of such molecules. In a subsequent study [266], the efficacy of a hexadecane-based lubricant with the same nanoparticles was studied. The number of molecules of the base lubricant was varied, and the effectiveness of the corresponding nanolubricant was investigated. It was found that when the number of base lubricant molecules was too small, it did not provide a significant contribution to lubrication, but simply filled the depressions on the rough surface. Certainly, strictly speaking, it should be borne in mind that when preparing a nanolubricant, the base fluid always has a specific density under given external conditions. This means that the number of molecules in the gap between the contacting planes cannot change significantly.

This brief review shows that considerable progress has been made in understanding the various mechanisms by which nanofluids influence the lubrication process. Nevertheless, numerous fundamental issues still remain unresolved. The most important of them are listed below.

- How does the viscosity of the lubricant vary with channel height?
- How does the viscosity of nanolubricant vary with channel height, nanoparticle size, and concentration?
- How does the rheology of nanofluids vary with channel height and shear rate?
- How do viscosity and rheology depend on temperature and pressure?
- How does thermal conductivity of lubricant and nanolubricant depend on channel height, nanoparticle size, and concentration?
- How does the friction coefficient depend on channel height and shear rate?

The questions formulated are in fact key questions for nanotribology. Moreover, nanofluids can be employed in microchannels and nanochannels for cooling specific devices. Here, too, the ability to determine and predict the thermophysical properties of nanofluids is needed. However, another important problem arises. Since implementation of a liquid–vapor phase transition is one of the most effective methods of enhancing heat transfer, it is essential to gain a deeper understanding of how this phase transition occurs in confined conditions, particularly, in channels of varying heights and in the presence of shear rates. This problem is also important for tribology, since nanofluids generally have a higher boiling point. Due to this, the lubricant reduces the likelihood of steam plugs formation.

8.2.5 *Hybrid Nanofluids*

This book is devoted to the study of the thermophysical properties of nanofluids, encompassing both conventional nanofluids with spherical nanoparticles and non-standard nanofluids with CNTs. The distinctive characteristics of the latter are primarily attributed to their unconventional morphology. Strictly speaking, CNT cannot be called a mesoobject. One of their sizes (length) is usually several micrometers, and in some cases tens of micrometers. Such radical differences in morphology also determine the significantly different thermophysical properties of these two types of nanofluids. However, so far, the so-called hybrid nanofluids have not been mentioned in this book. Two different types of hybrid nanofluids can be distinguished. The first is nanofluids, which use two (and sometimes three) types of conventional nanoparticles. The second is prepared using carbon nanotubes (CNTs) and conventional nanoparticles.

The properties of hybrid nanofluids have been actively studied for about a decade. Interest in hybrid nanofluids is huge. Just in the recent five to seven years, approximately two dozen specialized reviews have been dedicated to hybrid nanofluids. A select few of these reviews are listed below [13, 14, 49, 70, 76, 83, 106, 198, 225, 227, 259]. This is connected, on the one hand, with the amazing properties of these nanofluids, and on the other, with their practically infinite diversity (the number of combinations of base fluids, types of nanoparticles and CNTs in threes is very large). Such a great diversity has resulted in a huge number of experimental observations available in the literature indicating unusual properties of hybrid nanofluids. Unfortunately, the lack of systematicity in these data does not yet allow us to answer the question of what causes these unusual properties and whether they can be predicted. For this reason, these nanofluids have not been considered in this book.

The thermophysical properties of hybrid nanofluids will be analyzed in the future. To achieve this, it is necessary to carry out systematic experiments to measure their thermophysical properties for well-defined classes of nanofluids. It is necessary to systematically study not only the influence of the concentration of certain nanoparticles or CNTs, but also their size and material. Naturally, the base fluids should also be varied. The latter is not at all easy, since in some cases the base fluid contains also a surfactant. The mechanisms of transport processes in this case will be exceedingly complex. Accordingly, the experiments should be systematically supported by molecular dynamic simulations.

8.3 Concluding Remarks

The primary objective of this study was to study not only the thermophysical properties of nanofluids, but also the underlying mechanisms. Only in this case it is possible to talk about the possibility of creating nanofluids with predetermined properties, which is necessary from a practical standpoint. In this context, the issue of developing stable and persistent nanofluids immediately arises. This is an engineering and technological challenge. It is quite possible to solve it, and in some cases it is already solved, which paves the way for the extensive practical deployment of nanofluids. On the other hand, nanofluids represent completely unusual dispersed fluids. Understanding the transport processes occurring in nanofluids advances the understanding of fluid physics in general. At the same time, a significant challenge is that currently there is no satisfactory and consistent theory of transport processes in nanofluids. Similarly, as with rarefied gases, fluids are disordered systems. However, in the rarefied gases only binary collisions of molecules occur. In a rarefied gas, all transport processes are caused by binary interactions of molecules. All gas properties (mass, momentum, energy) are transferred by molecules. Therefore, the diffusion, viscosity, and thermal conductivity coefficients are related by rather simple relations.

In a dense gas and in liquid, the interaction of molecules is always collective. Consequently, the transport processes that occur in these systems are governed by complex and disparate mechanisms. It is worth understanding the distinction between these processes in dense gases and liquids. A defining attribute of a liquid is the existence of short-range order. It is this factor that is responsible for both the self-diffusion of fluid molecules and the viscosity of the fluid.

The occurrence of thermal conductivity is also contingent upon the presence of short-range order. However, energy transport is also accompanied by the formation of certain macroscopic perturbation propagating at the speed of sound. The presence of large macroscopic particles in the liquid does not change the short-range order of the basic fluid. In a coarse dispersed fluid, the particles generate hydrodynamic perturbations (of velocity, pressure, and temperature) in the base fluid, thereby changing the stress tensor and the heat flux vector. Accounting for these perturbations results in the occurrence of the effective transport coefficients, which were brilliantly derived by J. C. Maxwell and A. Einstein.

In nanofluids there are mesoscale particles, in the metric of the current fluid these are material points. They cannot generate hydrodynamic disturbances; however, such particles change the structure of the dispersed fluid. In particular, the short-range order of the molecules of the base fluid is changed. In addition, the pseudo-gas of nanoparticles itself can be structured. At sufficiently high concentrations, nanoparticles can form their own short-range order. The transformation of the structure of the nanofluid in comparison with that of the base fluid and the corresponding coarse-dispersed fluid determines the specific mechanisms of transport processes. The transport of momentum (viscosity) is directly determined by the local destruction and formation of a new short-range order. The excess thermal conductivity of nanofluids can be attributed to the interaction between nanoparticles and microfluctuations of hydrodynamic fields that are generated around them. These micro-fluctuations propagate at the speed of sound and their scale is almost independent of the nanoparticle size. Furthermore, in nanofluids, the contribution due to energy transport by both the molecules of the base fluid and the nanoparticles is of considerable significance. In the case of nanofluids with CNTs, the heat conduction mechanisms become more intricate. In this case, it is necessary to take into account the formation of a percolation structure of CNTs as well as the direct heat transport by nanotubes.

There is another very important motive, which is usually not paid attention to when creating an appropriate orchestration in description of the transport processes. When solving ordinary problems of external or internal flow, there is always an open system, the presence of which is simply fixed by certain boundary conditions. However, an open system can significantly alter the nature of transport processes within a subsystem (for example, in a particular simulated flow). A good example of such an influence is provided by the laminar-turbulent transition. If the flow begins in a laminar regime, its transition to turbulent regime occurs as a result of the development of external disturbances coming from the environment. In the spectrum of these disturbances, there will always (or almost always!) be an unstable perturbation whose amplitude will rapidly increase. As a result, the flow will eventually become turbulent. It is important to underline that in majority of instances turbulence-inducing disturbances originate from external sources. Microscopic fluctuations occurring in any flow are typically damped. Apparently, the only example where internal microscopic disturbances can lead to the development of hydrodynamic instability is the flow around the end of a sharp needle. In this case, the scale of the detached shear layer is of the same order as the scale of the corresponding microscopic fluctuations. The resulting resonance will lead to the development of hydrodynamic instability. Such a flow is generally very unstable, and in this case, the Kelvin–Helmholtz instability develops.

Another important example of the influence of the external environment on the nature of the mechanisms of transport processes was given in Chap. 3 in the molecular dynamics modeling of the rheology of a simple liquid and a nanofluid. With increasing shear rate, the nanofluid structure changes, in particularly, the short-range order in the fluid. Here the degree of influence of the environment on the subsystem (nanofluid) is clearly visible. At sufficiently high shear rates, momentum transport in the subsystem becomes independent of the sizes of the internal structural elements of the nanofluid. As consequence the momentum transport processes are practically independent of the thermal microfluctuation of these elements.

The experimental study of transport processes in nanofluids is a technically challenging and labor-intensive endeavor. The intricacy of the issue is even compounded when examining the transport characteristics of hybrid nanofluids. It is evident that a comprehensive experimental investigation of all the issues that have sparked interest is unfeasible. It is therefore essential to develop the ability to predict the properties of specific nanofluids with a reasonable degree of accuracy. In order to achieve this, it is necessary to employ the molecular dynamics method. The application of this method to the simulation of real systems remains as yet a significant challenge. Nevertheless, the outcomes obtained through the application of this method are, at the very least, qualitatively accurate in their prediction of the dependence of transport coefficients on a range of parameters. In this regard, molecular dynamics data possess predictive capacity.

In practice, nanofluid flows are encountered in a variety of applications, where their experimental study is often rather challenging. In such cases, the use of simulation is invaluable. A single-fluid hydrodynamic description can be used to simulate nanofluid flows in almost all cases. The exception pertains to flows in rather narrow micro- and nanochannels. In such cases, molecular dynamics-based simulation can prove very useful. In the context of hydrodynamic simulation of nanofluids, it is essential to bear two key considerations in mind. The effective transport coefficients are not adequately described by classical formulae. Thus, only real experimental data should be used. One last thing: the standard use of similarity parameters, where dimensionless transport equations are solved simply by varying these parameters, is completely unacceptable. Such an approach precludes the possibility of obtaining data that can be physically interpreted and deemed reasonable. It is best to solve the transport equations in dimensional form, and then determine the dependencies on the desired similarity parameters. However, it should always be remembered that all similarity parameters are complex functions of the concentration of nanoparticles, their size and material. Therefore, adequate dependencies on the similarity parameters can be obtained only by fixing all the specified values sequentially.

I would like to express my gratitude to those who have demonstrated the fortitude to read this book to the end. Regardless of anything, you have touched a completely new and amazing page of classical physics.

References

1. Abbas F, Ali HM, Shah TR et al (2020) Nanofluid: potential evaluation in automotive radiator. J Mol Liquids 297:112014
2. Adiwibowo MT, Ibadurrohman M (2019) Slamet, synthesis and performance test of nanofluidic detergents from palm oil-based primary alkyl sulfates surfactant and zinc oxide. AIP Conf Proc 2019:020049
3. Ahamed M, Akhtar MJ, Alhadlaq HA (2015) Comparative cytotoxic response of nickel ferrite nanoparticles in human liver HepG2 and breast MFC-7 cancercells. Chemosphere 135:278–288
4. Ahmad SHA, Saidur R, Mahbubul IM et al (2017) Optical properties of various nanofluids used in solar collector: a review. Renew Sustain Energy Rev 73:1014–1030
5. Al-Gebory L, Mengüç MP (2018) The effect of pH on particle agglomeration and optical properties of nanoparticle suspensions. J Quant Spectrosc Radiation Transfer 219:46–60
6. Al-Waeli AHA, Chaichan MT, Sopian K et al (2019) Modeling and experimental validation of a PVT system using nanofluid coolant and nano-PCM. Sol Energy 177:178–191
7. Alder BJ, Wainwright TE (1957) Phase transition for hard spheres system. J Chem Phys 27:1208–1209
8. Ali AS (2020) Application of nanomaterials in environmental improvement. In: Sen M (ed) Nanotechnology and the environment. IntechOpen
9. Ali I (2012) New generation adsorbents for water treatment. Chem Rev 112(10):5073–5091
10. Amanullah M, Al-Afraj MK, Al-Abdullatif ZA (2011) Preliminary test results of nano-based drilling fluids for oil and gas field application. In: Proc SPE/IADC drilling conference and exhibition, Amsterdam, p 139534
11. Asadi A, Asadi M, Rezaniakolaei A et al (2018) An experimental and theoretical investigation on heat transfer capability of $Mg(OH)_2$/MWCNT-engine oil hybrid nano-lubricant adopted as a coolant and lubricant fluid. Appl Thermal Eng 129:577–586
12. Aytas S, Yusan S, Sert S et al (2021) Preparation and characterization of magnetic graphene oxide nanocomposite (GO-Fe_3O_4) for removal of strontium and cesium from aqueous solutions. Character Appl Nanomat 4:63–76
13. Azim MKM, Arifutzzaman A, Saidur R et al (2022) Recent progress in emerging hybrid nanomaterials towards the energy storage and heat transfer applications: a review. J. Mol Liquids 360:119443
14. Babar H, Ali HM (2019) Towards hybrid nanofluids: preparation, thermophysical properties, applications, and challenges. J Mol Liquids 281:598–633
15. Bair S, McCabe C, Cummings PT (2002) Comparison of nonequilibrium molecular dynamics with experimental measurements in the nonlinear shear-thinning regime. Phys Rev Lett 88(5):058302
16. Bangale MS, Mitkare SS, Gattani SG et al (2012) Recent nanotechnological aspects in cosmetics and dermatological preparations. Int J Pharmacy Pharmaceutical Sci 4(2):88–97

17. Bankole MT, Abdulkareem AS, Mohammed IA et al (2019) Selected heavy metals removal from electroplating wastewater by purified and polyhydroxybutyrate functionalized carbon nanotubes adsorbents. Sci Rep 9:4475
18. Battez AH, González R, Viesca JL et al (2008) CuO, ZrO2 and ZnO nanoparticles as antiwear additive in oil lubricants. Wear 265(3–4):422–428
19. Bellos E, Tzivanidis C (2019) Alternative designs of parabolic trough solar collectors. Prog Energy Combust Sci 71:81–117
20. Bhanvase B, Barai D (2021) Introduction to nanofluids. Nanofluids Heat Mass Transfer. Elsevier, Amsterdam
21. Bloom EYa, Mayorov MM, Tsebers AO (1989) Magnetic fluids. Znanie, Riga
22. Bo P, Daohui L, Hamid M et al (2008) Adsorption and hysteresis of bisphenol A and 17α-ethinyl estradiol on carbon nanomaterials. Environ Sci Techn 42(15):5480–5485
23. Bondareva NS, Buonomo B, Manca O et al (2019) Heat transfer performance of the finned nano-enhanced phase change material system under the inclination influence. Int J Heat Mass Transfer 135:1063–1072
24. Boungiorno J, Hu L-W, Kim SJ et al (2008) Nanofluids for enhanced economics and safety of nuclear reactors: an evaluation of the potential features issues, and research gaps. Nucl Technol 162:80–91
25. Boyou NV, Ismail I, Sulaiman WRW et al (2019) Experimental investigation of hole cleaning in directional drilling by using nano-enhanced water-based drilling fluids. J Petroleum Sci Eng 176:220–231
26. Bucholz EW, Sinnott SB (2013) Computational investigation of the mechanical and tribological responses of amorphous carbon nanoparticles. J Appl Phys 113(7):073509
27. Bukhtiyarov VI, Slinko MG (2001) Metallic nanosystems in catalysis. Adv Chem 70(2):167–181
28. Cheraghian G, Rostami S, Afrand M (2020) Nanotechnology in enhanced oil recovery. Processes 8:1073
29. Choudhary S, Sachdeva A, Kumar P (2020) Investigation of the stability of MgO nanofluid and its effect on the thermal performance of flat plate solar collector. Renew Energy 147:1801–1814
30. Chungsying L, Huantsung C, Chunti L (2006) Removal of zinc(II) from aqueous solution by purified carbon nanotubes: kinetics and equilibrium studies. Industrial Eng Chem Res 45(8):2850–2855
31. Contreras EMC, Oliveira GA, Filho EPB (2019) Experimental analysis thermohydraulic performance of graphene and silver nanofluids in automotive cooling systems. Int J Heat Mass Tran 132:375–387
32. Cui ST, Cummings PT, Cochran HD et al (1998) Nonequilibrium molecular dynamics simulation of the rheology of linear and branched alkanes. Int J Thermophys 19(2):449–459
33. Dai W, Kheireddin B, Gao H et al (2016) Roles of nanoparticles in oil lubrication. Tribol Int 102:88–98
34. Daneshipour M, Rafee R (2017) Nanofluids as the circuit fluids of the geothermal borehole heat exchangers. Int Com Heat Mass Transfer 81:34–41
35. Das SK, Putra N, Roetzel W (2003) Pool boiling of nano-fluids on horizontal narrow tubes. Int J Multiphase Flow 29:1237–1247
36. Dedkov GV (2000) Nanotribology: experimental facts and mathematical models. Physics Uspekhi 43(6):541–572
37. Dehaj MS, Mohiabadi MZ (2019) Experimental investigation of heat pipe solar collector using MgO nanofluids. Sol Energy Mater Sol Cells 191:91–99
38. Dhahad HA, Chaichan MT (2020) The impact of adding nano-Al_2O_3 and nano-ZnO to Iraqi diesel fuel in terms of compression ignition engines' performance and emitted pollutants. Therm Sci Eng Prog 18:100535
39. Dhale LP, Wadhave PB, Kanade DV et al (2015) Effect of nanofluid on cooling system of engine. Int J Eng Appl Sci 2(10):8–10

40. Ding M, Liu C, Rao Z (2019) Experimental investigation on heat transfer characteristic of TiO_2-H_2O nanofluid in microchannel for thermal energy storage. Appl Therm Eng 160:114024
41. Doan D-Q, Fang T-H, Chen T-H (2020) Influences of grain size and temperature on tribological characteristics of CuAlNi alloys under nanoindentation and nanoscratcher. Int J Mech Sci 185:105865
42. Donzelli G, Cerbino R, Vailati A (2009) Bistable heat transfer in a nanofluid. Phys Rev Lett 102(10):104503
43. Duan H (2020) Analysis on the extinction properties of nanofluids for direct solar absorption. Physica E 120:114046
44. Duangthongsuk W, Wongwises S (2017) An experimental investigation on the heat transfer and pressure drop characteristics of nanofluid flowing in microchannel heat sink with multiple zigzag flow channel structures. Exp Therm Fluid Sci 87:30–39
45. Duarah S, Pujari K, Durai RD et al (2016) Nanotechnology-based cosmeceuticals: a review. Int J Appl Pharmaceutics 8:8–12
46. Ebrahimian M, Ansarifar GR (2016) Investigation of the nano fluid effects on heat transfer characteristics in nuclear reactors with dual cooled annular fuel using CFD (computational fluid dynamics) modeling. Energy 98:1–14
47. Eder SJ, Feldbauer G, Bianchi D et al (2015) Applicability of macroscopic wear and friction laws on the atomic length scale. Phys Rev Lett 115(2):025502
48. Elsheikh HA, Sharshir SW, Mostafa ME et al (2017) Applications of nanofluids in solar energy: a review of recent advances. Renew Sustain Energy Rev 82:3483–3502
49. Eshgarf H, Kalbasi R, Maleki A et al (2021) A review on the properties, preparation, models and stability of hybrid nanofluids to optimize energy consumption. J Thermal Anal Calorim 144(5):1959–1983
50. Ewen JP, Heyes DM, Dini D (2018) Advances in nonequilibrium molecular dynamics simulations of lubricants and additives. Friction 6(4):349–386
51. Ewen JP, Spikes HA, Dini D (2021) Contributions of molecular dynamics simulations to elastohydrodynamic lubrication. Tribol Lett 69:24
52. Fal J, Mahian O, Zyła G (2018) Nanofluids in the service of high voltage transformers: breakdown properties of transformer oils with nanoparticles, a review. Energies 11:2942
53. Fan X, Chen H, Ding Y et al (2008) Potential of 'nanofluids' to further intensify microreactors. Green Chem 10(6):670–677
54. Gandomkar A, Saidi MH, Shafii MB et al (2017) Visualization and comparative investigations of pulsating ferro-fluid heat pipe. Appl Therm Eng 116:56–65
55. Gehrke I, Geiser A, Somborn-Schulz A (2015) Innovations in nanotechnology for water treatment. Nanotechnol Sci Appl 8:1–17
56. Ghosh P, Han G, De M et al (2008) Gold nanoparticles in delivery applications. Adv Drug Delivery Rev 17:1307
57. Glasgow MDK, Chougule MB (2015) Recent developments in active tumor targeted multifunctional nanoparticles for combination chemotherapy in cancer treatment and imaging. J Biomed Nanotech 11:1859–1898
58. Gorji TB, Ranjbar AA (2017) A review on optical properties and application of nanofluids in direct absorption solar collectors (DASCs). Renew Sustain Energy Rev 72:10–32
59. Goshayeshi HR, Safaei MR, Goodarzi M et al (2016) Particle size and type effects on heat transfer enhancement of ferro-nanofluids in a pulsating heat pipe. Powder Technol 301:1218–1226
60. Gotovac S, Yang CM, Hattori Y et al (2007) Adsorption of polyaromatic hydrocarbons on single wall carbon nanotubes of different functionalities and diameters. J Colloid Interface Sci 314:18–24
61. Guimarey MJG, Comucas MJP, Lypez ER et al (2019) Thermophysical properties of polyalphaolefin oil modified with nanoadditives. J Chem Thermodynamics 131:192–205
62. Gupta V, Agarwal S, Saleh TA (2011) Chromium removal by combining the magnetic properties of iron oxide with adsorption properties of carbon nanotubes. Water Res 45(6):2207–2212

63. Gupta V, Mohapatra S, Mishra H et al (2022) Nanotechnology in cosmetics and cosmeceuticals—a review of latest advancements. Gels 8:173
64. He Y, Jin Y, Chen H et al (2007) Heat transfer and flow behavior of aqueous suspensions of TiO2 nanoparticles (nanofluids) flowing upward through a vertical pipe. Int J Heat Mass Transfer 50(11–12):2272–2281
65. Heiz A, Landman U (eds) (2007) Nanocatalysis. Springer, Heidelberg
66. Heris SZ, Shokrgozar M, Poorpharhang S et al (2014) Experimental study of heat transfer of a car radiator with CuO/ethylene glycol-water coolant. J Dispersion Sci Technol 35(5):677–684
67. Hossain MS, Saidur R, Sabri MFM et al (2015) Spotlight on available optical properties and models of nanofluids: a review. Renew Sustain Energy Rev 43:750–762
68. Hu CZ, Bai ML, Lv JZ et al (2014) Molecular dynamics investigation of the effect of copper nanoparticle on the solid contact between friction surfaces. Appl Surf Sci 321:302–309
69. Hussein AK, Ghodbane M, Said Z et al (2022) The effect of the baffle length on the natural convection in an enclosure filled with different nanofluids. J Therm Anal Calorim 147:791–813
70. Said Z (ed) (2022) Hybrid nanofluids. Preparation, characterization and applications. Elsevier, Amsterdam
71. Jalal R, Goharshadi EK, Abareshi M et al (2010) ZnO nanofluids: green synthesis, characterization, and antibacterial activity. Mat Chem Phys 121:198 201
72. Islam M, Shabani B, Rosengarten J et al (2015) The potential of using nanofluids in pem fuel cell cooling systems: a review. Renew Sustain Energy Rev 48:523–539
73. Jadhao V, Robbins MO (2019) Rheological properties of liquids under conditions of elastohydrodynamic lubrication. Tribology Lett 67:66
74. Jang JT, Lee J, Seon J et al (2018) Giant magnetic heat induction of magnesium-doped γ-Fe_2O_3 superparamagnetic nanoparticles for completely killing tumors. Adv Mater 30:1704362
75. Jang SP, Choi SUS (2006) Cooling performance of a microchannel heat sink with nanofluids. Appl Thermal Eng 26(17–18):2457–2463
76. Javvadi M, Nikhare C, Gurajala NK et al (2024) A review on hybrid nanofluids: preparation methods, thermophysical properties and applications. Iranian J Sci Tech: Trans Mech Eng. https://doi.org/10.1007/s40997-024-00772
77. Jayhooni SMH, Rahimpour MR (2013) Effect of different types of nanofluids on free convection heat transfer around spherical mini-reactor. Superlattices Microstruct 58:205–217
78. Ji C, Sun S, Wang B et al (2019) Molecular dynamic simulations of the roles of nanoparticles in sliding friction process. Chem Phys Lett 728:44–49
79. Jia L, Chen Y, Mo S 92013) Solid–liquid phase transition of nanofluids. Int J Heat Mass Transfer 59:29–34
80. Jiang G, Ni X, Yang L et al (2019) Synthesis of superamphiphobic nanofluid as a multifunctional additive in oil-based drilling fluid, especially the stabilization performance on the water/oil interface. Colloids Surf A Physicochem Eng Asp 588:124385
81. Jones N, Ray B, Ranjit KT et al (2008) Antibacterial activity of ZnO nanoparticle suspensions on a broad spectrum of microorganisms. FEMS Microbiol Lett 279:71–76
82. Jothy MA, Shanmuganathan S, Nagalakshmi (2015) An overview on niosome as carrier in dermal drug delivery. J Pharm Sci Res 7:923
83. Junankar AA, Purohit JK, Bhende NV (2022) A review on hybrid nanofluid: current research and sustainable development for turning operation. Int J Environment Sustain Develop 21:97–115
84. Jurado-Sánchez B, Sattayasamitsathit S, Gao W et al (2015) Self-propelled activated carbon Janus micromotors for efficient water purification. Small 11(4):499–506
85. Kabeel AE, El-Said EMS (2014) Applicability of flashing desalination technique for small scale needs using a novel integrated system coupled with nanofluid-based solar collector. Desalination 333:10–22
86. Kampa M, Castanas E (2008) Human health effects of air pollution. Environ Pollut 1519(2):362–367
87. Karami M, Bahabadi AMA, Delfani S et al (2014) A new application of carbon nanotubes nanofluid as working fluid of low-temperature direct absorption solar collector. Sol Energy Mater Sol Cells 121:114–118

88. Kamatchi R, Venkatachalapath S (2015) Parametric study of pool boiling heat transfer with nanofluids for the enhancement of critical heat flux: a review. Int J Thermal Sci 87:228–240
89. Kao MJ, Chang H, Wu YY et al (2007) Producing aluminum-oxide brake nanofluids using plasma charging system. J Chinese Soc Mech Eng 28(2):123–131
90. Kao MJ, Lo CH, Tsung TT et al (2007) Copper-oxide brake nanofluid manufactured using arc submerged nanoparticle synthesis system. J Alloys Compounds 434–435:672–674
91. Kaushik A, Mujawar M, Khan S (2020) Nanosensors for environmental applications. In: Han B, Tomer V, Nguyen TN, Ali F, Singh PK (eds) Nanosensors for smart cities. Elsevier, Amsterdam, p 143
92. Khaligh NG, Johan MR (2020) Recent advances in water treatment using graphene-based materials. Mini-Rev Organic Chem 17:74–90
93. Keshavarz Moraveji M, Razvarz S (2012) Experimental investigation of aluminum oxide nanofluid on heat pipe thermal performance. Int Commun Heat Mass Transfer 39:1444–1448
94. Khan MS, Sisodia MS, Gupta S et al (2019) Measurement of tribological properties of Cu and Ag blended coconut oil nanofluids for metal cutting. Eng Sci Technol Int J 22:1187–1192
95. Khlebtsov N, Dykman l (2011) Biodistribution and toxicity of engineered gold nanoparticles: a review of in vitro and in vivo studies. Chem Soc Rev 40:1647–1671
96. Kholy AEl, Alyan A, Fakhry K et al (2019) Thermal hydraulic analysis of supercritical water reactor cooled by TiO_2 nanofluid. J Nucl Sci Technol 56:291–299
97. Khooshehchin M, Mohammadidoust A, Ghotbinasa S (2020) An optimization study on heat transfer of pool boiling exposed ultrasonic waves and particles addition. Int Commun Heat Mass Transfer 114:104558
98. Khooshehchin M, Mohammadidoust A, Fathi S (2023) Experimental investigation of stabilizers of nanofluid in the pool boiling process. Heat Transfer Eng 44(5):442–460
99. Khoshvaght-Aliabadi M, Hassani SM, Mazloumi SH et al (2017) Effects of nooks configuration on hydrothermal performance of zigzag channels for nanofluid-cooled microelectronic heat sink. Microelectron Reliab 79:153–165
100. Kim K, Lee J-B (2007) High aspect ratio tapered hollow metallic microneedle arrays with microfluidic interconnector. Microsyst Technol 13(3–4):231–235
101. Kim SJ, Bang IC, Buongiorno J et al (2007) Surface wettability change during pool boiling of nanofluids and its effect on critical heat flux. Int J Heat Mass Transfer 50(19–20):4105–4116
102. Kim SK, Girard A, Jun S et al (2018) Effect of surface roughness on pool boiling heat transfer of water on hydrophobic surfaces. Int J Heat Mass Transfer 118:802–811
103. Kiran S, Ashraf A, Afzal G et al (2022) Green synthesis of magnesium oxide nanoparticles using leaves of Iresine herbstii for remediation of reactive brown 9 dye. Global NEST J 24(2):291–296
104. Kong L, Sun J, Bao Y (2017) Preparation, characterization and tribological mechanism of nanofluids. RSC Adv 7:12599–12609
105. Kotsmar C, Yoon KY, Yu H et al (2010) Stable citrate-coated Iron citrate-coated Iron oxide superparamagnetic nanoclusters at high salinity. Ind Eng Chem Res 49(24):12435–12443
106. Kshirsagar DP, Venkatesh MA (2021) A review on hybrid nanofluids for engineering applications. Mat Today: Proc 44:744–755
107. Kulkarni DP, Das DK, Vajjha RS (2009) Application of nanofluids in heating buildings and reducing pollution. Appl Energy 86(12):2566–2573
108. Kumar A, Hassan MA, Chand P (2020) Heat transport in nanofluid coolant car radiator with louvered fins. Powder Technol 376:631–642
109. Labhasetwar V, Leslie-Pelecky DL (2007) Biomedical applications of nanotechnology. Wiley, New York
110. Lahouij I, Bucholz EW, Vacher B et al (2012) Lubrication mechanisms of hollow-core inorganic fullerene-like nanoparticles: coupling experimental and computational works. Nanotechnology 23(37):375701
111. Lavanya M, Shalini S, Suvardhan K et al (2019) Nanotechnology-based water quality management for wastewater treatment. Env Chem Lett 17:65–121

112. Lee K, Arnot T, Mattia D (2011) A review of reverse osmosis membrane materials for desalination–development to date and future potential. J Membr Sci 370:1–22
113. Lee MH, Cho K, Shah AP et al (2005) Nanostructured sorbents for capture of cadmium species in combustion environments. Environ Sci Technol 39(21):8481–8489
114. Li X, Cheung SCP, Tu J (2014) Nucleate boiling of dilute nanofluids—mechanism exploring and modeling. Int J Thermal Sci 84:323–334
115. Li XF, Zhu DS (2009) Experimental study of cold storage characteristics of nanofluids as the phase change material. Mater Rev 23(3):11–16
116. Lin L, Kedzierski MA (2018) Specific heat of aluminum-oxide nanolubricants. Int J Heat Mass Transfer 126:1168–1176
117. Liu PZ, Lu J, Yu HL et al (2017) Lubricant shear thinning behavior correlated with variation of radius of gyration via molecular dynamics simulations. J Chem Phys 147:084904
118. Liu Z, Wang H, Liu C et al (2012) Magnetic cellulose-chitosan hydrogels prepared from ionic liquids as reusable adsorbent for removal of heavy metal ions. Chem Commun 48(59):7350–7352
119. Lohani A, Verma A, Joshi H et al (2014) Nanotechnology-based cosmeceuticals. D ISRN Dermatol 2014:843687
120. Lu CS, Chiu II, Liu CT (2006) Removal of zinc(II) from aqueous solution by purified carbon nanotubes: kinetics and equilibrium studies. Ind Eng Chem Res 45(8):2850–2855
121. Lu M-C, Chen R, Srinivasan V et al (2011) Critical heat flux of pool boiling on Si nanowire array-coated surfaces. Int J Heat Mass Transfer 54:5359–5367
122. Lv J, Bai M, Cui W et al (2011) The molecular dynamic simulation on impact and friction characters of nanofluids with many nanoparticles system. Nanoscale Res Lett 6:200
123. Maity SK, Ancheyta J, Marroquin G (2010) Catalytic aquathermolysis used for viscosity reduction of heavy crude oils: a review. Energy Fuel 24(5):2809–2816
124. Martín M, Villalba A, Fernández AI et al (2019) Development of new nano-enhanced phase change materials (NEPCM) to improve energy efficiency in buildings: lab-scale characterization. Energy Build 192:75–83
125. Mathas D, Holweger W, Wolf M et al (2021) Evaluation of methods for viscosity simulations of lubricants at different temperatures and pressures: a case study on PAO-2. Tribol Trans 64:1138–1148
126. Maurer-Jones MA, Gunsolus IL, Murphy CJ et al (2013) Toxicity of engineered nanoparticles in the environment. Anal Chem 85(6):3036–3049
127. Mehmood A, Waqas A, Said Z et al (2019) Performance evaluation of solar water heating system with heat pipe evacuated tubes provided with natural gas backup. Energy Rep 5:1432–1444
128. Meibodi SS, Kianifar A, Mahian O et al (2016) Second law analysis of a nanofluid-based solar collector using experimental data. J Therm Anal Calorim 126:617–625
129. Merum S, Veluru JB, Kranthi KS et al (2018) A review on nanomaterial revolution in oil and gas industry for EOR (enhanced oil recovery) methods. Res Dev Mater Sci 4:339–350
130. Milanese M, Colangelo G, Cretм A et al (2016) Optical absorption measurements of oxide nanoparticles for application as nanofluid in direct absorption solar power systems—part II: ZnO, CeO_2, Fe_2O_3 nanoparticles behavior. Sol Energy Mater Sol Cells 147:321–326
131. Minakov AV, Lobasov AS, Rudyak VYa et al (2014) Measuring of critical density of heat flow during boiling of nanoliquids on a cylindrical heater. Tech Phys Lett 40(7):582–584
132. Minakov AV, Mikhienkova EI, Zhigarev VA et al (2018) A study of the influence of nanoparticles on the properties of drilling fluids. Colloid J 80:418–426
133. Minakov AV, Mikhienkova EI, Pryazhnikov MIAD et al (2022) Experimental study of the rheological properties and stability of highly-concentrated oil-based emulsions. J Mol Liquids 349:11812
134. Minakov AV, Mikhienkova EI, Matveev AV et al (2019) Experimental study of antifriction properties of fluids with nanoparticles. J Friction Wear 40:420–424
135. Minakov AV, Mikhienkova EI, Neverov AL et al (2020) Comprehensive numerical study of the effect of nanoparticle additives on the cutting transport performance in horizontal boreholes. J Comp Design Eng 8:283–297

136. Minakov AV, Mikhienkova EI, Voronenkova YO et al (2019) Systematic experimental investigation of filtration losses of drilling fluids containing silicon oxide nanoparticles. J Natural Gas Sci Eng 71:102984
137. Minakov AV, Pryazhnikov MI, Guzei DV et al (2017) The experimental study of nanofluids boiling crisis on cylindrical heaters. Int J Thermal Sci 116:214–223
138. Minakov AV, Zhigarev VA, Mikhienkova EI et al (2018) The effect of nanoparticles additives in the drilling fluid on pressure loss and cutting transport efficiency in the vertical boreholes. J Petroleum Sci Eng 171:1149–1158
139. Mishra A, Tripathy P, Ram S et al (2009) Optical properties in nanofluids of gold nanoparticles in polyvinylpyrrolidone. J Nanosci Nanotechnology 9(7):4342–4347
140. Mohan K, Sundararaj S, Kannan KG et al (2020) Experimental analysis on refrigeration system using CNT, gold & HAUCL4 nanofluids. Mater Today Proc 33:360–366
141. Molinari J-F, Aghababaei R, Brink T et al (2018) Adhesive wear mechanisms uncovered by atomistic simulations. Friction 6(3):245–259
142. Mori S, Utaka Y (2017) Critical heat flux enhancement by surface modification in a saturated pool boiling: a review. Int J Heat Mass Transfer 108:2534–2557
143. Moshfegh AZ (2009) Nanoparticle catalysis. J Phys 42D:233001
144. Mu L, Sprando RL (2010) Application of nanotechnology in cosmetics. Pharmaceutical Res 27:1746–1749
145. Murugesan K, Srinivasan P, Mahadeva R et al (2020) Tuftsin-bearing liposomes co-encapsulated with doxorubicin and curcumin efficiently inhibit EAC tumor growth in mice. Int J Nanomedicine 15:10547–10559
146. Muzakkir SM, Hirani H (2022) Effect of using zinc oxide nanoparticles as antiwear and antifriction additive in mineral oil for a tribo-pair operating in mixed lubrication regime. Tribol Ind 44(4):592–598
147. Najafpoor A, Norouzian-Ostad R, Alidadi H et al (2020) Effect of magnetic nanoparticles and silver-loaded magnetic nanoparticles on advanced wastewater treatment and disinfection. J Mol Liquids 303:112640
148. Nanda S (2018) Nanocosmeceuticals: retrospect, precepts and prospects. Pharma Times 50(2):73–78
149. Naphon P, Nakharintr L, Wiriyasart S (2018) Continuous nanofluids jet impingement heat transfer and flow in a micro-channel heat sink. Int J Heat Mass Transfer 126:924–932
150. Navarrete N, Mondragyn R, Wen D et al (2019) Thermal energy storage of molten salt – based nanofluid containing nano-encapsulated metal alloy phase change materials. Energy 167:912–919
151. Nebykov D, Panov AO, Razvalyaeva AV et al (2023) Alcohols catalyzed by γ-Al_2O_3-supported nickel and copper nanoparticles. Russian J General Chem 93(8):1931–1940
152. Neumann RV (2010) Nanotechnology and the environment. Nova Science Publishers, Inc., New York
153. Neupert T, Bartel D (2023) Evaluation of various shear-thinning models for squalane using traction measurements, TEHD and NEMD simulations. Lubricants 11:178
154. Okonkwo EC, Wole-Osho I, Almanassra IW et al (2021) An updated review of nanofluids in various heat transfer devices. J Therm Anal Calorim 145:2817–2872
155. Qi C, Zhao N, Cui X et al (2018) Effects of half spherical bulges on heat transfer characteristics of CPU cooled by TiO_2-water nanofluids. Int J Heat Mass Transfer 123:320–330
156. Otanicar TP, Phelan PE, Prasher RS et al (2010) Nanofluid-based direct absorption solar collector. J Renew Sustain Energy 2(3):33102
157. Ovsianikov A, Chichkov B, Mente P et al (2007) Two photon polymerization of polymer-ceramic hybrid materials for transdermal drug delivery. Int J Appl Ceramic Technol 4:22–29
158. Pan B, Lin D, Mashayekhi H et al (2008) Adsorption and hysteresis of bisphenol A and 17a-ethinyl estradiol on carbon nanomaterials. Environ Sci Technol 42(15):5480–5485
159. Pandya NS, Desai AN, Tiwari AK et al (2021) Influence of the geometrical parameters and particle concentration levels of hybrid nanofluid on the thermal performance of axial grooved heat pipe. Therm Sci Eng Prog 21:100762

160. Pang S, Zhao L, Yuxiu A (2024) Advanced developments in nanotechnology and nanomaterials for the oil and gas industry: a review. Geoenergy Sci Eng 238(6):212872
161. Parmon V (2008) Nanomaterials in catalysis. Mater Res Innov 12(2):60–61
162. Patel SK, Modi KV (2020) Techniques to improve the performance of enhanced condensation area solar still: a critical review. J Cleaner Prod 268:122260
163. Patil BBT (2015) Waste water treatment using nanoparticles. J Adv Chem Eng 5(3):10–4172
164. Paul G, Hirani H, Kuila T et al (2019) Nanolubricants dispersed with graphene and its derivatives: an assessment and review of the tribological performance. Nanoscale 11:3458–3483
165. Philip J, Jaykumar T, Kalyanasundaram P et al (2003) A tunable optical filter. Measure Sci Technol 14:1289–1294
166. Pereira J, Moita A, Moreira A (2023) Fewer dimensions for higher thermal performance: a review on 2D nanofluids. Appl Sci 13:4070
167. Pitoniak E, Wu CY, Mazyck DW et al (2005) Adsorption enhancement mechanisms of silica-titania nanocomposites for elemental mercury vapor removal. Environ Sci Technol 39(5):1269–1274
168. Poddar AK (2023) Nanotechnology in ecology preservation. Eur Chem Bull 12(S3):2785–2795
169. Pollshetivar V, Varna RS (2010) Green chemistry by nano-catalis. Creen Chem 12:743–754
170. Popov YuV, Mokhov VM, Latyshova SE et al (2017) Colloid and nanosized catalysts in organic synthesis: XVIII. Disproportionation and cross-coupling of amines during catalysis with immobilized nickel nanoparticles. Russ J General Chem 87:2757–2761
171. Popov YuV, Mokhov VM, Latyshova SE et al (2018) Colloidal and nanosized catalysts in organic synthesis: XX. Continuous hydrogenation of imines and enamines catalyzed by nickel nanoparticles. Russ J General Chem 88:2035–2038
172. Pordanjani AH, Aghakhani S, Afrand M et al (2019) An updated review on application of nanofluids in heat exchangers for saving energy. Energy Conversion Management 198:111886
173. Pourpasha H, Heris SZ, Mahian O et al (2020) The effect of multi-wall carbon nanotubes/turbine meter oil nanofluid concentration on the thermophysical properties of lubricants. Powder Technol 367:133–142
174. Pryagnikov MI, Minakov AV, Rudyak VYa (2015) The influence of the size of nanoparticles and the heater size on the critical heat flux density in boiling nanofluids. Tech Phys Lett 60(6):798–804
175. Pustovalov VK, Astafyeva LG (2018) Optical properties of nanoparticles and nanofluids for direct absorption of solar radiation. Nanotechnol Environ Eng 3:15
176. Putri RAV, Ibadurrohman M, Slamet S (2020) Development of nanofluid biodegradable detergent from palm kernel oil and TiO_2. AIP Conf Proc 2237:020026
177. Pyrzynska K (2023) Preconcentration and removal of Pb(II) ions from aqueous solutions using graphene-based nanomaterials. Materials 16:1078
178. Rahmani AR, Athey AE, Ahmadian M et al (2015) Crosswell magnetic sensing of superparamagnetic nanoparticles for subsurface application. Soc Petroleum Eng 20:1–16
179. Rakhi MS, Suresh BG, Premalatha M (2016) Applications of nanotechnology in waste water treatment. A review. Imp J Interdiscip Res 2(11):1500–1511
180. Ramasamy US, Bair S, Martini A (2015) Predicting pressure viscosity behavior from ambient viscosity and compressibility: challenges and opportunities. Tribol Lett 57(2):11
181. Ramasamy US, Lichter S, Martini A (2016) Effect of molecular scale features on the polymer coil size of model viscosity index improvers. Tribol Lett 62(2):23
182. Ramasamy US, Len M, Martini A (2017) Correlating molecular structure to the behavior of linear styrene–butadiene viscosity modifiers. Tribol Lett 65(4):147
183. Rativa D, Gymez-Malagyn LA (2015) Solar radiation absorption of nanofluids containing metallic nanoellipsoids. Sol Energy 118:419–425
184. Rehman T-U, Ali HM, Saieed A et al (2018) Copper foam/PCMs based heat sinks: an experimental study for electronic cooling systems. Int J Heat Mass Transfer 127:381–393

185. Routbort J, Singh D, Timofeeva E et al (2010) Development, characterization, production, and demonstration of nanofluids for industrial cooling applications. Quarterly report #7. https://doi.org/10.2172/983764
186. Rosensweig RE (1985) Ferrohydrodynamics. Cambridge University Press, Cambridge
187. Rubbi F, Das L, Habib K et al (2021) State-of-the-art review on water-based nanofluids for low temperature solar thermal collector application. Solar Energy Mat Solar Cells 230:111220
188. Rudyak VYa, Belkin AA (1999) Liquid-solid phase transition in a heterogeneous system of solid spheres. J Exp Theor Phys 89:1086–1088
189. Rudyak VYa, Belkin AA (2014) Fluid viscosity under confined conditions. Doklady Phys 59(12):604–606
190. Rudyak VYa, Belkin AA (2018) Molecular-dynamics simulation of fluid viscosity in nanochannels. Nanosystems: Phys Chem Math 9:349–355
191. Saadati H, Hadad K, Rabie A (2018) Safety margin and fuel cycle period enhancements of VVER-1000 nuclear reactor using water/silver nanofluid. Nucl Eng Technol 50:639–647
192. Sadeghi G, Nazari S, Ameri M et al (2020) Energy and exergy evaluation of the evacuated tube solar collector using Cu_2O/water nanofluid utilizing ANN methods. Sustain Energy Technol Assess 37:100578
193. Sadiq IO, Sharif S, Suhaimi MA et al (2018) Enhancement of thermo-physical and lubricating properties of SiC nanolubricants for machining operation. Procedia Manuf 17:166–173
194. Said Z, Saidur R, Rahim NA (2014) Optical properties of metal oxides based nanofluids. Int Commun Heat Mass Transfer 59:46–54
195. Said Z, Arora S, Bellos E (2018) A review on performance and environmental effects of conventional and nanofluid-based thermal photovoltaics. Renew Sustain Energy Rev 96:302–316
196. Said Z, Sundar LS, Tiwari AK et al (2022) Recent advances on the fundamental physical phenomena behind stability, dynamic motion, thermophysical properties, heat transport, applications, and challenges of nanofluids. Phys Rep 946(2):1–94
197. Saidur R, Meng T, Said Z et al (2012) Evaluation of the effect of nanofluid-based absorbers on direct solar collector. Int J Heat Mass Transfer 55:5899–5907
198. Sajid MU, Ali HM (2018) Thermal conductivity of hybrid nanofluids: a critical review. Int J Heat Mass Transfer 126:211–234
199. Sajid MU, Ali HM (2019) Recent advances in application of nanofluids in heat transfer devices: a critical review. Renew Sustain Energy Rev 103:556–592
200. Seil JT, Webster TJ (2012) Antimicrobial applications of nanotechnology: methods and literature. Int J Nanomedicine 7:2767–2781
201. Salimi-Yasar H, Heris SZ, Shanbedi M et al (2017) Experimental investigation of thermal properties of cutting fluid using soluble oil based TiO_2 nanofluid. Powder Technol 310:213–220
202. Samsuri A, Hamzah A (2016) Water based mud lifting capacity improvement by multiwall carbon nanotubes additive. Int J Petroleum Gas Eng 3(8):1–9
203. Santhosh C, Velmurugan V, Jacob G et al (2016) Role of nanomaterials in water treatment applications: a review. Chem Eng J 306:1116–1137
204. Sharma A, Kumar S, Mahadevan N (2012) Nanotechnology: a promising approach for cosmetics. Int J Recent Adv Pharmaceutical Res 2(2):54–61
205. Sharma D, Pandey KM, Debbarma A et al (2017) Numerical investigation of heat transfer enhancement of SiO_2-water based nanofluids in light water nuclear reactor. Mater Today Proc 4:10118–10122
206. Sharma T, Reddy ALM, Chandra TS et al (2008) Development of carbon nanotubes and nanofluids based microbial fuel cell. Int J Hydrogen Energy 33(22):6749–6754
207. Sharafeldin MA, Gróf G, Abu-Nada E et al (2019) Evacuated tube solar collector performance using copper nanofluid: energy and environmental analysis. Appl Therm Eng 162:114205
208. Sharma H, Kushwah P, Saxena G (2021) Review on nanofluids as potential heat transfer fluid. Int J Eng Res Appl 11(7):17–28

209. Shawgo RS, Grayson ACR, Li Y et al (2002) BioMEMS for drug delivery Curr Opin Solid State Mat Sci 6(4):329–334
210. Sheikholeslami M, Rezaeianjouybari B, Darzi M et al (2019) Application of nanorefrigerant for boiling heat transfer enhancement employing an experimental study. Int J Heat Mass Transfer 141:974–980
211. Shojaeian M, Koşar A (2015) Pool boiling and flow boiling on micro-and nanostructured surfaces. Exp Thermal Fluid Sci 63:45–73
212. Siddique ARM, Muresan H, Majid SH et al (2019) An adjustable closed-loop liquid-based thermoelectric electronic cooling system for variable load thermal management. Therm Sci Eng Prog 10:245–252
213. Singh AP, Dwivedi RK, Suhane A (2021) Influence of nano particles on the performance parameters of lube oil—a review. Mater Res Express 8:102001
214. Singh AP, Tripathi A, Shukla MN et al (2021) Morphological parameters of nanoparticles used in nano lubrication—a review. IOP Conf Series Mat Sci Eng 1136:012029
215. Singh AP, Dwivedi RK, Suhane A (2022) In the context of nano lubrication, do nanoparticles exhibit favorable impacts on all tribo surfaces? A review. Prot Met Phys Chem Surf 58:325–338
216. Singh R, Lillard JW (2009) Nanoparticle-based targeted drug delivery. Experim Mol Pathol 86(3):215–223
217. Siricharoenpanich A, Wiriyasart S, Srichat A et al (2019) Thermal management system of CPU cooling with a novel short heat pipe cooling system. Case Stud Therm Eng 15:100545
218. Skomorokhova EA, Sankova TP, Orlov EI et al (2020) Size-dependent bioactivity of silver nanoparticles: antibacterial properties, influence on copper status in mice, and whole-body turnover. Nanotech, Sci Appl 13:137–157
219. Soltani S, Kasaeian A, Sarrafha H et al (2017) An experimental investigation of a hybrid photovoltaic/thermoelectric system with nanofluid application. Sol Energy 155:1033–1043
220. Song D, Hatami M, Wang Y et al (2016) Prediction of hydrodynamic and optical properties of TiO_2/water suspension considering particle size distribution. Int J Heat Mass Transfer 92:864–876
221. Spikes H (2015) Friction modifier additives. Tribol Lett 60:5
222. Srivastava I, Kotia A, Ghosh SK et al (2021) Recent advances of molecular dynamics simulations in nanotribology. J Mol Liquids 335:116154
223. Srivastava S, Khare E (2021) Nano-cosmeceuticals for health care. Int J Pharmacy Pharmaceut Sci 22(3):426–461
224. Sun B, Liu H (2017) Flow and heat transfer characteristics of nanofluids in a liquid-cooled CPU heat radiator. Appl Therm Eng 115:435–443
225. Sundar LS, Sharma KV, Singh MK et al (2017) Hybrid nanofluids preparation, thermal properties, heat transfer and friction factor—a review. Renew Sustain Energy Rev 68(3):185–198
226. Sundar LS, Said Z, Saleh B et al (2020) Combination of Co3O4 deposited rGO hybrid nanofluids and longitudinal strip inserts: thermal properties, heat transfer, friction factor, and thermal performance evaluations. Therm Sci Eng Prog 20:100695
227. Suneetha S, Subbarayudu K, Reddy PBA (2022) Hybrid nanofluids development and benefits: a comprehensive review. J Thermal Eng 8(3):445–455
228. Tabassum H, Sailaja A, Afreen H et al (2023) Sewage treatment using nanoparticles. In: Taşeli BK (ed) Sewage management. IntechOpen
229. Taketomi S, Chikazumi S (1993) Magnetic fluids. Mir, Moscow
230. Taylor RA, Phelan PE (2009) Pool boiling of nanofluids: comprehensive review of existing data and limited new data. Int J Heat Mass Transfer 52:5339–5347
231. Thamilselvi V, Radha KV (2017) Silver nanoparticle loaded silica adsorbent for wastewater treatment. Kor J Chem Eng 34(6):1801–1812
232. Tong T, Elimelech M (2016) The global rise of zero liquid discharge for wastewater management: drivers, technologies, and future directions. Critical review. Environ Sci Technol 50:6846–6855

233. Tong Y, Lee H, Kang W et al (2019) Energy and exergy comparison of a flat-plate solar collector using water, Al_2O_3 nanofluid, and CuO nanofluid. Appl Therm Eng 159:113959
234. Tyagi H, Phelan P, Prasher R (2009) Predicted efficiency of a low-temperature nanofluid based direct absorption solar collector. J Solar Energy Eng 131(4):0410041–0410047
235. Tyagi I, Gupta VK, Sadegh H et al (2017) Nanoparticles as adsorbent; a positive approach for removal of noxious metal ions: a review. Sci Technol Development 34(3):195–214
236. Tsaia CY, Chiena HT, Dingb PP et al (2004) Effect of structural character of gold nanoparticles in nanofluid on heat pipe thermal performance. Materials Lett 58:1461
237. Vafaei S, Borca-Tasciuc T, Podowski MZ et al (2006) Effect of nanoparticles on sessile droplet contact angle. Nanotechnology 17(10):2523–2527
238. Wong KV, Leon OD (2010) Applications of nanofluids: current and future. Adv Mech Eng 2010:1652–1660
239. Valizade M, Heyhat MM, Maerefat M (2019) Experimental comparison of optical properties of nanofluid and metal foam for using in direct absorption solar collectors. Sol Energy Mater Sol Cells 195:71–80
240. Vanossi A, Manini N, Urbakh M et al (2013) Colloquium: modeling friction: from nanoscale to mesoscale. Rev Modern Phys 85(2):529–552
241. Venu H, Madhavan V (2016) Effect of Al_2O_3 nanoparticles in biodiesel-diesel-ethanol blends at various injection strategies: performance, combustion and emission characteristics. Fuel 186:176–189
242. Venu H, Subramani L, Raju VD (2019) Emission reduction in a DI diesel engine using exhaust gas recirculation (EGR) of palm biodiesel blended with TiO_2 nano additives. Renew Energy 140:245–263
243. Venu H, Raju VD, Lingesan S et al (2020) Influence of Al2O3 nano additives in ternary fuel (diesel-biodiesel-ethanol) blends operated in a single cylinder diesel engine: performance, combustion and emission characteristics. Energy 215:119091
244. Vékás L, Bica D, Avdeev MV (2007) Magnetic nanoparticles and concentrated magnetic nanofluids: synthesis, properties and some applications. China Particuology 5:43–49
245. Vicaria JM, Gonz6lez-Beneded R, Sierra AA et al (2019) Use of α-amylase/silica particle suspensions to optimize cleaning in a simulated cleaning-in-place system. J Food Eng 247:64–73
246. Vinardell MP, Mitjans M (2015) Nanocarriers for delivery of antioxidants on the skin. Cosmetics 2(4):342–354
247. Vo TT, Nguyen TTN, Huynh TTT et al (2019) Biosynthesis of silver and gold nanoparticles using aqueous extract from *Crinum latifolium* leaf and their applications forward antibacterial effect and wastewater treatment. J Nanomat 2019:1–14
248. Vonarbourg A, Passirani C, Saulnier P et al (2006) Parameters influencing the stealthiness of colloidal drug delivery systems. Biomaterials 27(24):4356–4373
249. Wang X, Tao S, Xing B (2009) Sorption and competition of aromatic compounds and humic acid on multiwalled carbon nanotubes. Environ Sci Technol 43(16):6214–6219
250. Wulandari PP, Adiwibowo MT, Redjeki AS et al (2019) Synthesis of eco-friendly detergent based on crude palm oil and titania nanoparticles. Asian J Chem 31:2394–2396
251. Xu P, Zeng GM, Huang DL et al (2012) Use of iron oxide nanomaterials in wastewater treatment: a review. Sci Total Environment. 424:1–10
252. Wang S, Liu Z, Wang W et al (2017) Fate and transformation of nanoparticles (NPs) in municipal wastewater treatment systems and effects of NPs on the biological treatment of wastewater: a review. Roy Soc Chem 7:37065–37075
253. Wang Z, Dai L, Yao J et al (2021) Enhanced adsorption and reduction performance of nitrate by Fe–Pd–Fe_3O_4 embedded multi-walled carbon nanotubes. Chemosphere 281:130718
254. Wu JH, Phillips BS, Jiang W et al (2008) Bio-inspired surface engineering and tribology of MoS2 overcoated CBN-TiN composite coating. Wear 261:592–599
255. Wu JS, Zhang XX, Shen JC et al (2017) A review of thermal absorbers and their integration methods for the combined solar photovoltaic/thermal (PV/T) modules. Renew Sustain Energy Rev 75:839–854

256. Wu N, Hu N, Zhou G et al (2018) Tribological properties of lubricating oil with micro/nanoscale WS2 particles. J Experimental Nanosci 13:37–38
257. Yang K, Wang X, Zhu L et al (2006) Competitive sorption of pyrene, phenanthrene, and naphthalene on multiwalled carbon nanotubes. Environ Sci Technol 40(18):5804–5810
258. Yapar EA (2012) Nanomaterials and cosmetics. J Pharm Inst Univ 42:43–70
259. Yasmin H, Giwa SO, Noor S et al (2023) Experimental exploration of hybrid nanofluids as energy-efficient fluids in solar and thermal energy storage applications. Nanomaterials 13:278
260. You SM, Kim JH, Kim KH (2003) Effect of nanoparticles on critical heat flux of water in pool boiling heat transfer. Appl Phys Lett 83:3374–3376
261. Zakaria I, Azmi W, Mohamed WG et al (2015) Experimental investigation of thermal conductivity and electrical conductivity of Al_2O_3 nanofluid in water-ethylene glycol mixture for proton exchange membrane fuel cell application. Int Commun Heat Mass Transfer 61:61–68
262. Zakaria I, Mohamed WANW, Azmi WH et al (2018) Thermo-electrical performance of PEM fuel cell using Al_2O_3 nanofluids. Int J Heat Mass Transfer 119:460–471
263. Zhang L, Jiang Y, Ding Y et al (2007) Investigation into the antibacterial behavior of suspensions of ZnO nanoparticles (ZnO nanofluids). J Nanoparticle Res 9(3):479–489
264. Zhang XJ, Wu P, Qiu LM et al (2010) Analysis of the nucleation of nanofluids in the ice formation process. Energy Convers Manage 51:130–134
265. Zheng X, Su L, Deng G et al (2021) Study on lubrication characteristics of C4-alkane and nanoparticle during boundary friction by molecular dynamics simulation. Metals 11:1464
266. Zheng X, Su L, Deng G (2024) Influence of nanoparticles in lubricant on sliding contact of atomic rough surfaces—a molecular dynamics study. Lubricants 12:160
267. Zonouzi SA, Aminfar H, Mohammadpourfard M (2019) A review on effects of magnetic fields and electric fields on boiling heat transfer and CHF. Appl Therm Eng 151:11–25

The manufacturer's authorised representative in the EU is Springer Nature Customer Service Centre GmbH, Europaplatz 3, 69115 Heidelberg, Germany. If you have any concerns regarding our products, please contact ProductSafety@springernature.com

Printed and bound by CPI Group (UK) Ltd, Croydon, CR0 4YY
05/10/2026
02228133-0008